Foundations of Chemistry

Foundations of Chemistry
An Introductory Course for Science Students

Philippa B. Cranwell

University of Reading
Reading, UK

and

Elizabeth M. Page

University of Reading
Reading, UK

This edition first published 2021
© 2021 John Wiley & Sons Ltd

The right of Philippa B. Cranwell and Elizabeth M. Page to be identified as the authors of this work has been asserted in accordance with law.

Registered Office(s)
John Wiley & Sons, Inc., 111 River Street, Hoboken, NJ 07030, USA
John Wiley & Sons Ltd, The Atrium, Southern Gate, Chichester, West Sussex, PO19 8SQ, UK

Editorial Office
The Atrium, Southern Gate, Chichester, West Sussex, PO19 8SQ, UK

For details of our global editorial offices, customer services, and more information about Wiley products visit us at www.wiley.com.

Wiley also publishes its books in a variety of electronic formats and by print-on-demand. Some content that appears in standard print versions of this book may not be available in other formats.

Library of Congress Cataloging-in-Publication Data

Names: Cranwell, Philippa B., 1985– author. | Page, Elizabeth (Lecturer in Chemistry Education), author.
Title: Foundations of chemistry : an introductory course for science students / Philippa B. Cranwell, University of Reading, Reading, UK and Elizabeth M. Page, University of Reading, Reading, UK.
Description: Hoboken, NJ : Wiley, 2021. | Includes index.
Identifiers: LCCN 2020028483 (print) | LCCN 2020028484 (ebook) | ISBN 9781119513872 (paperback) | ISBN 9781119513919 (adobe pdf) | ISBN 9781119513902 (epub)
Subjects: LCSH: Chemistry.
Classification: LCC QD31.3 .C73 2021 (print) | LCC QD31.3 (ebook) | DDC 540–dc23
LC record available at https://lccn.loc.gov/2020028483
LC ebook record available at https://lccn.loc.gov/2020028484

Cover Design: Wiley
Cover Image: © alice-photo/Shutterstock

Set in 10/12pt Sabon by Straive, Pondicherry, India

10 9 8 7 6 5 4 3 2 1

This book is dedicated to the joy of a new life, and the passing of a well-lived and loved one.

Contents

Preface

Foundations of Chemistry is a concise course in advanced general chemistry specifically designed for students studying at Level 3 (A level or equivalent). It is especially relevant for students enrolled upon one-year foundation programmes catering for physical and life sciences provided by UK universities, and is intended to introduce students to the core elements of physical, organic, and inorganic chemistry. The text outlines the basic principles in each area of chemistry and builds on prior knowledge from GCSE (or equivalent) courses to quickly expand and develop students' knowledge and understanding. Each chapter contains worked examples that showcase core concepts, and includes practice questions with fully worked answers available on the Wiley website. Margin comments signpost students to knowledge that is covered elsewhere in the book and are used to highlight key learning objectives when appropriate. A summary is given at the end of each chapter that lists the main concepts and learning points.

The authors recognise that many students on foundation programmes study chemistry as a subsidiary subject. We hope that this text outlines and clearly explains the information and knowledge required at foundation level so that these students can progress smoothly onto their parent programmes.

Students are first introduced to the structure of the atom and how the periodic table is built up. The different types of bonding are then described, followed by the way in which the properties of materials are affected by bonding within and between molecules. The fundamentals of physical chemistry including thermodynamics, equilibria, acids and bases, reaction rates, redox, and electrochemistry are explained. Elements and typical compounds from some key groups in the periodic table are considered in more detail, with specific reference to the underlying principles of periodicity and energetics where relevant. Organic chemistry takes a mechanistic approach, so students are able to rationalise reactivity and chemical behaviour; the intention is not for students to learn by rote, but to be able to apply their knowledge and understanding to derive an answer. This skill can be developed by students in a chapter dedicated to synoptic questions in organic chemistry. Finally, a chapter dedicated to spectroscopy demonstrates how modern analytical techniques are used in organic chemistry to determine structure.

The book is concise and focusses on key ideas, yet points students to areas that they may meet if they continue to study chemistry.

Philippa Cranwell and Elizabeth Page
Reading, UK

Acknowledgements

The authors would like to thank Dr Chris Smith and Dr Jenny Eyley of the University of Reading for patiently reading and commenting on drafts of the manuscript. We are indebted to them for their advice and suggestions for clarification. We would also like to thank our families for the support that they have given us during the preparation of this book, and for their belief that there would be an eventual conclusion!

Contributors

Philippa Cranwell is Associate Professor of Organic Chemistry at the University of Reading. She has extensive experience of teaching students chemistry, ranging from A-level to Foundation level and higher. She has co-authored several texts relating to both practical and theoretical organic chemistry. She actively undertakes research in the field of chemistry education and regularly publishes her works. She was awarded a University of Reading Teaching Fellowship in 2016 for her contribution to teaching and learning.

Elizabeth Page is Emeritus Professor of Chemistry Education at the University of Reading. She has over 30 years' experience teaching chemistry at Foundation level and higher. She is author of several textbooks for life-sciences and chemistry students. Elizabeth has been an examiner for A-level chemistry and helped in the design of the revised A-level specifications in chemistry. During her time at Reading she established a strong network of chemistry teachers, providing a forum for discussions and guidance in teaching GCSE and A-level chemistry.

Elizabeth was awarded the Royal Society of Chemistry Education prize for her work with chemistry teachers and is a National Teaching Fellow.

About the companion website

This book is accompanied by a companion website:

www.wiley.com/go/Cranwell/Foundations

The website includes:

- Extended answers to the end of chapter questions
- Digital images of the in-text figures

0

Fundamentals

At the end of this chapter, students should be able to:

- Recognise the base SI quantities used in chemistry and state their symbols and units

- Convert between commonly used SI units

- Write numbers using scientific notation

- Recognise metric prefixes used in expressing large or small numbers.

- Understand the use of significant figures when expressing quantities and measurements and be able to round values to the correct number of significant figures

- Write chemical formulae and equations and balance equations

- Use the appropriate symbol to indicate the physical state of a substance in a chemical equation

0.1 Introduction to chemistry

Chemistry is a subject that underpins many other disciplines. At the heart of chemistry is the study of the elements that make up the periodic table, the reactions they undergo, and the new compounds that are formed.

Water is a compound that we are all familiar with, and most people know the formula for water is H_2O even if they know nothing else about chemistry. The formula of water tells us that it is a molecule made up of two atoms of the element hydrogen and one atom of the element oxygen. In your course, you will learn that the elements in the periodic table are composed of atoms and that atoms are made up of smaller particles called protons, neutrons, and electrons. It is the specific combination of protons, neutrons, and electrons that gives each element its particular properties.

Foundations of Chemistry: An Introductory Course for Science Students, First Edition.
Philippa B. Cranwell and Elizabeth M. Page.
© 2021 John Wiley & Sons Ltd. Published 2021 by John Wiley & Sons Ltd.
Companion website: www.wiley.com/go/Cranwell/Foundations

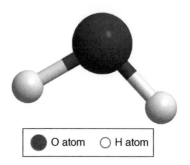

Figure 0.1 The shape of a water molecule, H_2O.

Hydrogen and oxygen are two of the smallest and lightest elements in the periodic table. Both hydrogen and oxygen are gases at room temperature, whereas water is a liquid. During your studies, you will learn why certain substances are gases or liquids, with low melting points and boiling points, and why other substances are solids that are very difficult to break down or melt. These properties depend to a certain extent upon how the atoms are arranged in molecules of the substances. For example, a water molecule has a bent shape (Figure 0.1). The bent shape of the water molecule is one of the factors that determines the melting and boiling point of water and ensures that it is a liquid at ambient temperatures. If the atoms in water were arranged in a straight line, water would have a much lower boiling point and would likely be a gas at room temperature. Clearly, this would have a major impact on life on earth. In this course, you will learn how to predict the shapes of small molecules such as water and see how important shape is in chemistry.

In this chapter, we will introduce some of the fundamental tools necessary for studying, understanding, and applying chemical principles. You may have met some of these rules before in other subject areas, and you will probably meet them again later in the book, but this chapter should act as a toolbox from which you can select the information and guidance you need for the rest of the course.

0.2   Measurement in chemistry and science – SI units

Chemistry is a practical subject, and our present knowledge of chemical properties and principles is based on experiments. Unfortunately, we don't have space in this book to describe many of the amazing experiments that early investigators carried out to enhance our understanding and knowledge of chemistry. The majority of experiments require making and recording measurements. The laws of science operate across the globe, so it's important that scientists make measurements that can be compared with each other. Therefore, measurements must be recorded in a universal and standard way. For this reason, the *metric system* was developed to establish a standardised set of units. The metric system has its origins in the eighteenth century. More recently, a revised metric system was introduced and adopted by scientists across the world. This system is known as the Système Internationale d'Unités, and the units in the system are known as SI units.

There are seven fundamental SI units. Six of these are commonly used in chemistry, and you will meet all of them in this course. The six **base units** used in chemistry are listed in Table 0.1 and are the units of mass, length, time, electrical current, temperature, and amount of substance. All other units for physical

Table 0.1 Base SI quantities used in chemistry with symbols and units.

Physical quantity	Symbol	Base unit	Unit symbol
Mass	m	kilogram	kg
Length	l	metre	m
Time	t	second	s
Electrical current	I	ampere	A
Temperature	T	kelvin	K
Amount of substance	n	mole	mol

Table 0.2 Commonly used derived units.

Quantity	Unit name	Symbol and base units
Area	square metre	m^2
Volume	cubic metre	m^3
Velocity (speed)	metre per second	$m\,s^{-1}$
Acceleration	metre per second squared	$m\,s^{-2}$
Density	kilogram per cubic metre	$kg\,m^{-3}$
Concentration	mole per cubic metre	$mol\,m^{-3}$
Energy	joule	$J = kg\,m^2\,s^{-2}$
Force	newton	$N = J\,m^{-1} = kg\,m\,s^{-2}$
Pressure	pascal	$Pa = N\,m^{-2} = kg\,m^{-1}\,s^{-2}$
Frequency	hertz	$Hz = s^{-1}$

quantities can be reduced to these base units. For example, speed is defined as the distance or length travelled divided by the time taken, so:

$$\text{speed} = \frac{\text{length}}{\text{time}} = \frac{\text{metre}}{\text{second}} = \frac{m}{s} = m/s = m\,s^{-1}$$

This defines the unit of speed as metre per second or m/s. In chemistry, as in most other scientific subjects, this would be written as: $m\,s^{-1}$, where the superscript '–1' means 'per second'.

There are many other units you will come across that can be reduced down to these base units. Units of this type are called **derived units** and usually have their own symbol. A list of derived units and their symbols is given in Table 0.2.

It is very important when carrying out calculations to keep track of the units, as you will be expected to quote the units of your answers. Units of each value in a calculation multiply or cancel with each other, as shown in the example here:

Acceleration is defined as the change in velocity (speed) divided by the time taken and can be represented by this equation:

$$\text{acceleration} = \frac{\text{change in velocity}}{\text{time}}.$$

Replacing the quantities with their units, we obtain:

$$\text{acceleration} = \frac{m\,s^{-1}}{s}.$$

The unit $\frac{1}{s}$ can also be written as s^{-1}, so the units for acceleration are $m\,s^{-1} \times s^{-1}$.

To multiply s^{-1} by s^{-1}, we simply add the –1 superscripts:
$$\text{acceleration} = m\,s^{-1} \times s^{-1} = m\,s^{-2}.$$

Dealing with exponents
Exponents tell us how many times a number should be multiplied by itself. For example:

$a^1 = a$

$a^2 = a \times a$

$a^{-1} = \frac{1}{a}$

$a^{-2} = \frac{1}{a^2}$

A special case is $a^0 = 1$

To multiply quantities with exponents just add the exponents together:

$a^2 \times a = a^{(2+1)} = a^3$

$a^{-1} \times a^2 = a^{(-1+2)} = a$

To divide quantities with exponents subtract the exponents:

$\frac{a^2}{a} = a^{(2-1)} = a$

$\frac{a^{-1}}{a^3} = a^{(-1-3)} = a^{-4}$.

If we know the speed of an object and wish to determine how far it travels in a certain time, we multiply the speed by the time:
distance = speed × time.

Inserting the units for speed and time gives us the unit for distance = m s^{-1} × s.

To multiply s^{-1} by s, again we add the superscripts: distance = m s^{-1} × s = m. This gives us the answer for distance in the correct units of metres. You should always check your answer when doing calculations to make sure that the value you obtain *and* the units are sensible.

Worked Example 0.1   Determine the derived units for the following quantities using the definitions given:

a. $\text{density} = \dfrac{\text{mass}}{\text{volume}}$

b. $\text{entropy} = \dfrac{\text{energy}}{\text{temperature}}$

c. $\text{pressure} = \dfrac{\text{force}}{\text{area}}$

d. $\text{power} = \dfrac{\text{energy}}{\text{time}}$

Solution

a. $\text{density} = \dfrac{\text{mass}}{\text{volume}} = \dfrac{\text{kg}}{\text{m}^3} = \text{kg m}^{-3}$

b. $\text{entropy} = \dfrac{\text{energy}}{\text{temperature}} = \dfrac{\text{J}}{\text{K}} = \dfrac{\text{kg m}^2\text{s}^{-2}}{\text{K}} = \text{kg m}^2\,\text{s}^{-2}\,\text{K}^{-1}$

c. $\text{pressure} = \dfrac{\text{force}}{\text{area}} = \dfrac{\text{N}}{\text{m}^2} = \dfrac{\text{kg m s}^{-2}}{\text{m}^2} = \text{kg m s}^{-2} \times \text{m}^{-2} = \text{kg m}^{-1}\,\text{s}^{-2}$

d. $\text{power} = \dfrac{\text{energy}}{\text{time}} = \dfrac{\text{J}}{\text{s}} = \dfrac{\text{kg m}^2\text{s}^{-2}}{\text{s}} = \text{kg m}^2\,\text{s}^{-2} \times \text{s}^{-1} = \text{kg m}^2\,\text{s}^{-3}$

0.3   Expressing large and small numbers using scientific notation

When studying science, you will likely come across numbers that are either extremely large or very small. It often isn't convenient (or practical) to write out the numbers 'longhand'. An example of this is in using the constant for the speed of light in vacuum. The value of this constant is approximately three hundred million metres per second (m s^{-1}), which can be written as 300 000 000 m s^{-1}. This method of writing numbers is sometimes called **non-exponential form**. A much simpler way of expressing this number is in **exponential form**, so 300 000 000 m s^{-1} is written as 3×10^8 m s^{-1}. The number 10^8 represents one hundred million or 100 000 000. When is it written as $\times 10^8$ this means there are eight zeroes after the 3. Some numbers are very small: for example, the Planck constant is 6.6×10^{-34} J s. The number 10^{-34} indicates there are 33 zeroes

Take care!
$3 \times 10^2 = 300$ (i.e. two zeroes after the number before the decimal point)
$3 \times 10^{-2} = 0.03$ (i.e. just one zero after the decimal point before the number)

after the decimal point and before the number itself. Clearly, it would be very tedious to write the number out in full!

When a number is written in exponential form, it is called **scientific notation**. The standard format is that the number is written with one digit *before* the decimal point and the multiplier 10^n is added after the number. The following examples show how numbers can be rewritten using scientific notation:

4 is written as 4×10^0.

42 is written as 4.2×10^1.

242 is written as 2.42×10^2.

2424 is written as 2.424×10^3.

To express numbers smaller than 1 using scientific notation, i.e. a decimal number, the number is converted to a number between 1 and 10 and the exponential factor 10^{-n} used to indicate the position of the decimal point, as in these examples:

0.4 is written as 4×10^{-1}.

0.042 is written as 4.2×10^{-2}.

0.00242 is written as 2.42×10^{-3}.

Any number raised to the power of zero is equal to 1, so $10^0 = 1$.

It isn't usual to use the factors 10^0 and 10^1 unless specifically required to give an answer in scientific notation.

Box 0.1 When converting numbers that are less than 1 expressed in scientific notation back to decimal numbers, move the decimal point to the left by the value of the exponent, adding zeroes. So 2.42×10^{-3} moves the decimal point by three places to the left, requiring two zeroes after the decimal point.

$$\begin{array}{c} 3\ \ 2\ \ 1 \\ 2.42 \times 10^{-3} = .0\ 0\ 2.42 \times 10^{-3} = 0.00242 \end{array}$$

Worked Example 0.2 Write the following numbers expressed in non-exponential form using scientific notation:

a. 675

b. 1 000 000

c. 0.98

d. 0.00355

Solution

a. To write 675 in scientific notation, we must move the decimal point (currently after the 5) two spaces to the left (i.e. divide by 100) and then multiply the remaining number by 100 or 10^2.
   675 becomes 6.75×10^2.

b. To write 1 000 000 in scientific notation, we must move the decimal point (currently after the last 0) six spaces to the left (i.e. divide by 1 000 000) and then multiply the remaining number by 1 000 000 or 10^6.
1 000 000 becomes 1×10^6.

c. To write 0.98 in scientific notation, we must move the decimal point one space to the right (i.e. multiply by 10) and then multiply the remaining number by 0.1 or 10^{-1} (i.e. divide by 10).
So 0.98 becomes 9.8×10^{-1}.

d. To write 0.00355 in scientific notation, we must move the decimal point three spaces to the right (i.e. multiply by 1 000) and then multiply the remaining number by 0.001 or 10^{-3} (i.e. divide by 1 000).
So 0.00355 becomes 3.55×10^{-3}.

Worked Example 0.3   Convert the following numbers, expressed in scientific notation, to non-exponential form.

a. 7.01×10^2

b. 6.912×10^5

c. 8.05×10^{-1}

d. 2.310×10^{-4}

Solution

a. Move the decimal point to the **right** by the number of spaces indicated by the exponent, i.e. by 2. 7.01 becomes 701.

b. Move the decimal point to the **right** by the number of spaces indicated by the exponent, i.e. by 5. 6.912×10^5 becomes 691 200.

c. Move the decimal point to the **left** by the number of spaces indicated by the exponent, i.e. by 1 space. 8.05×10^{-1} becomes 0.805.

d. Move the decimal point to the left by the number of spaces indicated by the exponent, i.e. by 4 spaces. 2.310×10^{-4} becomes 0.0002310.

0.4   Using metric prefixes

Scientists also have a shorthand for writing some large and small numbers using a prefix that represents a certain quantity. So the quantity 1 000 g, or 1×10^3 g in scientific notation, can be written as 1 kg. The letter 'k' represents the factor 'kilo', or 10^3. The quantity one-thousandth of a gram or 0.001 g (1×10^{-3} g) can be written as 1 mg, where the prefix 'm' represents the factor 'milli', or 10^{-3}. Table 0.3 gives some common prefixes, along with their symbols.

Note that the 'k' in kg is lowercase. The case of the unit can be very important, so make sure you pay attention to it!

Table 0.3 Some common prefixes and their values with quantities and symbols.

Factor	Number	Name	Symbol	Factor	Number	Name	Symbol
10^9	1 000 000 000	giga	G	10^{-9}	0.000 000 001	nano	n
10^6	1 000 000	mega	M	10^{-6}	0.000 001	micro	μ
10^3	1 000	kilo	k	10^{-3}	0.001	milli	m
				10^{-2}	0.01	centi	c
				10^{-1}	0.1	deci	d

0.4.1 Units of mass and volume used in chemistry

Mass

The **mass** of a substance is the amount we measure when we weigh the substance. Mass is a measure of the quantity of a substance. In science, we use the term **mass** as opposed to **weight** of a substance to describe the amount because the weight of a substance is related to the gravitational force acting on the substance. We often use the words 'mass' and 'weight' interchangeably, but they are not really the same thing scientifically.

The SI unit of mass is the kilogram. However, in the laboratory, chemicals and other objects are usually weighed in units of grams as the gram is a far more convenient amount to work with. Even though we use a balance to **weigh** a substance, we report its **mass** in grams, not its weight.

One kilogram is equivalent to one thousand grams, 1 000 g. 1 kg = 1 000 g. Therefore $1 \text{ g} = \dfrac{1}{1\,000} \text{ kg}$ or 0.001 kg. This can also be written as 1×10^{-3} kg.

Another common measure of mass is the milligram, mg. One milligram is one-thousandth of a gram:

$$1 \text{ mg} = \frac{1}{1\,000} \text{ g or } 1 \times 10^{-3} \text{ g} = 0.001 \text{ g}.$$

> You should always choose a balance appropriate to the level of accuracy of the mass required. If you need 'about two grams', for example, a two-figure balance is perfectly appropriate. If you are told to weigh 'accurately approximately two grams', this implies you need to use a four-figure balance to weigh around two grams of substance but record its weight accurately to four decimal places, e.g. 0.0001 g.

Volume

The volume of a substance is the space that it occupies. In chemistry, we are generally most concerned with the volumes of liquids and gases.

Liquid volumes are measured relatively easily, and the type of measuring device used will depend upon the accuracy with which we need to know the volume. There is always a certain amount of uncertainty when making a measurement. Different types of measuring equipment have different levels of uncertainty associated with them. For less accurate work, chemists generally use a measuring cylinder (Figure 0.2a). These can be obtained in different sizes, but the uncertainty in the volume when using a measuring cylinder depends on the size of the measuring cylinder and the graduations. The uncertainty on a measuring cylinder is half the volume of the smallest graduations. For a 25 cm^3 measuring cylinder where the graduations are 0.5 cm^3, the uncertainty is ±0.25 cm^3. If 25 cm^3 was measured into a 25 cm^3 measuring cylinder, the percentage uncertainty would be $\dfrac{0.25}{25} \times 100\% = 1\%$. If a 250 cm^3 measuring cylinder was used to measure out 25 cm^3 where the uncertainty was ±0.5 cm^3, the percentage uncertainty would be $\dfrac{0.5}{25} \times 100\% = 2\%$. It is therefore more accurate to use a smaller measuring cylinder for this volume of liquid.

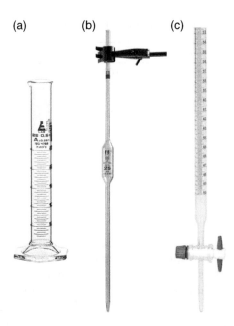

(a)   (b)   (c)

Figure 0.2 (a) $25\,cm^3$ measuring cylinder; (b) $25\,cm^3$ pipette; (c) $50\,cm^3$ burette. *Source:* Eisco Labs.

If we need to know the volume of liquid measured more accurately than this, either a pipette or a burette can be used. A pipette (Figure 0.2b) measures a fixed volume of liquid, and pipettes can be obtained in various sizes. A burette (Figure 0.2c) can be used to deliver variable volumes of liquids; although burettes can be obtained in various sizes, the $50\,cm^3$ burette is the size most commonly used. The uncertainty on a volume measurement when using a laboratory pipette is $\pm 0.06\,cm^3$. This means the uncertainty on measuring a volume of $25\,cm^3$ using a pipette is $\dfrac{0.06}{25} \times 100\% = 0.2\%$. For a standard burette, the uncertainty associated with each reading is $\pm 0.05\,cm^3$. If making one reading, then the percentage uncertainty in measuring $25\,cm^3$ is $\dfrac{0.05}{25} \times 100\% = 0.2\%$. When recording the amount added from a burette, we make two readings, so the total uncertainty associated with the readings is $\pm 0.1\,cm^3$. Thus the percentage uncertainty on the burette reading is $\dfrac{0.1}{25} \times 100\% = 0.4\%$.

The derived SI unit for volume is the cubic metre or m^3 because the volume of an object is obtained from the product of its height, width, and depth, which are all units of length (Figure 0.3a). As the SI unit of length is the m, the derived unit for volume has units of m^3. However, a volume of $1\,m^3$ is very large, so the m^3 is not a very useful unit.

In the chemistry laboratory, we often work with volumes a lot smaller than a cubic metre. A volume you will often encounter is the litre (Figure 0.3b). The symbol that represents the litre is L. However, strictly speaking, this is a non-SI unit, although it is in common use. Chemists usually work in volumes of cm^3 or dm^3 where:

$$1\,L = 1\,000\,mL = 1\,000\,cm^3 = 1\,dm^3 \ \left(\text{because } 1\,mL = 1\,cm^3\right)$$

(a)

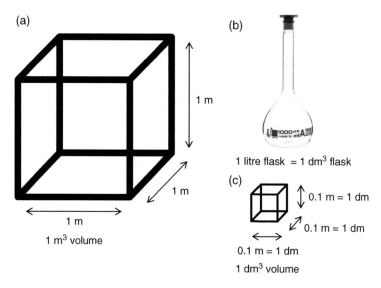

1 m

(b)

1 litre flask = 1 dm³ flask

(c)

0.1 m = 1 dm
0.1 m = 1 dm
0.1 m = 1 dm
1 dm³ volume

1 m
1 m
1 m³ volume

Figure 0.3 (a) Cube of volume 1 m³. (b) Flask of volume 1 L. *Source:* Eisco Labs. (c) Cube of volume 1 dm³, not to scale.

Therefore: 1 L = 1 dm³.

1 dm (decimetre) = 0.1 m (1×10^{-1} m) (Figure 0.3c).

1 dm³ (decimetre cubed) = $(0.1$ m$)^3$ = $(1 \times 10^{-1}$ m$)^3$ = 1×10^{-3} m³.

1 dm³ is therefore one thousandth of a m³. There are therefore 1 000 dm³ or 1 000 L in 1 m³.

Imagine the size of a 1 litre carton of orange juice. 1 000 cartons of 1 litre orange juice would fit in a cubic metre.

Remember

1 dm³ = 1 L = 1×10^{-3} m³
1 000 dm³ = 1 000 L = 1 m³.

Source: Dr Elizabeth Page.

Worked Example 0.4   Convert each of the following masses to grams. You may need to refer to Table 0.3 to obtain the numerical values of the prefixes used. Give your answers in scientific notation:

a. 2.32 Mg

b. 1 000 mg

c. 400 μg

Solution

    a. $2.32\,Mg = 2.32 \times 10^6\,g$

    b. $1\,000\,mg = 1\,000 \times 10^{-3}\,g = 1\,g$

    c. $400\,\mu g = 400 \times 10^{-6}\,g = 4.00 \times 10^{-4}\,g$

Worked Example 0.5   Express the following quantities in the units stated:

    a. $1\,cm^3$ in m^3

    b. $1\,000\,m^3$ in dm^3

    c. $1\,g\,m^{-3}$ in $g\,dm^{-3}$

Solution

    a. $1\,cm$ is the same as $0.01\,m$ or $1 \times 10^{-2}\,m$.
       Therefore $1\,cm^3 = (0.01\,m)^3 = (1 \times 10^{-2})^3\,m^3 = 1 \times 10^{-6}\,m^3$

    b. $1\,m$ is 10 times bigger than $1\,dm$: $1\,m = 10\,dm$.
       Therefore $1\,m^3 = 1 \times 10^3\,dm^3$.
       $1\,m^3$ is 1×10^3 (one thousand) times bigger than $1\,dm^3$.
       Therefore $1\,000\,m^3 = 1\,000 \times 10^3\,dm^3 = 1 \times 10^6\,dm^3$.

    c. $1\,dm^3$ is one-thousandth of a m^3.
       If the concentration is equal to $1\,g$ in $1\,m^3$, there will be $1\,000$ times fewer grams in $1\,dm^3$.
       So $1\,g\,m^{-3} = 1 \times 10^{-3}\,g\,dm^{-3}$.

0.5   Significant figures

When carrying out calculations in science, the answer must be given to the same level of accuracy as the values in the question. For example, if a balance weighs to two decimal places, the mass of substance weighed may be read as 5.02 g. The total number of significant figures in this number is three. This means we know the mass accurate to 3 significant figures.

The number can be rounded to a smaller number of significant figures as shown:

Three significant figures = 5.02 g

Two significant figures = 5.0 g

One significant figure = 5 g

An answer should be given to the same number of significant figures as the number with fewest significant figures in the calculation. For example, if the mass of solid is 5.02 g and the mass of water the solid is added to is 50 g (the same as 50 mL or 50 cm^3), the total mass of the solid and water would be 5.02 g + 50 g = 55 g. The answer can only be given to two significant figures as the mass of water is only known to this level of accuracy. In fact, the mass of water could actually be any mass between 49.50 g and 50.49 g, which are both equivalent to 50 g when given to two significant figures.

There are some rules for determining how many significant figures are in a number:

i. Any zeroes before a digit are not significant. For example, 0.005 is only accurate to 1 significant figure.

ii. Any zeroes after a digit are significant. For example, 0.00500 is accurate to 3 significant figures.

iii. Digits below the number 5 are always rounded downwards, and digits equal to or above the number 5 are rounded upwards, as in the following examples:

- The number 0.544 becomes 0.54 when written to two significant figures.

- The number 0.545 becomes 0.55 when written to two significant figures.

- The number 0.546 becomes 0.55 when written to two significant figures.

iv. When performing a calculation, determine the required number of significant figures and round up or down at the end of the calculation, not at the steps in between.

v. Always give your answer to the same accuracy as that of the value known to the least number of significant figures in the calculation.

Worked Example 0.6   Convert the following numbers to values with two significant figures, and write the answer in scientific notation:

a. 9 495 g

b. 0.00940 g

c. 0.09056 g

d. 19.005 g + 1.515 g

Solution

a. 9 495 g has four significant figures and becomes 9 500 g to two significant figures. In scientific notation, this is written as: 9.5×10^3 g.

b. 0.00940 g has three significant figures and becomes 0.0094 g to two significant figures. In scientific notation, this is written as: 9.4×10^{-3} g.

c. 0.09056 g has four significant figures and becomes 0.091 g to two significant figures when rounded up. In scientific notation, this is written as: 9.1×10^{-2} g.

d. The sum of 19.005 g + 1.515 g is 20.520 g, but this value has five significant figures. When it is rounded to two significant figures, it becomes 21 g as 0.52 rounds up to 1.0.

0.6 Calculations using scientific notation

0.6.1 Adding and subtracting

When adding numbers expressed using scientific notation, it is often useful to write the numbers using non-standard coefficients to simplify the mathematical process.

For example, the number 4 242 can be written in any of the following ways and retains its original numerical value:

$$4\ 242 = 4\ 242 \times 10^0$$

$$4\ 242 = 424.2 \times 10^1$$

$$4\ 242 = 42.42 \times 10^2$$

$$4\ 242 = 4.242 \times 10^3$$

Only the last figure is standard scientific notation. However, if we were required to add the number 4 242 to 5.00×10^2 and give the answer in standard scientific notation, the following procedure could be used:

$$4\ 242 + 5.00 \times 10^2 = 42.42 \times 10^2 + 5.00 \times 10^2 = 47.42 \times 10^2 = 4.742 \times 10^3$$

The example shows that the numbers are converted such that the exponent is common, i.e. 10^2 in this case, and then the numbers can be added.

Similarly, with numbers smaller than 1, the values are converted such that the exponents are common and added or subtracted in the normal manner. The exponent remains unchanged.

For example:

$$8.57 \times 10^{-2} - 1.23 \times 10^{-3} = 8.57 \times 10^{-2} - 0.123 \times 10^{-2} = 8.447 \times 10^{-2}$$
$$= 8.45 \times 10^{-2}\ \text{(after rounding).}$$

In this example the numbers are exact, that means we know them precisely. For example the number of people in a room. We therefore don't need to give the answer to a specific number of significant figures as, effectively, the number of significant figures is infinite.

This answer must be rounded, as we only know the original values to an accuracy of three significant figures.

Worked Example 0.7   Calculate the values of the following, giving your answers in scientific notation:

a. $102 + 1.310 \times 10^3 =$

b. $0.057\ 90 + 1.3 \times 10^{-4} =$

c. $3.120 \times 10^{-2} - 5.7 \times 10^{-4} =$

d. $6.375 \times 10^3 - 0.103 \times 10^2 =$

Solution

a. Express both numbers in a format that has the same exponent, and then add:

$$0.102 \times 10^3 + 1.310 \times 10^3 = 1.412 \times 10^3$$

b. Express both numbers in a format that has the same exponent, and then subtract:

$$579.0 \times 10^{-4} + 1.3 \times 10^{-4} = 580.3 \times 10^{-4} = 5.803 \times 10^{-2}$$

$$= 5.8 \times 10^{-2} \text{ (to 2 sig figs)}$$

c. Express both numbers in a format that has the same exponent, and then subtract:

$$3.120 \times 10^{-2} - 0.057 \times 10^{-2} = 3.063 \times 10^{-2}$$

$$= 3.1 \times 10^{-2} \text{ (to 2 sig figs)}$$

d. Express both numbers in a format that has the same exponent, and then subtract:

$$6.375 \times 10^3 - 0.0103 \times 10^3 = 6.3647 \times 10^3 = 6.36 \times 10^3 \text{ (to 3 sig figs)}$$

0.6.2   Multiplying and dividing numbers

When multiplying numbers written in scientific notation, the coefficients (i.e. the numbers in front of the 10^n) are multiplied and the exponents are added. So, for example:

$$(2.42 \times 10^2) \times (4.54 \times 10^2) = (2.42 \times 4.54) \times 10^{(2 + 2)} = 10.9868 \times 10^4$$

$$= 1.098\,68 \times 10^5$$

When dividing numbers written in scientific notation, the coefficients are divided and the exponents subtracted. So, for example:

$$\frac{2.42 \times 10^{-2}}{4.54 \times 10^{-4}} = \frac{2.42}{4.54} \times 10^{(-2-(-4))} = 0.533 \times 10^2 = 5.33 \times 10^1 \text{ or } 53.3$$

Worked Example 0.8   Calculate the answers to the following expressions, and present your result in scientific notation:

a. $7.5 \times 5.7 =$

b. $(6.4 \times 10^2) \times (1.30 \times 10^4) =$

c. $(1.751 \times 10^{-3}) \times (59.0 \times 10^{-2}) =$

d. $1.435 \times 10^{-2} \div 2.9 \times 10^{-4} =$

Solution

a. $7.5 \times 5.7 = 42.75 = 43 = 4.3 \times 10^1$

The smallest number of significant figures is two, so this becomes 43 after reducing the number of significant figures and rounding.

b. $(6.4 \times 10^2) \times (1.30 \times 10^4) = 8.32 \times 10^6 = 8.3 \times 10^6$

The smallest number of significant figures is two, so 8.32 becomes 8.3 after reducing the number of significant figures.

c. $(1.751 \times 10^{-3}) \times (59.0 \times 10^{-2}) = 103.309 \times 10^{-5} = 1.033\ 09 \times 10^{-3} = 1.03 \times 10^{-3}$

The smallest number of significant figures is three, so 1.03309 becomes 1.03 after reducing the number of significant figures.

d. $\dfrac{1.435 \times 10^{-2}}{2.9 \times 10^{-4}} = 0.4948 \times 10^{(-2-(-4))} = 0.4948 \times 10^2 = 4.948 \times 10^1 = 4.9 \times 10^1$

The smallest number of significant figures is two, so 4.948×10^1 becomes 4.9×10^1 after reducing the number of significant figures and rounding down. This would normally be written as 49.

0.7 Writing chemical formulae and equations

0.7.1 Writing chemical formulae

A chemical compound has a formula that represents the type and number of atoms in that compound. The plural of 'formula' is 'formulae' as the word derives from Latin.

Most people are familiar with H_2O and CO_2 as the formulae that represent water and carbon dioxide, respectively.

The letters in the formulae represent the elements in the periodic table, and the numbers given as subscripts after the letters indicate how many atoms of each element are in one unit of the compound. The formula H_2O tells us there are two atoms of hydrogen (H) and one atom of oxygen (O) in one molecule of the compound. In CO_2, there is one atom of carbon and two atoms of oxygen.

The terms 'atom' and 'molecule' will be explained fully in Chapter 1.

In the formulae of some compounds, there may be atoms enclosed in brackets (or parentheses): for example, $Ca(NO_3)_2$, the formula for calcium nitrate. The formula indicates that in one formula unit of calcium nitrate, there is one atom of calcium (Ca), two atoms of nitrogen (N), and six atoms of oxygen (O). The brackets act as a multiplier, multiplying all the atoms inside the bracket by the number in subscript outside the bracket.

Substances such as calcium nitrate, $Ca(NO_3)_2$, are known as **salts**. These are composed of **ions**. Ions are atoms, or groups of atoms, that have an overall charge. The charge can be positive or negative. A positively charged ion is known as a **cation**. A negatively charged ion is known as an **anion**. Table 0.4 lists some common cations and anions. You will see that most cations are actually metal ions.

Salts are neutral substances made up of positive cations and negative anions in a ratio that ensures the overall charge on the salt is zero. For example, aluminium sulfate is composed of aluminium (Al^{3+}) ions and sulfate (SO_4^{2-}) ions. Because the salt is neutral overall, the formula must be $Al_2(SO_4)_3$.

Table 0.4 Symbols and charges for some common cations and anions.

Common cations			Common anions		
Symbol	Name	Charge	Symbol	Name	Charge
Na^+	Sodium	+1	F^-	Fluoride	−1
K^+	Potassium	+1	Cl^-	Chloride	−1
Mg^{2+}	Magnesium	+2	Br^-	Bromide	−1
Ca^{2+}	Calcium	+2	O^{2-}	Oxide	−2
Al^{3+}	Aluminium	+3	S^{2-}	Sulfide	−2
Cu^{2+}	Copper	+2	$CO_3{}^{2-}$	Carbonate	−2
Fe^{2+}	Iron	+2	$SO_4{}^{2-}$	Sulfate	−2
Ag^+	Silver	+1	$NO_3{}^-$	Nitrate	−1
Zn^{2+}	Zinc	+2	$PO_4{}^{3-}$	Phosphate	−3
$NH_4{}^+$	Ammonium	+1	OH^-	Hydroxide	−1

0.7.2 Writing and balancing chemical equations

A chemical equation provides a shorthand method for describing the process taking place in a chemical reaction. The chemical formulae for the reactants are written on the left-hand side of the equation and the formulae for the products on the right-hand side. The number of each type of atom on the left-hand side of the equation must be the same as the number of each type of atom on the right-hand side. This involves **balancing** the equation by writing the number of moles of each substance in front of the formula for the substance. These numbers are called **stoichiometric coefficients**.

To balance a chemical equation, first write the reactants and products separated by a reaction arrow. For example, in the reaction of hydrogen gas with oxygen gas to form water, the reactants hydrogen and oxygen should be written on the left-hand side and the product water on the right-hand side of the reaction arrow. Note that both hydrogen and oxygen exist as molecules whose formulae are H_2 and O_2, respectively. The formula for water is H_2O.

$$H_2 + O_2 \rightarrow H_2O \qquad \text{UNBALANCED EQUATION}$$

The numbers of atoms of hydrogen are balanced on both sides, but the equation shows 2 atoms of oxygen on the left and only 1 atom on the right. To balance the equation, we therefore need to increase the number of oxygen atoms on the right to 2 by including 2 moles (or units) of H_2O, as shown in red:

$$H_2 + O_2 \rightarrow 2H_2O \qquad \text{UNBALANCED EQUATION}$$

However, we now have 4 atoms of hydrogen on the right and only 2 on the left. In order to fully balance the equation, we must increase the number of atoms of hydrogen on the left to 4, as shown:

$$2H_2 + O_2 \rightarrow 2H_2O \qquad \text{BALANCED EQUATION}$$

The equation is now balanced. The total number of atoms on the left of the equation is equal to the number on the right.

As another example, consider the reaction of calcium hydroxide, $Ca(OH)_2$, with nitric acid, HNO_3. The products in this reaction are calcium nitrate, $Ca(NO_3)_2$, and water.

The **mole** is the term used in chemistry to describe a specific amount of a material. It will be explained fully in Chapter 3.

Calcium has a charge of +2 and so requires two hydroxide ions of charge −1 for the compound to remain neutral. Calcium also requires two nitrate ions, $NO_3{}^-$, to form a neutral formula unit of calcium nitrate, $Ca(NO_3)_2$.

Again, write the formulae for the reactants on the left-hand side of the equation and the products on the right-hand side, as shown:

$$Ca(OH)_2 + HNO_3 \rightarrow Ca(NO_3)_2 + H_2O \qquad \text{UNBALANCED EQUATION}$$

In this case, it is simpler to treat the nitrate ion as a unit with formula NO_3^- rather than separately balancing nitrogen and oxygen atoms on both sides of the equation. It can be seen that there is one nitrate ion on the left-hand side but there are two on the right. Therefore we need to add 2 moles of nitric acid to the left-hand side:

$$Ca(OH)_2 + 2HNO_3 \rightarrow Ca(NO_3)_2 + H_2O \qquad \text{UNBALANCED EQUATION}$$

This has balanced the nitrate ions but not the hydrogen and oxygen atoms. There are 4 hydrogen atoms on the left-hand side but only 2 on the right. There are also 2 oxygen atoms (apart from in the NO_3^- ion) on the left-hand side but only 1 on the right. We therefore need to increase the number of water molecules to 2:

$$Ca(OH)_2 + 2HNO_3 \rightarrow Ca(NO_3)_2 + 2H_2O \qquad \text{BALANCED EQUATION}$$

The equation is now balanced.
The word equation for this reaction is spoken as:
1 mole of calcium hydroxide reacts with 2 moles of nitric acid to give 1 mole of calcium nitrate and 2 moles of water.

If a formula unit is written with a number in front of it in a chemical equation, for example, $2Ca(NO_3)_2$, this means that there are 2 units or lots of $Ca(NO_3)_2$ taking part in the reaction. The scientific term for this quantity is a *mole*. You will meet the mole and its definition in Chapter 3.

Worked Example 0.9   Balance the following equations:

    a. $Mg + O_2 \rightarrow MgO$

    b. $NaOH + H_2SO_4 \rightarrow Na_2SO_4 + H_2O$

    c. $CO + NO \rightarrow CO_2 + N_2$

    d. $C_2H_5OH + O_2 \rightarrow CO_2 + H_2O$

Solution

    a. $Mg + O_2 \rightarrow MgO$
      This requires 2 atoms of O on the right-hand side: $Mg + O_2 \rightarrow 2MgO$.
      And the left-hand side now requires 2 atoms of Mg:
      $2Mg + O_2 \rightarrow 2MgO$.

    b. $NaOH + H_2SO_4 \rightarrow Na_2SO_4 + H_2O$
      The left-hand side requires 2 atoms of Na:
      $2NaOH + H_2SO_4 \rightarrow Na_2SO_4 + H_2O$.
      Now the right-hand side needs 2 moles of H_2O:
      $2NaOH + H_2SO_4 \rightarrow Na_2SO_4 + 2H_2O$.

    c. $CO + NO \rightarrow CO_2 + N_2$
      The left-hand side needs 2 atoms of N: $CO + 2NO \rightarrow CO_2 + N_2$.
      The left-hand side now has 3 atoms of O and the right-hand side has 2, so increase the number of moles of CO_2 on the right:
      $CO + 2NO \rightarrow 2CO_2 + N_2$.
      Now the left-hand side needs one more atom of C and one more atom of O: $2CO + 2NO \rightarrow 2CO_2 + N_2$.

d. $C_2H_5OH + O_2 \rightarrow CO_2 + H_2O$

When a hydrocarbon is burnt in oxygen, the products are carbon dioxide and water. First balance the number of atoms of carbon on each side of the equation:
$C_2H_5OH + O_2 \rightarrow 2CO_2 + H_2O$.
Now balance the number of atoms of hydrogen by increasing the number of moles of water to 3:
$C_2H_5OH + O_2 \rightarrow 2CO_2 + 3H_2O$.
Next balance the number of atoms of oxygen by adding 3 moles of O_2 on the left-hand side:
$C_2H_5OH + 3O_2 \rightarrow 2CO_2 + 3H_2O$.

0.7.3 Indicating the physical state of reactants and products in chemical equations

When writing chemical equations, it is sometimes helpful to include the physical state of the reactants and products in the equation. This is especially important in equations relating to energy changes, as the energy required or released in the reaction depends upon the physical state of the products.

The common symbols used to describe the physical state of a reactant or product are given in Table 0.5.

Table 0.5 State symbols commonly used in chemical equations.

State	Symbol	Example Formula	Substance
Gas	g	$H_2O(g)$	Water vapour or steam
Liquid	l	$H_2O(l)$	Liquid water
Solid	s	$H_2O(s)$	Ice
Solid[1]	gr	$C(gr)$	Graphite
Solid[1]	d	$C(d)$	Diamond
Aqueous[2]	aq	$NaCl(aq)$	Sodium chloride dissolved in water

[1]The forms (called *allotropes*) of carbon, i.e. graphite and diamond, have specific symbols as energy changes are slightly different depending upon the form of carbon used in the reaction.
[2]Aqueous means that the substance is dissolved in water or is in aqueous solution.

Quick-check summary

- The measurement system used in chemistry is the SI system. There are six base units commonly used in chemistry. All units used in chemistry can be reduced to these six base units.

- Very large or very small numbers used in chemistry are expressed using scientific notation. In scientific notation, a number is expressed as a coefficient (a number between 1 and 10) multiplied by an exponential factor (10 raised to a whole number power). For numbers less than 1, the power is a negative number; and for numbers greater than 1, the power is a positive number.

- Metric prefixes can be used with units to indicate multiples of 10. These prefixes are letters that denote the factor used to multiply the unit.

- The properties of mass and volume are met very often in chemistry. The mass of an object is a measure of the amount of matter in the object. The volume of an object is a measure of the amount of space occupied by a substance. The litre (L) and millilitre (mL) are often used to denote volume, although these are not standard SI units.

- The number of significant figures in a measurement controls the number of significant figures obtained in any calculation involving that measured quantity. The number of significant figures in the result of a calculation cannot be expressed with a greater number of significant figures than that of the smallest number of significant figures in the measured quantity.

- When adding and subtracting values written using scientific notation, all exponents should be converted to the same size.

- When multiplying numbers written using scientific notation, the exponents should be added. When dividing numbers written using scientific notation, the exponents in the denominator should be subtracted.

- The chemical formula of a substance expresses the number of each type of atom in the substance.

- Ions are single atoms or groups of atoms that have a small whole number positive or negative charge. Positive ions are called cations, and negative ions are called anions. Cations and anions combine to form salts. The overall charge on the formula unit of a salt is zero.

- A chemical equation provides a shorthand method for describing the process taking place in a chemical reaction. The number and types of each atom on the left-hand side of a chemical equation must be the same as the number and types of each atom on the right-hand side of the equation. When this is the case, the equation is said to be balanced.

- The physical state of a material at the temperature of the reaction can be included in a chemical equation using a symbol in parentheses following the formula of the substance in the equation.

End-of-chapter questions

1 Determine the derived units of the following quantities:

   a. Acceleration, a (a = change in velocity/time)

   b. Force, F ($F = m \times a$, where m = mass and a = acceleration)

   c. Gravitational potential energy, PE ($PE = m \times g \times h$, where m = mass, g = acceleration due to gravity, and h = height)

2 Express the following numbers using scientific notation:

   a. 0.00849

   b. 84 265

   c. 354

   d. 0.0000218

3   Convert the following numbers to non-exponential form:

   a. 2.3×10^{-2}

   b. 2.8×10^{2}

   c. 4.95×10^{-4}

   d. 5.759×10^{5}

4   Write the following quantities in a more convenient form by changing the units, and give the answer in scientific notation:

   a. 6 300 m (change to km)

   b. 1 540 ms (change to s)

   c. 0.000456 kg (change to g)

   d. 639 500 J (change to kJ)

   e. 6 500 ns (change from ns to μs)

   f. 250 cm³ (change to dm³)

   g. 0.5 L (change to cm³)

5   How many significant figures are in each of the following numbers?

   a. 0.0945

   b. 83.220

   c. 106.0

   d. 0.00314

6   Determine the results of the following calculations, taking care to give the correct number of significant figures *and* units in the answers.

   a. 12.786 cm + 1.23 m =

   b. 3.21 cm × 150.091 mm =

   c. 1.090 m divided by 10 s =

   d. 2 100 J + 5.13 kJ + 6 MJ =

7   Determine the results of the following calculations, taking care to give the correct number of significant figures *and* units in the answers. Give your answer in scientific notation.

   a. 3.2×10^{3} m $\times 3.1 \times 10^{-5}$ kg m^{-1} =

   b. 9.47×10^{-3} km $\div 2.3 \times 10^{-1}$ m s^{-1} =

   c. 30.6010×10^{-3} kg $- 1.040 \times 10^{3}$ mg =

8   Using the information in Table 0.4, write the formulae of the following compounds:

   a. potassium nitrate

   b. calcium chloride

c. sodium sulfate

d. ammonium phosphate

e. aluminium sulfate

9 Balance the following equations by adding appropriate stoichiometric coefficients:

a. $N_2 + H_2 \rightarrow NH_3$

b. $Fe + O_2 \rightarrow Fe_2O_3$

c. $C_5H_{12} + O_2 \rightarrow CO_2 + H_2O$

d. $NaOH + H_2SO_4 \rightarrow Na_2SO_4 + H_2O$

e. $Li + H_2O \rightarrow LiOH + H_2$

10 Write balanced chemical equations for the following reactions, including state symbols, assuming they are carried out at room temperature. Use Table 0.4 to obtain the formulae of the salts included.

a. Hydrogen and oxygen to give water

b. Graphite (an allotrope of carbon) and oxygen to give carbon dioxide

c. Calcium carbonate (a solid) with hydrochloric acid (an aqueous solution of HCl) to give calcium chloride (in aqueous solution), water, and carbon dioxide

d. An aqueous solution of silver nitrate with an aqueous solution of sodium chloride to give a precipitate of silver chloride and sodium nitrate in solution

e. The thermal decomposition of solid lithium nitrate to solid lithium oxide, oxygen, and nitrogen dioxide (NO_2) gas

1

Atomic structure

At the end of this chapter, students should be able to:

- Draw a representation of an atom, and determine how many protons, neutrons, and electrons are present
- Explain how electrons are arranged in atoms
- Explain and draw s, p, and d orbitals
- Give the ground state electronic configuration of an atom
- Explain isotopes, and calculate the average atomic mass of an element from given isotopic ratios
- Explain the process of radioactive decay, and give relevant equations

1.1 Atomic structure

This section will outline the structure of the atom, as well as introduce isotopes and radioisotopes of the elements.

1.1.1 Subatomic particles

Understanding the atom is fundamental to being able to understand, and subsequently master, chemistry. All elements are made up of atoms, and the exact structure of each atom determines which element it is. There are two main regions inside an atom: the nucleus and the outer shells surrounding the nucleus. Within the nucleus are two types of subatomic particles: *protons* and *neutrons*. The number of protons in an atom determines the actual element. The third main type of subatomic particle is found in the outer shells, and these are called *electrons*. Electrons are very small and are negatively charged, and they orbit the

Foundations of Chemistry: An Introductory Course for Science Students, First Edition.
Philippa B. Cranwell and Elizabeth M. Page.
© 2021 John Wiley & Sons Ltd. Published 2021 by John Wiley & Sons Ltd.
Companion website: www.wiley.com/go/Cranwell/Foundations

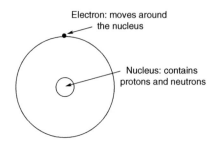

Figure 1.1 Simplified structure of the atom (not to scale).

Table 1.1 Properties of subatomic particles.

Particle	Symbol	Relative mass	Charge
Proton	p	1	+1
Neutron	n	1	0
Electron	e	$\dfrac{1}{1836}$ or 0.00055	−1

nucleus rather like planets orbit the sun. The exact manner in which an electron orbits the nucleus is defined by the *orbital* the electron occupies. This will be discussed in Section 1.2.4.

Protons and neutrons have the same mass and are much larger and heavier than electrons. Protons have a positive charge, and neutrons have no charge and so are neutral. Both protons and neutrons are found in the nucleus of the atom (Figure 1.1). Electrons have a much smaller mass than protons and neutrons and have a negative charge. Electrons are found in the region of space surrounding the nucleus. Table 1.1 gives the properties of the three main subatomic particles.

1.1.2   Mass number (*A*) and atomic number (*Z*)

Each element is described by the number of protons, neutrons, and electrons it possesses. These are represented by two quantities: the mass number (*A*) and the atomic number (*Z*) The mass number indicates the total number of protons and neutrons in the nucleus of the atom (p + n). The atomic number gives the number of protons in a neutral atom of the element (p). A neutral atom must have the same number of protons as electrons, so *Z* also indicates the number of electrons (e).

For any element X, we can represent the information about the mass number, *A*, and atomic number, *Z*, as shown in Figure 1.2a. As you can see, the mass number is written as a superscript and the atomic number as a subscript in front of the symbol for the element.

Information about the element beryllium is shown in Figure 1.2b using this convention.

Beryllium (Be) has an atomic number, *Z*, of 4; therefore, in a neutral molecule, there are 4 protons and 4 electrons. Beryllium has a mass number, *A*, of 9; therefore it has 5 neutrons (because the mass number is 9 and we already know that it has 4 protons from the atomic number, so 9 – 4 = 5 neutrons). The mass

The mass number gives the total number of protons and neutrons: $A = p + n$. The atomic number gives the total number of protons or electrons present in a neutral atom: $Z = p$ or e.

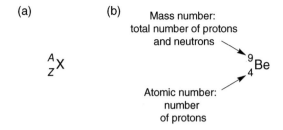

Figure 1.2 (a) General representation of mass number and atomic number for the element X. (b) The mass number and atomic number for beryllium.

number is written as a superscript and the atomic number as a subscript before the symbol for the element.

Worked Example 1.1   How many protons, electrons, and neutrons are present in an atom of aluminium, $^{27}_{13}$Al?

Solution
Aluminium exists as $^{27}_{13}$Al. The mass number is 27, and the atomic number is 13. The atomic number tells us how many protons there are, so in a neutral atom of aluminium, there are 13 protons, and there must be 13 electrons to balance the charge from the protons. If there are 13 protons, there must be 14 neutrons, because the number of protons and neutrons adds up to the mass number, 27.

Worked Example 1.2   How many protons, electrons, and neutrons are present in an oxide (O^{2-}) ion? The symbol for an oxygen atom is $^{16}_{8}$O.

Solution
Oxygen exists as $^{16}_{8}$O. An oxide ion is an oxygen atom with a 2– charge, so it has 2 extra electrons compared to a neutral oxygen atom. Therefore: oxygen contains 8 protons (due to the mass number) and 8 neutrons (due to the difference between the mass number and the atomic number); and, because it has a 2– charge, it has to contain 2 more electrons than it has protons (because each electron has a negative charge), so it has 10 electrons.

Remember from the previous chapter that an ion is an atom or a group of atoms that has a charge.

1.1.3   Isotopes

Many elements possess isotopes. An *isotope* is an atom of an element that has a different mass number. Because all isotopes of the same element have the same atomic number (Z), the number of protons must be the same. This is always the case. If the number of protons has changed, the element has also changed! Isotopes differ in the number of neutrons in the nucleus of the atom. An example of an element that has isotopes is bromine, which naturally exists in two forms:

- $^{79}_{35}$Br contains 35 protons, 35 electrons, and 44 neutrons.

- $^{81}_{35}$Br contains 35 protons, 35 electrons, and 46 neutrons.

An isotope is an atom of an element that has the same atomic number but differs in the number of neutrons and therefore mass number.

You will meet the average relative atomic mass of an element in Chapter 3.1.1

These isotopes are sometimes written as bromine-79 (^{79}Br) and bromine-81 (^{81}Br).

In naturally occurring bromine, the ratio of ^{79}Br to ^{81}Br is 50.5 : 49.5. This means that in 100 bromine atoms, 50.5 will be ^{79}Br and 49.5 will be ^{81}Br. The proportion naturally occurring of each isotope is called the **relative abundance** or **isotopic abundance**. Knowing the relative abundance of each isotope and its mass number allows us to calculate the average mass of one atom of the element as shown here for bromine:

$$\text{Average mass of one atom of Br} = \frac{(50.5 \times 79) + (49.5 \times 81)}{100} = 79.9$$

Note that the answer calculated has been given no units as A, the mass number, is unitless. The actual unit for the answer is the *atomic mass unit* (amu), which is an extremely small quantity. However, the unitless answer obtained is equal to the average **relative atomic mass**, A_r, of the element which is covered in Chapter 3.

Worked Example 1.3 Uranium exists as three isotopes: $^{238}_{92}$U, $^{235}_{92}$U, and $^{234}_{92}$U with approximate relative abundance 99.27%, 0.72%, and 0.01%, respectively. What is the average relative atomic mass of a uranium atom to three significant figures?

Solution

$$\text{Average relative atomic mass} = \frac{(238 \times 99.27) + (235 \times 0.72) + (234 \times 0.01)}{100}$$

$$= 237.99 = 238$$

1.1.4 Radioisotopes

Some isotopes are unstable and emit radiation; they are radioactive. Three types of emission are produced spontaneously from radioactive nuclei: alpha (α), beta (β), and gamma (γ). The exact species that is emitted is dependent upon the type of emission.

An alpha particle, α, has two protons and two neutrons – it is effectively the nucleus of a He atom. Compared to a helium atom, the alpha particle, α, has lost both of its electrons and therefore has a 2+ charge. We can represent it as: $^4_2$He or $^4_2\alpha$. Alpha (α) emission occurs when an alpha particle or helium nucleus is lost from the radioisotope. This is usually restricted to elements with atomic numbers over 83 that have a large proton-to-neutron ratio. This emission causes a change in the composition of the nucleus of the element from which it was emitted, generating a new element. Alpha radiation is not very penetrating and can be stopped by either paper or skin.

Alpha (α) particles are emitted by $^{210}_{84}$Po. An equation can be written for this radioactive decay process:

$$^{210}_{84}\text{Po} \rightarrow ^{206}_{82}\text{Pb} + ^4_2\text{He}$$

or

$$^{210}_{84}\text{Po} \rightarrow \,^{206}_{82}\text{Pb} + \,^{4}_{2}\alpha$$

In this case, because 2 protons are emitted from the nucleus, the atomic number of the new element formed is 82, which is the atomic number for lead (Pb). Therefore the element itself changes from polonium to lead. The atomic mass also alters because 2 protons and 2 neutrons have been lost. The new mass number (A) is 210 – 4 = 206, which tells us the new element has $A = 206$.

β^- (beta) emission arises when an electron ($^{0}_{-1}\text{e}$) is released from the atom and a proton is added to the nucleus. This sounds confusing, as a nucleus does not contain electrons, but what actually happens is that a neutron splits to give a proton and an electron. The proton remains in the nucleus, and the electron is emitted. The atomic number of the atom formed therefore increases by one. β^- radiation that generates a proton is usually seen in an element that has a neutron-rich nucleus. Beta radiation is more penetrating than alpha radiation and can be stopped by a sheet of aluminium foil.

An example of beta emission can be seen with the $^{40}_{19}\text{K}$ isotope:

$$^{40}_{19}\text{K} \rightarrow \,^{40}_{20}\text{Ca} + \,^{0}_{-1}\text{e}$$

The addition of a proton requires the atomic number to increase by one. In this case, potassium changed to calcium with an atomic number of 20.

Gamma (γ) radiation (or gamma rays) is penetrating electromagnetic radiation released from the nucleus of an atom. It consists of high-energy photons. Gamma radiation is often released alongside alpha (α) or beta (β) radiation. Gamma radiation is the most strongly penetrating and ionising of the three types of radiation and can only be stopped by using a lead sheet.

Electromagnetic radiation is a continuous range of wavelengths and hence includes the wavelength of visible light. A *photon* is the smallest discrete amount, or quantum, of electromagnetic radiation.

1.2 Electronic structure

This section will outline the structure of the atom in more detail. It describes how electrons are arranged in atoms and how this determines the way the periodic table of the elements is built up.

1.2.1 The periodic table

Figure 1.3 shows the arrangement of all 118 known elements in the periodic table. The elements are distributed over 18 columns and 7 rows. The elements are arranged according to atomic number, which increases from left to right and top to bottom. The atomic number equals the number of protons present in a neutral atom and therefore determines the position of the element in the periodic table.

The rows across the periodic table are called *periods,* and there are seven periods numbered from one to seven. The columns down the periodic tables are called *groups,* and there are 18 groups in total, but their naming usually runs from one to eight. All of the elements underneath hydrogen (H) are in Group 1, and beneath beryllium (Be) they are in Group 2. The groups in the middle of the periodic table, from scandium (Sc) to zinc (Zn), are called the *d-block* elements and are sometimes referred to by group numbers (i.e. Groups 3 to 12 in the 1-18 numbering system). The elements below boron (B) are in Group 3 (or Group 13), below C are in Group 4 (or Group 14), etc. up to those below helium (He), which are called Group 8 (sometimes called Group 0 or Group 18). Elements that exist in the same group have broadly similar chemical properties, due to having the same number of electrons in their outermost shell.

It is best to approach reading the periodic table like a book: start at the top left, and finish at the bottom right.

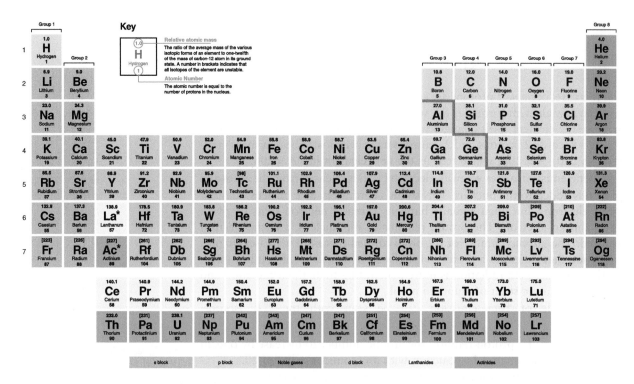

Figure 1.3 The standard modern form of the periodic table. *Source:* University of Reading. Public Domain.

In some textbooks, you may see Groups 1–8 numbered as Groups 1–18, where the d-block elements are numbered 3–12.

The periodic table is roughly divided by a zig-zag line that splits the table into two parts. The elements to the right of the zig-zag line are non-metals, and those to the left of the line are metals. There are many more metals than non-metals in the periodic table. Whether an element is a metal or non-metal has an impact upon its properties. In addition, the periodic table is split into four distinct areas: the s block, the p block, the d block, and the f block (Figure 1.3). These areas relate to the orbital in which the outermost electron resides, and will be discussed further in Section 1.2.4.

Metals tend to be lustrous, malleable, sonorous when struck, ductile, and good conductors of heat and electricity. Metals are often solid, with a high melting and boiling point, although the one notable exception is mercury (Hg). Metals usually have a high density and can be heavy. In comparison, non-metals tend to be dull, brittle, and poor conductors of heat and electricity, and they generally have lower melting and boiling points than metals. Elements that are close to the zig-zag line exhibit atypical properties and are sometimes referred to as *metalloids*. For example, silicon (Si) exhibits properties that would be expected for both a metal and a non-metal.

The bonding properties of metals and non-metals will be discussed in Chapter 2.

1.2.2 Electron energy levels

Electrons are arranged outside the nucleus of the atom in *energy levels* that are sometimes called *principal quantum shells*. You can imagine these as spherical layers extending out from the nucleus as in Figure 1.4. The energy levels are numbered from 1 to 7, with the level closest to the nucleus being level 1. These

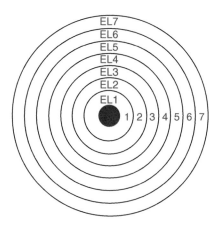

Figure 1.4 The energy levels (EL) in an atom. The integers represent the principal quantum number, n.

numbers are referred to as the *principal quantum number* of the electrons in that energy level. The principal quantum number has the symbol n. As the distance from the nucleus gets bigger, the volume of the layers increases. In addition, as the distance from the nucleus increases, the energy of the level increases along with the energy of the electrons in the level. The seven principal quantum levels relate to the seven rows across the periodic table.

1.2.3   Simple electronic configurations

The arrangement of electrons in the outer shells or principal quantum levels of the atom is called the *electronic configuration*. Electrons fill the lowest energy levels in an atom first. As n increases, the energy levels increase in both size and energy and can hold more electrons.

The maximum number of electrons that can be held in the first three energy levels is shown in Table 1.2. In this course, we will not be concerned with electrons in energy levels beyond $n = 4$.

The first shell can contain only two electrons. The elements that only have electrons in the first shell are hydrogen and helium, with one and two electrons, respectively. Once the first shell is full, at helium, the next electron (in lithium) must be placed in the second shell. The second shell can hold up to eight electrons. Once a shell is full, an element is said to be *stable*. The noble gases (Group 8, also called Group 18) all have a full outer shell, so they are stable and are very unreactive. Table 1.3 shows the arrangement of the electrons in shells for the first 11 elements in the periodic table.

You can see from the table that each consecutive element has one more electron, and these fill energy levels from $n = 1$ upwards. The first energy level is full at helium, and so the next electron (in lithium) enters the $n = 2$ shell. The second

Table 1.2 The maximum number of electrons in the first four energy levels.

Energy level or shell	n	Maximum number of electrons
First	1	2
Second	2	8
Third	3	18
Fourth	4	32

Table 1.3 Arrangement of electrons in the first 11 elements of the periodic table.

Element	Number of electrons or Z	Number in 1st shell ($n = 1$)	Number in 2nd shell ($n = 2$)	Number in 3rd shell ($n = 3$)
H	1	1	0	0
He	2	2	0	0
Li	3	2	1	0
Be	4	2	2	0
B	5	2	3	0
C	6	2	4	0
N	7	2	5	0
O	8	2	6	0
F	9	2	7	0
Ne	10	2	8	0
Na	11	2	8	1

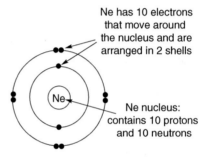

Ne has 10 electrons that move around the nucleus and are arranged in 2 shells

Ne nucleus: contains 10 protons and 10 neutrons

Figure 1.5 The structure of neon.

energy level is full at neon ($Z = 10$), and so the next electron (in sodium) enters the $n = 3$ energy level.

Electron configurations can be represented as in the diagram in Figure 1.5 for neon. This shows that the first and second shells are both full.

You will notice that there are four paired electrons and two single electrons in the second energy level of the O atom in Worked Example 1.4. Electrons in the same energy level tend to remain unpaired unless they have to pair up with another electron.

Worked Example 1.4   Draw the arrangement of electrons in oxygen.

Solution
The position of oxygen in the periodic table and its atomic number ($Z = 8$) tell us that a neutral atom of oxygen has 8 protons and therefore must have 8 electrons. However, oxygen is in the second row of the periodic table; therefore the electrons are arranged into two shells. The first shell contains two electrons because there is only space for two electrons, and the second shell contains six electrons. We can draw this as shown:

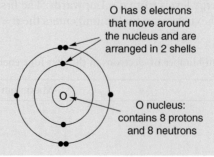

O has 8 electrons that move around the nucleus and are arranged in 2 shells

O nucleus: contains 8 protons and 8 neutrons

We have seen that each successive shell in an atom can hold an increasing number of electrons. The reason for the precise maximum number of electrons that each shell can hold is due to the arrangement of the electrons in each shell. The next section will explain how electrons are arranged in shells or energy levels.

1.2.4   Sub-shells and atomic orbitals

Within each energy level, there are distinct areas in space that the electrons occupy. These distinct areas are *sub-shells*, which are further arranged into *atomic orbitals*. A sub-shell is a group of atomic orbitals that have the same properties, such as shape. The sub-shells are called s, p, d, and f. Atomic orbitals are regions within a sub-shell that can contain up to two electrons and have a particular orientation in space. For example, the p sub-shell actually comprises three atomic orbitals, which are called p-orbitals. To understand about atomic orbitals and sub-shells, it is useful to spend a little time considering the nature of electrons.

We have seen that electrons are very small, they have a negative charge, and they move very quickly. The early twentieth century was an exciting time for physicists and chemists. Many ground-breaking experiments were carried out that helped determine the properties of the three main fundamental atomic particles: the proton, the neutron, and the electron. Unfortunately, there isn't time or space to describe these experiments here, but two important experiments gave scientists unexpected and potentially conflicting information about the properties of electrons. These experiments showed that electrons can behave as both particles and waves, and this theory became known as *wave-particle duality*. A consequence of wave-particle duality is Heisenberg's Uncertainty Principle, which states that it is impossible to accurately know both the position and velocity of an electron at the same time. Around the same time, a physicist named Schrödinger produced a mathematical model of the atom, which allowed for the wave-like behaviour of electrons and produced equations that described the position of an electron in the outer shells of an atom. The equations that Schrödinger derived led to the idea that there are regions in space around the nucleus where there is a high probability of finding an electron of a given energy. These regions in space are called *atomic orbitals* and can be thought of as sub-shells within the principal quantum shells.

When considering the occupancy of electrons in atomic orbitals or the behaviour of electrons in chemical reactions, it is often more intuitive to think of electrons as small negatively charged particles – and this is fine. However, it is important to remember that electrons also have wave-like properties, which prevents us from stating exactly where the electron is at any one time; we can only say where there is a good probability of finding the electron.

As stated earlier electrons in each principal quantum shell of an atom are arranged in orbitals. An *orbital* is an area in space where there is a high probability of finding an electron of a specific energy. An orbital is actually a mathematical function that describes the wave-like behaviour of an electron in an atom. When this function is plotted in three dimensions for an electron of a certain energy, a three-dimensional shape is obtained. The shape represents the space in which there is a high probability of finding an electron with that energy. An atomic orbital can hold a maximum of two electrons.

Any atomic orbital can hold a maximum of two electrons.

s orbitals

The hydrogen atom has just one electron, which occupies the first principal quantum shell. There is just one atomic orbital in this shell, called a 1s orbital. The number 1 tells us the orbital is in the first principal quantum shell ($n = 1$).

An s orbital is spherical in shape, and in any shell, there is only one s orbital. A 1s orbital is shown in Figure 1.6a. Also shown in this figure are a 2s orbital and a 3s orbital (Figures 1.6b and c). They are all spherical in shape and get larger as the value of n increases. Within an atom, all the s orbitals are concentric – this means they all have the nucleus as their centre and lie on top of each other. The shading in the figure represents the electron density or the probability of finding an electron in each orbital. You can see that in the 1s orbital, there is a high probability of finding the electron close to the nucleus. However, in the higher s orbitals, there are areas that are unshaded and therefore are unlikely to contain electrons. These areas are called *nodes*.

Elements that either have fully or partially filled s orbitals in their outer shells are called s-block elements. These elements occupy Groups 1 and 2 of the periodic table.

Each s orbital can contain a maximum of two electrons. The second element in the periodic table, helium, has a total of two electrons. Both of the electrons in helium are therefore in the 1s orbital because the s orbital can hold up to two electrons and this is the lowest-energy orbital. To allow both negatively charged electrons to occupy the same space and not fly apart, the electrons have opposite spins. This creates a very tiny magnetic field about each electron, causing them to behave as magnets with opposite polarity, and so holds them together in the orbital.

The concept of spin is beyond the scope of this textbook. But when representing electrons in orbitals, if there are two electrons in one orbital, make sure they are represented by arrows (or half arrows) pointing in opposite directions: i.e. one pointing up and one pointing down, as in Fig. 1.10.

p orbitals

The second principal quantum shell has one s orbital and three p orbitals. p orbitals are shaped like a dumb-bell or figure of eight (see Figure 1.7). There are three

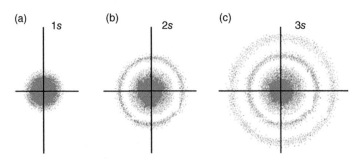

Figure 1.6 (a) The 1s orbital; (b) the 2s orbital; (c) the 3s orbital. *Source:* Redrawn from https://www.quora.com/What-is-a-simple-explanation-of-the-Quantum-mechanical-model-of-the-atom.

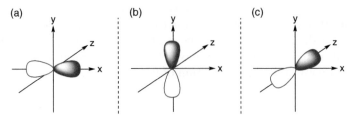

Figure 1.7 (a) The $2p_x$ orbital; (b) the $2p_y$ orbital; (c) the $2p_z$ orbital.

p orbitals in each quantum shell, where n is greater than 1, and each p orbital points in a different direction and is at 90° to the other two. For simplicity, they are generally drawn aligned along the x, y, and z axes. The p orbitals in the second shell are named $2p_x$, $2p_y$, and $2p_z$.

Just like s orbitals, p orbitals can only hold a maximum of two electrons. However as there are three p orbitals per energy level, there can be a maximum of six p electrons in total in a p sub-shell. Elements that have filled or partially filled p orbitals in their outer shells are found in the p block of the periodic table, which consists of Groups 3–8 (also known as Groups 13–18).

d orbitals

The third principal quantum shell in an atom contains an s orbital (3s), three p orbitals (3p), and five d orbitals (3d). The shapes of the d orbitals are shown in Figure 1.8. These are slightly more complex than p orbitals and again point in different directions along the x, y, and z axes. Four of the d orbitals have a similar shape (although three have lobes that point between the axes and the fourth has lobes that point along the axes). The fifth d orbital (d_{z^2}) has a dumb-bell shape with a donut ring around the middle. Again, each d orbital can hold a maximum of 2 electrons, so the maximum number of d electrons in any one energy level is 10. Elements that have outer electrons in d orbitals are found in the d block of the periodic table.

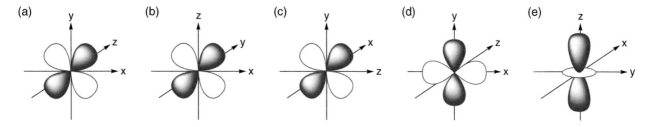

Figure 1.8 (a) The d_{xy} orbital; (b) the d_{xz} orbital; (c) the d_{yz} orbital; (d) the $d_{x^2-y^2}$ orbital; (e) the d_{z^2} orbital.

1.2.5   Describing electronic configurations

Now that you are familiar with the concept of principal quantum levels and atomic orbitals within these levels, the arrangement of electrons in atoms can be seen to follow a logical procedure.

As we move further away from the nucleus, the electrons and energy levels have higher energies. Within a principal quantum shell, the electrons in different types of orbitals have different energies. Electrons in s orbitals have a lower energy than electrons in p orbitals, which have a lower energy than electrons in d orbitals. So the order of energies of electrons in atomic orbitals is $s < p < d$. This is shown in Figure 1.9.

Electrons fill atomic orbitals according to the following rules:

1. Electrons always go into the lowest energy orbital possible. This is called the *Aufbau principle*, sometimes known as the *building-up* principle.

2. If there is more than one orbital of the same energy available, electrons always fill an unoccupied orbital first: Hund's rule.

Atomic orbitals are regions in space where there is a high probability of finding an electron. The $n = 1$ shell has one s orbital. The $n = 2$ shell has one s orbital and three p orbitals. The $n = 3$ shell has one s orbital, three p orbitals, and five d orbitals. Each atomic orbital can hold a maximum of two electrons.

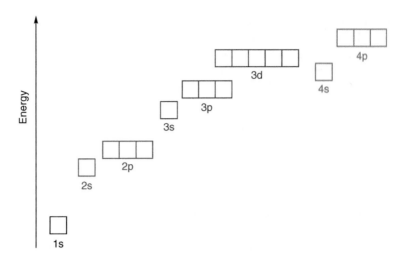

Figure 1.9 The relative energies of the orbitals in an atom. Relative energies not to scale.

3. If two electrons occupy the same orbital, they will have opposite spin: the Pauli exclusion principle.

We have already seen that the electrons for hydrogen and helium obey these rules. The lowest energy orbital is the 1s orbital, and this is filled at helium. The electrons are forced to have opposite spins in helium. The electron configuration for the atom is a shorthand that describes the occupancy of the orbitals. For hydrogen, the electron configuration is $1s^1$. For helium, the electron configuration is $1s^2$. The superscripted number represents the number of electrons in the 1s orbital for each element.

Once the 1s orbital is full, the next electron at lithium ($Z = 3$) must enter the second energy level, as this is the next lowest in energy. Within this energy level, the 2s orbital is of lower energy than the 2p orbitals. Thus the electron occupies the 2s orbital. The electron configuration of lithium is $1s^2 2s^1$.

The next element, beryllium ($Z = 4$), has four electrons, and the fourth electron must pair up with the electron in the 2s orbital, as this is of lower energy than the 2p orbitals. The electron configuration is $1s^2 2s^2$. The two electrons in the 2s orbital have opposite spins.

At boron ($Z = 5$), the 2s orbital is full, and so the next electron must occupy a 2p orbital. All are empty and of the same energy, and so we arbitrarily place the electron in the $2p_x$ orbital, although it could equally occupy $2p_y$ or $2p_z$. The electron configuration for boron is $1s^2 2s^2 2p^1$.

The next two electrons at carbon and nitrogen enter the other empty 2p orbitals. Nitrogen has the electron configuration $1s^2 2s^2 2p^3$. As all 2p orbitals are half-filled at nitrogen, the next electron of oxygen must pair with another p electron to give one fully occupied and two half-occupied 2p orbitals: $1s^2 2s^2 2p^4$. This can be visualised more easily by using the representation with electrons in boxes, as in Figure 1.10 for oxygen.

The next electrons complete the remaining two 2p orbitals so that at neon ($Z = 10$), we have a filled second shell of electrons: $1s^2 2s^2 2p^6$. We will see that this is a very stable arrangement of electrons and has significant consequences for the chemical reactivity of the element.

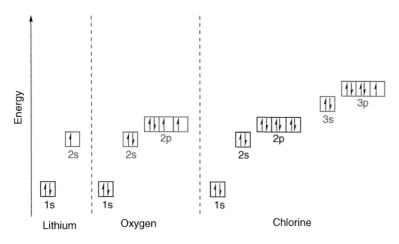

Figure 1.10 Electron arrangements in lithium, oxygen, and chlorine. Outer shell electrons are shown in red boxes.

Once the second shell is full at neon, the next electron at sodium ($Z = 11$) enters the third shell and occupies the 3s orbital. The filling process is then repeated as for the second row of elements. The electron arrangement for chlorine ($Z = 17$) is shown in Figure 1.10. Table 1.4 shows how electrons fill atomic orbitals for the first 36 elements – i.e. up to the end of the fourth row of the periodic table.

Within Table 1.4, there are various points to note. Firstly, electron filling appears to follow the rules just listed up to element 18, argon, when the 3p orbitals are full. The next electron might be expected to occupy the first 3d orbital; however, it can be seen that the electron in potassium actually occupies the 4s orbital. The reason for this is that the energy of the electrons in a 4s orbital is slightly lower than the energy of the electrons in the 3d orbitals, because the 4s orbital is larger and more diffuse. This means the electron in potassium occupies the 4s orbital preferentially, giving potassium a single outer s electron and placing the element in the s block. The electron configuration of potassium is $1s^2 2s^2 2p^6 3s^2 3p^6 4s^1$.

Secondly, the next electron at calcium also occupies the 4s orbital, giving an electron configuration of $1s^2 2s^2 2p^6 3s^2 3p^6 4s^2$. However, once the 4s orbital is filled, the next electron at scandium occupies the first 3d orbital, as this is now lower in energy than the 4p orbitals.

Finally, you may notice another anomaly in filling atomic orbitals at Cr and Cu. These two elements have just a single 4s electron, whereas the preceding elements have a filled 4s shell. Cr has one electron in each of its 3d orbitals, and copper has two electrons in each 3d orbital, leaving just one 4s electron in each case. The usual reason given for this relates to the extra stability associated with having a filled or half-filled set of d orbitals.

A shorthand way to write electron configurations is to use the symbol for the noble gas element (Group 8) to represent electrons in filled shells. So the electron configuration for potassium, $1s^2 2s^2 2p^6 3s^2 3p^6 4s^1$ can be written as $[Ar]4s^1$, where $[Ar]$ represents the configuration $1s^2 2s^2 2p^6 3s^2 3p^6$.

1.2.6 Electronic structures and the periodic table

The original organisation of elements in the periodic table was proposed by Mendeleev in 1869. This was before experiments had been carried out that proved the existence of atomic orbitals and energy levels in atoms. He organised the elements based on their atomic masses and other properties. It is now known

Table 1.4 Electron configurations for the first 36 elements.

Element symbol	Atomic number	$n = 1$ shell	$n = 2$ shell		$n = 3$ shell			$n = 4$ shell	
		1s	2s	2p	3s	3p	3d	4s	4p
H	1	1							
He	2	2							
Li	3	2	1						
Be	4	2	2						
B	5	2	2	1					
C	6	2	2	2					
N	7	2	2	3					
O	8	2	2	4					
F	9	2	2	5					
Ne	10	2	2	6					
Na	11	2	2	6	1				
Mg	12	2	2	6	2				
Al	13	2	2	6	2	1			
Si	14	2	2	6	2	2			
P	15	2	2	6	2	3			
S	16	2	2	6	2	4			
Cl	17	2	2	6	2	5			
Ar	18	2	2	6	2	6			
K	19	2	2	6	2	6		1	
Ca	20	2	2	6	2	6		2	
Sc	21	2	2	6	2	6	1	2	
Ti	22	2	2	6	2	6	2	2	
V	23	2	2	6	2	6	3	2	
Cr	24	2	2	6	2	6	5	1	
Mn	25	2	2	6	2	6	5	2	
Fe	26	2	2	6	2	6	6	2	
Co	27	2	2	6	2	6	7	2	
Ni	28	2	2	6	2	6	8	2	
Cu	29	2	2	6	2	6	10	1	
Zn	30	2	2	6	2	6	10	2	
Ga	31	2	2	6	2	6	10	2	1
Ge	32	2	2	6	2	6	10	2	2
As	33	2	2	6	2	6	10	2	3
Se	34	2	2	6	2	6	10	2	4
Br	35	2	2	6	2	6	10	2	5
Kr	36	2	2	6	2	6	10	2	6

that chemical reactivity depends upon electronic structure, so our current periodic table agrees with Mendeleev's. The periodic table consists of four main areas, which are generally shown shaded in different colours.

On the left hand of the periodic table are Groups 1 and 2. The elements in these groups are known as *s block* elements as their highest energy electrons are in s orbitals.

On the right of the periodic table are the *p block* elements. These constitute the elements in Groups 3–8 (or Groups 13–18) where the p orbitals are being filled.

In the middle section are the *d block* elements. This area includes 10 groups of elements from Sc to Zn in the first row of the d block. These elements are all metals and those in Groups 3-11 are often referred to as *transition elements*. The elements in the group headed by zinc (Group 12) are not transition elements.

At the bottom of the periodic table are the two rows of elements that form the *f block*. The elements in these two rows are called the *lanthanides* and *actinides* and have electrons filling the f orbitals. In total, f orbitals can hold 14 electrons, as there are 7f orbitals. A discussion of these orbitals is beyond the scope of this book.

Worked Example 1.5   Give the ground state electron configuration for $_{16}^{32}$S.

Solution

$_{16}^{32}$S has atomic number 16 and mass number 32. Therefore, a neutral atom of S has 16 protons, 16 electrons, and 16 neutrons. When putting the electrons into the orbitals, it is important to always start with the lowest-energy orbital.

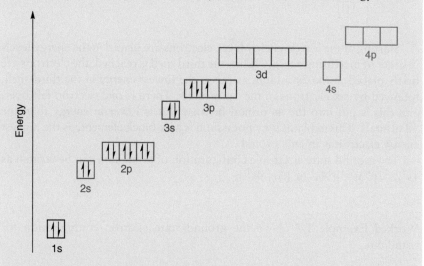

The reason we say 'ground state' electron configuration is because 'excited states' can also exist. The discussion of excited states is beyond the scope of this book but is discussed at degree level.

Electrons are placed in the energy levels starting at the lowest energy level (1s) in order of increasing energy (Aufbau principle). When filling the third shell, electrons are firstly placed in the 3s orbital, as this has the lowest energy in the third shell, followed by four electrons in the 3p orbitals. However, when putting the remaining four electrons in the 3p orbitals, each p orbital is filled with one electron first (Hund's rule). When each p orbital is half full, there is one electron left over, and it is at this point that the electrons are paired. When they are paired, they have opposing spins, which in this case is represented by drawing each half-arrow in opposite directions, fulfilling the requirements of the Pauli exclusion principle.

The ground state electronic configuration of sulfur can be written as $1s^2 2s^2 2p^6 3s^2 3p^4$.

Worked Example 1.6   Give the ground state electron configuration for $_{19}^{39}$K.

Solution

$_{19}^{39}$K has atomic number 19 and mass number 39, therefore in a neutral atom there are 19 protons, 19 electrons, and 20 neutrons.

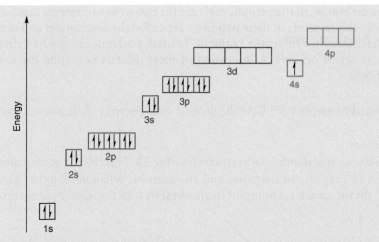

Starting at the lowest energy level, electrons are placed in the energy levels in order of increasing energy. When the third shell is reached, the electrons are firstly placed in the 3s orbital, as this is the lowest energy in the third shell, followed by six electrons in the 3p orbitals. There is one electron left over, and this is put into the 4s orbital because this is lower in energy than the 3d orbitals. This explains why potassium is an s-block element, as the highest energy electron is in an s orbital.

The ground state electronic configuration of potassium can be written as $1s^2 2s^2 2p^6 3s^2 3p^6 4s^1$ or [Ar] $4s^1$.

Worked Example 1.7   Give the ground state electron configuration for scandium.

Solution

$^{45}_{21}$Sc has atomic number 21 and mass number 45. Therefore, in a neutral atom of Sc, there are 21 protons, 21 electrons, and 24 neutrons. As with the other examples, and following the Aufbau principle, Hund's rule, and the Pauli exclusion principle, the ground state electronic configuration of scandium can be written as $1s^2 2s^2 2p^6 3s^2 3p^6 4s^2 3d^1$ or [Ar]$4s^2 3d^1$. Remember, the 4s orbital is filled before the 3d because it is lower in energy.

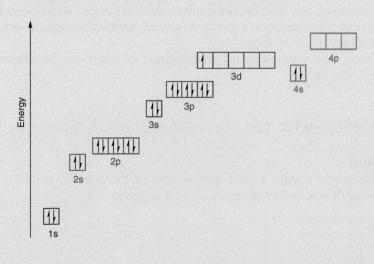

Quick-check summary

- Atoms are made up of a central, dense region containing neutrons and protons surrounded by electrons in outer shells with different energies.

- Protons and neutrons have the same relative masses as each other whereas electrons are almost 2000 times lighter. Protons have a positive charge, electrons have a negative charge and neutrons are neutral. Isotopes of an element have the same number of protons (Z) but different numbers of neutrons and therefore different mass number (A).

- There are three main types of radioactive decay namely alpha-, beta- and gamma-.

- Electrons are arranged in orbitals according to their energies in the outer shells, or energy levels of the atom.

- The ground state electronic configuration of an atom can be defined by using the Aufbau principle, the Pauli exclusion principle, and Hund's rule.

- Each successive element has one more proton in the nucleus and one more electron in its outer shell. The different areas of the periodic table (s block, p block etc) are characterised by the type of orbitals in the outer shells being filled by electrons.

End-of-chapter questions

1 What are the relative mass and charge of a proton, a neutron, and an electron?

2 How many protons, neutrons, and electrons do the following atoms and ions contain?

  a. $^{24}_{12}\text{Mg}$

  b. $^{74}_{32}\text{Ge}$

  c. $^{107}_{47}\text{Ag}$

  d. $^{90}_{40}\text{Zr}^{4+}$

  e. $^{127}_{53}\text{I}^{-}$

3 Define the term *isotope*. Include an example in your answer.

4 Zinc exists as five naturally occurring isotopes: ^{64}Zn, ^{66}Zn, ^{67}Zn, ^{68}Zn, and ^{70}Zn, with abundances 49.2%, 27.7%, 4.0%, 18.5%, and 0.6%, respectively. Calculate the average relative atomic mass of zinc.

5 What are the three types of radiation that can be emitted by radioisotopes? Explain the difference between each type, and give any relevant equations.

6 ^{222}Rn undergoes alpha decay. Give the equation for this process.

7 Draw a clearly annotated diagram showing the shells of electrons in the following elements:

a. Chlorine

b. Calcium

c. Neon

d. Carbon

8 On a clearly annotated diagram, draw and label:

a. A 1s orbital

b. Three 2p orbitals

c. Five 3d orbitals

9 Write the ground state electronic configurations for the following atoms or ions:

a. S

b. Sr

c. Se^{2-}

d. Co

e. Mn

f. Mg^{2+}

10 State the elements that have the following ground state electronic configurations:

a. $1s^2 2s^2 2p^2$

b. $1s^2 2s^2 2p^6 3s^2 3p^3$

c. $1s^2 2s^2 2p^6 3s^2 3p^6 3d^{10} 4s^2 4p^6$

d. $[Ar] 3d^5 4s^1$

2

Chemical bonding

At the end of this chapter, students should be able to:

- Describe different types of bonding using dot-and-cross diagrams where appropriate, namely:
  - Simple molecular and giant covalent bonding
  - Ionic bonding
  - Metallic bonding
- Deduce and explain the shapes of simple molecules
- Explain polarity and intermolecular forces
- Identify dipoles
- Deduce the strength of intermolecular forces and therefore some properties of a material

2.1 Bonding

This section will outline the type of bonding that exists in covalent molecules. It will also describe ionic bonding and metallic bonding.

2.1.1 Atoms and molecules

In the first chapter, we saw that an **atom** is an individual entity that contains electrons, protons, and neutrons. A **molecule** is formed when two (or more) atoms bond together. Some elements exist naturally as molecules, such as chlorine, Cl_2, nitrogen, N_2, and hydrogen, H_2. For example, Figure 2.1, shows eight hydrogen atoms and the four hydrogen molecules they form by bonding together.

Foundations of Chemistry: An Introductory Course for Science Students, First Edition.
Philippa B. Cranwell and Elizabeth M. Page.
© 2021 John Wiley & Sons Ltd. Published 2021 by John Wiley & Sons Ltd.
Companion website: www.wiley.com/go/Cranwell/Foundations

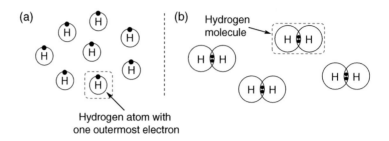

Figure 2.1 (a) Eight hydrogen atoms; (b) four hydrogen molecules.

An *atom* contains electrons, protons, and neutrons. A **molecule** is formed when two (or more) atoms bond together.

Atoms of different elements can also bond together to form molecules. When two or more elements are chemically joined together a **compound** is formed. Water, H_2O, carbon dioxide, CO_2, and ethanol, C_2H_5OH, are all examples of compounds.

Within chemistry, there are three main ways in which atoms can bond together to make larger units. These bonding types are called *metallic*, *ionic*, and *covalent*. The first part of this chapter explains the different types of bonding that can occur and the types of compounds formed.

When considering bonding, the driving factor that causes an atom to lose or gain electrons is the extra stability the atom gains when it has a full outer shell of electrons. For example, the noble gases in Group 8 (Group 18) are especially stable because all of their orbitals are filled with electrons. Other atoms in the s and p blocks of the periodic table react by losing, gaining or sharing electrons in order to have the same electron configuration as the nearest noble gas element. This tendency of elements to attain a full outer shell of electrons is called the *octet rule* because the noble gases Ne and Ar have eight outer electrons.

The word *octet* is derived from the Latin for the number eight, which is *octo*. Hence the octet rule refers to eight electrons.

2.1.2 Metallic bonding

Metallic bonding occurs in metals and alloys when metal atoms are bonded to other metal atoms. When metal atoms join together, their outer (valence shell) electrons are lost, leaving positively charged metal ions. The electrons form a 'sea' of negative charge surrounding the positive metal ions. The interaction between the sea of electrons and the positive metal centres holds the atoms together in the bulk metal. A representation of the bonding in sodium is shown in Figure 2.2. Sodium is in Group 1 and has only one outer electron that is easily

The valence shell is the outermost shell in an atom that contains electrons.

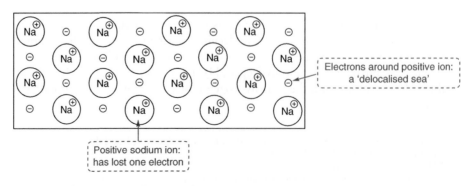

Figure 2.2 Bonding in sodium metal.

lost. The positive sodium ions form a regular ordered array surrounded by the sea of negative electrons.

An ion is an atom that has lost or gained one or more electrons to form a charged species. A positive ion is called a cation, and a negative ion is called an anion.

Worked Example 2.1   Explain and draw the bonding present in Magnox, an alloy of magnesium and aluminium that is used in the cladding of nuclear power reactors.

Solution
Magnesium is in Group 2, so it can lose two electrons. Therefore, each magnesium atom contributes two electrons to the sea of charge. Aluminium is in Group 3 and has three valence electrons that contribute to the sea of electrons. If we use this information, we can suggest what the bonding in the alloy may look like.

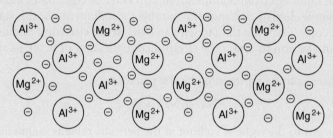

2.1.3   Ionic bonding

Ionic bonding occurs between a metal and a non-metal. An ionic bond is an *electrostatic* interaction between two oppositely charged ions that are attracted to each other. Ions are formed when atoms lose or gain electrons, so they have either a positive or a negative charge. The metal ion always has a positive charge and the non-metal ion a negative charge.

An example of this behaviour is shown by fluorine, which has the electron configuration $1s^2 2s^2 2p^5$. When a fluorine atom gains one more electron, it has the electron configuration of neon, i.e. $1s^2 2s^2 2p^6$. It now has a charge of −1 because it has gained one negatively charged electron and becomes a fluoride ion, F^-.

The process can be represented by the equation:

$$F + e^- \rightarrow F^-$$

The electron arrangements in the fluorine atom and fluoride ion are shown in Figure 2.3.

Note that the neutral F atom is called *fluorine*, as in the element, and the negatively charged F^- ion is called *fluoride*. This is the form of the element found in toothpaste, for example.

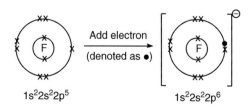

Figure 2.3 Addition of an electron to a fluorine atom to generate a fluoride ion. Electrons in the fluorine atom are represented by x.

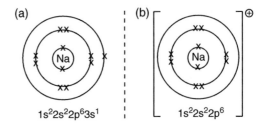

Figure 2.4 Arrangement of electrons in (a) a sodium atom; (b) a sodium ion.

To form a positive ion, the atom must lose an electron. Sodium in Group 1 has just one outer electron. Sodium attains an octet of eight outer electrons by losing this single electron. It is more favourable for sodium to lose this single electron and become a Na^+ ion than to gain an extra seven electrons. Sodium therefore loses an electron to attain the electron configuration of neon, as shown in Figure 2.4.

The process can be represented by the equation:

$$Na \rightarrow Na^+ + e^-$$

The element to the right of sodium in the periodic table is magnesium, Mg, and an atom of magnesium has two outer electrons ($1s^2 2s^2 2p^6 3s^2$). To attain a full outer shell, magnesium must lose both its 3s electrons. When this occurs, a Mg^{2+} ion is formed:

$$Mg \rightarrow Mg^{2+} + 2e^-$$

$$1s^2 2s^2 2p^6 3s^2 \rightarrow 1s^2 2s^2 2p^6 + 2e^-$$

The process of losing or gaining an electron to form an ion is called *ionisation*.

Metal atoms always lose electrons on ionisation to form positive ions, whereas non-metals gain electrons to form negative ions.

Elements in the first row of the periodic table (i.e. hydrogen and helium) cannot obtain an octet of electrons as they have a single s orbital that can only hold a maximum of two electrons. They therefore have a full outer shell when they have just two electrons in the 1s orbital. A helium atom has a full outer shell and so has no tendency to lose or gain electrons and therefore does not readily form ions. However, hydrogen has just one outer electron. It can either lose this electron and form a H^+ ion or gain an electron to form a H^- ion. The H^- ion is called a *hydride* ion.

$$H \rightarrow H^+ + e^-$$

$$H + e^- \rightarrow H^-$$

When a positively charged cation and a negatively charged anion are formed, an electrostatic interaction exists between the oppositely charged ions. This interaction is the basis of ionic bonding, and the attractive force between the ions is similar to that experienced between the north and south poles of a magnet. These attractive forces extend among all the cations and anions in a structure to give an *ionic lattice*. An ionic lattice is a regular three-dimensional array of positive and negative ions extending throughout the structure. The attractive forces between the oppositely charged ions are very large; therefore, ionic bonds are very strong, and ionic lattices are difficult to break down. The bonding in sodium chloride can be used as an example of ionic bonding.

Sodium chloride is an ionic compound that contains sodium ions (Na^+) and chloride ions (Cl^-). Sodium is in Group 1 and therefore has one electron in its outer shell. Chlorine is in Group 7 (Group 17) and has seven outer electrons. To have a full outer shell, sodium must lose one electron and chlorine must gain one more electron, as shown in Figure 2.5.

In Figure 2.5, sodium's outer electrons are represented by dots and chlorine's outer electrons by crosses. By losing one electron to chlorine, sodium now has one fewer electron than protons and so has an overall positive charge. In contrast, chlorine has one more electron than protons, so it has an overall negative charge. The resulting ions are shown in square brackets, where the charge of each ion is at the top right, on the outside of the bracket.

The positive sodium ions and negative chloride ions are attracted to each other by electrostatic forces and are arranged in a regular lattice array. Figure 2.6 shows the arrangement of ions in the sodium chloride lattice. Many millions of these units are linked together in a grain of salt (Figure 2.7).

When chlorine has a negative charge, the name changes to *chloride*. The same is true for fluorine (fluoride), bromine (bromide), and iodine (iodide).

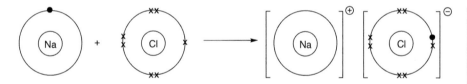

Figure 2.5 Bonding in NaCl. Note: only outer-shell electrons are shown for clarity.

This type of diagram, where electrons are represented by dots and crosses, is called a *dot-and-cross diagram* and sometimes a *Lewis structure*. Note that all electrons are identical to each other, regardless of whether they are drawn as dots or crosses.

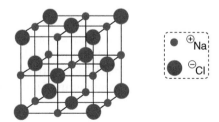

Figure 2.6 The sodium chloride lattice. *Source:* Based on https://www.chemguide.co.uk/atoms/structures/ionicstruct.html.

Figure 2.7 Salt, sodium chloride. *Source:* Dr Philippa Cranwell.

Ionic bonding is very strong, and, as a result, ionic compounds or salts have similar physical properties. Ionic compounds usually have high melting and boiling points and are solids at room temperature. They form hard crystalline structures. When solid, they do not conduct heat or electricity; but when molten, they do, as the ions can move and carry a charge or transfer heat energy.

Ionic bonding occurs between metals and non-metals, and there is a strong electrostatic interaction between the bonding partners.

Worked Example 2.2   Give the electronic configuration of the following atoms and the likely electron configuration and charge of the species formed by each atom upon ionisation. State the noble gas atom that has the same electron configuration as the ion formed.

  a. potassium

  b. chlorine

Solution

  a. Potassium is an s block element in Group 1 of the periodic table. It is a metal with one outer electron. The electron configuration of potassium is $1s^2 2s^2 2p^6 3s^2 3p^6 4s^1$. To gain a full outer shell, potassium tends to lose its outer electron and form a K^+ ion with electron configuration $1s^2 2s^2 2p^6 3s^2 3p^6$. This electron configuration is the same as argon, Ar.

  b. Chlorine is in Group 7 (Group 17) of the periodic table and has seven outer electrons with electron configuration $1s^2 2s^2 2p^6 3s^2 3p^5$. It is a non-metal and tends to gain an electron to form the chloride, Cl^-, ion with electron configuration $1s^2 2s^2 2p^6 3s^2 3p^6$. This electron configuration is the same as argon, Ar.

Worked Example 2.3   Draw the Lewis structure (dot-and-cross diagram) for calcium fluoride, CaF_2.

Solution

Calcium fluoride contains two elements: calcium and fluorine. Calcium is a metal on the left of the periodic table, in Group 2. Fluorine is a non-metal and resides in Group 7 (Group 17). Because we have a metal and a non-metal in the compound, the type of bonding present is ionic. Therefore, the main interaction between atoms will be electrostatic: i.e. a positive ion with a negative ion.

As stated earlier, calcium is in Group 2. Therefore, to gain a full outer shell, it needs to lose two electrons. In doing this, it will have a 2+ charge.

Fluorine is in Group 7. Therefore, to gain a full outer shell, it needs to gain one electron. By doing this, it has a negative charge. To balance the two electrons given up by calcium, there need to be two fluorine atoms present that can each take one electron.

When asked a question about the type of bonding in a compound, the first thing to check is the types of elements that are bonded and whether they are metals or non-metals.

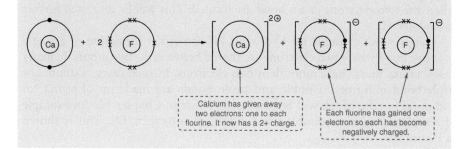

Calcium has given away two electrons: one to each flourine. It now has a 2+ charge.

Each fluorine has gained one electron so each has become negatively charged.

2.1.4   Covalent bonding

Covalent bonding occurs between two non-metal atoms. In a covalent bond, each atom contributes one or more electrons to share with the other atom. The driving force behind covalent bonding is similar to that in ionic bonding: each element wants to achieve a full outer shell of electrons. In the case of covalent bonding, electrons are shared rather than lost or gained. The covalent bond itself is very strong; a large amount of energy is required to break a covalent bond.

We will use the formation of the H—H bond in molecular hydrogen to explain covalent bonding. Hydrogen is an element that can undergo covalent bonding. In an atom of hydrogen, there is only one electron in a 1s orbital (the outermost shell). To achieve a full outer-shell, hydrogen needs to have two electrons; therefore, one hydrogen atom shares its electron with that of another atom of hydrogen. The 1s orbitals containing the electrons overlap. This is shown in Figure 2.8a. In this figure, the electrons on each hydrogen are shown as a dot or a cross. When the electrons are shared, they are shown within the overlapping rings of the outermost shell. When two electrons are shared, they are said to form a *bonding pair* between the two atoms. One pair of electrons is one bond and can be represented by a single line, as in Figure 2.8b. In the case of hydrogen, H_2,

Covalent bonding occurs between non-metals. A covalent bond is a shared pair of electrons.

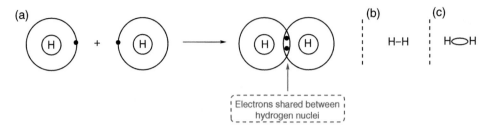

Figure 2.8 (a) Formation of a single bond in hydrogen, H_2; (b) an alternative representation of a hydrogen molecule with a single bond; (c) a representation of the location of the electrons in a σ bond.

The symbol σ is the Greek letter sigma. A σ bond consists of a pair of electrons shared between two covalently bonded atoms.

Although the electrons in the hydrogen atoms are shown as dots and crosses, all the electrons are identical.

The symbol π is the Greek letter pi. A π bond consists of a pair of electrons located between two covalently bonded atoms in a plane at 90° to the bond axis.

In covalent bonding, the first two electrons to be shared form a sigma bond. If more than two electrons are shared, the electrons occupy pi bonds.

there is one pair of electrons that is located between the two nuclei. This is called a (sigma) σ bond. Figure 2.8c is a representation of the electron cloud showing where the two electrons in a σ bond are located. This will be discussed further in Chapter 12.

The previous example for hydrogen, H_2 shows the formation of a single covalent bond where two electrons are shared between the two atoms. In many cases, atoms must share more than two electrons. In such cases, a double or triple bond is formed. Double and triple bonds are made up of sigma (σ) and pi (π) bonds, which will be discussed further in Chapter 12. An example of a molecule where a double bond is present is oxygen, O_2. This is shown in Figure 2.9.

The oxygen atom has six outer electrons. The electron configuration is $1s^2 2s^2 2p^4$. An oxygen atom requires two additional electrons to fill its outer shell and attain a complete octet of electrons as in the atom neon, the closest noble gas. Thus, in the formation of an oxygen molecule, each oxygen atom shares two of its outer electrons with another oxygen atom. There is therefore a total of four electrons bonding the oxygen atoms together. When four electrons are shared between two atoms, a double bond is formed, which is represented by a double horizontal line between the atoms in the bond.

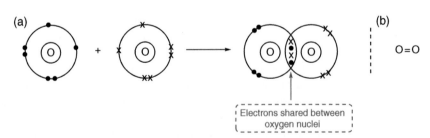

Figure 2.9 (a) Formation of a double bond in oxygen, O_2; (b) an alternative representation of an oxygen molecule with a double bond.

Worked Example 2.4 Draw a dot-and-cross diagram to show the bonding present in carbon dioxide, CO_2.

Solution

Carbon dioxide contains two elements: carbon and oxygen. Both are non-metals; therefore, the type of bonding between them will be covalent, i.e. shared pairs of electrons. Carbon has four electrons in its outer shell and therefore needs to gain another four electrons to fill its outer shell. Oxygen has six electrons, so it needs to gain two more electrons. In addition, the name *carbon dioxide* gives a hint about the structure, with the 'di' showing that there are two oxygen atoms.

This information allows us to deduce that if carbon shares two electrons with each oxygen, each oxygen will have eight electrons in its outer shell and so will have a complete octet. Conversely, if oxygen shares two electrons each with carbon, carbon will have gained a share of four electrons in total, so it will also have a full outer shell. In the case of carbon dioxide, there are two pairs of electrons between each nucleus (a). Thus, there is a double bond between each carbon and oxygen atom. This is represented by a double line between the atoms (b).

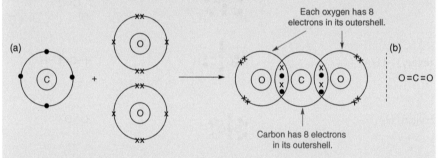

For all structures containing covalent bonds, it is possible to work out the bonding and how the atoms are arranged. Some common molecules are shown in Table 2.1.

To ensure that you understand this topic, try to draw each of the molecules in Table 2.1 using dot-and-cross diagrams and starting from the component atoms, accounting for all of the electrons shown.

Dative covalent bonding

In some of the molecules shown in Table 2.1, a pair of electrons is 'left over' or not involved in covalent bond formation. Examples of molecules where this occurs include ammonia (one pair of electrons on the nitrogen), water (two pairs of electrons on the oxygen), and nitrogen (one pair of electrons on each nitrogen). This pair of electrons is called a *lone pair of electrons* or, more colloquially, a *lone pair*. Such pairs of electrons can also undergo bonding to other species that can accept a pair of electrons, such as a positive hydrogen ion, H^+. When both electrons in a covalent bond originate from the same atom, a *dative covalent bond* is formed. An example of this is ammonia (NH_3), which can form a dative covalent bond to a proton (H^+) to form an ammonium ion (NH_4^+). The dot-and-cross diagrams showing the formation of a dative covalent bond in the ammonium ion are given in Figure 2.10a. A dative covalent bond can be shown either as a line or as an arrow. The arrow is drawn such that it starts on the atom that has donated the electrons and finishes on the atom that has gained the pair of electrons, as shown in Figure 2.10b. Once formed, the dative covalent bond between the nitrogen and hydrogen atoms is chemically no different to the other nitrogen–hydrogen bonds in the ammonium ion.

A dative covalent bond is formed when both electrons in a covalent bond are provided by the same atom.

Table 2.1 The names, molecular formulae, dot-and-cross diagrams, and display formulae of some commonly encountered covalently bonded molecules.

Name and molecular formula	Dot-and-cross diagram	Display formula
Fluorine, F_2	F ˟• F	F−F
Water, H_2O	H O H	H−O−H
Ammonia, NH_3	H N H	H−N−H (with H below N)
Methane, CH_4	H C H	H−C−H (with H above and below C)
Ethene, C_2H_4	H C C H	H₂C=CH₂
Ethyne, C_2H_2	H C C H	H−C≡C−H
Oxygen, O_2	O O	O=O
Nitrogen, N_2	N N	N≡N

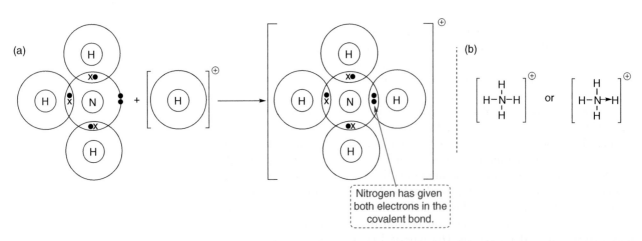

Figure 2.10 (a) Bonding in the ammonium ion, NH_4^+. Note: only outer-shell electrons are shown for clarity. (b) An alternative representation of bonding in the ammonium ion.

Nitrogen has given both electrons in the covalent bond.

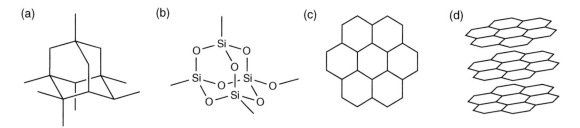

Figure 2.11 (a) Bonding in diamond; (b) bonding in silicon dioxide; (c) two-dimensional bonding in graphite, looking down from above; (d) three-dimensional bonding in graphite, looking through the layers.

Simple molecular covalent bonding and giant covalent bonding

Covalent bonding can lead to two different types of structures. These are called *simple covalent* (or simple molecular) or *giant covalent* (or giant molecular) structures. The main difference between simple molecular covalent bonding and giant covalent bonding is that a simple molecule is just that: a small molecule that contains covalent bonds. Examples of molecules that have simple molecular bonding are oxygen (O_2), methane (CH_4), chlorine (Cl_2), and ethanol (C_2H_6O). Giant covalent bonding leads to extended repeated units such as in silicon dioxide (SiO_2) and graphite and diamond (both elemental forms of carbon). A substance that has a giant covalent structure is much more complex and contains an extended array of bonds. This type of bonding does not result in discrete molecules because the bonding is repeated throughout the structure. The structures formed are not molecules as the bonding continues repeatedly in three dimensions. The structures of diamond, silicon dioxide, and graphite are shown in Figure 2.11.

2.2 Valence Shell Electron Pair Repulsion Theory (VSEPR)

The shape of a covalently bonded molecule or ion is determined by the number and arrangement of the pairs of electrons around the central atom. Electron pairs are negatively charged centres and are repelled by neighbouring centres of electron density. The structures formed are most stable when the centres of electron density are as far apart as possible from each other. If you remember this, you will be able to tackle any problems involving the shapes of simple molecules. The process for working out the shape of a molecule is called valence shell electron pair repulsion theory, or VSEPR for short. It is much simpler than the name suggests!

In this model, a single area of electron density or charge is defined as:

The arrangement of atoms around a bonding centre is entirely based upon the repulsion of areas of negative charge.

- A lone pair of electrons

- A single bond – composed of one pair of electrons

- A double bond – composed of two pairs of electrons

- A triple bond – composed of three pairs of electrons

2.2.1 How to determine the number of areas of electron density around a central atom

The first stage in working out the shape of a molecule is to determine the number of areas of electron density around the central atom. This can be done by drawing a dot-and-cross diagram that shows the number of bonded electrons and the number of lone pairs. The number of areas of electron density is equal to the number of bonded atoms plus the number of lone pairs.

Once we know the number of areas of electron density, there are various fundamental shapes that the molecule can have that minimise the repulsive forces between these negatively charged areas.

In the following section, examples of molecules adopting these fundamental shapes are given so you can see how the overall shape is obtained.

Box 2.1   Depicting the three-dimensional shape of a molecule
It is difficult on a flat piece of paper or screen to draw a three-dimensional shape. Most molecules are three-dimensional with bonds that point in different directions. Chemists use many different ways of showing three-dimensional structures, but in this book we will use single lines, wedges, and dashed lines. Each type of line means something slightly different and gives important information about the location and direction of the bond in space. The single line means that the bond is in the plane of the page; the wedge means that the bond is coming out of the page towards you; and the dash means that the bond is going into the page, away from you.

The different representations of bonds in a molecule.

2.2.2 Two electron centres around the central atom: linear molecules

At centres where there are two areas of electron density, the bonded atoms arrange themselves at 180° to each other, generating a linear structure. One of the simplest molecules that adopts this shape is beryllium chloride, $BeCl_2$. Beryllium has electron configuration $1s^2 2s^2$ with just two electrons in its outer shell. Even by sharing electrons with two chlorine atoms, the beryllium atom cannot gain a full outer shell of electrons but shares each of its valence electrons with the unpaired electron of a chlorine atom. In this way, the chlorine atom gains an octet of electrons in its outer shell, and the beryllium atom has just four, as shown in Figure 2.12. Each chlorine atom forms a single bond to the central beryllium atom, and as there are just two centres of electron density, these arrange themselves at 180° to give a linear structure.

The Be centre is said to be **electron deficient** because it has only four electrons in its outer shell when it has bonded to the chlorine atoms. In the solid state the Be centres accept lone pairs of electrons from chlorine atoms in neighbouring molecules to form long polymer chains. Single molecules of $BeCl_2$ are only present in the gas phase.

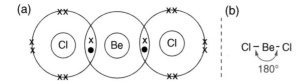

Figure 2.12 Dot-and-cross diagram for beryllium chloride, $BeCl_2$ showing two centres of electron density and linear shape.

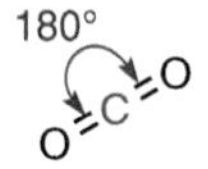

$$\underset{O}{\overset{180°}{\curvearrowright}} \underset{O}{\overset{}{C}} O$$

Figure 2.13 A linear centre in carbon dioxide, CO_2.

Carbon dioxide is another linear molecule. We saw in Section 2.1.4 that each central carbon atom has a double bond joining it to an oxygen atom. These two areas of electron density (the two double bonds) spread themselves as far apart as possible in a linear O=C=O arrangement, as shown in Figure 2.13.

2.2.3 Three electron centres around the central atom: trigonal planar molecules

Centres that have three areas of electron density adopt a trigonal planar structure, with angles of approximately 120° between each of the centres of electron density. A simple molecule with this structure is BCl_3.

Boron has three electrons in its outer shell, and each chlorine atom has seven. Thus boron shares each of its valence electrons with one unpaired electron from chlorine. This gives the boron atom six outer electrons, and the chlorine atoms each have eight outer electrons. The six electrons around boron form three single bonds to the chlorine atoms, thus creating three areas of electron density. These three bonds spread so that they are 120° apart and form a trigonal planar (or triangular planar) shape, as shown in Figure 2.14.

The boron centre is electron deficient because it has only six electrons in its outer shell once it has bonded to the chlorine atoms.

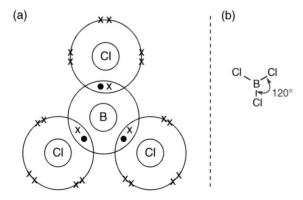

Figure 2.14 (a) Bonding in boron trichloride, BCl_3 with a trigonal planar centre; (b) showing the bonding angles in boron trichloride, BCl_3.

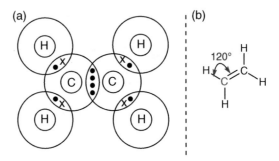

Figure 2.15 (a) Bonding in C_2H_4, ethene with trigonal planar centre; (b) showing the bonding angles in ethene.

Another molecule that has a trigonal planar arrangement around its central atoms is ethene, C_2H_4. Each carbon atom has four electrons in its outer shell; to attain a complete octet, carbon shares two of these electrons with the single electrons of two hydrogen atoms and shares the remaining two electrons with two electrons from the other bonded carbon atom. Thus, each carbon atom in C_2H_4 has two single C–H bonds and one double C=C bond, giving it three distinct areas of electron density. Again, the atoms around each carbon atom spread so that they are approximately 120° apart in a trigonal planar arrangement, as shown in Figure 2.15.

2.2.4   Four electron centres around the central atom: tetrahedral, pyramidal, and bent molecules

When a central atom has four sets of electron density surrounding it, the areas of electron density are based upon a tetrahedron, but the final shape of the molecule depends on the number of bonded pairs. These shapes are called tetrahedral, pyramidal and, bent (or angular). Figure 2.16 shows examples of molecules that adopt each of the three shapes. When there are four single bonds to the central atom, e.g. in methane, the molecule adopts a tetrahedral shape (Figure 2.16a). When there are three single bonds to the central atom and one lone pair of electrons, e.g. in ammonia, NH_3, the molecule adopts a pyramidal shape (Figure 2.16b). When there are two bonds and two pairs of electrons around the central atom, e.g. water, H_2O, the molecule adopts a bent or angular shape (Figure 2.16c). The lone pairs have an impact on the bond angles between the bonded atoms and the central atoms because the lone pairs occupy more space than the bonding pairs. In a symmetrical tetrahedral molecule, the angle between

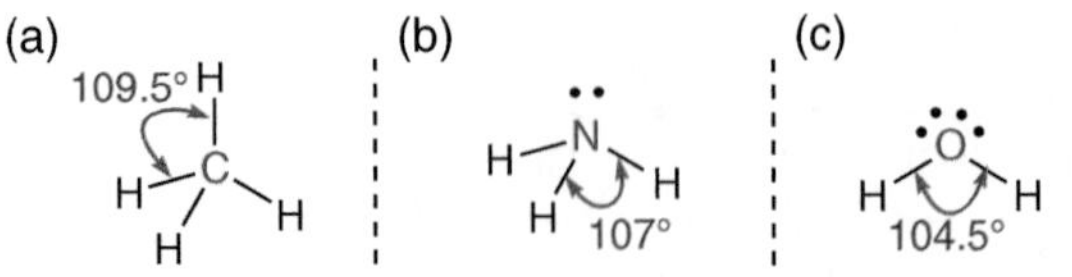

(a)   (b)   (c)

Figure 2.16 (a) Bonding angles in a tetrahedral bonding centre; (b) bonding angles in a pyramidal bonding centre; (c) bonding angles in a bent (or angular) bonding centre.

each of the covalent bonds is 109.5°; in a pyramidal compound, the angle is 107°; and in a bent compound, it is 104.5°. This reflects the larger space that the lone pairs occupy, pushing the bonding electrons closer together; see Figure 2.16.

The rule of thumb for deciding the order of interactions between areas of electron density in a molecule is:

Lone pair–lone pair > lone pair–bonded pair > bonded pair–bonded pair

2.2.5 Five electron centres around the central atom: trigonal bipyramidal molecules

If there are five areas of electron density around the central atom, the molecule adopts a trigonal bipyramidal shape, as shown in Figure 2.17 for phosphorus pentachloride, PCl_5. The angle in the plane around the phosphorus atom between each of the equatorial bonds is 120°, and the angle between the equatorial bonds and the axial bonds is 90°.

Box 2.2 You will come across the terms *axial* and *equatorial* throughout your studies in chemistry. An axial bond is one that runs vertically up and down along a single axis, whereas an equatorial bond is located horizontally across the page or in the equatorial plane of the molecule.

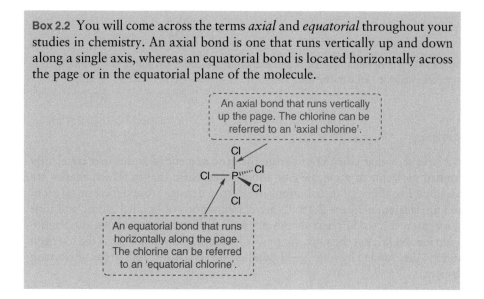

An axial bond that runs vertically up the page. The chlorine can be referred to an 'axial chlorine'.

An equatorial bond that runs horizontally along the page. The chlorine can be referred to an 'equatorial chlorine'.

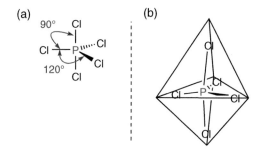

Figure 2.17 (a) Bonding angles in a trigonal bipyramidal molecule; (b) phosphorus pentachloride, PCl_5 superimposed into a trigonal bipyramid.

2.2.6   Six electron centres around the central atom: octahedral molecules

When there are six areas of electron density around the central atom or six single covalent bonds, the molecule adopts an octahedral shape. If all six bonded atoms are identical, the angle between each of the bonds is 90°, as this is the furthest apart that all six components can arrange themselves around the central atom. Figure 2.18 depicts the shape of the sulfur hexafluoride, SF_6 molecule: the six fluorine atoms arrange themselves as far apart as possible around the central sulfur atom. The reason for the name *octahedral* is that the shape made by the six bonds from a central atom has eight faces, which is an octahedron.

SF_6 is used extensively in insulating high-voltage electrical transmittance cables and switching gear. It is now known to be the most potent green-house gas and is banned in all applications apart from in the electrical industry.

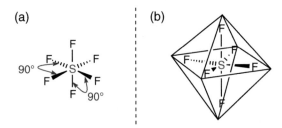

Figure 2.18 (a) Bonding angles in an octahedral molecule; (b) sulfur hexafluoride, SF_6 superimposed into an octahedron.

Summary

We have seen that when considering the arrangement of atoms in a covalently bonded molecule, a molecule can adopt seven basic shapes. These shapes are based on five different distributions of electron density around the central atom and are shown in Figure 2.19. Other combinations of bonded atoms and lone pairs can result in different shapes not included here: for example, the ammonium ion, NH_4^+, has the same shape as a methane, CH_4 molecule as the nitrogen atom is surrounded by four N—H bonds which constitute four areas of electron density.

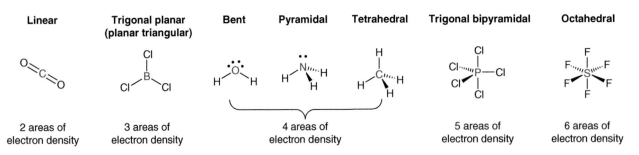

Figure 2.19 Common shapes of simple covalent molecules.

Worked Example 2.5   Describe the bonding present in silane, SiH_4, and determine the shape of the molecule.

Solution
The first thing to determine is the number of areas of electron density around the central silicon atom. Both silicon and hydrogen are non-metals; therefore, the bonding is covalent. A dot-and-cross diagram can be used to describe the bonding. There are four bonds from the silicon atom to the hydrogen atoms, and no lone pairs; thus there are four areas of electron density, so the shape of silane is tetrahedral. The bonding and shape of the molecule are exactly analogous to methane, CH_4, as both carbon and silicon are in the same group of the periodic table.

Worked Example 2.6   Describe the bonding present in hydrogen sulfide, H_2S, and determine the shape of the molecule.

Solution
Sulfur, the central atom, is in Group 6 (Group 16), so it has six valence electrons and needs to gain two more to complete the octet. This is achieved by bonding to hydrogen. There are four areas of electron density around sulfur consisting of two single bonds and two lone pairs of electrons. The four areas of electron density arrange themselves as far apart as possible and form a tetrahedral shape. However, because two of these areas of electron density are lone pairs, the actual shape of the molecule appears to be bent (i.e. we can't 'see' the lone pairs!). As discussed previously, lone pairs of electrons occupy more space than bonding pairs of electrons, and push the hydrogen atoms more closely together. The H—S—H bond angle is therefore slightly less than 109.5°. So although the arrangement of electron centres is based on a tetrahedron, it is slightly distorted, and hydrogen sulfide is therefore bent (also called *v-shaped* or *angular*).

Worked Example 2.7   Describe the bonding present in formaldehyde, CH_2O, and determine the shape of the molecule.

Solution
Firstly, decide on the central atom. Carbon is the only element in the molecule that can have more than two bonds, so carbon must be the central atom. The carbon atom requires four more electrons to complete its octet.
   To satisfy the octet rule, oxygen needs an extra two electrons and so shares two of its electrons with two from carbon, making a double bond. The two hydrogen atoms share one electron each with carbon to form two single bonds. Using a dot-and-cross diagram, it can be seen that there are three areas of electron density (remember, a double bond counts as one area). Therefore, to achieve minimum repulsion, the three areas spread out to be roughly 120° apart from each other, and the shape is trigonal planar or planar triangular.

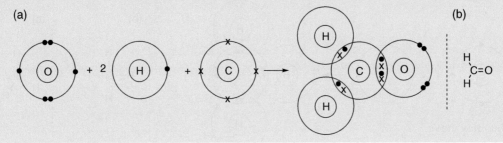

Worked Example 2.8   Describe the bonding present in hydrogen cyanide, HCN, and determine the shape of the molecule.

Solution
In HCN, the central carbon atom is bonded to one hydrogen atom and one nitrogen atom. Carbon requires four more electrons to complete its octet. Hydrogen requires one more electron to complete its outer shell, so carbon shares one of its electrons with it to form a single C—H bond. Nitrogen has five valence electrons and so needs a further three electrons to complete its octet. It obtains these by sharing three of its electrons with the carbon atom to form a triple bond (having six electrons). From the dot-and-cross diagram, there are two areas of electron density at the carbon atom (remember, a triple bond counts as one area), so to achieve minimum repulsion, the two areas need to be 180° apart from each other. The shape is therefore linear.

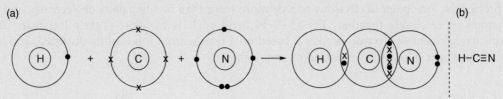

Worked Example 2.9   Explain and describe the bonding present in phosphorus pentafluoride, PF_5.

Solution
The first thing to determine is how many areas of electron density are around the central phosphorus atom using a dot-and-cross diagram. Phosphorus is in Group 5 (Group 15) and therefore has five outer-shell electrons. Each fluorine has seven outer-shell electrons and therefore wants to gain one electron to complete the octet. Fluorine

achieves this by sharing each of its unpaired electrons with one from phosphorus, generating a pentavalent central phosphorus atom. There are no lone pairs of electrons, so the five single bonds arrange themselves in a trigonal bipyramidal shape.

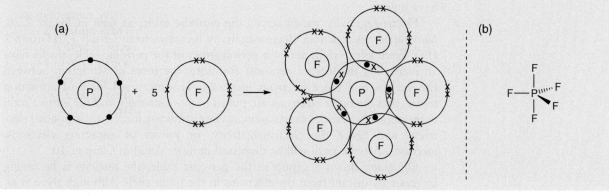

Box 2.3 You may have noticed the phosphorus atom now has 10 outer electrons, which is two more than the accepted octet. Phosphorus is said to 'expand the octet' by accommodating more than eight electrons in its outer shell. The term for this behaviour is *hypervalency*, and several theories have been used to explain it. A discussion of hypervalency is beyond the scope of this textbook, but it should be noted that you could also encounter hypervalency when looking at bonding in the elements sulfur, chlorine, and iodine.

2.3 Polar bonds and polar molecules

We have seen that atoms bond to each other in different ways to form covalent or ionic compounds. These compounds have very different physical and chemical properties to each other. The main factor that determines whether two atoms will form an ionic or covalent bond is the difference in *electronegativity* between the elements concerned. In a covalently bonded material, the main factor that determines the physical properties of the material is whether the molecule is *polar* or *non-polar*. These features of a molecule are of fundamental importance and will be explained in the following section.

2.3.1 Electronegativity

The electronegativity of an element is the power of an atom in a molecule to pull electrons towards itself when the electrons are in a bond. Linus Pauling developed a scale that compares the relative electronegativities of elements with each other. Pauling's scale has values from 0.7 to 4.0. The symbol for electronegativity is the Greek letter chi: χ. The more electronegative an element, the higher the value of χ. Fluorine is the most electronegative element with a value of 4.0; and francium, at the bottom of Group 1, is the least electronegative with a value of 0.7. Note that electronegativities are relative values, which means they don't have units.

Electronegativity varies across the periodic table, as seen in Figure 2.20. Moving across a period, electronegativity increases from Group 1 to Group 7 (17). From left to right across a period (row) of the periodic table, nuclei have an increasing number of protons; therefore, the force of attraction between the nucleus and outer electrons increases. This increased force of attraction means that the outer electrons are pulled in more strongly and the atomic radii decrease; so the nucleus exerts an increasing attractive force on any bonded electrons, and hence the electronegativity, or power of attracting electrons, increases. Atomic radii will be discussed in more detail in Chapter 10.

On going down a group in the periodic table, the nucleus is becoming increasingly distant from the electrons in the outer shell. Although there is an increase in nuclear charge, the nucleus is screened by more and more layers of filled electron shells, so the outer electrons experience less and less of the nuclear charge. Thus, the elements become less electronegative and more electropositive. The least electronegative element is francium at the bottom of Group 1. Notice that no values are given for the electronegativities of some noble gas elements as they rarely form bonds.

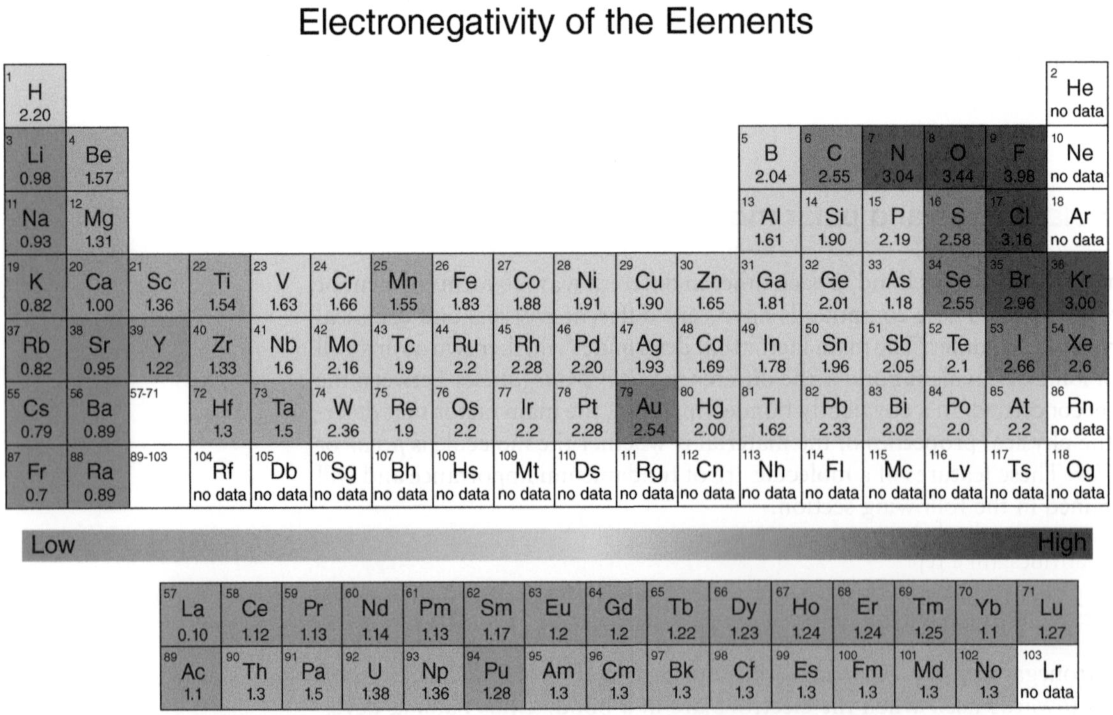

Electronegativity of the Elements

Figure 2.20 Periodic table showing Pauling electronegativity of most of the elements. You are unlikely to be expected to memorise these values. *Source:* By Anne Helmenstine. Retrieved from https://sciencenotes.org/electronegativity-definition-and-trend/

2.3.2 Polar bonds

Consider a covalent bond formed between two atoms A and B. The single bond is composed of a pair of electrons. If atoms A and B have the same electronegativity, then on average, the pair of electrons will be located evenly between the two atoms, as shown in Figure 2.21. This type of bond is called a *pure covalent* bond and is formed when the elements at the end of the bond are the same: for example, Cl_2 or H_2.

However, if one of the atoms (say, atom B) is more electronegative than the other, the electrons will be pulled towards that atom, and the distribution will no longer be evenly spread. Atom B will have a greater share of electrons than atom A, as shown in Figure 2.22. The result of this is that atom B becomes slightly negatively charged compared to atom A. We represent this charge by a δ– sign next to atom B. Because atom B takes up a slight negative charge, atom A must become slightly positively charged to balance the overall charge, and this is represented by the symbol δ+.

A bond where there is an uneven distribution of charge, such as that shown in Figure 2.23, is called a *polar covalent* bond. It occurs between atoms where there is a reasonable difference in electronegativity. Examples of polar covalent bonds are found in hydrogen chloride, HCl, where the chlorine atom is more electronegative than hydrogen; and carbon dioxide, CO_2, where oxygen is more electronegative than carbon.

If there is a large difference in electronegativity between the atoms in a bond, then the electrons are concentrated on the more electronegative element. In the extreme case, the more electronegative element pulls the electrons completely towards itself and becomes a negatively charged ion (an anion), and the bonding is ionic. This generally occurs when one of the atoms in the bond is a metal and

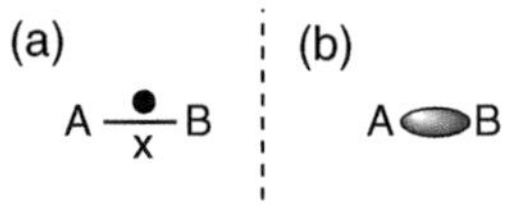

> The Greek letter δ is pronounced 'delta' and in maths and chemistry means 'a little bit' or 'slightly'. So δ– means slightly negative, and δ+ means slightly positively charged.

(a)   (b)

A $-\bullet\!\!/\!\!x-$ B   A◗B

Figure 2.21 (a) A pair of electrons shared evenly between two atoms with the same electronegativity; (b) a representation of the cloud of electrons.

(a)   (b) δ+   δ–

A $-\bullet\!\!/\!\!x-$ B   A◗ B

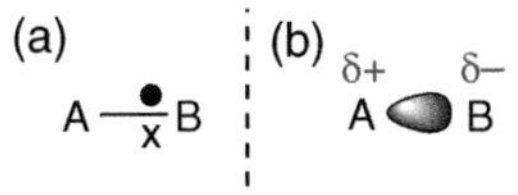

Figure 2.22 (a) A pair of electrons shared unevenly between two atoms with different electronegativities; (b) a representation of the cloud of electrons, showing the charge distribution.

(a)   (b)

δ+ δ–   δ– δ+ δ–
H–Cl   O=C=O

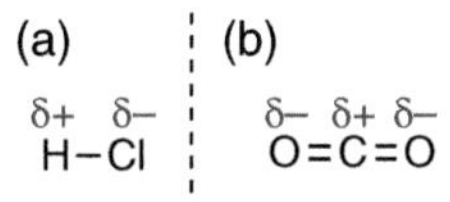

Figure 2.23 Polar covalent bonds in (a) hydrogen chloride, HCl and (b) carbon dioxide, CO_2, showing the charge distribution.

> • *Pure covalent* bonds are formed when the atoms in a bond have the same electronegativity values.
> • *Polar covalent* bonds are formed when the atoms in a bond have a small difference in electronegativity values.
> • *Ionic* bonds are formed between metals and non-metals that typically have a large electronegativity difference.

the other a non-metal. The non-metal becomes an anion and the metal a cation, and now electrostatic interactions hold the ions together in the lattice.

2.3.3   Polar molecules

If a molecule has polar bonds, it may be polar overall and have an overall dipole moment, which is important as this can affect how one molecule interacts with another. For example, the molecule H—Cl has a polar bond and an overall dipole moment. This is depicted by an arrow drawn beside the molecule with a line through the flat end. The direction of the arrowhead shows the direction of electron charge in the molecule, and the arrow points from the less electronegative end to the more electronegative end of the molecule, as shown in Figure 2.24.

Another example is the molecule chloromethane, CH_3Cl, which has a tetrahedral shape. The C—Cl bond is polar because the chlorine atom is more electronegative than the carbon atom (C has $\chi = 2.5$ and Cl has $\chi = 3.0$). The C—H bonds are usually considered to be non-polar because the electronegativities of carbon and hydrogen are similar (H has $\chi = 2.2$). The molecule therefore has a dipole moment in the direction of the chlorine atom, as shown in Figure 2.25.

However, not all molecules with polar bonds have a dipole moment. We have seen that in carbon dioxide, CO_2, both C=O double bonds are polar. However, the molecule is linear, and Figure 2.26 shows that the two polar bonds pull charge from the carbon atom in opposite directions to each other. As the size of the dipole is equal in both C=O bonds, there is no net overall dipole moment. You can think of this like two teams in a tug of war pulling against each other. If both teams have exactly the same strength, neither team will win.

Consider the series of molecules based on methane, CH_4, where the hydrogen atoms are replaced in turn to give fluoromethane, CH_3F, difluoromethane, CH_2F_2, trifluoromethane, CHF_3, and tetrafluoromethane, CF_4, shown in Figure 2.27. The molecule fluoromethane, Figure 2.27a, has a permanent dipole

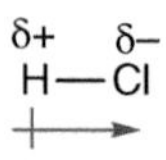

Figure 2.24 Hydrogen chloride molecule showing the charge separation and direction of overall dipole moment.

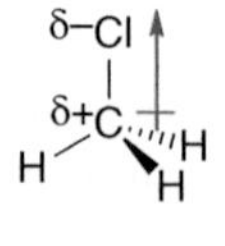

Figure 2.25 Chloromethane, CH_3Cl, is a polar molecule.

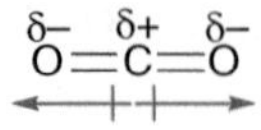

Figure 2.26 Carbon dioxide has polar bonds but no overall dipole moment.

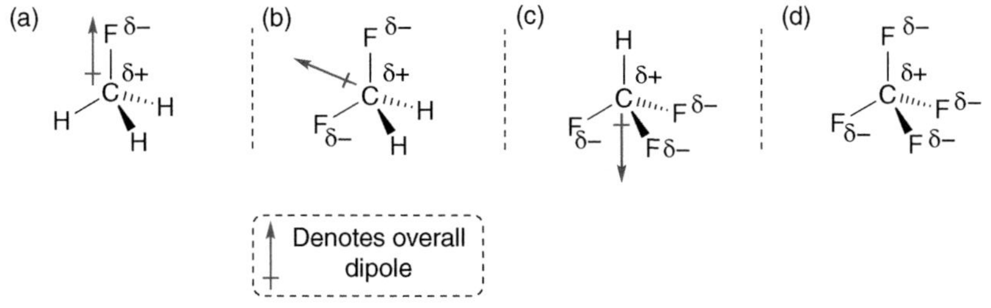

Figure 2.27 (a) Overall molecular dipole in fluoromethane; (b) overall molecular dipole in difluoromethane; (c) overall molecular dipole in trifluoromethane; (d) tetrafluoromethane with no overall dipole but polar bonds.

moment. It has the same shape as chloromethane, shown in Figure 2.25. The dipole moment is along the C—F bond from the carbon atom to the fluorine atom. In difluoromethane, there are two C—F bonds, and each is polar. The molecule is polar overall as the two dipoles along the bonds add together and give an overall dipole moment oriented between the C—F bonds. Trifluoromethane, CHF_3, also has an overall dipole in the molecule that points in the opposite direction to the remaining C—H bond: Figure 2.27c. Finally, tetrafluoromethane, CF_4, does not have an overall dipole because there is no overall charge separation in the molecule. The polar C—F bonds pull charge in opposite directions and cancel each other out: Figure 2.27d.

Worked Example 2.10   Determine if the molecule propan-2-one (acetone), C_3H_6O, is polar and has an overall dipole moment.

Solution
Once the structure of acetone has been determined, using a dot-and-cross diagram, it is possible to see that it is trigonal planar in shape; there are two single carbon-to-carbon bonds, and there is a double bond between oxygen and carbon. The carbon-carbon and carbon-hydrogen bonds are not polar, because carbon and hydrogen have similar electronegativities, but the carbon-oxygen bond is polar because oxygen is more electronegative than carbon. This molecule is also polar overall, as charge is pulled towards the oxygen atom, leaving the central carbon atom slightly positively charged.

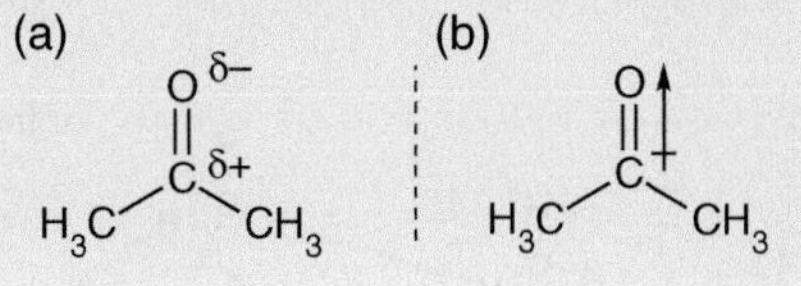

2.4   Intermolecular forces

The bonds that hold atoms together in a molecule are known generally as *intramolecular forces* because they are *inside* a molecule. They are very strong and not easily broken. For example, in the molecule water, H_2O, the covalent O—H bonds are the intramolecular forces. It is very difficult to split water into its component atoms, oxygen and hydrogen. In fact, it takes roughly 51.5 kJ energy to split 1 g of water into its elements, oxygen and hydrogen. If we could easily separate water molecules into gaseous hydrogen and oxygen, we would have a low-cost way of producing hydrogen as a fuel stock.

However, simple covalent molecules such as water don't exist in isolation; they are surrounded by many millions of other similar molecules with much weaker forces between them. The forces that exist *between* molecules are called *intermolecular* forces. They are much weaker than the forces holding atoms together but are very important. The intermolecular forces holding water molecules together in liquid water are around 1.3 kJ per gram, which is about one-twentieth the strength of an O—H bond. If it weren't for these intermolecular forces, we wouldn't have any liquid water to drink – water would exist as a gas on the earth's surface.

Intermolecular forces determine the properties of a covalently bonded compound, such as its melting and boiling point. It is important to understand how they arise in order to have an idea of the strength of these forces.

Figure 2.28 shows the strong intramolecular forces within the methane, CH_4, molecule and the weaker intermolecular forces between the methane molecules.

There are three main types of intermolecular force between molecules:

- Instantaneous dipole–induced dipole or London dispersion forces

- Permanent dipole–permanent dipole

- Hydrogen bonding

The first two types of intermolecular forces that involve dipole-to-dipole interactions (both permanent and instantaneous) are called *van der Waals forces*. Van der Waals forces are attractive forces between slightly positively and slightly negatively charged areas of a molecule. The term *van der Waals* is reasonably general and does not take into account the type of dipoles that are interacting. The term *London dispersion forces* is more specific, and this name is used for instantaneous dipole to induced dipole interactions. The third type of intermolecular force, hydrogen bonding, is a special type of dipole–dipole interaction.

The origins of these interactions will be discussed in the following sections.

It is important that you understand the difference between inter- and intramolecular forces, as it is subtle but very important. Intermolecular forces are between two *different* molecules (like international flights are between two different countries); intramolecular forces are between the atoms that form a bond and are *within* a molecule

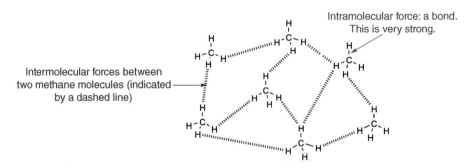

Intramolecular force: a bond. This is very strong.

Intermolecular forces between two methane molecules (indicated by a dashed line)

Figure 2.28 Comparison of inter- and intramolecular forces.

2.4.1 Permanent dipole–permanent dipole interactions

Permanent dipole–permanent dipole interactions are stronger than London dispersion forces but weaker than hydrogen bonds. These interactions occur through space and are purely electrostatic with a δ+ charge in one molecule interacting with a δ– charge in another molecule. Figure 2.29 shows the permanent dipoles set up in HCl molecules, where the chlorine atom is more electronegative than the hydrogen atom. Permanent dipole–permanent dipole interactions are formed between the oppositely charged ends of molecules.

2.4.2 London dispersion forces (instantaneous dipole–induced dipole)

An instantaneous dipole can occur in a bond between any two elements, regardless of the electronegativities of the bonded atoms. As the name suggests, they are fleeting and so do not last very long, but they can have an impact upon other molecules that are nearby. The common name for an instantaneous dipole to induced dipole interaction is *London dispersion forces*.

If the atoms in a bond have similar electronegativities, the electron charge is evenly distributed between them. However, because electrons are constantly moving, there is still a chance that at any one moment, the electrons may suddenly be at one end of the bond, rendering that end of the bond slightly negatively charged (δ–) and the other end of the bond slightly positively charged (δ+) in comparison. This forms an instantaneous dipole. Once an instantaneous dipole has been set up, it induces a dipole in another bond in a nearby molecule: Figure 2.30.

This is because the shift of electrons generating a δ– charge forces the electrons in a nearby bond to be repelled, so a dipole is formed in the nearby molecule, as shown in Figure 2.30c. London dispersion forces are reasonably weak because they are short-lived, but they are important nevertheless.

Instantaneous dipoles (or dispersion forces) increase with increasing polarisability of the molecule. The more readily polarisable the molecule, the larger the instantaneous dipole and induced dipole. *Polarisability* is a measure of how

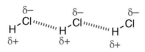

Figure 2.29 Permanent dipoles in the hydrogen chloride molecule and resultant permanent dipole–permanent dipole interactions.

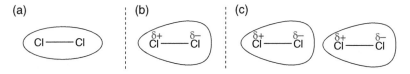

Figure 2.30 (a) Chlorine molecule with even distribution of charge; (b) Chlorine molecule with instantaneous dipole showing charge distribution; (c) neighbouring chlorine molecule with an induced dipole, showing charge distribution.

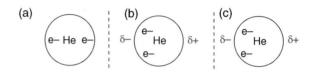

Figure 2.31 (a) Helium atom ($Z = 2$) showing even distribution of electrons; (b) helium atom with instantaneous dipole due to temporary movement of charge; (c) induced dipole in a neighbouring helium atom.

easily the charge distribution in an atom or molecule can be distorted by the application of an external electrical field or charge. The greater the number of electrons, the more readily polarisable the molecule is, so instantaneous dipoles increase with increasing molecular mass. In addition, the larger the surface area of a molecule or the larger the area of possible contact between two molecules, the stronger the intermolecular forces.

Instantaneous dipoles are responsible for the very weak intermolecular forces formed between noble gas atoms. The noble gases consist of monatomic molecules: single atoms of neon, argon, etc. Clearly, there are no intramolecular forces here, as there are no bonds, and the intermolecular forces are extremely weak. Instantaneous dipoles are formed by the random movement of electron density from one side of the atom to another, which then induces a dipole in a neighbouring atom, as shown in Figure 2.31. Helium has only two electrons, so the size of the dipole is very small. The heavier noble gases possess more electrons and so have larger dipoles.

> A monatomic molecule is composed of just one atom. A diatomic molecule such as Cl_2 has two atoms. A triatomic molecule such as H_2O has three atoms. A polyatomic atom has several atoms.

2.4.3   Hydrogen bonding

The final type of interaction is hydrogen bonding, which is the strongest type of intermolecular force and has about one-fifth the strength of a typical covalent bond. Hydrogen bonding is actually a special type of dipole–dipole interaction. Hydrogen bonds are formed between molecules that contain a hydrogen atom bonded to a small, strongly electronegative element such as nitrogen, oxygen, or fluorine.

The hydrogen atom is very small and has only one electron in its 1s orbital. When bonded to an electronegative element such as nitrogen, N, oxygen, O, or fluorine, F, a dipole exists, and the hydrogen atom adopts a partial positive charge ($\delta+$) while the other atom becomes slightly negatively charged ($\delta-$). The partial positive charge is therefore concentrated over a very small volume and makes the hydrogen atom strongly polarising. This allows the hydrogen atom from one molecule to attract electron density from a small electronegative atom (such as nitrogen, oxygen, or fluorine) in a neighbouring molecule, and the hydrogen atom becomes sandwiched between the two more electronegative atoms. This can be seen in Figure 2.32, which depicts the formation of a hydrogen bond between two molecules of water. The highly polarising hydrogen atom in one water molecule ($\delta+$) is able to attract electron density in the form of a lone pair of electrons from a neighbouring oxygen atom ($\delta-$). The lone pair is now 'shared' between the hydrogen atom and the oxygen atom of a neighbouring molecule, and a hydrogen bond is formed. Because hydrogen bonding requires both the bonded and the neighbouring atom to be highly electronegative and small, hydrogen bonding generally only takes place between hydrogen and nitrogen, oxygen, and fluorine.

> Hydrogen bonding can occur when a hydrogen atom in a molecule is bonded to a strongly electronegative element such as nitrogen, oxygen, or fluorine. A hydrogen bond is formed between the H atom of one molecule and the N, O, or F atom of a neighbouring molecule.

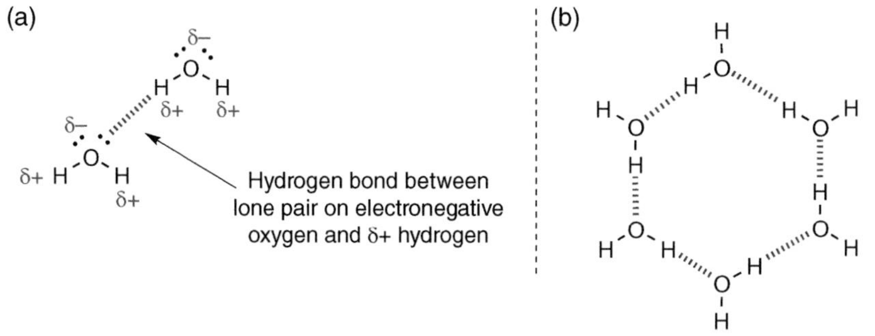

Figure 2.32 (a) Formation of a hydrogen bond between two molecules of water where molecules are in constant motion; (b) formation of hydrogen bonds in ice, where molecules are in a fixed position.

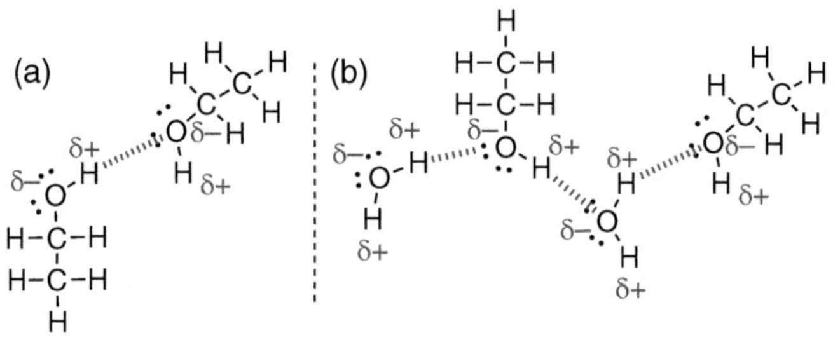

Figure 2.33 (a) Hydrogen bonding between two ethanol, C_2H_5OH, molecules; (b) hydrogen bonding between ethanol, C_2H_5OH, and water molecules.

In liquid water, hydrogen bonds are constantly forming and breaking, as in Figure 2.32a, whereas in ice (Figure 2.32b), the molecules are held together in fixed positions by hydrogen bonds. Each oxygen atom can form two hydrogen bonds to neighbouring molecules as it possesses two lone pairs of electrons.

Hydrogen bonds are also formed between different types of molecules. The group of organic molecules known as alcohols contains an —OH bond. Ethanol, C_2H_5OH, is a well-known alcohol. Hydrogen bonding can occur between molecules of ethanol (Figure 2.33a). Hydrogen bonding can also occur between molecules of ethanol and molecules of water, as shown in Figure 2.33b.

2.4.4 Summary of strengths of intermolecular forces

Intermolecular forces are the weak forces of attraction formed between simple covalently bonded molecules. There are three types of intermolecular forces:

- London dispersion forces: instantaneous dipole–induced dipole forces
- Permanent dipole–permanent dipole forces
- Hydrogen bonding

Table 2.2 Approximate strengths of different types of bonds and intermolecular forces.

Bond type/intermolecular force type	Strength (approx.)/kJ mol^{-1}
Ionic bond (e.g. NaCl)	771
Single covalent bond (e.g. Cl–Cl)	242
Instantaneous dipole–induced dipole	0–10
Permanent dipole–permanent dipole	5–20
Hydrogen bond	20–45

The order of strengths of these forces is:

instantaneous–induced dipole forces < permanent dipole–permanent dipole forces < hydrogen bonding. Table 2.2 gives approximate values for the strengths or energies of different types of intermolecular forces and bond types.

Worked Example 2.11   Determine the types of intermolecular forces present in the following molecules: (a) silane (SiH_4); (b) methanol (H_3COH); and (c) dichloromethane (CH_2Cl_2).

Solution

a. Silane is a tetrahedral molecule with four Si—H bonds. The electronegativities of silicon and hydrogen are very similar and the molecule is symmetrical, so there is no overall dipole present. This means that the most likely type of interaction between two silane molecules is instantaneous dipole–induced dipole or London dispersion forces.

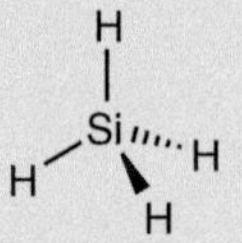

b. Methanol contains an O—H bond, so there is a large dipole between the carbon and the oxygen and an even bigger dipole between the hydrogen and the oxygen atoms. This can create permanent dipole to permanent dipole interactions. More specifically, because the hydrogen of the O—H bond is bonded to oxygen, methanol can undergo hydrogen bonding with other molecules of methanol, CH_3OH.

c. Dichloromethane contains two carbon-chlorine bonds, which both have a permanent dipole due to the different electronegativities of carbon and chlorine; carbon is δ+ and chlorine δ−. This is a permanent dipole; therefore, the dominant intermolecular force in dichloro-methane is permanent dipole–permanent dipole.

2.4.5   A special case: ion-dipole intermolecular forces

One final intermolecular force should be considered, although it follows the same rules and conventions as the intermolecular forces already discussed: the ion–dipole force. The ion can be either positively charged (a cation) or negatively charged (an anion). The dipole that it interacts with can be induced, instantaneous, or permanent. It really depends upon the molecules involved.

For example, a sodium ion (Na^+) could interact with ethane (a non-polar molecule) through either of the following:

- An instantaneous dipole, where the ethane itself could develop an instantaneous dipole due to random electron movement within the bonds in the ethane molecule

- An induced dipole, where the ethane itself could develop an induced dipole due to the close proximity of the sodium ion to it

Another example could be the sodium ion (Na^+) ion interacting with methanol (a polar molecule). The methanol itself has a permanent dipole, so the interaction would be between the δ– of the oxygen on methanol and the Na^+ ion itself.

Quick-check summary

- There are three main types of chemical bonding: metallic, covalent, and ionic. Metallic bonding can be described by a lattice of positive metal ions held together by a sea of negative electrons.

- Ions are formed when atoms lose or gains electrons.

- Ionic bonding occurs between positive and negative ions held together in a regular lattice by electrostatic forces.

- Covalent bonding occurs when atoms share one or more pair of electrons.

- Dative covalent bonding occurs when one atom provides both electrons to another atom capable of accepting them.

- Covalent bonding results in simple molecular and giant molecular structures.

- VSEPR is a theory that allows the prediction of shapes of molecules based upon the arrangement of pairs of electrons around the central atom.

- The type of bonding in a compound determines its physical properties.

- A polar bond is one that has a permanent dipole moment due to a difference in electronegativities between the atoms in the bond.

- A polar molecule possesses an overall dipole moment.

- Intramolecular forces are the strong forces within a molecule that hold the atoms together.

- Intermolecular forces are weak forces between molecules that hold them together.

- There are three main types of intermolecular forces: instantaneous dipole to induced dipole, permanent dipole to permanent dipole, and hydrogen bonding.

End-of-chapter questions

1 Draw a dot-and-cross diagram to show the bonding present in a molecule of nitrogen, N_2.

2 Explain the difference between ionic, covalent, and metallic bonding, giving an example of each type of bonding.

The term 'species' is used to represent a compound such as a molecular substance or an ionic material or a metal.

3 State the most likely type of bonding present in the following species:

   a. Aluminium

   b. NaCl

   c. Formaldehyde, CH_2O

   d. $MgBr_2$

   e. Bronze (an alloy of copper and tin)

   f. Ethyne, C_2H_2

   g. Borane, BH_3

4 Draw the bonding present in the following species:

   a. Strontium

   b. $CaCl_2$

   c. Formaldehyde, CH_2O

   d. $AlCl_3$

   e. Ethyne, C_2H_2

   f. Borane, BH_3

   g. $NiCl_2$

5 Determine whether the following species contain a permanent dipole:

   a. Ethyne, C_2H_2

   b. Ethane, C_2H_6

   c. $AlCl_3$

   d. CO_2

   e. CH_3OH

   f. 1,3-difluorobenzene

   g. 1,4-difluorobenzene

6 List the most likely intermolecular forces that exist in each of the following compounds. There may be more than one type of intermolecular force in each molecule.

   a. Ammonia, NH_3

   b. Iodine, I_2

   c. Formaldehyde, CH_2O

   d. Ethanoic acid, CH_3COOH

   e. Methane, CH_4

   f. Silane, SiH_4

7 Draw the hydrogen bonding between the following molecules:

   a. Ammonia, NH_3 and methanol (CH_3OH)

   b. Ethanoic acid (CH_3COOH) and water

   c. Two hydrogen fluoride, HF, molecules

8 Suggest the shapes of the following molecules:

   a. BF_3

   b. $CHCl_3$

   c. PCl_3

   d. H_2S

   e. $BeCl_2$

3

Amount of Substance

At the end of this chapter, students should be able to:

- Determine relative atomic and molecular masses using the periodic table of the elements

- Explain the meaning of the mole and use Avogadro's number to calculate numbers of atoms of elements or molecules of substances in a certain amount in moles

- Use the equation relating the amount of substances in moles to mass and molar mass to calculate any one of these parameters knowing the other two

- Calculate theoretical and actual percentage yields in chemical reactions

- Calculate percentage composition by mass of elements in compounds from the molecular formula and relative atomic masses

- Calculate percentage purity of compounds from analytical data relating to percentage composition

- Determine the empirical and molecular formula of a compound from its percentage composition by mass

- Calculate the concentration of a solution in various units, knowing its composition

- Perform calculations to determine how to dilute a concentrated solution to a given molarity

- Carry out calculations using titration data to determine the concentration of a solution

- Use the molar gas volume to work out the mass of a certain volume of a known gas under standard conditions of temperature and pressure

Foundations of Chemistry: An Introductory Course for Science Students, First Edition.
Philippa B. Cranwell and Elizabeth M. Page.
© 2021 John Wiley & Sons Ltd. Published 2021 by John Wiley & Sons Ltd.
Companion website: www.wiley.com/go/Cranwell/Foundations

3.1 Masses of atoms and molecules

Atoms join together to make molecules, and molecules react together to make different molecules or join together to make bigger molecules. In making these new substances, scientists need to know how much of one molecule or compound will react with another and how much product will be obtained as a result. To achieve this, we need to be able to count molecules or groups of molecules. And to use the right amount of reacting substance, we need to be able to measure it by determining its mass or volume or by knowing its concentration. This means we need to know about the relative masses of the atoms and molecules in reacting substances.

3.1.1 Relative atomic mass, A_r

In Chapter 1, we saw that different atoms have different numbers of neutrons, protons, and electrons, and this means that atoms of different elements have different masses. Because atoms are so small, it is impossible to weigh just one, so we measure their masses relative to each other. The stable isotope of carbon-12, $^{12}_{6}C$, has been chosen as the standard by which to compare the masses of other atoms. One atom of carbon-12 has been assigned a mass of exactly 12 atomic mass units (amu). So one amu is equal to one-twelfth the mass of an atom of carbon-12, which is equivalent to 1.66×10^{-27} kg. The masses of other atoms are obtained by comparing the average mass of one atom of the element to one-twelfth the mass of a carbon-12 atom. This value is called the **relative atomic mass** of the element and has the symbol, A_r.

One twelfth the mass of a carbon-12 atom is a non-SI unit of mass equal to the atomic mass constant, u. It has a mass of approximately $1.660\,540 \times 10^{-27}$ kg.

$$A_r = \frac{\text{average mass of one atom of element}}{\frac{1}{12}\,\text{mass of one atom of carbon-12}}$$

The relative atomic mass of an element, A_r, is defined as the average mass of one atom of the element compared to the mass of one-twelfth of an atom of carbon-12.

Using this scale, the relative atomic masses as shown in a periodic table of the elements are obtained. Hydrogen, the lightest element, has a relative atomic mass of 1.008 when given to three decimal places (four significant figures). Nitrogen, which is slightly heavier than carbon, has a A_r of 14.007. For most calculations in this book, we will use relative atomic masses to just one decimal place because these A_r values are close to whole numbers for most elements. One exception to this is chlorine, which has a relative atomic mass of 36.5.

3.1.2 Relative molecular mass, M_r

The relative molecular mass of a compound, M_r, is the mass of one molecule of the compound compared to the mass of one-twelfth of an atom of carbon-12. The mass of one molecule of the compound is therefore obtained by adding together the relative atomic masses of all the atoms in the compound.

The relative molecular mass of a molecule, M_r, is defined as the mass of one molecule compared to the mass of one-twelfth of an atom of carbon-12.

Worked Example 3.1   Obtain the relative molecular mass of glucose, $C_6H_{12}O_6$.

Solution
Total number of atoms in the molecule = 6 × carbon (C) + 12 × hydrogen (H) + 6 × oxygen (O)

Find the relative atomic mass of each atom:

$$C = 12.0 \qquad H = 1.0 \qquad O = 16.0$$

Multiply the A_r of each atom by the number of each atom in one molecule of glucose:

$$C : 6 \times 12.0 = 72.0 \qquad H : 12 \times 1.0 = 12.0 \qquad O : 6 \times 16.0 = 96.0$$

Add together the relative atomic masses of all the atoms:

$$72.0 + 12.0 + 96.0 = 180.0$$

So the relative molecular mass of glucose, $M_r = 180.0$

3.1.3   Relative formula mass

Not all compounds exist as molecules. In Chapter 2, we saw that some substances are made up of ions held together by electrostatic forces. Ionic substances don't contain single discrete molecular units but repeating arrays of oppositely charged ions. The term **relative formula mass** is given to the mass of the formula unit of the substance. The same symbol, M_r, is also used to represent the relative formula mass.

Worked Example 3.2   Obtain the relative formula mass of one unit of ammonium sulfate, $(NH_4)_2SO_4$.

Solution
The procedure here is exactly the same as when determining the relative molecular mass. First, use the formula to determine the number of atoms of each element in the formula unit.

Total number of atoms of each element:

$$N = 2 \quad H = 8 \quad S = 1 \quad O = 4$$

Find the relative atomic mass of each element:

$$N = 14.0, H = 1.0, S = 32.0, O = 16.0$$

Multiply the A_r of each atom by the number of each atom:

$$N : 2 \times 14.0 = 28.0 \quad H : 8 \times 1.0 = 8.0$$

$$S : 1 \times 32.0 = 32.0 \quad O : 4 \times 16.0 = 64.0$$

Add together the relative atomic masses of all the atoms:

$$28.0 + 8.0 + 32.0 + 64.0 = 132.0$$

So the relative formula mass of $(NH_4)_2SO_4$, $M_r = 132.0$.

In the formula for a molecule or ionic substance, a subscript indicates the number of atoms of that element in the formula. If the subscript follows a parenthesis (bracket), then all the atoms in the parenthesis are multiplied by the number.

3.2   Amount of substance

Atoms and molecules are extremely small, and it is impossible to count or weigh them using standard laboratory equipment. When molecules react together they do so in fixed amounts, but the numbers involved are so huge that it's not practical to write down the exact figure. Instead, it was decided to collect a certain specific number of atoms and molecules together and call this number of atoms or molecules a **mole**.

3.2.1   The mole

The number of atoms or molecules in this fixed amount known as a mole was defined as $6.022\ 140\ 76 \times 10^{23}$. The name given to this number is the *Avogadro number*, and it has the symbol L or N_A. It is usually rounded to 6.02×10^{23}. The quantity of substance that contains 6.02×10^{23} atoms, molecules, or ions is called a **mole,** and its symbol is ***mol***. Clearly, this number is huge and incredibly difficult to imagine. To grasp an idea of its magnitude, one mole of footballs would form a planet the size of the earth, and one mole of doughnuts would cover the surface of the earth to a depth of five miles! The reason this number was chosen is that if we take any element and weigh out an amount equal to its relative atomic mass in grams, we will have 6.02×10^{23} atoms of the element: for example, in 12.0 g of carbon-12, there are 6.02×10^{23} atoms of carbon-12; in 32.0 g of sulfur, there are 6.02×10^{23} atoms of sulfur; and in 180.0 g of glucose, there are 6.02×10^{23} molecules of $C_6H_{12}O_6$. One mole of any substance contains the same number of particles but has a different mass depending on the masses of the atoms making it up. In the same way, six bananas are heavier than six satsumas, which are heavier than six blueberries (Figure 3.1).

The Avogadro number, $6.022\ 140\ 76 \times 10^{23}$ is known extremely accurately but we will simply use the number to three significant figures in this text, i.e. 6.02×10^{23}.

Figure 3.1 The same number of different types of fruit have different masses, as do different atoms. *Source:* Elizabeth Page.

A mole, unit *mol*, is the amount of substance that contains exactly 6.022 140 76 $\times 10^{23}$ (the Avogadro constant) elementary entities. The term *elementary entities* can refer to particles such as atoms, molecules, electrons, ions, or any other entity.

It is important to define the type of particle being measured. One mole of nitrogen molecules, N_2, contains 6.02×10^{23} molecules of N_2. However, because every nitrogen molecule consists of two nitrogen atoms, one mole of nitrogen molecules contains $2 \times 6.02 \times 10^{23}$ nitrogen atoms.

Worked Example 3.3   Calculate the number of:

    a. molecules of water, H_2O

    b. atoms of hydrogen, H

    c. atoms of oxygen, O

in one mole of water.

Solution

    a. One mole of water by definition contains 6.02×10^{23} molecules of water, H_2O.

    b. One molecule of water contains two atoms of hydrogen (H), so one mole of water contains $2 \times 6.02 \times 10^{23}$ or 1.204×10^{24} atoms of hydrogen (H).

    c. One molecule of water contains 1 atom of oxygen (O), so one mole of water contains $1 \times 6.02 \times 10^{23}$ atoms of oxygen (O).

3.2.2   Converting between moles and masses of substances – molar mass

The mass of one mole of a substance in grams is called its **molar mass,** which has the symbol M and unit $g\,mol^{-1}$. This gives us an equation that we can use to calculate the amount in moles of a substance if we have a certain quantity in grams:

$$\text{amount in moles } (n) = \frac{\text{mass of substance in grams } (m)}{\text{molar mass } (M)}$$

The amount in moles is often given the symbol n, so the equation becomes:

$$n = \frac{m}{M}$$

The following equation includes the units for each quantity. You can see that we obtain the unit of $g\,mol^{-1}$ for molar mass (M) by dividing the mass (m) by the amount in moles (n) of the substance:

$$M\,g\,mol^{-1} = \frac{m\,g}{n\,mol}$$

If we require a certain number of moles of a substance and need to know what mass this equates to, then we can rearrange the equation to get an expression for the mass of substance:

mass of substance in grams (m) = amount in moles (n) × molar mass (M)

or $m = n \times M$

The molar mass of a substance has the same numerical value as the relative molecular mass or relative formula mass of the substance, but it has a unit of $g\,mol^{-1}$ (grams per mole). One mole of water has a relative molecular mass, M_r, of 18 but a molar mass, M, of $18\,g\,mol^{-1}$. In the same way, the mass of one mole of an element – for example, copper – has the same numerical value as the relative atomic mass of copper $(A_r = 63.5)$ but an actual mass, m, of 63.5 g. The mass of one mole of nitrogen atoms is 14.0 g because the A_r of N = 14.0, but nitrogen exists as N_2 molecules, so the mass of one mole of molecules is $14.0 \times 2 = 28.0$ g.

Worked Example 3.4   Calculate the amount in moles of sodium chloride, NaCl, in 117 g of sodium chloride.

Solution
To determine the number of moles, we divide the mass of substance by the molar mass using this equation:

$$\text{amount in moles } (n) = \frac{\text{mass of substance in grams } (m)}{\text{molar mass } (M)}$$

First of all, we need to find the molar mass of sodium chloride by adding the relative atomic masses of sodium and chlorine and converting to grams: A_r (Na) = 23.0, A_r (Cl) = 35.5, so M_r (NaCl) = 23.0 + 35.5 = 58.5, and molar mass NaCl = $58.5\,g\,mol^{-1}$.

$$\text{Amount in moles, } n(\text{NaCl}) = \frac{m}{M} = \frac{117\,g}{58.5\,g\,mol^{-1}} = 2.00\text{ mol.}$$

Worked Example 3.5   Calculate the mass of 0.5 mol of glucose, $C_6H_{12}O_6$.

Solution
This can be found by using ratios. If one mole of $C_6H_{12}O_6$ has a mass of 180 g, then half a mole of glucose has a mass of 180 g/2 = 90 g.
Using the earlier equation:

mass of substance in grams (m) = amount in moles (n) × molar mass (M)

$$= 0.5\text{ mol} \times 180\,g\,mol^{-1} = 90\,g$$

3.3   Calculations with moles

A chemical equation tells us the number of moles of each reactant (on the left-hand side) that react together to give products (on the right-hand side). The number of moles of product and the formulae of the products are given on the right-hand side of the equation.

3.3.1 Reacting masses

When powdered lead oxide is heated with charcoal, the lead oxide is converted to metallic lead and carbon dioxide is produced. The equation for the reaction is:

$$2PbO + C \rightarrow 2Pb + CO_2$$

The numbers in front of each of the reactants and products tell us the number of moles of reactants required and the number of moles of products formed.

This equation tells us that one mole of carbon is required to react with every two moles of lead oxide. These are the reactants. For every two moles of lead oxide used, two moles of metallic lead are formed along with one mole of carbon dioxide as products. These numbers give us the **stoichiometry** of the reaction equation. These numbers are sometimes called the **stoichiometric numbers** or **stoichiometric coefficients,** and they are needed to balance the equation – i.e. to ensure that we have the same number and type of atoms on the left-hand side as on the right-hand side of the equation.

Once we know the balanced chemical equation, we can calculate the number of moles and the mass of each reactant required.

> The stoichiometric coefficients of a reaction indicate the number of moles of each reactant and product in the balanced chemical equation. A chemical equation must be balanced, as the total number of atoms of reactants must be the same as the total number of atoms of products – even if they are arranged differently.

Worked Example 3.6   Calculate the mass of charcoal required to react with 10 g of lead oxide (PbO) to produce metallic lead. How much lead would be obtained, assuming the reaction is 100% efficient?

Solution

Step 1: Write the balanced chemical equation for the reaction:
$2PbO + C \rightarrow 2Pb + CO_2$.

Step 2: Calculate the number of moles of lead oxide, PbO, in 10 g. To do this, we use the equation:

$$\text{amount in moles } (n) = \frac{\text{mass of substance in grams}(m)}{\text{molar mass}(M)}$$

The equation requires the molar mass of PbO, which can be calculated to be $223 \, \text{g mol}^{-1}$.
Inserting the values in the equation, we obtain:

$$\text{amount in moles of PbO } (n) = \frac{m}{M} = \frac{10 \, \text{g}}{223 \, \text{g mol}^{-1}} = 0.045 \, \text{mol}$$

Step 3: Use the stoichiometric coefficients from the equation to calculate the number of moles of charcoal (carbon) required and hence the mass of charcoal:
Two moles of PbO react with one mole of C.
One mole of PbO reacts with half a mole of C.
The ratio of PbO to C is 2:1.
0.045 moles of PbO react with 0.045/2 moles of C = 0.0225 mol.
A_r (C) = 12.0 g, therefore 0.0225 moles of C = 0.0225 × 12.0 g = 0.27 g.

Step 4: Use the stoichiometry of the equation to determine the number of moles of lead produced for each mole of lead oxide used.

Convert this number of moles of lead to mass of lead using the atomic mass of lead:

Two moles of PbO produce two moles of Pb. The ratio of PbO to Pb is 1:1.

0.045 moles of PbO produce 0.045 moles of Pb.

0.045 moles of Pb has a mass of $0.045 \, \text{mol} \times 207.2 \, \text{g mol}^{-1} = 9.3 \, \text{g}$ (A_r (Pb) = 207.2).

Worked Example 3.7   Magnesium metal burns brightly in air, reacting with oxygen to produce magnesium oxide. A chemist had a supply of only 16 g of pure oxygen. How much magnesium would react, and how much magnesium oxide would be formed?

Solution

Step 1: Write the balanced chemical equation for the reaction.
To balance the equation, first write down the formulae of the reactants and products as:

$$Mg + O_2 \rightarrow MgO \qquad \text{UNBALANCED EQUATION}$$

Note that oxygen exists as molecules of O_2, so we include O_2 (not O) on the left-hand side. Because we have two atoms of oxygen on the left-hand side, we need to add two atoms to the right-hand side, so we insert a 2 in front of MgO. The equation becomes:

$$Mg + O_2 \rightarrow 2MgO \qquad \text{UNBALANCED EQUATION}$$

Now, to balance the equation, we need one more atom of Mg on the left-hand side, so the final equation is:

$$2Mg + O_2 \rightarrow 2MgO \qquad \text{BALANCED EQUATION}$$

Step 2: Calculate the number of moles of oxygen molecules available.
To do this, we have to calculate the molar mass of oxygen, O_2:
$2 \times 16.0 = 32.0 \, \text{g mol}^{-1}$.

$$\text{amount in moles of } O_2 = \frac{16 \, \text{g}}{32.0 \, \text{g mol}^{-1}} = 0.50 \, \text{mol}$$

Step 3: Use the stoichiometric coefficients from the equation to calculate the number of moles of magnesium that will react with 0.50 mol oxygen and hence the mass of magnesium.
The equation tells us 1 mole of oxygen molecules reacts with 2 moles of magnesium. Therefore:

0.50 moles O_2 react with 1 mole magnesium.

1 mole of Mg has a mass of 24 g. Therefore 24 g of magnesium would react.

Step 4: Use the stoichiometry of the reaction equation to calculate the amount in moles of magnesium oxide produced.
From the equation, 2 moles of Mg is converted to 2 moles of MgO. Therefore 1 mole of Mg would be converted to 1 mole of MgO. As MgO has a M_r of 40, this would produce 40 g of magnesium oxide.

3.3.2 Percentage yield

Generally, when a chemical reaction is carried out, not all of the reactants are converted to products. In the previous examples, it was assumed that there was 100% conversion; but in practice, some of the reactants or products may be lost or converted to side products by unwanted reactions, so the **yield** of the reaction is less than 100%. The percentage yield in a reaction is a measure of the actual yield compared to the theoretical yield multiplied by 100%.

$$\text{percentage yield of product} = \frac{\text{moles of product obtained}}{\text{theoretical moles of product expected}} \times 100\%$$

The percentage yield can also be calculated as:

$$\text{percentage yield of product} = \frac{\text{mass of product obtained}}{\text{theoretical mass of product expected}} \times 100\%$$

The percentage yield is an important property of a reaction as it tells us how efficient the reaction is in converting reactants to products. The higher the percentage yield, the more efficient the reaction, which is important in industrial chemical reactions.

In a chemical reaction, one of the reactants usually runs out first or is present in a limited quantity. This is called the *limiting reagent*. The amount of the limiting reagent determines the maximum amount of product that can be formed.

> **Worked Example 3.8**   7.50 g of methanol (CH_3OH) reacted with excess ethanoic acid (CH_3COOH) to give 14.2 g of methyl ethanoate (CH_3COOCH_3) and water according to the following equation:
>
> $$CH_3OH + CH_3COOH \rightarrow CH_3COOCH_3 + H_2O$$
>
> Calculate the percentage yield of the reaction.
>
> *Solution*
> The first step in a calculation of this type is to write the chemical equation for the reaction to determine the number of moles of reactant required to produce one mole of product. Then find the actual amount in moles of the reactant that is the limiting reagent used and the amount in moles of product obtained. Next, find the theoretical maximum amount of product that could be obtained from this amount of starting reagent, and calculate the percentage yield.
>
> **Step 1:** The chemical equation for the reaction is:
>
> $$CH_3OH + CH_3COOH \rightarrow CH_3COOCH_3 + H_2O$$
>
> This equation is balanced, so 1 mole of methanol produces 1 mole of methyl ethanoate.
>
> **Step 2:** Calculate the amount in moles of limiting reagent used. As we are told that the ethanoic acid is in excess, the amount of methanol must determine the maximum amount in moles of product that

It is safer when carrying out calculations of percentage yield to work in moles rather than masses of reactants and products. Working in moles generally leads to fewer mistakes as you are less likely to omit the *stoichiometry* of the reaction.

The *atom economy* is another way of defining the efficiency of a reaction. The atom economy is a measure of the percentage of reactants that become useful products. It is important in a green economy to have sustainable processes where raw materials and expensive chemicals are not wasted. In chemical reactions, the efficiency and yield depend upon many factors including the conditions under which the reaction is carried out, the techniques and processes used, and any side products produced. In practice, reactions are never 100% efficient.

The **theoretical yield** in a reaction is the amount of product predicted by the stoichiometric equation based on the amount of reactants used. This assumes 100% conversion of reactants to products. The **percentage yield** is the relative amount of products obtained by mass or moles compared to that expected from the stoichiometric equation.

can be obtained. We say that the methanol is the limiting reagent.

Use the equation $n = \dfrac{m}{M}$ to find the amount in moles of methanol.

Molar mass of methanol, CH_3OH, $= (12.0 + 4 \times 1.0 + 16.0)\,g\,mol^{-1}$

$$= 32.0\,g\,mol^{-1}$$

Amount in moles of methanol $(n(CH_3OH))$ used $= \dfrac{7.50\,g}{32.0\,g\,mol^{-1}}$

$$= 0.234\,mol$$

Therefore the theoretical maximum amount in moles of methyl ethanoate that could be formed is 0.234 mol. The molar mass of methyl ethanoate is $74.0\,g\,mol^{-1}$ so this would have a mass of $0.234\,mol \times 74.0\,g\,mol^{-1} = 17.3\,g$.

The percentage yield of the reaction

$$= \frac{\text{mass of product obtained}}{\text{theoretical mass of product expected}} \times 100\%$$

Substituting for the masses in the equation, we obtain

$$\text{Percentage yield} = \frac{14.2\,g}{17.3\,g} \times 100 = 82.1\%.$$

Percentage yields for chemical reactions can also be calculated in terms of the number of moles of product obtained compared to the maximum theoretical number of moles possible. In this case, the equation for percentage yield is:

$$\%\text{yield} = \frac{\text{amount in moles of product obtained}}{\text{theoretical amount in moles of product expected}} \times 100\%$$

We should of course get the same result for the yield whichever equation is used.

If this equation is applied to the problem in Worked Example 3.8, the number of moles of product obtained must be calculated from the mass and the molar mass of the product.

The amount in moles of product obtained:

$$n(\text{product}) = \frac{\text{mass of methyl ethanoate}}{\text{molar mass of methyl ethanoate}} = \frac{14.2\,g}{74.0\,g\,mol^{-1}} = 0.192\,mol$$

The percentage yield of the reaction $= \dfrac{0.192\,mol}{0.234\,mol} \times 100\% = 82.1\%.$

As we have seen in Worked Example 3.8, it is important to know the limiting reagent in a reaction when calculating the percentage yield. The limiting reagent is the one that is totally consumed in a reaction and therefore determines the maximum amount of product that can be obtained. Worked Example 3.9 demonstrates how to determine the limiting reagent when this information isn't stated in the question.

Worked Example 3.9   In the synthesis of potassium iodate, KIO_3, 3.1 g of potassium chlorate, $KClO_3$, was dissolved in warm water and 3.6 g of powdered iodine, I_2, added to the solution. After neutralisation with potassium hydroxide and evaporation, 4.1 g of potassium iodate was obtained. Calculate the percentage yield of the reaction. The equation for the reaction is:

$$2KClO_3 + I_2 \rightarrow 2KIO_3 + Cl_2$$

Solution

We are given the equation for the reaction and the masses of each starting material. The first step is to calculate the amount in moles of each material used and the amount in moles of product formed. This requires calculating the molar mass of each reagent and the product. This is done in a step-wise fashion here:

	$KClO_3$	I_2	KIO_3
Mass, m	3.1 g	3.6 g	4.1 g
Amount in moles, n	$\dfrac{3.1\,g}{122.5\,g\,mol^{-1}}$ $= 0.025\,mol$	$\dfrac{3.6\,g}{254\,g\,mol^{-1}}$ $= 0.014\,mol$	$\dfrac{4.1\,g}{214\,g\,mol^{-1}}$ $= 0.019\,mol$

The chemical equation states that 1 mole of I_2 requires 2 moles of $KClO_3$. As we have 0.014 moles of I_2, this would require 0.028 moles $KClO_3$ (i.e. 2 × 0.014 mol) for complete reaction. From the calculations, we have only 0.025 moles of $KClO_3$, so the $KClO_3$ is the limiting reagent.

Alternatively, we can prove that $KClO_3$ is the limiting reagent by stating that we have 0.025 moles $KClO_3$ and that this would require 0.0125 moles I_2 for complete reaction. As we have 0.014 moles I_2, we have sufficient I_2 to react with all the $KClO_3$.

Because we have 0.025 moles $KClO_3$, we can't obtain more than 0.025 moles product, KIO_3. The equation tells us that for 100% yield, 2 moles of $KClO_3$ produces 2 moles KIO_3.

The actual amount in moles of KIO_3 obtained is 0.019, so the percentage yield is obtained from:

$$\%yield = \frac{amount\ in\ moles\ of\ product\ obtained}{theoretical\ amount\ in\ moles\ of\ product\ expected} \times 100\%$$

$$= \frac{0.019\,mol}{0.025\,mol} \times 100\% = 76\%$$

In practice, some of the I_2 is lost as vapour when it sublimes from the warm water, and not all the I_2 is available for reaction, which is why it is added in excess.

3.3.3 Percentage composition by mass

The formula of a compound gives us the numbers and types of each atom in the compound. We can use these to determine the percentage by mass of a particular element in a compound. This is the theoretical percentage composition by mass of the element, and assumes the compound is pure.

Percentage composition by mass of element X:

$$\%(X) = \frac{mass\ of\ element\ X\ in\ one\ mole\ of\ compound}{mass\ of\ one\ mole\ of\ the\ compound} \times 100\%$$

By using the atomic masses of each of the atoms in the compound, we can determine the total mass of each element in one mole of the compound and thus the theoretical percentage composition by mass of each element in the compound.

The following example shows how this technique can be used.

Worked Example 3.10   A molecule of glucose, $C_6H_{12}O_6$, has a molar mass of $180\,g\,mol^{-1}$. Calculate the theoretical percentage by mass of each element in the compound.

Solution
A sample of 180 g of pure glucose would be made up of $6 \times 12.0\,g = 72.0\,g$ of carbon, $12 \times 1.0\,g = 12.0\,g$ of hydrogen, and $6 \times 16.0\,g = 96.0\,g$ of oxygen.

The percentage composition by mass of carbon in glucose:
%(C) = 72.0 g/180 g × 100 = 40.0%.

The percentage composition by mass of hydrogen in glucose:
%(H) = 12.0 g/180 g × 100% = 6.7%.

The percentage composition by mass of oxygen in glucose:
%(O) = 96.0 g/180 g × 100 = 53.3%.

Worked Example 3.11   Calculate the percentage composition by mass of silver in a sample of pure silver nitrate, $AgNO_3$.

Solution
The mass of 1 mole of silver nitrate
= 108 g (Ag) + 14.0 g (N) + 48.0 g (3 × O) = 170 g.

Percentage composition of silver by mass in $AgNO_3$
$= \dfrac{108\,g}{170\,g} \times 100 = 63.5\%$

If a chemical substance is analysed and the percentage composition by mass of any of the elements in it is lower than that expected from the molecular formula, then the substance is likely to be impure. The percentage purity of a compound can therefore be defined as:

Percentage purity of a compound
$$= \frac{\text{Percentage composition of element found by experiment}}{\text{Theoretical percentage composition (from molecular formula)}} \times 100\%$$

Worked Example 3.12   Methyl butanoate ($C_3H_7COOCH_3$) is one ester present in fresh strawberries. A sample of commercially prepared methyl butanoate, to be used in strawberry flavouring, was determined to have a percentage composition by mass of carbon of 57.2%. Calculate the percentage purity of the substance.

Solution
The molecular formula indicates the number and type of each atom in one molecule of methyl butanoate and allows us to calculate the molar mass.

Molar mass of $C_3H_7COOCH_3$
= 5 × 12.0 g (C) + 10 × 1.0 g (H) + 2 × 16.0 g (O) = 102.0 g

Theoretical percentage by mass of carbon in methyl butanoate
$= \dfrac{60.0\,g}{102.0\,g} \times 100\% = 58.8\%$

Percentage composition by mass of carbon analysed for the compound
= 57.2%

Thus the percentage purity of the compound

$$= \frac{\text{Percentage composition of element found by experiment}}{\text{Theoretical percentage composition (from molecular formula)}} \times 100\%$$

$$= \frac{57.2\%}{58.8\%} \times 100\% = 97.3\%.$$

3.3.4 Empirical formula

The **empirical formula** of a compound gives us the simplest whole-number ratio of the elements present in one molecule. The molecular formula of a compound tells us how many of each type of atom are present in one molecule of the compound. The empirical formula is sometimes the same as the molecular formula.

For example, the molecular formula of glucose is $C_6H_{12}O_6$. If we divide the number of each atom by 6, we get the empirical formula CH_2O. The strawberry ester, methyl butanoate, has a molecular formula of $C_5H_{10}O_2$. We can't make this formula any simpler by dividing through by a small whole number, so in this case the molecular formula is the same as the empirical formula.

The empirical formula for a compound is obtained by knowing the mass of each element present in the compound, or from the percentage composition by mass of the elements.

For example, if we take a known mass of magnesium metal and burn it in excess oxygen, the compound magnesium oxide is formed. By weighing the magnesium oxide formed, we can work out the mass of oxygen in the compound and so find the ratio of the number of moles of magnesium to oxygen in the compound.

Some typical results for this experiment are given in Worked Example 3.13 along with the calculation to determine the empirical formula.

Worked Example 3.13   A sample of 1.2 g of magnesium metal was burnt completely in excess oxygen, taking care not to lose any of the product. The mass of white powder obtained was 2.0 g and was assumed to be composed of magnesium oxide. Determine the empirical formula of magnesium oxide.

Solution
First write down the information you have from the question:

$$\text{Mass of magnesium metal} : 1.2 \text{ g}$$

$$\text{Mass of magnesium oxide formed} : 2.0 \text{ g}$$

Therefore, the mass of oxygen in 2.0 g of magnesium oxide
$$= 2.0 - 1.2 \text{ g} = 0.8 \text{ g}$$

So 2.0 g of magnesium oxide contains 1.2 g of magnesium and 0.8 g of oxygen. Convert the masses of the elements to moles:

When magnesium metal burns in air, it forms both magnesium oxide and magnesium nitride, as magnesium can react with nitrogen in the air. In this example, we are assuming that no magnesium nitride is formed.

Note that the oxygen in magnesium oxide is present as oxide ions (O^{2-}) combined with the magnesium ions. (Refer to Chapter 2 for a reminder of ionic bonding.) Thus we use the atomic mass of oxygen, O, (and not the mass of O_2) to obtain the amount in moles.

Amount in moles magnesium = $1.2\,g/24.0\,g\,mol^{-1}$ = 0.05 mol

Amount in moles oxygen = $0.8\,g/16.0\,g\,mol^{-1}$ = 0.05 mol

The ratio of oxygen to magnesium in the compound is therefore $=\dfrac{0.05\,mol}{0.05\,mol} = \dfrac{1}{1}$. So the empirical formula of magnesium oxide is MgO.

This type of calculation can be carried out using a table, as shown here for MgO. This method allows you to keep track of the mass, molar mass, and amount in moles of each element.

Check! We know the charge on a magnesium ion is +2 and the charge on an oxide ion is −2, so this formula seems logical as the charges balance.

Element	Mg	O
Mass, (m)/g	1.2	0.8
Molar mass, (M)/g mol^{-1}	24.0	16.0
Amount in moles, (n)/mol	0.05	0.05
Divide through by smallest number of moles	1	1

The empirical formula is therefore MgO.

Worked Example 3.14 A sample of 1.50 g of a hydrocarbon undergoes complete combustion to give 4.40 g of CO_2 and 2.70 g of H_2O. What is the empirical formula? The molar mass of the hydrocarbon is $30.0\,g\,mol^{-1}$. Determine the molecular formula.

Solution
The key to these questions is to use the CO_2 and H_2O generated to work out how much carbon was in the starting hydrocarbon. To do this, we need to use the mass of the original sample (1.50 g), the proportion of carbon or hydrogen in the combustion product, and the amount of combustion product produced.

To determine the mass of carbon in the starting hydrocarbon, we know that CO_2 has M of $44.0\,g\,mol^{-1}$, and M of carbon is $12.0\,g\,mol^{-1}$; therefore we can work out the proportion of carbon in CO_2:

$$\text{Proportion carbon in } CO_2 = \frac{12.0\,g\,mol^{-1}}{44.0\,g\,mol^{-1}} = 0.27$$

As 4.40 g CO_2 was produced by combustion of 1.50 g of hydrocarbon, the mass of carbon in the hydrocarbon can be calculated.

Mass of carbon in 1.50 g hydrocarbon = 0.27 × 4.40 g = 1.20 g

As 2.70 g H_2O was produced by combustion of 1.50 g hydrocarbon, a similar calculation can be carried out to determine the mass of hydrogen in the hydrocarbon. However, in this case there are two hydrogen atoms in one water molecule. The mass of hydrogen, H, in 1.50 g of the hydrocarbon can be obtained from the following equation:

$$\text{Mass hydrogen in 1.50 g hydrocarbon} = \frac{2.0\,\text{g mol}^{-1}}{18.0\,\text{g mol}^{-1}} \times 2.70\,\text{g} = 0.30\,\text{g}$$

Once we have this information, we can continue as before.

Element	C	H
Mass (m)/g	1.20	0.30
Molar mass, (M)/g mol^{-1}	12.0	1.0
Amount in moles, (n)/mol	0.10	0.30
Divide through by smaller number of moles	1	3
Answer (empirical formula)	CH$_3$	
Molar mass of empirical unit	15.0 g mol^{-1}	
Molar mass of compound	30.0 g mol^{-1}	
Number of empirical units per molecule	2	

The molar mass (M) of the compound is 30.0 g mol^{-1}, and the empirical unit mass is 15.0 g mol^{-1}. Therefore:

$$\text{Number of empirical units in molecule} = \frac{30.0\,\text{g mol}^{-1}}{15.0\,\text{g mol}^{-1}} = 2$$

This means that there are two empirical units in the compound, so its molecular formula is CH$_3 \times 2$, giving C$_2$H$_6$.

Worked Example 3.15   Ethyl ethanoate is an ester with a smell of pear drops and is often used in nail polish remover. Analysis of a sample of ethyl ethanoate showed that it contained 54.5% by mass carbon and 9.10% by mass hydrogen. The remainder was oxygen. Determine the empirical formula of ethyl ethanoate. If the molar mass of ethyl ethanoate is 88 g mol^{-1}, calculate the number of empirical units in the molecular formula and hence the molecular formula.

Solution

The steps required to solve a problem such as this involve first finding the number of moles of each element in 100 g of the compound and then comparing the ratio of the number of moles of each to find the smallest whole-number ratio. To find the percentage oxygen in the compound first subtract the percentage of carbon and hydrogen from 100%:

$$54.5\% + 9.10\% = 63.6\%$$
$$\text{Percentage of oxygen} = 100 - 63.6 = 36.4\%$$

Step 1: Find the amount in moles of each element in 100 g of the compound by dividing the mass of each element in 100 g by its atomic mass.

Step 2: Find the ratio of the number of atoms of each element by dividing the amount in moles of each element by the smallest number of moles, which is that of oxygen. Round the answers if close to a whole number. This gives the number of atoms of each element in the empirical unit.

Step 3: Calculate the molar mass of the empirical formula unit = $2 \times 12.0 + 4 \times 1.0 + 16.0 = 44.0\,\text{g mol}^{-1}$.

Step 4: Divide the molar mass of the compound by the molar mass of the empirical formula unit to get the number of empirical formula units in the molecule.

	Element	C	H	O
	Mass (m) in 100 g compound/g	54.5	9.10	36.4
	Atomic mass, (A)/g mol^{-1}	12.0	1.0	16.0
Step 1	Amount in moles, (n)/mol	4.54	9.1	2.28
Step 2	Divide through by smallest number of moles	1.99	3.99	1.00
	Answer (empirical formula)	C_2H_4O		
Step 3	Molar mass of empirical unit/g mol^{-1}	44.0		
	Molar mass of compound/g mol^{-1}	88.0		
Step 4	Number of empirical units per molecule	2		

The molecular formula for ethyl ethanoate is $C_4H_8O_2$, and it is written as $CH_3COOC_2H_5$.

In this example, the ratios of each element obtained in Step 2 are close to a whole number, i.e. 2 in the case of carbon and 4 in the case of hydrogen. We therefore convert the numbers to the closest integer. In some cases, you may find that the ratio of one element to another is not close to an integer value. For example, in the compound Al_2O_3, the ratio of oxygen to aluminium obtained would be 1.5, giving the formula $AlO_{1.5}$. This is clearly not possible, so the ratios obtained should be multiplied up to obtain whole numbers. In this case, we would multiply by 2 to obtain Al_2O_3.

3.4   Solutions; concentrations and dilutions

When a substance such as sugar dissolves in water, a **solution** is formed. The substance that is being dissolved is called the **solute**, and the liquid it is dissolving in is called the **solvent**.

Frequently, chemists and biologists carry out reactions and analytical determinations in aqueous solution, i.e. water. But the same principle holds, that they need to know how much of each substance is reacting.

This section describes the various ways in which the concentration of a solution can be expressed and the procedures and calculations involved in preparing solutions of a specific concentration by dilution.

3.4.1   Measuring and expressing concentrations

For a reminder on units of volume, see Chapter 0

When a substance is dissolved in a solvent such as water, a solution is formed. The concentration of the solution is a measure of how much of the substance is dissolved per unit volume of solvent (Figure 3.2).

It is always safer to convert volumes to dm^3 to ensure the resulting concentration is obtained in mol dm^{-3}. A volume in cm^3 or mL should be converted to dm^3 by multiplying by 10^{-3}. A volume in litres can be changed directly to dm^3 as 1 L is the same as 1 dm^3.

There are many different ways of expressing the concentration of a solution, but chemists prefer to work in mol dm^{-3}. One dm^3 (cubic decimetre) is the same volume as a litre or 1000 cm^3. Therefore the unit of concentration is **mol per dm^3**, written as **mol dm^{-3}**, and a concentration of 1 mol dm^{-3} is obtained when one mole of a substance is dissolved in 1 dm^3 of water.

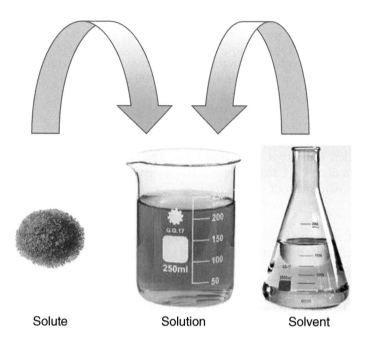

Solute   Solution   Solvent

Figure 3.2 A solute and a solvent are combined to form a solution.

$$\text{Concentration } \left(\text{mol dm}^{-3}\right) = \frac{\text{number of moles (mol)}}{\text{volume (dm}^3)} \text{ or } c = \frac{n}{V}$$

Using this equation, we can calculate concentrations of solutions but also work out the number of moles of a solute dissolved in a certain volume of solution, if we know its concentration. Because solutions are normally made up in the laboratory by weighing out a certain amount of solid, it's usually necessary to convert the mass of solute to an amount in moles to calculate the concentration. This is shown in the following example.

Concentrations in chemistry are normally expressed in units of mol dm^{-3}, but we often shorten this to M, which stands for *molar*. A bottle of sodium hydroxide solution that is labelled 0.1 M contains 0.1 mol sodium hydroxide per dm^3 of water. You will often see this unit used in the laboratory.

Worked Example 3.16 A solution of sodium hydroxide was made by dissolving 2.00 g of solid sodium hydroxide in 200 cm^3 of water. Calculate the concentration of the solution in mol dm^{-3}.

Solution
This type of problem must be done in two stages. The first step is to find the number of moles of sodium hydroxide used. The second step is to find the concentration of the solution when this amount in moles is dissolved in the solvent.

Step 1: Calculate the number of moles of sodium hydroxide. The formula of sodium hydroxide is NaOH.
The molar mass of NaOH is $23.0 + 16.0 + 1.0 = 40.0 \text{ g mol}^{-1}$.

$$\text{Amount in moles of NaOH} = \frac{2.00 \text{ g}}{40.0 \text{ g mol}^{-1}} = 0.05 \text{ mol}$$

Step 2: Now use the equation for concentration to determine the concentration of sodium hydroxide in the solution:

Sodium hydroxide dissolves readily in water. Although the concentration of the solution can be calculated in the way shown, the sodium hydroxide slowly reacts with carbon dioxide in the atmosphere, so the concentration changes over time.

$$c = \frac{n}{V} = \frac{0.05\ \text{mol}}{200 \times 10^{-3}\ \text{dm}^3} = 0.25\ \text{mol dm}^{-3}$$

Note that the volume of water has been converted from units of cm^3 to dm^3 by dividing by 1000, as there are $1000\ cm^3$ in $1\ dm^3$. Dividing by 1000 is the same as multiplying by 10^{-3}.

Worked Example 3.17 A technician has $450\ cm^3$ of silver nitrate solution left over from a practical class. The concentration of the solution is $0.135\ \text{mol dm}^{-3}$. Calculate the mass of silver nitrate in the solution and the equivalent mass of silver.

Solution
In this example, we know the concentration of the solution and its volume but need to calculate the number of moles of silver nitrate. Once we have the number of moles of silver nitrate, this can be converted to a mass of silver nitrate which can be used to calculate the mass of silver.

Step 1: Calculate the number of moles of silver nitrate by rearranging the equation $c = \frac{n}{V}$ to get an expression for n. The expression required is $n = c \times V$.

Insert the values for the concentration and the volume of solution to find the number of moles, n:

$$\text{number of moles},\ n = 0.135\ \text{mol dm}^{-3} \times 450 \times 10^{-3}\ \text{dm}^3$$
$$= 60.8 \times 10^{-3}\ \text{mol}$$

Step 2: Convert the amount in moles of silver nitrate to a mass of silver nitrate using the molar mass of silver nitrate ($AgNO_3$):

$$\text{Molar mass of } AgNO_3 = 108 + 14.0 + 48.0 = 170\ \text{g mol}^{-1}$$

$$\text{Mass of } AgNO_3 = \text{number of moles} \times \text{molar mass}$$
$$= 60.8 \times 10^{-3}\ \text{mol} \times 170\ \text{g mol}^{-1} = 10.3\ \text{g}$$

The mass of silver is obtained by multiplying the number of moles of silver (which is the same as the number of moles of silver nitrate) by the mass of 1 mole of silver.

$$\text{Mass of silver} = 60.8 \times 10^{-3}\ \text{mol} \times 108\ \text{g mol}^{-1} = 6.57\ \text{g}.$$

As the current price of silver is around 0.4 pence per gram, this is worth about £2.60, so it is probably not worth extracting such a small amount from the solution.

3.4.2   Solutions and dilutions

Many solutions we use in both the laboratory and everyday life are made by diluting a more concentrated solution. For example, a drink of orange squash is made by adding water to a small amount of the concentrated cordial. Stock solutions in the laboratory are made up to an accurately known concentration and then diluted to give solutions with lower concentrations. Dilutions are carried out by taking a known volume of the concentrated solution and then adding a specific volume of solvent – usually water. It is often necessary to either work out the new concentration of the diluted solution or the amount of water required to produce a specific concentration.

The concentration of the diluted solution can be calculated by keeping track of the number of moles and the volume of the concentrated solution.

Worked Example 3.18 $100 \, cm^3$ of a solution of hydrochloric acid of concentration $1.00 \, mol \, dm^{-3}$ is taken and $900 \, cm^3$ of water added to give $1000 \, cm^3$ of diluted HCl solution. Calculate the concentration of the new solution.

Solution
First calculate the number of moles of HCl in the $100 \, cm^3$ solution using $n = c \times V$:

Number of moles of HCl = $1.00 \, mol \, dm^{-3} \times 100 \times 10^{-3} \, dm^3 = 0.100 \, mol$

$0.100 \, mol$ of HCl in $100 \, cm^3$ of water is added to $900 \, cm^3$ of water. The new total volume is $900 \, cm^3 + 100 \, cm^3$ which is equal to $1000 \, cm^3$, so the new concentration can be obtained:

$$c = \frac{n}{V} = \frac{0.100 \, mol}{1000 \times 10^{-3} \, dm^3} = 0.100 \, mol \, dm^{-3}$$

Therefore the solution has been diluted by one-tenth from $1.00 \, mol \, dm^{-3}$ to $0.100 \, mol \, dm^{-3}$.

Typically, when preparing solutions in the laboratory, the concentration required is known, but the factor by which the concentrated stock solution must be diluted has to be calculated: i.e. the volume of concentrate required must be determined.

Worked Example 3.19   A technician has prepared $500 \, cm^3$ of sodium hydroxide solution of concentration $0.750 \, mol \, dm^{-3}$. He requires a concentration of $0.100 \, mol \, dm^{-3}$ for an experiment. Calculate the volume of water that must be added to the concentrate to obtain the required dilution.

Solution
The concentration required for the new solution is $0.100 \, mol \, dm^{-3}$, and the original concentration is $0.750 \, mol \, dm^{-3}$ so the original solution must be diluted by a factor of $(0.750 \, mol \, dm^{-3}/0.100 \, mol \, dm^{-3}) = 7.50$.

As the original volume is $500 \, cm^3$, this must therefore be diluted by a factor of 7.50. It must be made up to a new volume of $500 \, cm^3 \times 7.50 = 3750 \, cm^3$. Because the original volume is $500 \, cm^3$, we must add a total of $3750 - 500 \, cm^3 = 3250 \, cm^3$ water.

Alternatively, we can solve this problem by keeping track of the number of moles of sodium hydroxide. The total number of moles of sodium hydroxide does not change between the original solution and the diluted solution. If we calculate the original number of moles of sodium hydroxide and then equate this to the number of moles in the final diluted solution, we can calculate the new volume required for the final solution.

Amount in moles of NaOH in the original solution,

$$n = c \times V = 0.750 \, \text{mol dm}^{-3} \times 500 \times 10^{-3} \, \text{dm}^3 = 375 \times 10^{-3} = 0.375 \, \text{mol}$$

This must be the same as the number of moles in the diluted solution of concentration $0.100 \, \text{mol dm}^{-3}$. So we can use the equation $V = \dfrac{n}{c}$ to calculate the new volume of solution:

$$\text{Final volume of NaOH solution} = 0.375 \, \text{mol}/0.100 \, \text{mol dm}^{-3}$$

$$= 3.75 \, \text{dm}^3 = 3750 \, \text{cm}^3$$

If the initial volume of concentrated solution is $500 \, \text{cm}^3$, then the volume of water required $= 3750 - 500 \, \text{cm}^3 = 3250 \, \text{cm}^3$.

Clearly, the answers should be the same no matter which way you solve the problem.

3.4.3   Alternative units of concentration

In some types of work, the concentration of a solution is given in terms of the percentage of the solute (the dissolved substance) in the solution. There are two ways of expressing this. The concentration can be expressed as **weight per weight (w/w)** percentage composition or **weight per volume (w/V)** percentage composition.

The weight per weight percentage ($w/w\%$) is equal to the mass of solute divided by the total mass of solution:

$$w/w\% = \frac{\text{mass of solute}}{\text{mass of solution}} \times 100\%.$$

A solution that is made up from $10 \, \text{g}$ of glucose dissolved in $90 \, \text{g}$ of water would have a w/w percentage composition:

$$= \frac{10 \, \text{g}}{10 \, \text{g} + 90 \, \text{g}} \times 100\% = \frac{10}{100} \times 100\% = 10\% \text{ glucose}.$$

The weight per volume percentage ($w/V\%$) is equal to the mass of solute divided by the volume of solution. Because the process of dissolving a solid in a liquid is rarely accompanied by a significant change in volume, the volume used in this expression can be either the volume of solvent or the volume of solution. This gives the expression:

$$w/V\% = \frac{\text{mass of solute}}{\text{volume of solution}} \times 100\%$$

Again the mass of solute, glucose, is $10 \, \text{g}$. The volume of solution is obtained by converting the mass of water ($90 \, \text{g}$) using the density of $1.0 \, \text{g}$ cm^{-3}. The volume of solution is therefore $90 \, \text{cm}^3$. The w/V percentage composition is calculated as:

$$w/V\% = \frac{10 \, \text{g}}{90 \, \text{cm}^3} \times 100\% = 11\%$$

When calculating w/V percentages, a mass in grams is divided by a volume in cm^3 – you do not need to convert the volume to dm^3.

You can see from the calculations that the $w/w\%$ (10%) is slightly different from the $w/V\%$ (11%) even though the volume of the solution is the same.

Worked Example 3.20   11.2 g of glycine (M_r = 79) was dissolved in 100 cm^3 of water. Calculate the concentration of the solution in:

    a. mol dm^{-3}

    b. *w/w*%

    c. *w/V*%

The density of water is $1.0\,\text{g cm}^{-3}$.

Solution

    a. To calculate the concentration in mol dm^{-3}, we use the relationship $c = \dfrac{n}{V}$ and first calculate the number of moles of glycine dissolved.

$$\text{Number of moles of glycine} = \frac{m}{M} = \frac{11.2\,\text{g}}{79\,\text{g mol}^{-1}} = 0.142\,\text{mol}$$

Next, calculate the concentration using:

$$c = \frac{n}{V} = \frac{0.142\,\text{mol}}{100 \times 10^{-3}\,\text{dm}^3} = 1.42\,\text{mol dm}^{-3}$$

    b. To calculate the *w/w* percentage composition, use the expression:

$$w/w\% = \frac{\text{mass of solute}}{\text{mass of solution}} \times 100\%$$

The mass of solution is equivalent to the mass of water plus the mass of solute. The mass of water is 100 g as water has a density of $1\,\text{g cm}^{-3}$:

$$w/w\% = \frac{11.2\,\text{g}}{100\,\text{g} + 11.2\,\text{g}} \times 100\% = 10.1\%$$

    c. To calculate the *w/V* percentage composition, use the expression:

$$w/V\% = \frac{\text{mass of solute}}{\text{volume of solution}} \times 100\% = \frac{11.2\,\text{g}}{100\,\text{cm}^3} \times 100\% = 11.2\%$$

3.5   Titration calculations

Titrations are used to determine the amount of a substance present in a solution of an unknown concentration. Titrations are useful throughout analytical chemistry for determining many different factors in a whole range of products such as foodstuffs, personal products, agricultural products, pharmaceuticals, etc. Titrations are carried out by measuring the volume of a solution of known concentration (say, solution Y) that reacts exactly with a certain volume of the solution of unknown concentration, solution X. The colour change of an indicator is often used to determine when all the substance of unknown concentration (X) has been used up. At this point, the titration is stopped, and the volume of solution Y is measured exactly. By knowing the concentration and the volume of this solution used, we can find the number of moles that have reacted. We can then relate this to the number of moles of the unknown substance that have reacted using the chemical equation for the reaction.

Box 3.1   Titration calculations - relating reacting numbers of moles

Equation for the reaction: $Y + X \rightarrow$ products

1 mole X reacts with 1 mole Y

Amount in moles of known substance Y

$$n(Y) = V(Y) \times c(Y)$$

Amount in moles of unknown substance X

$$n(X) = V(X) \times c(X)$$

1 mole of X reacts with 1 mole of Y
Therefore n moles of X react with n moles of Y:

$$n(X) = n(Y)$$

$$V(X) \times c(X) = V(Y) \times c(Y)$$

$$c(X) = \frac{V(Y) \times c(Y)}{V(X)}$$

Equation for the reaction: $2Y + X \rightarrow$ products

1 mole of X reacts with 2 moles of Y

Therefore n moles of X react with $2n$ moles of Y:

$$n(X) = 2 \times n(Y)$$

$$V(X) \times c(X) = 2(V(Y) \times c(Y))$$

$$c(X) = 2 \times \frac{V(Y) \times c(Y)}{V(X)}$$

The apparatus used in titrations is accurate to four significant figures so, when carried out carefully, titrations can give very accurate results for concentrations and amounts of substances.

The balanced chemical equation must first be written for the reaction between X and Y so that the number of moles of X that react with a certain number of moles of Y can be determined. The example outlined in Box 3.1 shows how to relate the amount in moles of each of the reacting substances for chemical equations with different stoichiometries.

Worked Example 3.21   In a titration to determine the concentration of a solution of hydrochloric acid, 25.00 cm³ of hydrochloric acid solution was measured with a pipette and added to a conical flask, and a few drops of methyl orange indicator were added. Sodium hydroxide solution of concentration 0.0980 mol dm⁻³ was added from a burette until the indicator just changed colour and the acid was completely neutralised by the sodium hydroxide. The titration was repeated three times until consistent results were obtained. The average volume of sodium hydroxide solution required was found to be 23.25 cm³. Calculate the concentration of the hydrochloric acid solution.

Solution
Solving a titration problem such as this requires first writing a balanced chemical equation for the reaction to determine the number of moles of hydrochloric acid that react with one mole of sodium hydroxide. Following the steps as in the previous box, the number of moles of sodium hydroxide can be calculated and then equated to the number of moles of hydrochloric acid reacting.

Once the number of moles of hydrochloric acid is known, the concentration of the acid can be found.

Step 1: Write a balanced chemical equation for the reaction:

$$NaOH + HCl \rightarrow NaCl + H_2O$$

Therefore 1 mol NaOH reacts with 1 mol HCl.

Step 2: Identify the reagent for which you know both the volume and concentration. In this case, it is NaOH. Calculate the number of moles of NaOH reacting: $n(NaOH) = V(NaOH) \times c(NaOH)$.

$$n(NaOH) = 23.25 \times 10^{-3}\,dm^3 \times 0.0980\,mol\,dm^{-3}$$
$$= 2.279 \times 10^{-3}\,mol$$

Step 3: Equate the number of moles of NaOH to the number of moles of HCl:
$$n(HCl) = n(NaOH) = 2.279 \times 10^{-3}\,mol$$

Step 4: Rearrange the equation $n(HCl) = V(HCl) \times c(HCl)$ to find the concentration of hydrochloric acid:

$$c(HCl) = \frac{n(HCl)}{V(HCl)} = \frac{2.279 \times 10^{-3}\,mol}{25.00 \times 10^{-3}\,dm^3}$$
$$= 0.09114\,mol\,dm^{-3}$$

In many cases, the reactants in the balanced chemical equation for a titration are not in a simple 1 : 1 ratio. This means that you have to take care when carrying out calculations to correctly relate the number of moles of each reacting substance: for example, as in Worked Example 3.22.

Worked Example 3.22   A titration was carried out to determine the concentration of sulfuric acid using sodium hydroxide of concentration $0.0980\,mol\,dm^{-3}$. An average volume of $22.65\,cm^3$ of sodium hydroxide was required to neutralise $25.00\,cm^3$ of sulfuric acid. Calculate the concentration of the sulfuric acid.

Solution
Follow the steps as in Example 3.21:

Step 1: Write the balanced chemical equation for the reaction. Remember that an acid and a base react together to give a salt and water.

$H_2SO_4 + NaOH \rightarrow Na_2SO_4 + H_2O$ UNBALANCED EQUATION

$H_2SO_4 + 2NaOH \rightarrow Na_2SO_4 + 2H_2O$ BALANCED EQUATION

2 moles of NaOH react with 1 mole of H_2SO_4, and therefore 1 mole of NaOH reacts with half a mole of H_2SO_4.

The salt formed in this reaction is sodium sulfate (Na_2SO_4). Because the SO_4^{2-} ion has a double negative charge, we need two sodium (Na^+) ions per SO_4^{2-} ion to balance the charge. We therefore need to add 2 moles of NaOH to the right-hand side of the equation and 2 moles of water to the left.

Step 2: Identify the reagent for which you know both the volume and concentration. In this case, it is NaOH. Calculate the number of moles of NaOH reacting:

$$n(NaOH) = V(NaOH) \times c(NaOH)$$

$$n(\text{NaOH}) = 22.65 \times 10^{-3}\,\text{dm}^3 \times 0.0980\,\text{mol dm}^{-3}$$
$$= 2.220 \times 10^{-3}\,\text{mol}$$

Step 3: From the equation, we know that 1 mole of NaOH reacts with half a mole of H_2SO_4. So the number of moles of H_2SO_4 reacting is half the number of moles of NaOH:

$$n(\text{H}_2\text{SO}_4) = \tfrac{1}{2}\,n(\text{NaOH}) = \tfrac{1}{2} \times 2.220 \times 10^{-3}\,\text{mol} = 1.110 \times 10^{-3}\,\text{mol}$$

Step 4: Rearrange the equation $n(\text{H}_2\text{SO}_4) = V(\text{H}_2\text{SO}_4) \times c(\text{H}_2\text{SO}_4)$ to find the concentration of sulfuric acid:

$$c(\text{H}_2\text{SO}_4) = \frac{n(\text{H}_2\text{SO}_4)}{V(\text{H}_2\text{SO}_4)} = \frac{1.110 \times 10^{-3}\,\text{mol}}{25.00 \times 10^{-3}\,\text{dm}^3}$$
$$= 0.04440\,\text{mol dm}^{-3}$$

Box 3.2 It is possible to use a general equation when solving titration problems – but it is easy to make mistakes if you don't understand the theory properly. However, you may have been taught the following method.

For the general equation where n is the number of moles of Y reacting with m moles of X:

$$n\text{Y} + m\text{X} \rightarrow \text{products,}$$

the equation:

$$\frac{c(\text{Y}) \times V(\text{Y})}{n(\text{Y})} = \frac{c(\text{X}) \times V(\text{X})}{m(\text{X})}$$

equates the number of moles of X$(= m)$ to the number of moles of Y$(= n)$ taking into account the stoichiometry of the reaction.

This can be rearranged to:

$$c(\text{X}) = \frac{c(\text{Y}) \times V(\text{Y}) \times m(\text{X})}{n(\text{Y}) \times V(\text{X})}$$

The problem in Worked Example 3.22 can be solved using the general equation. If we substitute X for H_2SO_4 (the unknown concentration) and Y for NaOH, then the general equation becomes:

$$c(\text{H}_2\text{SO}_4) = \frac{c(\text{NaOH}) \times V(\text{NaOH}) \times m(\text{H}_2\text{SO}_4)}{n(\text{NaOH}) \times V(\text{H}_2\text{SO}_4)}$$

Therefore:

$$c(\text{H}_2\text{SO}_4) = \frac{0.0980\,\text{mol dm}^{-3} \times 22.65 \times 10^{-3}\,\text{dm}^3 \times 1\,\text{mol}}{2\,\text{mol} \times 25.00 \times 10^{-3}\,\text{dm}^3}$$

$$= 0.04440\,\text{mol dm}^{-3}$$

3.5.1 Back titration

A **back titration** is a titration where the concentration of the substance being analysed is determined by reacting it with a known amount of excess reagent. The remaining excess reagent is then titrated with another reagent. This process of back titration can be used, for example, to determine the amount of calcium carbonate in toothpaste.

Source: Elizabeth Page.

Calcium carbonate, $CaCO_3$, or chalk, is added as a mild abrasive to toothpaste. Carbonates react with acids to produce salts, carbon dioxide, and water. Carbonates are not soluble in water, but if a known quantity of acid is added in excess to an unknown quantity of calcium carbonate, the carbonate reacts completely with the acid. The remaining unreacted acid can then be titrated with a base such as sodium hydroxide of known concentration to find the amount in moles of unreacted acid. This can then be subtracted from the initial quantity of acid added to determine the amount in moles of acid that has reacted with the calcium carbonate. Knowing the ratio of moles of acid to moles of calcium carbonate, the amount in moles of calcium carbonate can be calculated.

Worked Example 3.23   In an experiment to determine the proportion of calcium carbonate in toothpaste, a 0.450 g sample of toothpaste was weighed and mixed with deionised water. 40.00 cm^3 of 0.100 mol dm^{-3} hydrochloric acid was added, and the mixture was warmed slightly to drive off carbon dioxide gas produced. On cooling, the mixture was titrated against sodium hydroxide solution. 23.20 cm^3 of sodium hydroxide of concentration 0.0980 mol dm^{-3} was required to completely neutralise the remaining hydrochloric acid. Calculate the percentage of calcium carbonate in the toothpaste.

Solution
As explained earlier, this type of procedure is called *back titration*. It is a two-stage process in which the analyte (substance whose concentration is being determined) is reacted with excess acid, and the remaining unreacted acid is titrated. The amount of acid that reacted can be obtained by subtraction. The calculation is therefore carried out in a number of steps:

Step 1: Calculate the total number of moles of HCl added ($n(HCl)_i$).

Step 2: Calculate the number of moles of HCl remaining ($n(HCl)_f$).

Step 3: Subtract $(n(HCl)_i) - (n(HCl)_f)$ to determine the number of moles of HCl that reacted with the calcium carbonate.

Step 4: Determine the number of moles of calcium carbonate that reacted and hence the mass and proportion in the toothpaste.

The values given in the Worked Example are used here to illustrate the steps 1-4.

Step 1: The total number of moles of HCl added is obtained from:

$$n(HCl)_i = c(HCl)_i \times V(HCl)_i = 40.00 \times 10^{-3} \, dm^3 \times 0.100 \, mol \, dm^{-3}$$
$$= 4.00 \times 10^{-3} \, mol$$

Step 2: The number of moles of HCl remaining is obtained from the titration with NaOH. The equation for this titration is:

$$HCl + NaOH \rightarrow NaCl + H_2O.$$

The equation tells us that 1 mole of NaOH reacts with 1 mole of HCl.
Calculate the number of moles of NaOH:

$$n(NaOH) = c(NaOH) \times V(NaOH)$$
$$= 0.0980 \, mol \, dm^{-3} \times 23.20 \times 10^{-3} \, dm^3$$
$$= 2.274 \times 10^{-3} \, mol$$

The number of moles of HCl remaining, $n(HCl)_f = 2.274 \times 10^{-3} \, mol$.

Step 3: Calculate the number of moles of HCl reacted by subtracting $n(HCl)_f$ from $n(HCl)_i$:

$$n(HCl)_i - n(HCl)_f = 4.00 \times 10^{-3} - 2.274 \times 10^{-3} \, mol$$
$$= 1.726 \times 10^{-3} \, mol$$

Step 4: We therefore know that 1.726×10^{-3} moles of HCl reacted with the calcium carbonate in the toothpaste. Write a balanced equation for the reaction to determine the number of moles of calcium carbonate in this sample:

$$CaCO_3 + 2HCl \rightarrow CaCl_2 + CO_2 + H_2O$$

The equation states that 2 moles of HCl react with 1 mole of calcium carbonate; therefore, each 1 mole of HCl reacts with 0.5 moles of calcium carbonate.
As the number of moles of HCl reacting = 1.726×10^{-3}, the number of moles of calcium carbonate present is $1.726 \times 10^{-3}/2 = 0.863 \times 10^{-3} \, mol$.

Next calculate the mass of calcium carbonate from the number of moles and the molar mass: $m(CaCO_3) = n(CaCO_3) \times M(CaCO_3)$

$$= 0.863 \times 10^{-3} \, mol \times 100 \, g \, mol^{-1} = 0.0863 \, g$$

The percentage of $CaCO_3$ in the toothpaste sample is obtained from:

$$\frac{mass \, CaCO_3}{mass \, of \, toothpaste} \times 100\% = \frac{0.0863 \, g}{0.450 \, g} \times 100 = 19.2\%$$

3.6   Calculations with gas volumes

Many chemical reactions take place between gases, and it is important to be able to calculate the number of moles of each gas reacting. Avogadro suggested that equal volumes of gases contain the same number of molecules, no matter what the chemical nature of the gas is. This hypothesis has been proved to be approximately true under conditions of normal pressures and temperatures. At room temperature (20 °C) and pressure (1 atm), one mole of any gas is said to have a volume of 24.0 dm^3. At 0 °C and a pressure of 1 atm, one mole of gas has a slightly smaller volume of 22.4 dm^3.

In Chapter 4, we will see how this volume can be calculated for any conditions of temperature and pressure. The volume occupied by one mole of gas at room temperature and pressure is sometimes referred to as the **molar gas volume, V_m**. Another way to think of the molar gas volume is to view it as the density of the gas. The smaller the volume occupied by one mole the more dense the gas. The important point is that at the same temperature and pressure, the same volume of any gas contains the same number of molecules. Because we know that one mole of any gas contains 6.02×10^{23} molecules, we can calculate the number of moles of a gas in a certain volume:

$$\text{Amount in moles of gas}, n = \frac{\text{volume of gas}, V}{\text{molar volume}, V_m}$$

By rearranging the equation, if we know the number of moles of gas in the sample, we can calculate the volume occupied by the gas using the molar volume:

$$\text{Volume of gas}, V = \text{Amount in moles}, n \times \text{molar volume}, V_m$$

Worked Example 3.24   Calculate the number of moles in 500 cm^3 of CO_2 at 20 °C and 1 atm pressure and hence the mass of CO_2 in the sample.

Solution
Assuming the molar volume, V_m of any gas at room temperature (20° C) and pressure (1 atm) is 24.0 dm^3, we can find the number of moles of gas in 500 cm^3.
Using the earlier equation:

$$\text{Amount in moles of gas}, n = \frac{\text{volume of gas}, V}{\text{molar volume}, V_m}$$

$$\text{Amount in moles}, n = \frac{500 \times 10^{-3} \text{ dm}^3}{24.0 \text{ dm}^3 \text{ mol}^{-1}} = 0.0208 \text{ mol}$$

The molar mass of CO_2 is 44.0 g mol^{-1} (12.0 + 16.0 × 2). Therefore, the mass of 0.0208 mol CO_2 = 0.0208 g × 44.0 g mol^{-1} = 0.915 g.

The molar volume, V_m, at 298 K and 1 atm is 24.4 dm^3 mol^{-1}. One mole of gas occupies 24.4 dm^3 at this temperature and pressure. Room temperature is generally considered to be 20° C (293 K). At this temperature the volume of one mole of a gas at 1 atm pressure is 24.0 dm^3. At 0° C (273 K) and 1 atm pressure the volume decreases to 22.4 dm^3.)

Worked Example 3.25   Calculate the volume occupied by 0.32 g of oxygen gas at room temperature and pressure.

Solution
Oxygen is a diatomic molecule with the formula O_2. The molar mass of oxygen is 32.0 g mol^{-1}. Therefore, in 0.32 g of O_2, there are $\dfrac{0.32 \text{ g}}{32.0 \text{ g mol}^{-1}} = 0.01$ mol oxygen molecules.

Using the earlier equation and the molar volume, find the volume occupied by the gas:

$$\text{Volume of gas}, V = \text{Amount in moles}, n \times \text{molar volume}, V_m$$
$$V = 0.01 \text{ mol} \times 24.0 \text{ dm}^3 \text{ mol}^{-1} = 0.24 \text{ dm}^3$$

A **diatomic molecule** has two atoms in one molecule of the compound: for example, O_2, N_2, Cl_2, etc. They are sometimes called dioxygen, dinitrogen, dichlorine, etc. to make this clear. The noble gases (Group 8 or 18) are **monatomic** molecules, which means they are composed of single atoms, e.g. He, Ar, etc. CO_2 is an example of a triatomic molecule.

Quick-check summary

- The relative atomic mass of an element is equal to the average mass of one atom of the element divided by one-twelfth of the mass of an atom of carbon-12.

- The relative molecular mass of a compound is obtained by summing the relative atomic masses of all the atoms in the compound.

- A mole of a substance is the quantity that contains Avogadro's number of particles, which can be atoms, ions, or molecules. The Avogadro constant is 6.02×10^{23} mol^{-1}.

- The amount in moles of a substance is obtained by dividing the mass of substance by its molar mass. The equation that relates the amount in moles (n) to mass (m) and molar mass (M) is $n = \dfrac{m}{M}$.

- The percentage yield of a reaction is an indication as to the success of the reaction in producing products and is obtained by dividing the actual yield by the expected yield in units of either mass or moles. The equation for percentage yield is:

$$\% \text{ yield of product} = \frac{\text{amount of product obtained}}{\text{theoretical amount of product expected}} \times 100\%$$

- The percentage composition by mass of an element in a compound is the mass of that element in one mole of the compound divided by the mass of one mole of the compound.

$$
\begin{aligned}
&\% \text{ composition by mass of X} \\
&= \frac{\text{mass of element X in one mole of compound}}{\text{mass of one mole of compound}} \times 100\%
\end{aligned}
$$

- The empirical formula of a compound gives the smallest whole-number ratio of the elements present in one molecule or formula unit of the compound.

- Concentration (c) is expressed as the amount in moles of compound (n) divided by the volume of solution (V). The unit of concentration is $mol\,dm^{-3}$: $c = \dfrac{n}{V}$.

- Concentration can also be expressed in terms of weight per weight percentage ($w/w\%$) or weight per volume percentage ($w/V\%$).

- Solutions of specific concentration can be made by diluting more concentrated solutions by a fixed amount. The dilution factor indicates by how much the more concentrated has been diluted.

- Titrations are used in quantitative analysis to determine the concentration of solutions. Calculations using the amount in moles of a known reactant are used to determine the amount in moles of a second reactant, and hence its concentration. The balanced stoichiometric equation for the reaction must be known.

- Avogadro's hypothesis states that equal volumes of gases contain the same number of molecules. At 20 °C and 1 atm pressure, one mole of any gas has a volume of 24.0 dm^3. This quantity is known as the molar gas volume. It is obtained by dividing the volume of gaseous sample by the amount in moles of the gas: $V_m = \dfrac{V}{n}$.

End-of-chapter questions

1 Calculate the mass of the following:

  a. 0.20 mol calcium carbonate, $CaCO_3$

  b. 0.50 mol bromine, Br_2

  c. 0.10 mol hydrated copper sulphate, $CuSO_4.5H_2O$

2 Calcium hydroxide is made by heating calcium carbonate strongly to form calcium oxide (quicklime) and then adding water to produce calcium hydroxide (slaked lime).

  a. What mass of water would be required to completely convert all the calcium oxide, formed from 10 kg calcium carbonate, to calcium hydroxide?

  b. What mass of calcium hydroxide would be produced from 10 kg calcium carbonate?

3 6.0 cm^3 of phenylamine was treated with 8.0 cm^3 of ethanoic anhydride, and 6.45 g of solid N-phenylethanamide was obtained according to the following equation:

$(CH_3CO)_2O$	+	$C_6H_5NH_2$	$\rightarrow$	$CH_3CONHC_6H_5$	+	CH_3COOH
Ethanoic anhydride		phenylamine		N–phenylethanamide		ethanoic acid

  Calculate the percentage yield in the reaction.
  (Density of phenylamine = 1.02 g cm^{-3}; density of ethanoic anhydride = 1.08 g cm^{-3}.)

4 Analysis of an organic compound containing C, H, and O only showed that it contained 66.7% C and 11.1% H by mass. The relative molecular mass of the compound was 72. Find the empirical formula and the molecular formula for the compound.

5 A sample of 0.85 g of silver nitrate ($AgNO_3$) was found to contain 0.51 g of silver. Calculate the actual percentage composition of silver in the sample and the percentage purity.

6 A standard saline drip used to replenish electrolytes in the body is 0.90% w/V sodium chloride. Calculate the molarity of the solution.

7 25.0 cm^3 of $0.101 \text{ mol dm}^{-3}$ sodium hydrogen carbonate solution was titrated with dilute sulfuric acid of unknown concentration. 19.5 cm^3 of acid was required to neutralise the sodium hydrogen carbonate solution. Find the concentration of the sulfuric acid in mol dm^{-3} and g dm^{-3}. The equation for the reaction is

$$2NaHCO_3 + H_2SO_4 \rightarrow Na_2SO_4 + 2H_2O + 2CO_2$$

8 In an experiment to determine the nature of an alkali metal element in a sample of metal carbonate, M_2CO_3, 2.25 g of the metal carbonate was dissolved in 100 cm^3 water. 25.0 cm^3 of the M_2CO_3 solution was titrated against hydrochloric acid of concentration $0.400 \text{ mol dm}^{-3}$. 20.5 cm^3 of HCl solution was required to completely neutralise the metal carbonate solution. Determine the identity of the alkali metal in the metal carbonate.

9 The percentage purity of a sample of aspirin prepared in the laboratory can be determined by back titration. In an experiment, a 1.500 g sample of aspirin was dissolved in 50 cm^3 of NaOH of concentration $0.500 \text{ mol dm}^{-3}$. The solution was boiled to ensure complete reaction of the aspirin according to this equation:

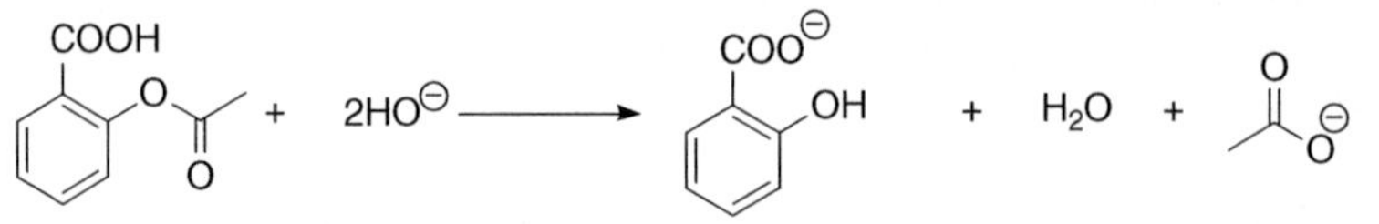

The resulting solution was titrated against $0.500 \text{ mol dm}^{-3}$ HCl, and 20.65 cm^3 of HCl was required to completely neutralise the excess sodium hydroxide. Calculate the percentage purity of the aspirin (aspirin: $C_9H_8O_4$, $M_r = 180.15$).

10 Answer the questions that follow:

a. What is the maximum volume of chlorine gas at room temperature and pressure that could be obtained by heating 5.0 g of manganese(IV) oxide with excess concentrated hydrochloric acid? The equation for the reaction is:

$$MnO_2 + 4HCl \rightarrow MnCl_2 + Cl_2 + 2H_2O$$

b. Potassium nitrate(V) decomposes on heating to potassium nitrite and oxygen. What mass of potassium nitrate would be required to produce 1.00 dm^3 of oxygen gas at room temperature and pressure? The equation for the decomposition reaction is:

$$2KNO_3 \rightarrow 2KNO_2 + O_2$$

(Assume 1 mole of gas has a volume of 24.0 dm^3 at room temperature and pressure.)

4

States of matter

At the end of this chapter, students should be able to:

- Describe the three main states of matter

- Compare the motion of particles in solids, liquids, and gases and the strengths of interactions between particles in these materials

- Describe the arrangement of particles in metals, ionic solids, and simple covalent and giant molecular structures

- Explain how the physical properties of simple materials such as melting and boiling points are determined by the strengths of intermolecular forces in the materials

- Understand the derivation of the ideal gas law and its relevance in defining the behaviour of gases

4.1 Introduction

Chapter 2 showed that the type of bonding between atoms and the forces between molecules depend upon the nature of the elements that are bonded together. In this chapter, we will see how bonding and intermolecular forces determine the properties of materials.

Most materials can be classified as solids, liquids, or gases – the three common states of matter. Most substances are made up of smaller particles such as atoms, molecules, or ions. In Chapter 2, it was shown that atoms within molecules have bonds between them called intramolecular forces, and molecules have weaker interactions between them known as intermolecular forces. It is the strength of the intermolecular forces that determines whether the substance exists as a solid, liquid, or gas at room temperature.

In solids, particles are relatively close to each other, usually in a regular arrangement. If a solid is heated, the particles gain kinetic energy and begin

Foundations of Chemistry: An Introductory Course for Science Students, First Edition.
Philippa B. Cranwell and Elizabeth M. Page.
© 2021 John Wiley & Sons Ltd. Published 2021 by John Wiley & Sons Ltd.
Companion website: www.wiley.com/go/Cranwell/Foundations

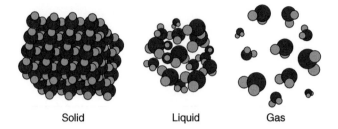

Solid   Liquid   Gas

Figure 4.1 Molecules of water in the solid, liquid, and gas states.
Source: Redrawn from NinetyEast, https://www.google.com/search?q=solids+
liquids+and+gases&client=firefox-b&source=lnms&tbm=isch&sa=
X&ved=0ahUKEwi5lK77987bAhVhMJoKHQhEAuEQ_AUICigB&
biw=960&bih=439#imgrc=gjLnfSjdY7gvKM.

to vibrate. Eventually, they vibrate so much that they break away from their
fixed positions and become mobile and move randomly in the liquid state, as
shown for water in Figure 4.1.

In the liquid state, the forces between particles are continually being broken
and reformed as the particles move away and then become closer to each other. If
the liquid is heated, the particles gain even more kinetic energy and eventually
escape from the liquid state into the space above. The particles are now in the
gaseous state, or gas phase.

In the gas phase, the particles are further from each other than in the solid (on
the molecular scale), so the intermolecular forces are now very weak. The forces
between the particles become weaker as the distance between the particles
increases. In a gas, the molecules move completely randomly in all directions
throughout the space that encloses the gas. Figure 4.2 illustrates the change from
solid to liquid to gas for water as it is heated.

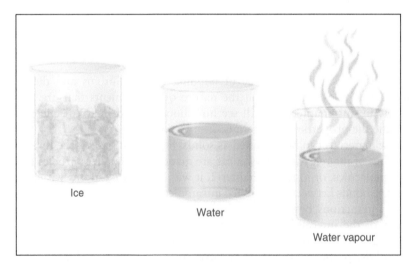

Ice

Water

Water vapour

Figure 4.2 The change in state from solid to liquid and then to gas for water on heating.
Source: Heba Soffar, 'The physical and chemical properties of water', 2017. Science online.
https://www.online-sciences.com/the-matter/the-physical-and-the-chemical-properties-
of-the-water/. Licensed under CC-BY-4.0.

In this chapter, we will look at the three states of matter and the forces holding particles together in each physical state. We will see how these forces differ between covalent and giant molecular materials, metals, and ionic solids, and how the forces impact the properties of the materials.

4.2 Solids

Chapter 2 introduced the three main types of materials: metals, ionic salts, and covalent compounds. In metals and ionic solids, the particles are arranged in regular arrays or *lattices*. Covalent compounds can be divided into two further types: simple covalent (also known as simple molecular) materials and giant covalent (or giant molecular) materials. Simple covalent materials are composed of small molecules, whereas giant covalent compounds have extended arrays of atoms.

4.2.1   Metallic lattices

The particles that make up metallic materials are atoms. In the solid state, metals consist of a regular arrangement of atoms, such as iron atoms or copper atoms. The valence (outer) electrons of the metal's atoms become delocalised, or lost, from the rest of the atom and behave as a 'sea' of electrons surrounding the metal ions, which have a resultant positive charge as depicted in Figure 4.3 for sodium metal. The sea of electrons holds the positive metal ions in a regular lattice with strong electrostatic forces between the metal ions and electrons. This results in the high melting and boiling points observed for most metals.

When an electric field is applied to a metal, the electrons are free to move to the positive electrode and hence conduct electricity. This makes metals good conductors of electricity.

Within the three-dimensional structure of a metal are layers of metal atoms in regular positions. If a mechanical force is applied to the material, the layers of atoms slide over each other, and this causes metals to be malleable or easily shaped.

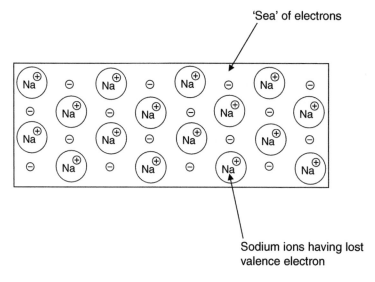

Figure 4.3 Metallic bonding in sodium.

4.2.2   Ionic lattices

Ionic compounds consist of regular arrays of positive cations and negative anions. The cations and anions are held together by strong electrostatic forces of attraction between the oppositely charged ions. Figure 4.4a shows a sodium chloride, NaCl, lattice; Figure 4.4b shows a calcium fluoride, CaF_2, lattice.

Although the ions are represented as spheres separated from each other in space, in practice, the oppositely charged ions touch each other, as in Figure 4.5. This method of depicting the ionic lattice where spheres or circles

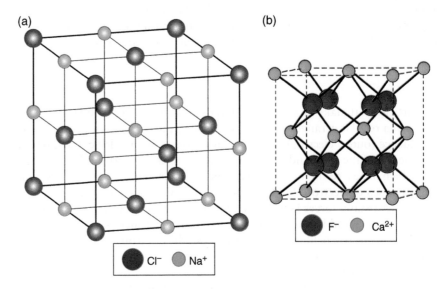

Figure 4.4 Ionic lattices. (a) Sodium chloride, NaCl. *Source:* Proline server, 'Ion lattice of a face centered cubic Bravais-lattice', 2007, Wikimedia Commons. https://commons. wikimedia.org/wiki/File:Ionlattice-fcc.svg. Licensed under CC-BY-SA-3.0. (b) Calcium fluoride, CaF_2. *Source:* 'Calcium the Archetypal Alkaline Earth Metal', https://cnx.org/ contents/9G6Gee4A@25.8:rllrYLPV@3/Solid-State-and-Solution-Chemistry; CC 3.0.

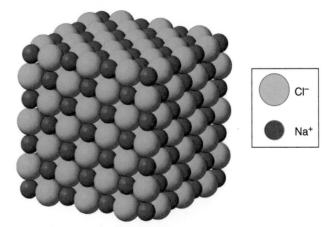

Figure 4.5 Space-filling diagram for sodium chloride showing space occupied by the ions. *Source:* The crystal structure of sodium chloride by Benjah-bmm27: own work, public domain, 2006. https://commons.wikimedia.org/w/index.php?curid=702423.

represent the spaces occupied by the atoms is called a *space-filling* diagram. Strong electrostatic forces hold the ions together in the lattice. To melt an ionic solid, these forces must be overcome by applying energy in the form of heat. The high strengths of these electrostatic forces mean that ionic solids have very high melting points and boiling points.

In the solid state, ionic materials do not conduct electricity as there are no freely moving charged particles that can move towards the electrodes. However, when an ionic material is melted so that it forms a liquid and an electric field is applied, the positive ions move towards the negative electrode and the negative ions move towards the positive electrode, and so a current flows. Ionic materials are hard and brittle, and they often form beautiful crystalline shapes. Ionic materials are not malleable as they break or shear under pressure.

4.2.3   Simple molecular solids and giant molecular structures

Two main types of materials are formed when atoms are joined by covalent bonds: *simple molecular* and *giant molecular* compounds.

Simple molecular solids consist of small molecules such as molecules of iodine, I_2, in crystalline iodine, or molecules of carbon dioxide, CO_2, in dry ice. The intermolecular forces between the molecules are weak and result in low melting and boiling points; therefore, simple covalent species are usually liquids or gases at room temperature. Some exceptions to this rule are iodine which is a solid at room temperature, and some waxes that are long-chain hydrocarbons.

Giant molecular solids have atoms bonded together by strong covalent bonds in a regular array that extends throughout the solid material. To melt the material, these strong covalent bonds must be broken, so the melting and boiling points of giant covalent materials are very high. Examples of giant molecular bonding are found in graphite, diamond, and silicon dioxide, the structures of which are shown in Figure 4.6.

Neither simple covalent nor giant covalent materials conduct electricity as there are no free electrons or ions to carry charge. The only exception is graphite,

> Covalent bonding occurs when atoms share their valence electrons to form strong bonds that hold them together.

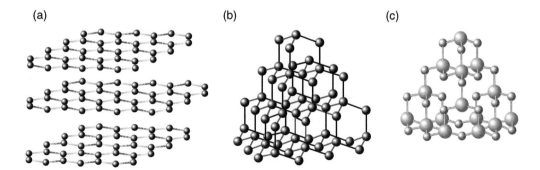

(a)   (b)   (c)

Figure 4.6 (a) The structure of graphite. *Source:* Modified from 'Giant covalent molecules', AQA, BBC. www.bbc.co.uk/schools/gcsebitesize/science/add_aqa_pre_2011/atomic/differentsubrev3.shtml. (b) The structure of diamond. (c) The structure of silicon dioxide. The lines represent the covalent bonds between the atoms of carbon in (a) and (b) and the atoms of Si and O in (c).

Table 4.1 A summary of the properties of ionic, covalent, and metallic compounds.

	Metallic	Ionic	Covalent	
			Simple	Giant
Intramolecular forces[a]	Strong	Strong	Strong	Strong
Intermolecular forces[a]			Weak	
Melting/boiling point	High	High	Low	High
Electrical conductivity	Good	Solid: poor Molten: good	None	Usually poor apart from certain materials
Malleability	Good	Poor	Poor	Poor

[a] The terms *intermolecular* force and *intramolecular* force have been used to describe the forces between and within particles of the material, respectively, even though there are no molecules in metals and ionic solids.

which has electrons between layers of carbon atoms that can move to the positive electrode if a voltage is applied.

The properties of ionic, covalent, and metallic compounds are summarised in Table 4.1.

Worked Example 4.1   Classify the following materials according to the type of *intra*molecular forces that exist within them:

a. Water, H_2O

b. Potassium chloride, KCl

c. Silicon dioxide, SiO_2

d. Gold, Au

e. Benzene, C_6H_6

Solution
Intramolecular forces are those forces between the atoms or ions in molecules or materials. The question is therefore asking about the type of bonding in the materials.

a. Water (H_2O) is a covalent liquid, and its atoms are therefore covalently bonded together.

b. Potassium chloride (KCl) is an ionic solid, and electrostatic forces hold the ions together in the lattice.

c. Silicon dioxide (SiO_2) has a giant covalent structure, and its atoms are held together by covalent bonds.

d. Gold (Au) is a metal, so its atoms are held together by metal–metal bonds.

e. Benzene (C_6H_6) is a covalently bonded liquid, so its atoms are held together by covalent bonds.

Worked Example 4.2   Comment on the likely type of *inter*molecular forces in the following materials, and predict the physical state of the material at room temperature and hence its approximate melting point:

  a. Carbon dioxide, CO_2

  b. Diamond, C

  c. Toluene, $C_6H_5CH_3$

  d. Aluminium sulfate, $Al_2(SO_4)_3$

  e. Polyethene, $(C_2H_4)_n$

Solution
Intermolecular forces are the forces between molecules that hold them together.

  a. Carbon dioxide (CO_2) is a covalently bonded molecule that does not have a dipole moment. It therefore has very weak intermolecular forces. It is a gas at room temperature and so has a low melting point.

  b. Diamond (C) is, of course, a solid at room temperature with a very high melting point. The atoms of carbon are held by strong covalent bonds, and the structure of diamond is giant covalent, so there are no intermolecular forces as such.

  c. Toluene, $C_6H_5CH_3$, is a hydrocarbon with covalent bonds between the atoms. It does not possess any electronegative atoms, and therefore it has very weak intermolecular forces. It is a liquid at room temperature with a low melting point (-95 C).

  d. Aluminium sulfate, $Al_2(SO_4)_3$, contains Al^{3+} and SO_4^{2-} ions. It has an ionic structure, with electrostatic forces between the oppositely charged ions. Its melting point is high.

  e. Polyethene $(C_2H_4)_n$ is a long-chain polymer. The long, straight hydrocarbon chains have relatively strong intermolecular forces between them, so the material is a solid at room temperature, although its melting point is quite low (around 110 C).

4.3 Liquids

We have seen that supplying energy to a solid in the form of heat makes the particles vibrate more vigorously, whether they are atoms, molecules, or ions. Eventually, the forces between the particles weaken, and the solid changes into the liquid state and melts. The melting point of the solid depends on the strength of the bonds between the particles. Liquids can flow because the particles can move past each other. The intermolecular forces between the particles are continually being broken and re-formed as the particles approach and then move away from each other. These forces are strong enough to keep the liquid in one place, so a liquid takes up the shape of the container that holds it.

In the case of a liquid, the term *particle* can mean an atom, a molecule, or an ion, depending upon the nature of the liquid. For example, liquid water contains water molecules, whereas molten NaCl would contain Na^+ ions and Cl^- ions.

4.3.1   Evaporation and condensation, vapour pressure, and boiling

As the temperature of a liquid is increased, the particles gain more kinetic energy and so become more separated from each other. Eventually, particles on the surface of the liquid gain sufficient kinetic energy to overcome the forces holding them in the liquid state and enter the space above the liquid to become a gas. The liquid is said to *evaporate* or *vaporise*. The space above the liquid that becomes occupied by particles is called the vapour. If the vapour is cooled, the particles lose kinetic energy and move around less quickly. The forces of attraction between the particles become stronger, and the particles will eventually join together and return to the liquid state. This process is called *condensation*. Evaporation and condensation are reversible processes, and you will have observed them many times. For example, when an open pan of water boils on a stove, the water evaporates. When the vapour hits a cold surface, it condenses back into liquid water. In an open container, the liquid will eventually all evaporate.

However, if the liquid is heated in a closed container, the space above the liquid in the container becomes more and more saturated with molecules that are forced closer together. The molecules with the least kinetic energy form new intermolecular forces with the nearest molecules and eventually return to the liquid state. Eventually, the rate at which molecules escape from the surface due to evaporation is equal to the rate at which molecules recombine and condense back to the liquid state. An *equilibrium* is established, as shown in Figure 4.7. The pressure exerted by the molecules in the gas phase above the liquid as they collide with the walls of the container is called the *vapour pressure*. As the temperature is increased, the vapour pressure of the liquid increases as more molecules have sufficient kinetic energy to escape into the gas phase. A new equilibrium is established at the higher temperature.

Boiling occurs when the pressure above the liquid is equal to the external atmospheric pressure and bubbles of vapour form within the body of the liquid. The temperature at which this occurs is called the boiling point, T_b, of the liquid.

When the vapour pressure is equal to the external pressure (usually around 1 atm), the liquid is said to boil. The temperature at which the vapour pressure of a liquid becomes equal to the external pressure is the *boiling point, T_b*, of the liquid. The stronger the intermolecular forces, the higher the boiling point of the material.

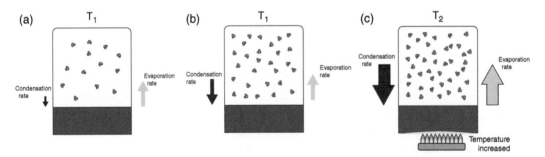

Figure 4.7 Movement of molecules from a liquid in a closed container. (a) The rate of evaporation is greater than the rate of condensation at temperature T_1. (b) Eventually at temperature T_1 the vapour phase has been come saturated with molecules, and the rate of evaporation is equal to the rate of condensation. An equilibrium is established. (c) The liquid is heated further to temperature T_2 and initially evaporation increases until eventually, the rate of condensation again becomes equal to the rate of evaporation and a new equilibrium is established. The vapour pressure in (c) is higher than in (b). *Source:* 'Evaporation Rates, Condensation Rates, and Relative Humidity'. https://www.e-education.psu.edu/meteo3/l4_p4.html.

Worked Example 4.3 Describe what happens in terms of the intermolecular forces in the liquid when water is heated from room temperature to its boiling point.

Solution

The question asks about the effect upon the intermolecular forces in water when it is heated to its boiling point. Start the answer by describing the intermolecular forces that exist between the molecules.

The water molecule has a large permanent dipole moment, which results in strong dipole-dipole interactions between neighbouring molecules. These dipole–dipole interactions are, more specifically, hydrogen bonds which is a very strong type of intermolecular force. When heated, the water molecules gain kinetic energy, which causes them to move about more rapidly, and the hydrogen bonds are weakened and eventually broken. As the hydrogen bonds are broken, the molecules on the surface of the liquid water escape. The space above the liquid becomes saturated with gaseous water molecules that exert a pressure, known as the vapour pressure of the water. When the vapour pressure of the water becomes equal to the external pressure, the water is said to boil. At the boiling point, the liquid water is in equilibrium with the vapour above it. The rate at which water molecules escape from the surface of the water is equal to the rate at which the intermolecular forces are re-formed and return to the liquid state.

4.3.2 Effect of intermolecular forces on melting and boiling points

One of the most significant effects of intermolecular forces is on the melting and boiling points of covalently bonded materials. The stronger the intermolecular forces, the more energy is required to overcome them, and the higher the melting and boiling point of the substance. In order for the substance to melt and then boil, the forces between molecules must be broken. So, in general, covalently bonded substances with permanent dipole–permanent dipole interactions have higher boiling points than those with induced dipole–induced dipole forces. This effect can be seen by comparing the molecules *cis*-dichloroethene ((Z)-CHClCHCl) and *trans*-dichloroethene ((E)-CHClCHCl) (Figure. 4.8).

(Z)-dichloroethene has a permanent dipole moment because both electronegative Cl groups are arranged on the same side of the double bond. There can be no rotation about a double bond. Therefore (Z)-dichloroethene has a higher boiling point (60.2 C) than the non-polar isomer, (E)-dichloroethene (48.5 C), in which the polar C—Cl bonds are diagonally opposite each other.

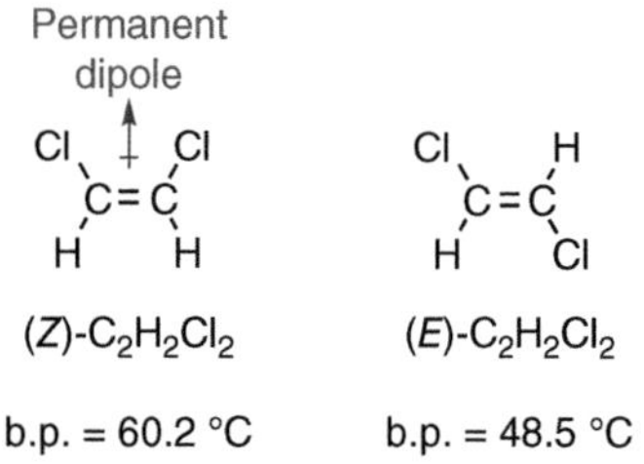

Permanent
dipole

Cl ↕ Cl   Cl   H
  C = C   C = C
H   H   H   Cl

(Z)-$C_2H_2Cl_2$ (E)-$C_2H_2Cl_2$

b.p. = 60.2 °C   b.p. = 48.5 °C

Figure 4.8 Comparison of boiling points in (Z)- and (E)-dichloroethene.

In molecules without permanent dipole moments, the melting and boiling points depend upon the strength of the instantaneous dipole–induced dipole interactions. The strength of these interactions depends largely upon the molecular mass of the substance and upon the arrangement of the atoms and groups in the molecules. The Group 7 (17) elements F_2, Cl_2, Br_2, and I_2 are non-polar and therefore have only instantaneous dipole–induced dipole (van der Waals) forces between the molecules. The strength of these forces can be readily observed by comparing the physical states of each of the elements at room temperature. Both fluorine and chlorine, the lighter elements, are gases at room temperature. Bromine is a liquid and iodine is a solid, indicating that the melting points and boiling points of the elements increase as we go down Group 7 (17) due to the stronger van der Waals forces in molecules with greater numbers of electrons. This information is shown in Table 4.2.

Noble gases are composed of single atoms and are known as *monatomic molecules*, posessing extremely weak instantaneous dipole–induced dipole forces. As a result, the boiling points of the noble gases are very low in comparison to those of other substances with similar molar masses, and they are very difficult to liquefy. In fact, the boiling point of helium is very close to absolute zero (0 K or −273 C). From Table 4.3, it can be seen that the boiling points of the Group 8 (18) elements increase down the group as the value of Z and the molar masses increase.

Non-polar molecules, such as hydrocarbons, that possess similar molar masses have melting and boiling points that depend upon the degree of branching in the molecule. Figure 4.9 shows the structures of hexane and its branched isomer 2,2-dimethylbutane. These molecules are structural isomers and have the same molar mass of 86 g mol^{-1}. The straight-chain hexane has a boiling point of 68 C, whereas the branched isomer has a lower boiling point of 49.7 C.

The higher boiling point in *n*-hexane is due to the larger induced dipole that is established in the straight-chain molecule. In the branched molecule, despite the

Structural isomers are molecules that have the same molecular formula but differ in the way the atoms are connected together and so have different structures.

Table 4.2 Physical properties of the halogens.

Element	State at room temperature	Molar mass/g mol^{-1}	Melting point/ C	Boiling point/ C
F_2	Gas	38	−220	−188
Cl_2	Gas	71	−101	−34
Br_2	Liquid	160	−7.2	60
I_2	Solid	254	114	184

Table 4.3 Properties of noble gases.

Noble gas	Z	Molar mass/g mol^{-1}	Boiling point/ C	Boiling point/K
He	2	4	−269 C	4
Ne	10	20	−246 C	27
Ar	18	40	−186 C	87
Kr	36	84	−152 C	121
Xe	54	131	−108 C	165
Rn	86	222	−62 C	211

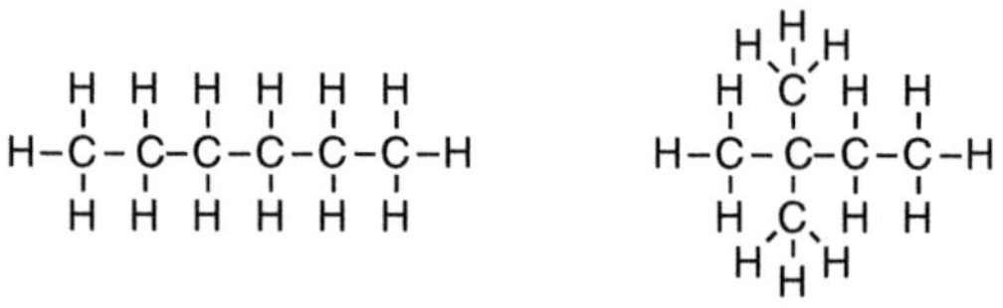

n-hexane
C_6H_{14}
b.p. = 68 °C

2,2-dimethylbutane
C_6H_{14}
b.p. = 49.7 °C

Figure 4.9 Comparison of boiling points of *n*-hexane and 2,2-dimethylbutane.

The *n* in *n*-hexane means the carbon atoms that make up the hexane molecule are all arranged in a line, with no branching.

same molar mass and therefore number of electrons, the induced dipole is smaller because of the shorter chain length for the dipole to operate along.

Worked Example 4.4 State whether the following species have permanent or induced dipole moments.

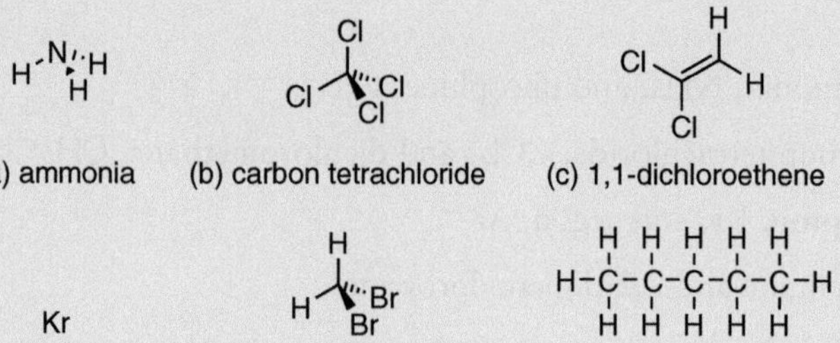

(a) ammonia (b) carbon tetrachloride (c) 1,1-dichloroethene

Kr

(d) krypton (e) dibromomethane (f) *n*-pentane

Solution

a. Ammonia has a trigonal pyramidal structure. The nitrogen atom has a lone pair of electrons and is more electronegative than hydrogen, so there is a permanent dipole towards the nitrogen atom.

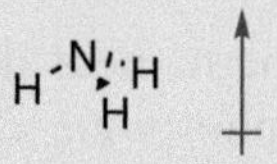

b. Carbon tetrachloride has a tetrahedral structure with a Cl atom at each of the four points of the tetrahedron. There is therefore no overall dipole moment, but instantaneous dipole–induced dipole forces exist.

c. In 1,1-dichloroethene, both Cl atoms are on the same C atom. The Cl atoms are more electronegative than the H atoms, so there is an overall permanent dipole moment in the direction shown.

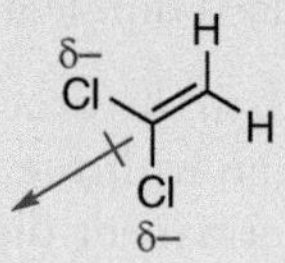

d. Krypton, Kr, is a noble gas. As such, it is a monoatomic molecule, with only instantaneous dipole–induced dipole forces.

e. Dibromomethane has a tetrahedral structure. The Br atoms are more electronegative than the H atoms, so there is an overall permanent dipole moment in the direction shown.

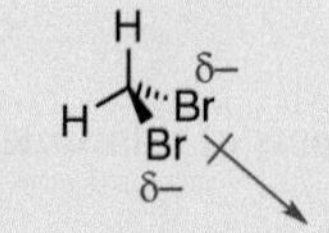

f. Pentane is a straight-chain hydrocarbon with no electronegative atoms. It therefore has an instantaneous dipole-induced dipole moment.

Worked Example 4.5 Using your knowledge from the previous example, state which of the following pairs of molecules has the higher boiling point and why:

a. Ammonia, NH_3, and phosphine, PH_3

b. Carbon tetrachloride, CCl_4, and dichloromethane, CH_2Cl_2

c. Krypton, Kr, and argon, Ar

d. *n*-Pentane and 2,2-dimethylpropane

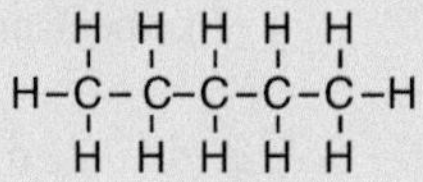

n-pentane 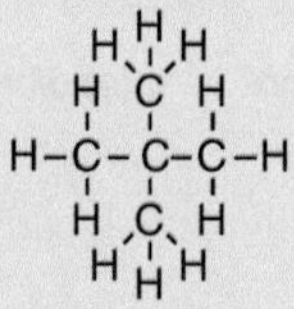   2,2-dimethylpropane

Solution

a. Ammonia has a greater permanent dipole moment and therefore a higher boiling point than phosphine as the N atom is more electronegative than P. Also, hydrogen bonding exists within ammonia, which is not possible in phosphine.

b. There is no permanent dipole in CCl_4, whereas there is a permanent dipole moment in CH_2Cl_2 (similar to CH_2Br_2). Therefore, CH_2Cl_2 has a higher boiling point than CCl_4.

c. Both Kr and Ar are noble gases with only instantaneous dipole–induced dipole forces. However, Kr has more electrons than Ar and therefore has a larger induced dipole and higher boiling point.

d. *n*-Pentane is a straight-chain hydrocarbon, whereas its structural isomer 2,2-dimethylpropane with the same molar mass is branched. Neither molecule has a permanent dipole moment, but pentane has a larger induced dipole as it has a longer unbranched chain than 2,2-dimethylpropane. Pentane therefore has a slightly higher boiling point than 2,2-dimethylpropane.

Worked Example 4.6   The structures of *cis*- and *trans*-but-2-ene are shown. State which isomer you would expect to have the higher boiling point, and explain why.

cis-but-2-ene   trans-but-2-ene

Solution
Both *cis*- and *trans*-but-2-ene have a double bond between the central two carbon atoms. This means that the two halves of the molecule are fixed and there can be no rotation about the bond. Therefore the methyl groups (CH_3) are in fixed positions with respect to each other. Although there are no electronegative atoms in either molecule, the presence of the two CH_3 groups on the same side of the double bond in *cis*-but-2-ene causes it to have a very small permanent dipole, whereas *trans*-but-2-ene has no dipole. The *cis*- (*Z*-) isomer therefore has a slightly higher boiling point than the *trans*- (*E*) isomer. The boiling point of *cis*-but-2-ene is 3.7 °C, whereas the boiling point of *trans*-but-2-ene is 0.9 °C.

4.3.3   The effect of hydrogen bonding on melting and boiling points of covalent compounds

Hydrogen bonding is the strongest type of intermolecular force. It occurs between molecules that have a hydrogen atom bonded to a small, strongly electronegative atom. The three most important electronegative atoms involved in hydrogen bonding are oxygen, nitrogen, and fluorine. Because hydrogen bonding is a very strong intermolecular force, it has a large impact on the melting and boiling points of compounds in which hydrogen bonding can take place.

Figure 4.10 shows the hydrogen bonding between molecules of water. The intramolecular bonding is shown by solid black lines, and the intermolecular hydrogen bonds are represented by dashed lines.

When a liquid boils, the intermolecular forces must be overcome so that the water molecules can escape into the vapour. Because they are very strong, the

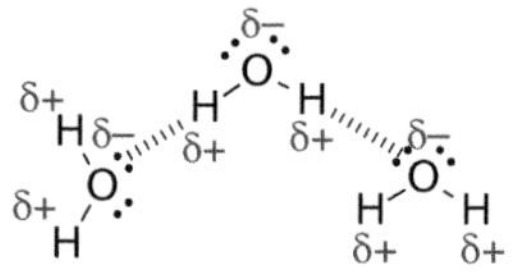

Figure 4.10 Hydrogen bonding between water molecules.

intermolecular forces in water are responsible for its high boiling point. Water, H_2O, is the hydride of oxygen – the first member of Group 6 (16). The lower members of Group 6 (16) form the following hydrides: H_2S, H_2Se, and H_2Te (Figure 4.11). Each of these molecules has a bent shape like the water molecule, as the central atom possesses two lone pairs of electrons. However, the electronegativity of the Group 6 (16) element decreases as we go down the group, and only the water molecule with its small electronegative oxygen atom can exhibit hydrogen bonding.

If hydrogen bonding did not exist between water molecules, the boiling points of the Group 6 (16) hydrides would be expected to increase steadily as we go down the group. This is the case for the Group 4 (14) hydrides, as shown in Figure 4.12, where the molecules (CH_4, SiH_4, GeH_4, SnH_4, and PbH_4) are non-polar and the intermolecular forces are very weak. Here there is a direct correlation between the size of the central atom and the boiling point of the hydride. The larger the central atom, the stronger the van der Waals forces, so the higher the boiling point. In Group 6 (16), the first member, H_2O, has a boiling point of 100 C and is a liquid at room temperature due to the strong intermolecular forces. If we compare H_2O with the hydride of the next element down Group 6 (16), H_2S, the boiling point is much lower at –60 C. There is no hydrogen bonding between molecules of H_2S, which means that the intermolecular forces are much weaker than in H_2O, and this results in H_2S being a gas at room temperature. The boiling points of the Group 6 (16) hydrides are shown by the grey line in Figure 4.12. The boiling points of the hydrides from H_2S to H_2Te increase steadily down the group as the van der Waals forces increase due to the increase

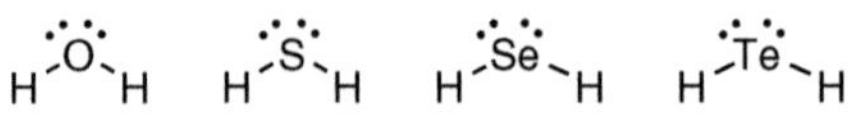

Remember, hydrogen bonding can only exist between hydrogen atoms and the elements N, O, and F.

$$H \overset{\cdot\cdot}{\underset{}{O}} H \quad H \overset{\cdot\cdot}{\underset{}{S}} H \quad H \overset{\cdot\cdot}{\underset{}{Se}} H \quad H \overset{\cdot\cdot}{\underset{}{Te}} H$$

Figure 4.11 Group 6 (16) hydrides.

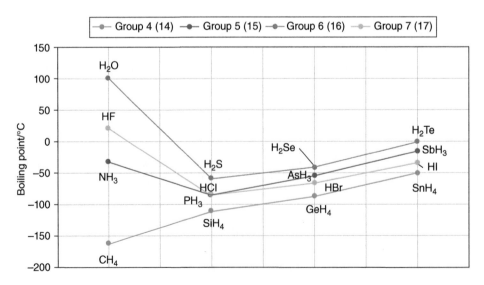

Figure 4.12 Boiling points of hydrides of Groups 4 (14), 5 (15), 6 (16), and 7 (17) elements.

in atomic number of the central atom. The unusually high boiling point of H_2O ensures that water is a liquid at ambient temperature. If this were not the case, it would be impossible for us to survive on earth. The plots of boiling point against period for the Group 5 (15) and Group 7 (17) hydrides are also shown in Figure 4.12. In each case, the boiling point of the first hydride (NH_3 or HF) is higher than the hydrides of the other group members. The only exception to this is SbH_3 (Group 5, Period 5), which has a higher boiling point (–17 C) than the first member of the group, NH_3 (–33 C) due to the relatively weaker hydrogen bonds in NH_3 and stronger van der Waals forces in SbH_3.

The strong intermolecular forces that arise due to hydrogen bonding in water give the molecule some very specific properties that underpin all life on earth.

Structure of Ice

Hydrogen bonding also exists between water molecules in the solid state, where water molecules are arranged in a regular lattice. The varied and beautiful shapes of ice crystals occur by the formation of a three-dimensional structure resulting from the directional properties of the hydrogen bonds between molecules of water. In ice, the water molecules are held further apart than they are in the liquid state where the molecules are moving randomly. The means that ice has a density that is lower than liquid water at its melting point of 0 C. As ice is warmed above its melting point and the hydrogen bonds begin to break, the density starts to increase to a maximum at 4 C. Once it reaches 4 C, the denser water sinks to the bottom of lakes and ponds, allowing aquatic creatures to survive even when the surface of the water is frozen (Figure 4.13).

The Heat Capacity of Water

The *heat capacity* of a material is a measure of the amount of heat energy required to raise the temperature of the material by one degree Celsius. Water has an unusually high heat capacity of $4.2\,J\,K^{-1}$ or $4.2\,J\,C^{-1}$. This high heat capacity is a result of the extensive hydrogen bonding that exists between water molecules in the liquid state. When liquid water is heated, some of the energy is used to weaken and break the hydrogen bonds. The temperature rise is therefore much lower than it would be if there were no hydrogen bonding present. The high heat capacity of water has a significant effect on regional climate, making coastal regions cooler during the day and warmer at night time. Oceans, seas,

The heat capacity of a material is a measure of the amount of heat energy required to raise the temperature of the material by one degree Celsius or one degree Kelvin. You will see this in more detail in Chapter 6.

Figure 4.13 Ice crystals. *Source:* ulleo/Pixabay.

Source: Pezibear/Pixabay.

Figure 4.14 Coastal regions are kept cool by the high heat capacity of water caused by hydrogen bonding. *Source:* Elizabeth Page.

and lakes can absorb and store heat energy during the day and lose it at night time, which makes coastal climates more temperate (Figure 4.14).

Hydrogen Bonding in Biological Molecules

The hydrogen bond has particular importance in biological molecules. The optimum hydrogen bond distance is around 2.8×10^{-10} m, and there is a linear relationship between the atoms or groups involved. Due to this strength and directionality, the distance and orientations between molecules that are hydrogen-bonded are constrained or fixed. For example, hydrogen bonds hold together the helical strands of DNA molecules. The hydrogen bonds between the nitrogen-containing bases purine and pyrimidine in opposite DNA strands constrain the rings to lie flat and at a fixed distance from each other. They also ensure that only thymine links to adenine and cytosine to guanine, as shown in Figure 4.15.

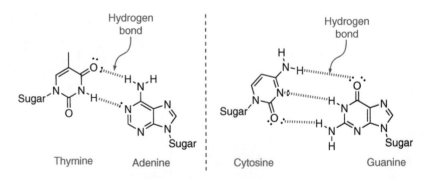

Figure 4.15 Hydrogen bonding between base pairs in DNA.

Worked Example 4.7   Compare the boiling points of the hydrides NH_3, H_2O, and HF given in Figure 4.12. Give some reasons why the boiling points of these hydrides are in the order $H_2O > HF > NH_3$.

Solution
All three hydrides experience hydrogen bonding between their molecules. Hydrogen bonding arises because of the high electronegativity and the lone pairs on the central elements (N, O, and F). Fluorine has the highest electronegativity of all three elements and the greatest number of lone pairs, so we might expect the boiling point order to follow $HF > H_2O > NH_3$. However, although F in HF has three lone pairs of electrons, it has only one available H atom for hydrogen bonding and so forms chains of HF molecules. Nitrogen in NH_3 has three H atoms but only one lone pair and so again forms chains of NH_3 molecules. However, H_2O has two lone pairs and two H atoms, which gives each H_2O molecule the possibility of forming four hydrogen bonds in a three-dimensional network. There are therefore more hydrogen bonds to overcome before boiling is achieved in H_2O compared to in NH_3 and HF.

Worked Example 4.8   Bromine, Br_2, is a reddish-brown corrosive liquid. A small amount of bromine is placed in a glass beaker fitted with a lid, and the beaker is placed in a fume hood. Some thermal data for bromine are given here:

$$T_m (Br_2) = -7.2 \ C$$

$$T_b (Br_2) = 58.8 \ C$$

T_m is the melting point, and T_b is the boiling point.

Describe the appearance of the bromine sample and the space above the bromine at the following temperatures. Use ideas about the movement of particles and intermolecular forces to explain your predicted observations.

a. −10 C

b. +10 C

c. 25 C

d. 60 C

Solution

a. Bromine will be solid at −10 C as the melting point (and freezing point) is −7.2 C. By cooling liquid bromine below its melting point, the kinetic energy of the molecules is reduced. Bromine has weak van der Waals forces between the molecules; on cooling, the kinetic energy of the molecules decreases, allowing stronger intermolecular forces to form in the solid state.

b. At +10 C, the bromine in the bottom of the beaker will be liquid, and the space above the bromine will be mainly colourless as few Br_2 molecules will have sufficient energy to escape into the vapour phase at this temperature.

c. On heating the bromine to 25 C, it should still be in the liquid phase, and the space above the liquid will become darker as some Br_2 molecules evaporate into the vapour phase.

d. The boiling point of bromine is given as 58.8 C, so at 60 C and atmospheric pressure, the liquid bromine will be boiling. The space above the liquid will be dark brown as it is saturated with bromine molecules. It is not recommended to carry out this procedure even in a fume hood, as boiling bromine is highly corrosive.

Worked Example 4.9   Classify the following solids into the main structural types according to the type of bonding that exists between the atoms, ions, or molecules. State whether the particles that make up the materials are atoms, molecules, or ions. Comment on the possible melting point of each solid material (very high/high/low/very low).

a. Ice

b. Calcium oxide

c. Solid methane

d. Tungsten, W

e. Solid argon

f. Pure (frozen) ethanoic acid (CH_3COOH)

g. Wax (waxes are composed of long chains of alkane molecules with the general formula C_nH_{2n+2})

h. Silicon

Solution
Take each of the materials in turn and consider the bonding.

a. Ice is made up of water (H_2O) molecules. It has a simple molecular structure. The atoms that make up the water molecules are joined together by strong covalent bonds. Hydrogen bonds are the main type of intermolecular forces between the molecules. Because ice has a simple covalent structure, the melting point is quite low (0 C). (See Figure 4.10 for hydrogen bonding in water.)

b. Calcium oxide (CaO) has an ionic structure. The Ca^{2+} and O^{2-} ions are held together by strong electrostatic forces. These must be broken if calcium oxide melts; and as the forces are very strong, the melting point will be very high (2572 C).

c. Solid methane is composed of molecules of CH_4 and therefore has a simple molecular structure. The carbon atoms are joined to the hydrogen atoms by strong covalent bonds. The CH_4 molecules are held together in the solid state by weak intermolecular forces (van der Waals forces), as CH_4 is non-polar. Therefore the melting point is very low (−182 C) and must be below room temperature as CH_4 is a gas at 25 C.

d. Tungsten (W) is a transition element and therefore a metal. It is composed of atoms of tungsten, and therefore has a metallic structure with a high melting point (3422 °C).

e. Argon is a noble gas and composed of single atoms of argon. The interatomic forces between argon atoms are very weak as argon has only 18 electrons per atom ($Z = 18$). The melting point of argon is very low (–189.4 °C).

f. Pure ethanoic acid (known as glacial ethanoic acid) is ethanoic acid that is free from water. This is also known as *anhydrous* ethanoic acid. Ethanoic acid has the formula CH_3COOH. The carbon, hydrogen, and oxygen atoms are joined together by strong covalent bonds to form molecules of CH_3COOH. Ethanoic acid has a simple molecular structure. However, it does have quite strong intermolecular forces due to hydrogen bonding between the –OH groups of one molecule and the C=O groups of a neighbouring molecule. Ethanoic acid is a liquid at room temperature with a melting point of 16.6 °C.

g. Wax is typically composed of molecules of long-chain alkanes of general formula C_nH_{2n+2}, where n is very large. The bonding between atoms of carbon and hydrogen is covalent, and the long-chain molecules are held together by van der Waals forces. Although these are not 'small' molecules like CO_2 and CH_3COOH, the structure is still classed as simple molecular. The intermolecular forces are fairly strong because the long carbon chains can stack closely together. The melting point of wax is relatively low, although the exact melting point will depend upon the nature of the long-chain alkanes making up the wax. Melting points of waxes are typically in the range 46 to 68 °C.

h. Silicon is an element ($Z = 14$) found beneath carbon in the periodic table. As such, it has a structure similar to diamond and is composed of silicon atoms joined together by strong covalent bonds. Each Si atom is bonded to four other Si atoms with a tetrahedral arrangement about each Si atom. To melt silicon, the strong covalent bonds between the atoms must be broken, so the melting point is high (1414 °C).

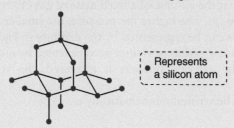

Structure of silicon.

4.4 Gases

4.4.1 Ideal gases

The behaviour of molecules in the gas phase is described by the *kinetic theory of gases*, which makes certain assumptions. The main points of kinetic theory are:

- Molecules in the gas phase move around quickly and randomly.

- Molecules in the gas phase are well separated from each other, and the distance between the molecules is large compared to the diameter of the molecules.

- There are no appreciable intermolecular forces between molecules in the gas phase.

- Molecules in the gas phase collide with each other in *elastic collisions* where no kinetic energy is lost upon collision.

- The average kinetic energy of a gas particle is directly proportional to the temperature.

- All gases at the same temperature have the same kinetic energy, and lighter molecules move faster than heavier molecules.

Any gas that behaves exactly according to these criteria is known as an *ideal gas*. Gases that come closest to ideal behaviour are the noble gases in Group 8 (18) of the periodic table. These gases have very weak intermolecular forces. Most *real gases* that we meet do not behave exactly according to the previous assumptions.

4.4.2 The ideal gas equation

The behaviour of an ideal gas can be described by the ideal gas equation, which is based upon the principles of the kinetic theory of gases. As real gases behave very similarly to ideal gases under normal conditions, we can use the ideal gas equation to represent the properties and behaviour of all gases.

Boyle's Law

The symbol $\propto$ is called the *proportionality symbol* and implies that the quantity on the left-hand side of the equation is directly proportional to the quantity on the right. In this case, the volume of the gas is proportional to the inverse (one over) of the pressure. Alternatively, we can say that the volume of a gas (*V*) is inversely proportional to the pressure (*p*).

Boyle's law states that the volume of a fixed mass of gas (*V*) is inversely proportional to its pressure (*p*). The higher the pressure, the smaller the volume occupied by the gas. This can be represented by the diagram in Figure 4.16, where a fixed amount of gas in a piston or syringe is compressed by increasing the external pressure. As the external pressure is increased, the volume of the gas decreases.

Boyle's law can be written mathematically as:

$$V \propto \frac{1}{p}.$$

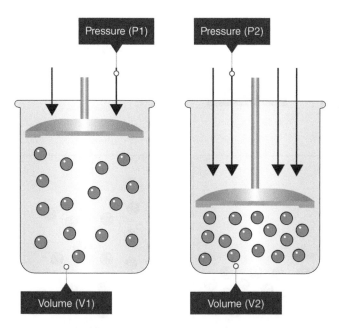

Figure 4.16 Boyle's law states that the volume of a gas is inversely proportional to the pressure. As the pressure is increased from *P1* to *P2*, the volume of a fixed number of moles of a gas decreases from *V1* to *V2*.

Charles's Law

Charles's law states that the volume of a fixed mass of gas (*V*) is directly proportional to the temperature (*T*). If the gas is held in a piston as in Figure 4.17, warming the gas increases the kinetic energy of the molecules, which causes them to spread apart and occupy more space. The law can be written mathematically as:

$$V \propto T$$

The Ideal Gas Law

Combining the expressions for Boyle's law and Charles's law gives:

$$V \propto \frac{T}{p} \text{ or } p\,V \propto T$$

By adding a constant of proportionality, k, to the equation, we can write:

$$p\,V = k\,T$$

This equation holds for a fixed mass of gas. A further gas law derived by Avogadro states that the volume of a gas depends upon the number of molecules or moles of gas present. The equation can be adapted slightly to include the dependence on the number of moles of gas, n, that are involved. The equation becomes:

$$p\,V = n\,R\,T$$

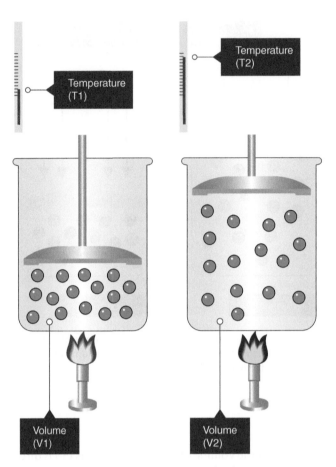

Figure 4.17 Charles's law states that as the temperature of a fixed number of moles of a gas is increased from *T1* to *T2*, the volume of the gas increases from *V1* to *V2*.

Notice that the equation now includes a new constant of proportionality, *R*. This is an important constant known as the *gas constant*. It is important because it is a constant for any gas, not just an ideal gas. The value of the gas constant, *R* is $8.31 \, \text{J K}^{-1} \, \text{mol}^{-1}$ or $8.31 \, \text{Pa m}^3 \, \text{K}^{-1} \, \text{mol}^{-1}$.

The equation is known as the *ideal gas equation*.

SI Units and the Ideal Gas Equation

Refer back to Chapter 0 for a reminder of SI units and how to apply them.

It is important that you make sure all quantities are given in the correct units when using the ideal gas equation. To ensure this, SI units should be used consistently when applying the equation.

Units of volume

The SI unit of volume is m^3. A volume of $1 \, \text{m}^3$ is 1000 times larger than the more commonly used value of $1 \, \text{dm}^3$, as shown in Figure 4.18.

Units of pressure

The SI unit of pressure is the pascal, Pa, where $1 \, \text{Pa} = 1 \, \text{N m}^{-2}$. Using units of pascals in the equation will ensure that the calculated volume is obtained in

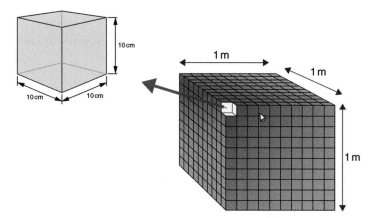

Figure 4.18 The large cube has a volume of 1 m^3. The sides of the small cube are one-tenth the length of the large cube (0.1 m or 1 dm or 10 cm). The volume of the small cube is 1 dm^3 and the large cube is made up of 1000 small cubes. *Source:* Redrawn from 'One metre cube is equal to how many litres?', https://www.quora.com/One-metre-cube-is-equal-to-how-many-litres.

Pressure is defined as force (in newtons) per unit area (in m^2). The SI unit of pressure is the Pa (pascal). 1 Pa = 1 N m^{-2}. Atmospheric pressure is measured in atmospheres (atm). A normal atmospheric pressure at sea level is around 1 atm or 1.01 × 10^5 Pa.

m^3, provided the value used for the gas constant is 8.31 J K^{-1} mol^{-1} or 8.31 Pa m^3 K^{-1} mol^{-1}. However, a pressure of 1 Pa is tiny, and pressure is more commonly measured in atmospheres (atm) or bar. 1 atm = 1.01 × 10^5 Pa, and 1 bar = 1 × 10^5 Pa.

Temperature
The SI unit of temperature is the kelvin (K). The Kelvin scale starts at 0 K, which is equivalent to –273 C. There are no negative values on the Kelvin scale, as 0 K is the coldest temperature theoretically possible to obtain. Most thermometers are calibrated to measure temperatures in Celsius (C); therefore, we often have to convert K to degrees Celsius (C) and vice versa. A temperature of 0 K is equivalent to –273 C, and a temperature of 0 C is equivalent to 273 K. To convert from degree Celsius to kelvin, 273 must be added to the temperature in degree Celsius. Although absolute values of temperature differ by 273 between the Kelvin scale and the Celsius scale, the size of a kelvin is the same as the size as a degree Celsius, as shown in Figure 4.19.

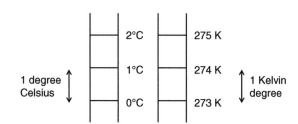

To convert from a temperature in degrees Celsius to kelvin, you should add 273. However, the magnitude of a degree Celsius is the same as that of a kelvin – i.e. the temperature difference, ΔT, is the same

Figure 4.19 The Celsius and Kelvin temperature scales.

Worked Example 4.10   20.0 g of carbon dioxide were placed in a vessel of volume 5.00 dm³ at a temperature of 0 C. Calculate the pressure exerted by the gas ($R = 8.31\,J\,K^{-1}\,mol^{-1}$).

Solution
To solve the equation, we assume CO_2 behaves as an ideal gas and use the equation $p\,V = n\,R\,T$. Rearrange the equation to get an expression for the pressure, p:

$$p = \frac{n\,R\,T}{V}$$

Values for n, T, and V must be inserted with the correct units:

$$n = \text{amount in moles of } CO_2 = \frac{m}{M} = \frac{20.0\,g}{44.0\,g\,mol^{-1}} = 0.455\,mol$$

$$T = \text{temperature in K} = 0\ C + 273 = 273\,K$$

$$V = \text{volume in m}^3 = 5.00 \times 10^{-3}\,m^3$$

Inserting the values into the equation,

$$p = \frac{0.455\,mol \times 8.31\,J\,K^{-1}\,mol^{-1} \times 273\,K}{5.00 \times 10^{-3}\,m^3} = 206 \times 10^3\,J\,m^{-3}$$

$$= 206\,000\,N\,m^{-2} = 206\,000\,Pa \text{ or } 2.06 \times 10^5\,Pa$$

Always include the units in your calculation and then cancel the units after calculating the result. After cancelling the units, we are left with $J\,m^{-3}$.

$1\,J = 1\,N\,m$, so substituting for J, the answer becomes $206\,000$ $N\,m \times m^{-3} = 206\,000\,N\,m^{-2}$. As 1 $N\,m^{-2} = 1\,Pa$, the answer can also be written as 2.06×10^5 Pa. The answer is given to three significant figures.

Worked Example 4.11   A cylindrical neon tube of diameter 12 mm and length 1.2 m contains a small amount of neon gas at a pressure of 1333 Pa. If the tube is operated at a temperature of 20 C, calculate the mass of neon in the tube ($R = 8.31\,J\,K^{-1}\,mol^{-1}$).

Solution
The ideal gas equation must be used to solve this problem. The equation is given by $p\,V = n\,R\,T$. Rearrange the equation to get an expression for the amount of gas in moles, n: $n = \dfrac{p\,V}{RT}$.

Values for p, V, R, and T must be inserted with the correct units:

$$p = 1333\,Pa$$

V, the volume of the cylinder, is obtained from the equation $V = \pi\,r^2\,h$, where r is the radius of the cylinder and h is the height. The radius, r, is half the diameter = 12 mm/2 = 6 mm. The height, $h = 1.2$ m.

Don't forget to ensure you convert all your units to the same dimensions so mm are converted to m here by dividing by 1000.

$$V = \pi r^2 h = 3.142 \times \left(6 \times 10^{-3}\,m\right)^2 \times 1.2\,m = 135.7 \times 10^{-6}\,m^3 = 1.36 \times 10^{-4}\,m^3$$

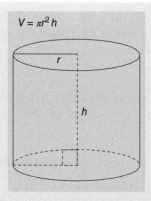

$$T = 20\ C + 273 = 293\ K$$

Inserting the previous values in the equation for the amount in moles,

$$n = \frac{1333\ \text{Pa} \times 1.36 \times 10^{-4}\ \text{m}^3}{8.31\ \text{J K}^{-1}\ \text{mol}^{-1} \times 293\ \text{K}} = 0.745 \times 10^{-4}\ \text{mol} = 7.45 \times 10^{-5}\ \text{mol}$$

As the molar mass of neon is $20.2\ \text{g mol}^{-1}$, the mass of gas required is

$$n \times M = 7.45 \times 10^{-5}\ \text{mol} \times 20.2\ \text{g mol}^{-1} = 150 \times 10^{-5}\ \text{g} = 1.50 \times 10^{-3}\ \text{g}$$

4.4.3   The molar gas volume, V_m

Avogadro's principle states that one mole of any gas at the same temperature and pressure has the same number of molecules. Therefore the volume of any gas at the same temperature and pressure depends upon the amount (in moles) of gas present. We can use the ideal gas equation to determine the volume occupied by 1 mole of a gas. This is called the *molar volume, V_m.*

Using the ideal gas equation, $p\,V = n\,R\,T$, the volume of gas can be expressed as $V = \dfrac{n\,R\,T}{p}$.

The volume of 1 mole of gas at 1 atm pressure $(1.01 \times 10^5\ \text{Pa})$ and 273 K $(0\ C)$ is:

$$\frac{1\ \text{mol} \times 8.31\ \text{J K}^{-1}\text{mol}^{-1} \times 273\ \text{K}}{1.01 \times 10^5\ \text{Pa}} = 2246 \times 10^{-5}\ \text{m}^3 = 2246 \times 10^{-2}\ \text{dm}^3 = 22.4\ \text{dm}^3$$

From the calculation, it can be seen that this volume does not depend upon the type of gas – it only depends upon the temperature and the pressure, so 1 mole of any gas at the same temperature and pressure has the same volume. Under these conditions $(1.01 \times 10^5\ \text{Pa}$ and 273 K$)$, this volume is $22.4\ \text{dm}^3$.

Note on units: $1\ \text{Pa} = 1\ \text{N m}^{-2}$, and $1\ \text{J} = \text{N m}$. When these derived units are inserted in the ideal gas equation, the units of volume are m^3.

$$V = \frac{\text{mol} \times \text{N m K}^{-1}\text{mol}^{-1} \times \text{K}}{\text{N m}^{-2}} = \text{m}^3.$$

Worked Example 4.12   Calculate the volume of 1 mole of carbon dioxide at 0 C and 1 atm pressure and at 25 C and 1 atm pressure. Comment on the difference in your values.

Solution
The calculation is carried out as shown previously using the ideal gas equation, $p\,V = n\,R\,T$. Rearranging this to get an expression for the volume gives $V = \dfrac{n\,R\,T}{p}$.

Inserting values for n, p, and R, the expression becomes:

$$V = \frac{1\,\text{mol} \times 8.31\,\text{J}\,\text{mol}^{-1}\text{K}^{-1} \times 273\,\text{K}}{101\,325\,\text{Pa}} = 0.0224\,\text{m}^3$$

$$= 22.4\,\text{dm}^3 \text{ (to three significant figures)}$$

At 25 C, the calculation is similar. The temperature must be converted to K: 25 C + 273 = 298 K. The pressure in pascals is the same:

$$V = \frac{n\,R\,T}{p} = \frac{1\,\text{mol} \times 8.31\,\text{J}\,\text{mol}^{-1} \times 298\,\text{K}^{-1}}{101\,325\,\text{Pa}} = 0.0244\,\text{m}^3 = 24.4\,\text{dm}^3$$

As expected, the volume at the higher temperature is slightly greater as a result of Charles's law.

Note on units: 1 atm = 101 325 Pa, and 1 Pa = $N\,m^{-2}$.

This value of 24 dm³ is often quoted as a constant for the volume of one mole of any gas at 1 atm pressure and 25 C (298 K).

Quick-check summary

- There are three common states of matter: solids, liquids, and gases.

- In metals and ionic solids, the atoms are arranged in a regular array.

- In metals, the atoms are held together by a sea of delocalised electrons. The strong electrostatic forces between the electrons and the positive metal ions result in high melting and boiling points.

- Metals are good conductors of electricity at room temperature due to the mobile electrons.

- Ionic materials are generally solids at room temperature due to the strong electrostatic forces between positive and negative ions. They have very high melting and boiling points. They do not conduct electricity in the solid state but will conduct electricity when molten.

- Covalent bonding leads to the formation of simple molecular and giant molecular materials.

- Simple molecular materials have low melting and boiling points due to the weak intermolecular forces between the molecules. They do not conduct electricity either as solids or as liquids.

- Giant molecular materials have very high melting and boiling points as they consist of vast three-dimensional arrays of bonded atoms.

- The behaviour of gases is described by the kinetic theory of gases.

- A set of laws describes the behaviour of gases.

- Boyle's law states that the volume of a fixed amount of gas at constant temperature is inversely proportional to the pressure.

- Charles's law states that the volume of a fixed amount of gas at constant pressure is directly proportional to the temperature.

- The volume of a gas at a fixed temperature and pressure is directly proportional to the amount of gas.

- The relationship between the volume, pressure, temperature, and amount of a gas is described by the ideal gas equation $pV = nRT$.

End-of-chapter questions

1 The following table gives some data on the physical properties of four substances A to D. Complete the gaps in the table using your knowledge of bonding between different types of materials.

Substance	Melting point	Boiling point	Electrical conductivity in the solid state	Electrical conductivity in the liquid state	Structure type
A	Low	Low	Poor	Poor	
B			Poor	Poor	Giant molecular
C	High	High	Poor	Good	
D	High	High	Good	Good	

2 The structures of diamond and CO_2 are shown here:

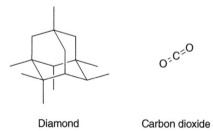

Diamond   Carbon dioxide

a. Name the structure types that diamond and carbon dioxide belong to.

b. Neither diamond nor solid carbon dioxide melts on heating. They *sublime*, which means they change directly from a solid to a vapour without going through the liquid phase. For diamond, this will only happen at very high pressures and temperatures of over 5000 C, whereas solid carbon dioxide will sublime readily at around −78 C. Explain the difference in their sublimation temperatures.

c. Explain why neither diamond nor carbon dioxide will conduct electricity.

3   Liquid bromine is a highly corrosive substance that is dark brown in colour.

  a. A small amount of bromine is transferred to a conical flask inside a fume hood at room temperature, and a stopper is inserted. Describe what you expect to happen in the flask.

  b. Use the kinetic theory of gases to describe what is happening on the molecular scale in part (a).

  c. The flask is warmed gently. What do you expect to observe in the space above the liquid bromine in the flask?

  d. The stopper is taken off the flask. Describe what you would expect to observe in the flask.

4   Oxygen tanks used in scuba diving are actually filled with compressed air, as breathing pure oxygen would be harmful to divers. A typical 'oxygen' tank has a volume of 12.0 L and is filled with air at a pressure of 200 bar and a temperature of 20 C. Calculate the mass of air in such a tank, assuming that air is composed of nitrogen (80%) and oxygen (20%). Note: 1 bar is 100 000 Pa.

5   A 3.65 g sample of pure magnesium carbonate, $MgCO_3$, decomposed on heating to release carbon dioxide gas.

  a. Calculate the volume of CO_2 produced at 60 C and 100 kPa (R = 8.31 J K^{-1} mol^{-1}).

  b. Calculate the mass of magnesium oxide that will be produced when 3.65 g of $MgCO_3$ completely decomposes.

6   Chlorine monofluoride (ClF) reacts with metallic tungsten at temperatures around 400 C to produce tungsten hexafluoride and chlorine according to the equation

$$6ClF(g) + W(s) \quad WF_6(g) + 3Cl_2(g)$$

  Tungsten hexafluoride is a gas at room temperature and is used in the semiconductor industry.

  a. Find the volume of tungsten hexafluoride that can be formed from 18.4 g of tungsten powder at 400 C and 100 kPa.

  b. The melting point of WF_6 is 2.3 C, and the boiling point is 17 C. What type of structure is solid WF_6 likely to possess?

7   For the following pairs of molecules, identify which has the higher boiling point. Explain your answer.

  a. Xe and Kr

  b. Decane and octane

  c. Pentane and 2-methylbutane

  d. 1,1-dichloroethene and (E)-1,2-dichloroethene

8   Convert the following temperatures to the temperature scale indicated:

   a. 25 °C to kelvin
   b. 100 K to degree Celsius
   c. −5 °C to kelvin
   d. 10 K to degree Celsius

9   a. On heating, the temperature of a beaker of water increased from 20 °C to 100 °C. Give the temperature difference in degrees kelvin.

   b. Antifreeze is added to water in a car's windscreen wash bottle to reduce the temperature at which the water freezes. On a certain morning, the external temperature was 3 °C, but addition of antifreeze prevented the windscreen liquid from freezing until it reached −5 °C. By how many degrees kelvin difference was the external temperature compared to the new freezing point?

   c. Explain why H_2O has a melting point of 0 °C, whereas H_2S, H_2Se, and H_2Te are all gases at room temperature.

10 a. If the volume of a fixed mass of argon gas at a pressure of 1.00×10^5 Pa is 0.0312 m^3 at 0 °C, calculate its volume at a pressure of 10.00×10^5 Pa and a temperature of 100 °C.

   b. A steel tank used for delivering propane fuel is fitted with a safety valve that opens if the internal pressure exceeds 0.990 atm. It is filled with propane at 18 °C and a pressure of 0.950 atm, but the temperature one summer day reaches 30 °C. Determine whether the safety valve will open.

5

Oxidation-reduction (redox) reactions

At the end of this chapter, students should be able to:

- Understand that the processes of oxidation and reduction involve the transfer of electrons

- Determine the oxidation numbers of elements in common chemical compounds

- Write redox half-equations for oxidation and reduction reactions, and balance half-equations to obtain an overall equation for the redox reaction

- Understand the action of oxidising and reducing agents in terms of electron transfer

- Use information from redox titrations to calculate concentrations and other relevant data about reacting species

- Recognise disproportionation reactions and identify the redox processes involved

5.1 Redox reactions

5.1.1 Electron transfer in redox reactions

The term **redox** is composed from the two words **reduction** and **oxidation**. Oxidation and reduction reactions involve the transfer of electrons. In fact, most chemical reactions occur with the transfer of electrons. As we have seen in Chapter 2, elements react together to attain a more stable arrangement of outer electrons.

Foundations of Chemistry: An Introductory Course for Science Students, First Edition.
Philippa B. Cranwell and Elizabeth M. Page.
© 2021 John Wiley & Sons Ltd. Published 2021 by John Wiley & Sons Ltd.
Companion website: www.wiley.com/go/Cranwell/Foundations

Oxidation and reduction reactions were previously thought to involve the gain or loss of oxygen. When an element combines with oxygen, such as the reaction of magnesium ribbon in oxygen to form magnesium oxide, it is said to be *oxidised*. The equation for this reaction is

$$2Mg(s) + O_2(g) \rightarrow 2MgO(s)$$

The chemical opposite of oxidation is reduction, which was originally defined as the removal of oxygen. Often this chemical process is brought about by hydrogen, which can combine with oxygen to produce water. For example, heating copper(II) oxide with hydrogen gas leaves metallic copper, and water is lost as steam:

$$CuO(s) + H_2(g) \rightarrow Cu(s) + H_2O(g)$$

The copper(II) oxide is said to be *reduced* as it has lost oxygen.

If we look more closely at these reactions, we can see that oxidation and reduction do not occur in isolation. In the first reaction, magnesium metal is oxidised to magnesium oxide. The element magnesium has the electron configuration $1s^2 2s^2 2p^6 3s^2$. In magnesium oxide, the magnesium atom is present as an ion with a 2+ charge and electron configuration $1s^2 2s^2 2p^6$. The magnesium atoms now have a full outer shell of electrons. Each magnesium atom has lost two electrons, which have been gained by the oxygen atoms. So, in being oxidised, the magnesium has lost electrons (Figure 5.1). The oxygen atoms are present as molecular oxygen, O_2. Each oxygen atom has the electron arrangement $1s^2 2s^2 2p^4$. The oxygen atoms gain the electrons the magnesium atom has lost. The oxygen atoms have been reduced to oxide (O^{2-}) ions by gaining electrons (Figure 5.2).

In the second example, where copper(II) oxide is being reduced to metallic copper by hydrogen, the copper 2+ ions (Cu^{2+}) are gaining two electrons to form elemental copper and so are being reduced. The hydrogen molecules are combining with oxygen to form water and so are being oxidised. Oxidation and reduction are occurring simultaneously.

Oxidation and reduction reactions are now defined as involving the transfer of electrons as electron transfer processes always accompany redox reactions.

Oxidation is the loss of electrons. Reduction is the gain of electrons.

The processes occur simultaneously as electrons are transferred between the species being oxidised and reduced.

Use this mnemonic to help you remember: OIL RIG: Oxidation Is Loss of electrons. Reduction Is Gain of electrons.

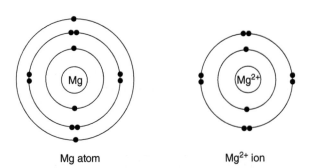

Mg atom   Mg²⁺ ion

Figure 5.1 Electron arrangements in a Mg atom and Mg^{2+} ion.

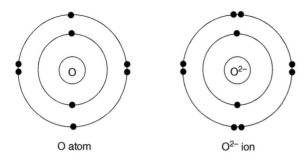

Figure 5.2 Electron arrangements in an O atom and O^{2-} ion.

The majority of chemical reactions involve oxidation and reduction. The corrosion of metals such as iron, the combustion of fuels, photosynthesis, and respiration are all types of redox reactions.

Worked Example 5.1 State which atoms are oxidised and which are reduced in the following reactions:

  a. $2Ca + O_2 \rightarrow 2CaO$

  b. $2Na + Cl_2 \rightarrow 2NaCl$

  c. $2Fe + \frac{3}{2} O_2 \rightarrow Fe_2O_3$

  d. $CuSO_4 + Zn \rightarrow ZnSO_4 + Cu$

Solution

  a. The calcium atoms have lost two electrons and are oxidised to Ca^{2+}. The oxygen atoms have each gained two electrons and are reduced to O^{2-}.

  b. The sodium atoms have lost one electron and are oxidised to Na^+. The chlorine atoms have gained one electron each and are reduced to Cl^-.

  c. The iron atoms have each lost three electrons and are oxidised to Fe^{3+}. The oxygen atoms have each gained two electrons and are reduced to O^{2-}.

  d. The zinc atoms lose two electrons and are oxidised to Zn^{2+}. The Cu^{2+} ions in $CuSO_4$ gain two electrons and are reduced to $Cu(0)$ in the metal.

5.1.2   Oxidation number

The **oxidation number** of an element (**ON**) provides a useful tool to help keep track of electrons in chemical reactions and indicate which species is being oxidised and which reduced. The oxidation number of an element in a compound tells us about the degree of oxidation of the element. Sometimes the term *oxidation state* is used in place of *oxidation number,* but both have the same meaning.

Atoms in elemental species such as magnesium metal (Mg) or oxygen molecules (O_2) have been neither oxidised nor reduced so their oxidation state is said to be zero.

In an ionic compound, the oxidation number is effectively the charge on the ion. In the compound magnesium oxide, MgO, the oxidation number of the magnesium ion is +2, as it is present as the Mg^{2+} ion, and so is in a higher oxidation state than in the metal.

The oxide ion has a charge of -2, so the oxidation number of the oxygen is therefore -2. The compound could be written as $Mg^{2+} O^{2-}$.

There are a few important principles to note here:

1. The oxidation number of an element in an ionic compound is the same as the charge on the ion.

2. An oxidation number can be positive, e.g. $+2$, or negative, e.g. -2. If the oxidation number is positive, the atom is more oxidised than in the element. If the oxidation number is negative, the atom is more reduced than in the element.

3. The oxidation number of an atom in the elemental state is zero. For example, the oxidation number of the magnesium atom in the metal is zero. The oxidation number of the oxygen atoms in oxygen gas is zero.

4. Because a compound is uncharged overall, the sum of the oxidation numbers in a neutral compound is zero. For example, in copper oxide, CuO, the oxidation state of the copper is $+2$, and the oxidation state of the oxygen is -2.

5. If the species has an overall charge, for example, the sulfate ion, SO_4^{2-}, or the carbonate ion, CO_3^{2-} the sum of the oxidation numbers of the elements in the ion is equal to the charge on the ion.

It is a little more difficult to assign oxidation numbers to atoms in covalent compounds as there are no ions formally present. For example, in a water molecule, H_2O, the atoms are covalently bonded by sharing electrons, and neither the oxygen atom nor the hydrogen atoms have lost or gained electrons. But we know, due to differences in electronegativity, that electrons are not distributed evenly between the two atoms. To assign oxidation numbers in covalent compounds, we treat them as if they were ionic and assign the electrons to the more electronegative atom and treat the compound as if it were ionic.

In H_2O, we would assign the oxygen atom the oxidation number -2 and the hydrogen atoms the oxidation number $+1$ to balance the charge. This doesn't mean that there are O^{2-} and H^+ ions present in water, it just provides us with a simple method of accounting for the transfer of electrons.

There are some general rules when assigning oxidation numbers to atoms in compounds:

1. The oxidation number of an atom in the elemental state is zero.

2. Atoms in Groups 1 and 2 have the same oxidation number (ON) as the group number. For example, the ON of sodium in $NaCl$ = $+1$, and the oxidation number of magnesium in MgO is $+2$. Aluminium in Group 3 (13) takes a $+3$ charge.

3. Certain atoms have fixed oxidation numbers:

   - Fluorine = -1

   - Hydrogen = $+1$ (apart from in metal hydrides, where it is -1 as the metal is more electropositive than hydrogen and has a positive oxidation number)

   - Oxygen = -2 (apart from in the peroxide ion, O_2^{2-}, where it is -1)

4. In covalent compounds, the more electronegative element takes the negative oxidation number.

For a reminder about electronegativity see Chapter 2.

Worked Example 5.2   Determine the oxidation numbers (ON) of each atom in the following species:

a. KBr

b. $AlCl_3$

c. NO_2

d. CO_3^{2-}

e. H_2SO_4

Solution

a. KBr: Potassium is in Group 1, so its ON = +1. The ON of bromine is therefore –1.

b. $AlCl_3$: Aluminium is in Group 3, so the ON of aluminium = +3. The compound is neutral, and there are three chlorine atoms, so the ON of each chlorine atom is –1.

c. NO_2: Oxygen is more electronegative than nitrogen and typically takes the ON –2. There are two oxygen atoms, so the total charge due to oxygen is $2 \times -2 = -4$. The ON of nitrogen is therefore +4 to maintain a neutral molecule.

d. CO_3^{2-}: Oxygen is more electronegative than carbon and typically takes the ON –2. The total charge due to oxygen is therefore $3 \times -2 = -6$. There is an overall charge of –2 on the ion; therefore, the charge on the carbon atom is +4. We can show this using algebra:

$$(ON\ of\ C) + 3 \times (ON\ of\ O) = -2$$

$$(ON\ of\ C) = -2 - (3 \times (ON\ of\ O)) = -2 - (3 \times -2) = -2 - (-6) = +4$$

e. H_2SO_4: Hydrogen has the ON +1, and oxygen has the ON –2. There are two hydrogen atoms $= 2 \times +1 = +2$ and four oxygen atoms $= 4 \times -2 = -8$. The sum of the oxidation numbers of oxygen and hydrogen is therefore $= (+2) + (-8) = -6$. The molecule is overall uncharged, so the charge on the sulfur atom must be +6.

$$2 \times (ON\ of\ H) + (ON\ of\ S) + 4 \times (ON\ of\ O) = 0$$

$$2 \times 1 + (ON\ of\ S) + 4 \times -2 = 0$$

$$2 + (ON\ of\ S) + (-8) = 0$$

$$(ON\ of\ S) = -2 + 8 = +6$$

5.1.3 Naming compounds based on the oxidation state of elements in the compound

Some elements can exist in a number of different oxidation states. This is specified by writing the oxidation number in Roman numerals after the symbol. For example, N_2O is nitrogen(I) oxide, whereas NO is nitrogen(II) oxide. Cu_2O is copper(I) oxide, whereas CuO is copper(II) oxide. In general conversation, the numbers often are not mentioned and are instead signified by the endings -ous and -ic. The ending -ous refers to the element in the lower oxidation state, whereas the ending -ic refers to the element in the higher oxidation state. For example, N_2O is nitrous oxide, and NO is nitric oxide; Cu_2O is cuprous oxide, and CuO is cupric oxide.

Variable oxidation states are often found in oxyanions (oxygen-containing ions) and can be signified using numerals in the same way: NO_2^- is nitrate(III), whereas NO_3^- is nitrate(V). SO_4^{2-} is sulfate(VI), whereas SO_3^{2-} is sulfate(IV).

However, in general conversation, it is more common to use different endings to refer to different oxyanions without specifying the oxidation state. The ending -ate is used for the highest oxidation state, whereas the ending -ite is used for the second highest. For example, NO_3^- and SO_4^{2-} are referred to as nitrate and sulfate, whereas NO_2^- and SO_3^{2-} are usually called nitrite and sulfite.

As you will see in Chapter 11, many transition metals can exist in different oxidation states, and the use of oxidation numbers when naming these compounds is common. Iron is a transition element that can form two oxides. Iron(II) oxide has iron in the +2 oxidation state with formula FeO. Iron(III) oxide has iron in the +3 oxidation state with formula Fe_2O_3. Iron(II) oxide is known as ferrous oxide and iron(III) oxide is ferric oxide.

> Anions that have the ending 'ate' or 'ite' generally contain oxygen and are derived from acids containing the anion. For example, sodium sulfate, Na_2SO_4, is derived from sulfuric acid, H_2SO_4, whereas sodium sulfite, Na_2SO_3, is obtained from the less common sulfurous acid, H_2SO_3.

> The Latin word for iron is ferrum, which gives us the names ferrous and ferric oxides.

Worked Example 5.3   Give the formulae of the following compounds:

a. Ammonium nitrate(V)

b. Potassium sulphite

c. Sodium chlorate(VII)

d. Cupric chloride

Solution

a. This contains the nitrate(V) ion, which has the formula NO_3^- as nitrogen must be in the +5 oxidation state. The formula of the salt is therefore NH_4NO_3.

b. This contains the sulfite ion, which has the formula SO_3^{2-} and has sulfur in the +4 oxidation state. The formula of the salt is therefore K_2SO_3.

c. Sodium chlorate(VII) contains the anion of the oxyacid of chlorine that has chlorine in the +7 oxidation state. The formula of the anion is therefore ClO_4^-. The formula of sodium chlorate(VII) is therefore $NaClO_4$. The ClO_4^- or chlorate(VII) ion is sometimes called 'perchlorate'.

d. Copper has two possible oxidation states: either +1 or +2. Cupric chloride contains copper in the higher (+2) oxidation state, so the formula is $CuCl_2$.

Worked Example 5.4   Give the names of the following compounds, ensuring that you include the appropriate oxidation state of the element given in bold:

a. $Ca(\mathbf{N}O_2)_2$

b. $Na\mathbf{Cl}O_3$

c. $\mathbf{Pb}O_2$

d. $\mathbf{P}_2O_5$

e. $K_2\mathbf{Cr}O_4$

Solution

a. The NO_2^- ion has nitrogen in the +3 oxidation state, so the name of this compound is calcium nitrite or calcium nitrate(III).

b. The ClO_3^- ion can be calculated to have chlorine in the +5 oxidation state, so its name is chlorate(V). The name of the salt is therefore sodium chlorate(V).

c. Lead (Pb) can have an oxidation state of +2 or +4. As this oxide has two atoms of oxygen per atom of lead, the lead must be in the +4 oxidation state. The name is therefore lead(IV) oxide.

d. The phosphorus atom in P_2O_5 can be calculated to have the oxidation state +5. The name of this compound is therefore phosphorus(V) oxide.

e. This salt contains the CrO_4^{2-} ion with chromium in the oxidation state +6. The name of the salt is therefore potassium chromate(VI).

5.1.4   Redox half-equations

We have seen that oxidation and reduction occur simultaneously with the transfer of electrons. Redox half-equations provide a convenient way of keeping track of the electrons when balancing equations.

During the reaction of magnesium metal in air, the magnesium is oxidised to Mg^{2+} and the oxygen reduced to O^{2-}. The two processes can be represented by two half-equations that include the electrons.

The half-equation for the oxidation of magnesium is:

$$Mg \rightarrow Mg^{2+} + 2e^- \text{ Oxidation Is Loss of electrons} \tag{5.1}$$

The magnesium atom (on the left of the equation) has lost its two outer electrons, which are written on the right of the equation along with the magnesium 2+ ion formed.

The half-equation for the reduction of oxygen is:

$$O_2 + 4e^- \rightarrow 2O^{2-} \text{ Reduction Is Gain of electrons} \tag{5.2}$$

Note that one oxygen atom requires two electrons to be reduced to an O^{2-} ion. The equation needs to be balanced with the same number of atoms and the same charge on either side. Therefore, two oxygen atoms in one molecule of O_2 require four electrons to be reduced to two O^{2-} ions. The electrons lost by the magnesium atom are gained by the oxygen molecule. As four electrons are required to reduce one oxygen molecule, two atoms of magnesium must be involved. We therefore multiply Eq. (5.1) by 2 to obtain Eq. (5.3):

$$2Mg \rightarrow 2Mg^{2+} + 4e^- \tag{5.3}$$

The left- and right-hand sides of Eqs. (5.3) and (5.2) can be added to obtain the overall equation. The electrons on each side of the equation will now cancel.

$$2Mg + O_2 + 4e^- \rightarrow 2Mg^{2+} + 2O^{2-} + 4e^-$$

$$2Mg + O_2 \rightarrow 2MgO$$

In this example, it is clear that the magnesium is being oxidised as it is reacting with oxygen. However, it is not always clear which species is being oxidised and which is being reduced. Oxidation numbers can help in determining this.

Worked Example 5.5   When chlorine gas is passed over heated iron filings, the iron reacts to produce iron(III) chloride. Write half-equations to represent the oxidation and reduction reactions, and combine the half-equations to obtain the overall equation for the reaction.

Solution
We know that the oxidation number of an atom in the elemental state is zero. In this reaction, both the iron filings and the chlorine gas are present as the elements. The iron reacts to form iron(III) chloride with an oxidation number of +3. The iron is therefore oxidised, and so the chlorine must be reduced.

When writing half-equations, always check that both the number and type of each atom *and* the overall charge are the same on each side of the half-equation.

The half-equation for the oxidation of iron is:

$$Fe \rightarrow Fe^{3+} + 3e^- \tag{5.4}$$

The half-equation for the reduction of one molecule of chlorine is:

$$Cl_2 + 2e^- \rightarrow 2Cl^- \tag{5.5}$$

The two half-equations must be combined and the electrons cancelled. This requires multiplying Eq. (5.4) by two and Eq. (5.5) by three:

$$2Fe \rightarrow 2Fe^{3+} + 6e^- \tag{5.6}$$

$$3Cl_2 + 6e^- \rightarrow 6Cl^- \tag{5.7}$$

Adding the left-hand sides and the right-hand sides of Eqs. (5.6) and (5.7) gives:

$$2Fe + 3Cl_2 + 6e^- \rightarrow 2Fe^{3+} + 6Cl^- + 6e^-$$

The electrons on each side cancel to leave us with a balanced equation for the overall reaction:

$$2Fe + 3Cl_2 \rightarrow 2FeCl_3$$

Worked Example 5.6   Aqueous potassium dichromate(VI), $K_2Cr_2O_7$, can be used to oxidise iron(II) ions to iron(III) in the presence of dilute hydrochloric acid. The dichromate(VI) is reduced to chromium(III), which is green. Use your knowledge of oxidation numbers and the information given to balance the chemical equation between dichromate(VI) ions and iron(II) ions.

Solution
Start by writing down the information you have been given about the reactants and the products in this process. We are told that the dichromate ions are converted to chromium(III) or Cr^{3+} ions. The reaction requires H^+ ions from the dilute HCl, which are converted to water molecules. The dichromate ions react with Fe^{2+}, which is converted to Fe^{3+}.

$$Cr_2O_7^{2-} + Fe^{2+} + H^+ \rightarrow Cr^{3+} + Fe^{3+} + H_2O \quad \text{UNBALANCED EQUATION}$$

We could add stoichiometric coefficients to each side of this equation until we managed to get the atoms and charges to balance, but it is much more efficient to use redox equations to balance both sides.

First identify the two redox half-equations.
$Cr_2O_7^{2-}$ is being reduced to Cr^{3+}, so write this half-equation first.
Each Cr(VI) ion is being reduced to Cr(III) and so gains three electrons. There are two Cr(VI) ions so we will need six electrons in total. The reaction is a reduction reaction, and so the electrons must be added to the left-hand side:

$$Cr_2O_7^{2-} + 6e^- \rightarrow 2Cr^{3+} + 7[O] \quad \text{UNBALANCED HALF-EQUATION}$$

The oxygen atoms from the dichromate ion are converted to water, so we need to add the appropriate number of H^+ ions to the left-hand side. As there are 7 atoms of oxygen, we will need 14 hydrogen ions to form 7 molecules of water:

$$Cr_2O_7^{2-} + 14H^+ + 6e^- \rightarrow 2Cr^{3+} + 7H_2O \quad \text{BALANCED HALF-EQUATION} \quad (5.8)$$

The iron(II) ions are being oxidised to iron(III), so each iron(II) ion must lose one electron:

$$Fe^{2+} \rightarrow Fe^{3+} + e^- \quad \text{BALANCED HALF-EQUATION} \quad (5.9)$$

We now have redox half-equations for the oxidation and reduction processes taking place. To combine the two half-equations and cancel the electrons, we need to multiply Eq. (5.9) by six and then add the two equations:

$$6Fe^{2+} \rightarrow 6Fe^{3+} + 6e^- \quad (5.10)$$

Adding Eqs. (5.8) and (5.10):

$$Cr_2O_7^{2-} + 14H^+ + 6Fe^{2+} + 6e^- \rightarrow 2Cr^{3+} + 6Fe^{3+} + 7H_2O + 6e^-$$

The electrons cancel to leave the overall ionic equation:

$$Cr_2O_7^{2-} + 14H^+ + 6Fe^{2+} \rightarrow 2Cr^{3+} + 6Fe^{3+} + 7H_2O$$

Again check that the atoms and charges are balanced on both sides.

An ionic equation is one that shows only the species that are changed on either side of the equation. For instance, here we have ignored the potassium ions (from $K_2Cr_2O_7$) and the chloride ions (from HCl) as they remain as K^+ ions and Cl^- ions on both sides of the equation.

5.1.5   Oxidising agents and reducing agents

Redox reactions involve the oxidation of one chemical species and the reduction of another. The chemical species that is responsible for bringing about the oxidation part of the redox reaction is called the **oxidising agent**. When a substance is oxidised, it **loses** electrons (OIL, Oxidation Is Loss), so the oxidising agent must be capable of accepting these electrons. By accepting the electrons, the oxidising agent is reduced (RIG, Reduction Is Gain). Its oxidation state or oxidation number decreases as it gains electrons. Oxidising agents typically have elements in high oxidation states. Examples of common oxidising agents are potassium manganate(VII), $KMnO_4$, sodium chromate(VI), Na_2CrO_4, and sodium dichromate(VI), $Na_2Cr_2O_7$. Many oxidising agents are classed as dangerous materials and are labelled as such during transportation and in the laboratory (Figure 5.3).

The chemical species that is responsible for causing the reduction process in a redox reaction is known as the **reducing agent**. When a substance is reduced, it gains electrons (RIG), and the reducing agent must be able to provide these electrons. Thus the reducing agent is oxidised because it has lost electrons. Its oxidation state or number increases.

Look at the reaction between potassium manganate(VII) and iron(II) chloride in sulfuric acid. The ionic equation is given here:

$$MnO_4^{-}(aq) + 5Fe^{2+}(aq) + 8H^+(aq) \rightarrow Mn^{2+}(aq) + 5Fe^{3+}(aq) + 4H_2O(l)$$

Here the manganate(VII) ions are acting as the oxidising agent. They are reduced to Mn^{2+}. The red text indicates the reduction reaction, and the blue text indicates the oxidation reaction. The half-equation for the reduction reaction is:

$$MnO_4^{-}(aq) + 8H^+(aq) + 5e^- \rightarrow Mn^{2+}(aq) + 4H_2O(l)$$

An oxidising agent is reduced by gaining electrons.

> An oxidising agent gains electrons and is itself reduced. A reducing agent loses electrons and is itself oxidised.

The Fe^{2+} ions are acting as the reducing agent. They are oxidised to Fe^{3+}. The half-equation for the oxidation reaction is:

$$Fe^{2+}(aq) \rightarrow Fe^{3+}(aq) + e^-$$

A reducing agent is oxidised by losing electrons.

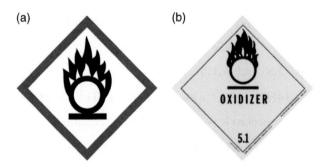

Figure 5.3 (a) The international pictogram for an oxidising agent; (b) safety label for an oxidising agent.

Worked Example 5.7 Identify the oxidising and reducing agents in the following redox reactions. Write half-equations for the oxidation and reduction reactions, clearly showing the transfer of electrons.

a. $Ni(s) + 2Fe^{3+}(aq) \rightarrow Ni^{2+}(aq) + 2Fe^{2+}(aq)$

b. $Cl_2(g) + 2I^-(aq) \rightarrow 2Cl^-(aq) + I_2(s)$

c. $MnO_2(s) + 4HCl(aq) \rightarrow MnCl_2(aq) + Cl_2(g) + 2H_2O(l)$

Solution
To identify the oxidising and reducing agents, first establish the species that are being oxidised and reduced. Do this by working out the oxidation states of each species on either side of the equation:

a. Here nickel metal with ON zero is being oxidised to Ni^{2+}. Nickel metal is therefore the reducing agent. The Fe^{3+} ions are being reduced to Fe^{2+}, so they are acting as the oxidising agent. The two half-equations are:

$$Ni(s) \rightarrow Ni^{2+}(aq) + 2e^- \quad \text{Metallic Ni is reducing agent}$$
$$Fe^{3+}(aq) + e^- \rightarrow Fe^{2+}(aq) \; Fe^{3+} \text{ ions are oxidising agent}$$

> Metallic elements are in the zero oxidation state and generally act as reducing agents as they are electropositive and are oxidised to positive oxidation states by loss of electrons.

b. On the left-hand side of the equation is elemental chlorine gas, which has chlorine atoms in the zero oxidation state. On the right-hand side of the equation are Cl^- ions in the −1 oxidation state. The Cl_2 is therefore reduced and acting as the oxidising agent:

$$Cl_2(g) + 2e^- \rightarrow 2Cl^-(aq)$$

The iodide ions on the left-hand side of the equation are in the −1 oxidation state and are oxidised by loss of electrons to molecular iodine, I_2. As the iodide ions are oxidised, they act as the reducing agent:

$$2I^-(aq) \rightarrow I_2(s) + 2e^-$$

c. First concentrate on the MnO_2 and Mn^{2+}. In MnO_2, the manganese atom is in the +4 oxidation state, whereas on the right-hand side of the equation it is in the +2 oxidation state. MnO_2 has therefore been reduced and acts as the oxidising agent:

$$MnO_2(aq) + 4H^+(aq) + 2e^- \rightarrow Mn^{2+}(aq) + 2H_2O(l)$$

The other half of the redox equation involves the Cl^- ions, which have been oxidised to Cl_2. The Cl^- ions are therefore acting as the reducing agent:

$$2Cl^-(aq) \rightarrow Cl_2(g) + 2e^-$$

5.2   Disproportionation reactions

The reverse of a disproportionation reaction is called a coproportionation reaction and occurs when two species with different oxidation states combine to form a single species with an intermediary oxidation state.

In some chemical reactions, one species is simultaneously oxidised and reduced. These types of reactions are called **disproportionation** reactions and are a specific type of redox reaction. They can be broken down into two half-equations in the same way as other redox reactions. Keeping track of oxidation numbers provides the best way to balance a disproportionation reaction.

The reaction of chlorine gas with aqueous sodium hydroxide provides an example of a disproportionation reaction. The molecular chlorine is oxidised to the chlorate(I) ion (or hypochlorite ion) and reduced to the chloride (Cl^-) ion. The overall equation for the reaction is given by:

$$Cl_2(g) + 2NaOH(aq) \rightarrow NaOCl(aq) + NaCl(aq) + H_2O(l)$$

The ionic equation for the reaction (omitting sodium ions) is:

$$Cl_2(g) + 2OH^-(aq) \rightarrow OCl^-(aq) + Cl^-(aq) + H_2O(l)$$

(ON = 0)   (ON = 1)   (ON = – 1)

Writing the half-equation for the oxidation of Cl_2, we have:

$$\tfrac{1}{2}Cl_2(g) \rightarrow OCl^-(aq) + e^- \qquad \text{UNBALANCED HALF-EQUATION}$$

The O in OCl^- is derived from the sodium hydroxide, so we add OH^- to the left-hand side to obtain the overall equation:

$$\tfrac{1}{2}\,Cl_2(g) + 2OH^-(aq) \rightarrow OCl^-(aq) + H_2O(l) + e^- \quad \text{BALANCED HALF-EQUATION}$$

The oxygen atom in OCl^- has a formal oxidation number of –2.

The half-equation for the reduction of Cl_2 is:

$$\tfrac{1}{2}\,Cl_2(g) + e^- \rightarrow Cl^-(aq)$$

Combining the two half-equations and cancelling the electrons gives the overall equation for the reaction:

$$Cl_2(g) + 2OH^-(aq) \rightarrow OCl^-(aq) + Cl^-(aq) + H_2O(l)$$

Worked Example 5.8   The decomposition of hydrogen peroxide, H_2O_2, results in the formation of water and oxygen. Write two half-equations to represent the oxidation and reduction reactions, and combine the half-equations to obtain the overall equation for the disproportionation reaction.

Solution
Start this problem by writing down the information you have been given in an unbalanced equation. We are told that hydrogen peroxide produces water and oxygen upon decomposition. Don't worry about balancing the equation just yet:

$$H_2O_2(aq) \rightarrow O_2(g) + H_2O(l) \qquad \text{UNBALANCED EQUATION}$$

Next, determine the oxidation numbers of the species being oxidised and reduced. It is likely that it is oxygen that is being both oxidised and reduced as we have hydrogen peroxide on the left and molecular oxygen on the right of the equation.

Write the oxidation numbers of oxygen in each species below the species in the unbalanced equation:

$$H_2O_2(aq) \rightarrow O_2(g) + H_2O(l) \qquad \text{UNBALANCED EQUATION}$$
$$\qquad -1 \qquad\quad 0 \qquad -2$$

Remember that oxygen takes an ON of –1 in the peroxide (O_2^{2-}) ion.

The above equation shows the oxygen atom is being both oxidised (in O_2) and reduced (in H_2O). Next, write two half-equations for the oxidation and reduction processes:

$$H_2O_2(aq) \rightarrow O_2(g) + 2[H^+](aq) + 2e^- \qquad \text{Oxidation}$$
$$H_2O_2(aq) + 2e^- \rightarrow 2[H^+](aq) + 2[O^{2-}](aq) \text{ Reduction}$$

The two half-equations can be added together to obtain the overall equation for the reaction and the electrons cancel:

$$2H_2O_2(aq) \rightarrow O_2(g) + 4[H^+](aq) + 2[O^{2-}](aq)$$

$$2H_2O_2(aq) \rightarrow O_2(g) + 2H_2O(l)$$

Although shown as ions, The $[H^+]$ and $[O^{2-}]$ are actually combined in molecular water. They do not exist as ions here as water is covalent, but the charges are used to keep account of the electrons.

5.3   Redox titrations

Redox reactions are frequently used in titrations to quantitatively determine the concentration or amount of a reacting substance. Many redox titrations involve a colour change when the reaction is complete as one or other of the reactants is used up or becomes in excess. They are often self-indicating.

Consider the reaction between potassium manganate(VII) and iron(II) ions in acidic solution. The equation for this reaction was given in Section 5.1.5. This redox reaction can be used to determine the amount of iron in a substance containing iron(II): for example, iron tablets available from pharmacists for preventing anaemia.

$$\underset{\text{purple}}{MnO_4^-}(aq) + 5Fe^{2+}(aq) + 8H^+(aq) \rightarrow \underset{\text{colourless}}{Mn^{2+}}(aq) + 5Fe^{3+}(aq) + 4H_2O(l)$$

Potassium manganate(VII) is deep purple in colour. When it reacts with the iron(II) ions in the flask, it is reduced to colourless Mn^{2+} ions. When all the iron(II) ions have reacted, the next drop of potassium manganate(VII) added turns the solution in the flask pale pink, indicating that all the iron(II) has been used up and the manganate ions are in excess. This is the endpoint of the reaction, and the volume of potassium manganate(VII) added can be recorded. Knowing the volume and concentration of the potassium manganate added to the iron solution, the number of moles of iron(II) can be calculated and hence the mass of iron in the tablets. The following example shows how the calculation is carried out.

Worked Example 5.9   Iron is present in pharmaceutical-grade iron tablets as an Fe(II) salt. A titration was carried out to determine the percentage of iron in an iron tablet. One tablet weighing 2.00 g was ground in a pestle and mortar and dissolved in 100 cm³ of hot sulfuric acid. The cool solution was transferred to a 250 cm³ volumetric flask and made up to the mark with deionised water. 25.00 cm^3 portions of the iron(II) solution were added to a conical flask and titrated with potassium manganate(VII) ($0.0095 \text{ mol dm}^{-3}$) from a burette. The endpoint was determined by the first permanent pink colour in the flask. The average titre required was 10.50 cm^3 $KMnO_4$ solution. Use the titration results to determine the percentage of iron(II) in the iron tablet.

Solution

The approach to solving a problem of this type is to first write the redox equation for the reaction occurring during the titration. This is given in the previous section:

$$MnO_4^-(aq) + 5Fe^{2+}(aq) + 8H^+(aq) \rightarrow Mn^{2+}(aq) + 5Fe^{3+}(aq) + 4H_2O(l)$$

From this, we can see that 1 mole of $KMnO_4$ solution reacts with 5 moles of Fe(II) solution.

From the information given, we can determine the average number of moles of $KMnO_4$ solution used to react with 25.00 cm^3 of Fe(II) solution:

$$n(KMnO_4) = V(KMnO_4) \times c(KMnO_4) = 10.50 \times 10^{-3} \text{ dm}^3 \times 0.0095 \text{ mol dm}^{-3}$$

$$= 9.975 \times 10^{-5} \text{ mol}$$

From the equation, one mole of MnO_4^- reacts with five moles of Fe^{2+}. Therefore, the amount in moles of Fe^{2+} in 25.00 cm^3 solution, $n(Fe(II))$

$$= 5 \times 9.975 \times 10^{-5} = 4.988 \times 10^{-4} \text{ mol}.$$

The amount in moles of Fe^{2+} in 250 cm^3 solution is 10 times this amount:

$$= 4.988 \times 10^{-4} \times 10 = 4.988 \times 10^{-3} \text{ mol}$$

The mass of Fe(II) in 250 cm^3 solution

$$= 4.988 \times 10^{-3} \text{ mol} \times 56.0 \text{ g mol}^{-1} = 279 \times 10^{-3} \text{ g or } 279 \text{ mg}$$

The percentage of iron in the tablet, $\%(Fe(II))$

$$= \left(\frac{\text{mass Fe(II)}}{\text{mass of tablet}} \right) \times 100 = \left(\frac{279 \times 10^{-3} \text{ g}}{2.00 \text{ g}} \right) \times 100 = 14.0\%$$

[Note that the atomic mass of Fe^{2+} ions is the same as the atomic mass of iron]

As the K^+ ions and SO_4^{2-} ions do not take part in the reaction we will omit them. They are called 'counterions'.

Don't forget to convert the volume in cm^3 to dm^3 by multiplying by 10^{-3}.

Remember $1 \times 10^{-3} \text{ g} = 1 \text{ mg}$

The following example is a titration in which one redox reaction using a solution of unknown concentration (KIO_3, in this case) is used to generate iodine which goes on to react with a solution of known concentration ($Na_2S_2O_3$) in a redox titration.

The amount of $Na_2S_2O_3$ required to react completely with the iodine is determined, and this allows the calculation of the unknown concentration of potassium iodate solution.

Worked Example 5.10   Potassium iodate(V), KIO_3, oxidises potassium iodide, KI, in acidic solution to form iodine, I_2. The liberated iodine reacts with sodium thiosulfate solution in a further redox reaction in which the iodine is reduced back to iodide ions by the thiosulfate ions. The equation for this reaction is:

$$I_2(s) + 2S_2O_3{}^{2-}(aq) \rightarrow 2I^-(aq) + S_4O_6{}^{2-}(aq)$$

In an experiment to determine the concentration of a solution of potassium iodate, $25.0\ cm^3$ of potassium iodate reacted with excess potassium iodide in acidic solution in a conical flask to produce iodine. The iodine was titrated against sodium thiosulfate solution of concentration $0.100\ mol\ dm^{-3}$. $17.0\ cm^3$ of sodium thiosulfate solution was required to completely react with all the iodine.

Use the results from the titration to work out the concentration of the potassium iodate solution used.

Solution

In this problem, two separate redox reactions are taking place. The first reaction is between the potassium iodate and potassium iodide to release iodine. Next, the iodine released reacts with sodium thiosulfate in the titration. We are asked to determine the concentration of potassium iodate solution used, so work backwards from the information given.

We can find the amount of iodine produced using the information that the iodine reacts with sodium thiosulfate according to:

$$I_2(s) + 2S_2O_3{}^{2-}(aq) \rightarrow 2I^-(aq) + S_4O_6{}^{2-}(aq)$$

Each 2 moles of sodium thiosufate reacts with 1 mole of iodine.
Amount in moles of sodium thiosulfate:
$$n(Na_2S_2O_3) = V(Na_2S_2O_3) \times c(Na_2S_2O_3)$$
$$= 17.0 \times 10^{-3}\ dm^3 \times 0.100\ mol\ dm^{-3} = 17.0 \times 10^{-4}\ mol.$$

Because each 1 mole of iodine reacts with 2 moles of sodium thiosulfate, the number of moles of iodine present is half the number of moles of sodium thiosulfate used:

$$\text{Amount in moles of iodine produced} = \frac{17.0 \times 10^{-4}\ mol}{2} = 8.5 \times 10^{-4}\ mol$$

Next work out the equation for the reaction of iodate and iodide ions to give iodine using half-equations.
The iodate(V) ions are reduced to elemental iodine, I_2, so each iodine atom requires five electrons:

$$IO_3{}^-(aq) + 6H^+(aq) + 5e^- \rightarrow \tfrac{1}{2}I_2(s) + 3H_2O(l) \qquad (5.11)$$

The iodide ions are oxidised by loss of electrons to iodine:

$$I^-(aq) \rightarrow \tfrac{1}{2}I_2(s) + e^- \qquad (5.12)$$

To obtain the overall redox equation, Eq. (5.12) must first be multiplied by five to allow the electrons to cancel:

$$5I^-(aq) \rightarrow \tfrac{5}{2}I_2(s) + 5e^- \qquad (5.13)$$

Now the left and right-hand sides of Eqs. (5.11) and (5.13) can be added and the electrons cancelled:

$$IO_3{}^-(aq) + 5I^-(aq) + 6H^+(aq) + \cancel{5e^-} \rightarrow 3I_2(s) + 3H_2O(l) + \cancel{5e^-} \quad (5.14)$$

This redox equation shows us that 1 mole of iodate(V) ions produces 3 moles of iodine upon reaction with excess iodide ions.

Iodine is a dark brown to yellow colour in aqueous solution, depending upon the concentration. This endpoint is self-indicating and can be detected when all the iodine has reacted with the sodium thiosulfate solution. Starch solution is generally added near the endpoint to make the colour change even more pronounced. Iodine turns deep blue in the presence of starch, and so the endpoint is when the blue colour has just disappeared.

Volumes in cm^3 must be converted to dm^3 by dividing by 1000 or multiplying by 10^{-3}.

Check that the charge and the number of atoms on each side of the half-equation are balanced.

The number of moles of iodate(V) = $\frac{1}{3} \times 8.5 \times 10^{-4} = 2.83 \times 10^{-4}$ mol IO_3^-.

Using $c = \dfrac{n}{V}$, we can calculate the concentration, c, of potassium iodate:

$$c = \frac{n}{V} = \frac{2.83 \times 10^{-4} \text{ mol}}{25.0 \times 10^{-3} \text{ dm}^3} = 0.113 \text{ mol dm}^{-3}$$

Quick-check summary

- Redox reactions are reactions in which electrons are transferred from one species to another.

- Redox reactions involve simultaneous oxidation and reduction.

- When a species loses electrons, it is said to be oxidised, and its oxidation number increases.

- When a species gains electrons, it is said to be reduced, and its oxidation number decreases.

- Oxidising agents are species that cause oxidation. They remove electrons from the species being oxidised and so are reduced in the process by gaining electrons.

- Reducing agents are species that cause reduction. They provide electrons to the species being reduced and so are oxidised in the process by losing electrons.

- Oxidation numbers (ON) or oxidation states indicate the number of electrons lost or gained by the species compared to when it is in its elemental state.

- All species in their elemental state are assigned an oxidation number of zero.

- A redox equation can be divided into two half-equations, one representing the oxidation reaction and the other representing the reduction reaction.

- Oxidation numbers can be used when balancing redox half-equations.

- Disproportionation reactions are reactions in which the same species is simultaneously oxidised and reduced.

- Redox reactions can be used quantitatively in redox titrations.

End-of-chapter questions

1 Identify the species being oxidised and reduced in the following redox reactions, and give the change in oxidation numbers of the oxidised and reduced species:

    a. $CuO + H_2 \rightarrow Cu + H_2O$

    b. $2Mg + CO_2 \rightarrow 2MgO + C$

    c. $2AgNO_3 + Cu \rightarrow 2Ag + Cu(NO_3)_2$

    d. $Cl_2 + 2KI \rightarrow 2KCl + I_2$

2 Determine the oxidation numbers of the atoms in bold in the following species:

a. Na**Cl**O

b. H$_2$**S**O$_4$

c. **N**$_2$H$_4$

d. K**Mn**O$_4$

e. **P**$_2$O$_5$

f. Na$_2$**Cr**$_2$O$_7$

3 Give the formulae of the following:

a. Sodium chlorate(I)

b. Iron(III) oxide

c. Potassium nitrate(V)

d. Sodium sulfite

e. Phosphorus(V) chloride

4 Write balanced equations for the overall redox reactions occurring between the following:

a. The reaction between metallic nickel and iron(III) ions in aqueous solution in which nickel is oxidised to nickel(II) and iron(III) is reduced to iron(II).

b. The reaction between bromide ions (Br^-) and bromate(V) (BrO_3^-) ions in acidic solution to form bromine (Br_2).

c. The reaction between potassium dichromate(VI) ($Cr_2O_7^{2-}$) and iron(II) ions in acidic solution in which dichromate is reduced to chromium(III) and iron(II) is oxidised to iron(III).

d. The reaction between manganate(VII) ions (MnO_4^-) and oxalate ($C_2O_4^{2-}$) ions in acidic solution in which manganate is reduced to manganese(II) and oxalate is oxidised to carbon dioxide, CO_2.

5 Write equations for the following redox reactions using half-equations:

a. The reaction of copper(II) with iodide ions to produce copper(I)iodide and iodine.

b. The reaction of manganate(VII) ions (MnO_4^-) with hydrogen peroxide (H_2O_2) in acidic solution to produce manganese(II) and oxygen.

6 State whether the following equations represent redox reactions or disproportionation reactions, and explain your answers.

a. $2CuCl(aq) \rightarrow CuCl_2(aq) + Cu(s)$

b. $IO_3^-(aq) + 5I^-(aq) + 6H^+(aq) \rightarrow 3I_2(s) + 3H_2O(l)$

c. $3NaClO(aq) \rightarrow 2NaCl(aq) + NaClO_3(aq)$

7   A redox titration was used to determine the percentage of iron in a sample of iron wire. A length of wire weighing 1.55 g was dissolved with heating in sulfuric acid where it was oxidised to Fe^{2+}. Once dissolved, the iron(II) solution was diluted with deionised water and transferred to a 250 cm^3 volumetric flask and made up to the mark with water. A 25.00 cm^3 portion of the solution was pipetted into a conical flask and titrated against $KMnO_4$ (concentration 0.01950 mol dm^{-3}) until the first permanent pink colour was obtained. 24.60 cm^3 of $KMnO_4$ solution was required to completely react with the Fe(II) solution. Construct a balanced equation for the reaction, and use the equation to determine the percentage of iron in the wire.

8   In an experiment to determine the percentage of copper in a hydrated copper(II) salt, 0.500 g of salt was weighed and dissolved in 250 cm^3 of deionised water. A 25.0 cm^3 of portion of this solution was taken, and excess potassium iodide was added to the solution to release iodine. The iodine was titrated against 0.0210 mol dm^{-3} sodium thiosulfate solution ($Na_2S_2O_3$), and 9.35 cm^3 of sodium thiosulfate was required to completely react with all the iodine liberated. Calculate the percentage of copper in the salt.

6

Energy, enthalpy, and entropy

At the end of this chapter, students should be able to:

- Explain the meaning of enthalpy change when applied to chemical reactions

- Understand the difference between exothermic and endothermic reactions and distinguish between them

- Describe and define some fundamental enthalpy changes that occur in chemical reactions

- Understand how calorimetry can be used to measure enthalpy changes experimentally

- Understand that Hess's law is another way to express the first law of thermodynamics

- Apply Hess's law to obtain lattice enthalpies using the Born–Haber cycle

- Use mean bond enthalpies to calculate enthalpy changes in reactions

- Explain how entropy changes occur in chemical reactions and predict the direction of entropy changes

- Explain what is meant by a spontaneous process and how spontaneous processes relate to the second law of thermodynamics

- Calculate entropy changes in chemical reactions

- Define the Gibbs free energy change and calculate free energy changes in chemical reactions at a specific temperature from thermodynamic data

- Use the Gibbs expression to predict spontaneity in chemical reactions

Foundations of Chemistry: An Introductory Course for Science Students, First Edition.
Philippa B. Cranwell and Elizabeth M. Page.
© 2021 John Wiley & Sons Ltd. Published 2021 by John Wiley & Sons Ltd.
Companion website: www.wiley.com/go/Cranwell/Foundations

6.1 Enthalpy changes

6.1.1 Energy and enthalpy

The terms *enthalpy* and *energy* are often interchanged. We will use the term *enthalpy change* predominantly in this course.

All chemical reactions take place with a change in energy, and there are many different types of energy changes. Many chemical reactions release energy in the form of heat. Some chemical reactions can release light and sound energy as well as heat: for example, fireworks (Figure 6.1). Others release electrical energy, such as the reactions in batteries, fuel cells, and biochemical pathways. Although most chemical reactions release energy overall, some reactions absorb energy.

The exchange of energy that takes place between a chemical reaction and its surroundings when at constant pressure is called the **enthalpy change**. The **enthalpy** of a reaction is the total amount of energy associated with the reactants, and the symbol used is H. However, it's not possible to measure absolute values of enthalpy, so we always refer to the enthalpy *change* in a chemical reaction. The enthalpy change has the symbol ΔH and is defined as the difference in enthalpy between the products and the reactants in a reaction:

The Greek letter Δ is used as a symbol meaning change in mathematics and chemistry. It is spoken as 'delta'. So ΔH is expressed as 'delta H'.

$$\Delta H = H_{products} - H_{reactants}$$

This equation can be written in words as:

Enthalpy change = enthalpy of products – enthalpy of reactants

In chemical reactions, the term *enthalpy change* normally means the amount of heat exchanged when the reaction is carried out at constant pressure. The units of

Figure 6.1 Fireworks release energy in the form of heat, light, and sound.
Source: Victoria Page.

enthalpy change are kJ mol^{-1}. The actual value of the enthalpy change for a specific reaction depends upon three variables: the amount of substance, the temperature, and the pressure at which the reaction is carried out. The standard enthalpy change is the enthalpy change that occurs when the reaction is carried out under standard conditions. This means at a pressure of 1 bar (1.00×10^5 Pa) with the reactants in their standard states. Values for enthalpy changes are usually quoted at a temperature of 298 K (25 °C). When referring to a standard enthalpy change, we use the symbol $\Delta H^{\ominus}$. We say this as 'delta H plimsoll'.

> The superscript $\ominus$ symbol means that the reaction is carried out under standard conditions, and the enthalpy change is measured under these conditions.

> The standard state of a substance is its physical state under standard conditions, and because standard conditions are very close to normal room temperature and pressure, you are probably familiar with most of these states. For example, water is a liquid, and we represent this by writing $H_2O(l)$. Iron is a solid, represented by $Fe(s)$, and oxygen is a gas, $O_2(g)$. When a substance such as sodium chloride is in solution in water, we write this as $NaCl(aq)$. The 'aq' stands for aqueous, which means 'in water'.

6.1.2 Exothermic and endothermic reactions

In many chemical reactions, heat energy is released to the surroundings. This type of reaction is called an **exothermic reaction**. We can detect the heat energy released because the reaction mixture gets hot, and this heat energy flows to the surroundings. The surroundings include any material around the reaction mixture – this can be the air in the room, the apparatus, and any solvent or insulation material. In some reactions, such as the combustion of magnesium wire or petrol, it is obvious that heat is given out as we can observe the material burning and measure a temperature increase. However, in some reactions, the enthalpy change is small and not as easy to detect or measure.

Some reactions absorb heat energy from the surroundings when they take place. The reaction mixture gets cooler and so heat energy flows from the surroundings to equilibrate the temperatures. These are known as **endothermic reactions**. The direction of heat flow in exothermic and endothermic reactions is shown in Figure 6.2.

Several ammonium salts dissolve in water endothermically, so the temperature of the solution decreases. Reusable hot packs and instant cold packs use exothermic and endothermic reactions to release or absorb heat and are used in various medical applications, as shown in Box 6.1.

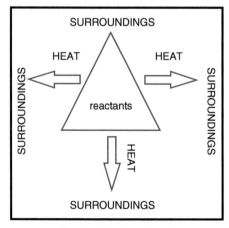

Energy is lost as heat in
an **exothermic** reaction
and the surroundings become warmer.

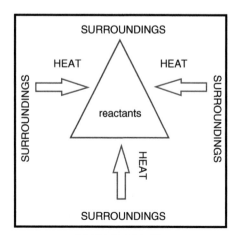

Energy is gained in
an **endothermic** reaction
and the surroundings become cooler.

Figure 6.2 Energy flow in exothermic and endothermic reactions.

Box 6.1 The principles of exothermic and endothermic reactions are used in hot and cold packs, which can bring relief for sports injuries and other similar medical problems. In the hot pack shown, a super-saturated solution of sodium acetate is forced to crystallise out. As it does so, it releases energy in an exothermic reaction. It can be re-used by warming the pack so that the salt re-dissolves and the procedure is repeated.

A cold pack such as the one shown here contains two bags: one containing water surrounded by another containing a salt such as ammonium nitrate. The salt dissolves in water in an endothermic reaction and absorbs heat, making the surroundings cooler.

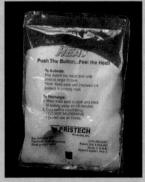

A hot pack containing sodium acetate. *Source:* Science Buddies

A cold pack containing ammonium acetate. *Source:* betterbraces.co.uk.

6.1.3   Reaction pathway diagrams

Diagrams that plot the enthalpy changes occurring in a reaction are known as **reaction pathway diagrams** or **enthalpy profile diagrams**. Example plots are shown in Figure 6.3. In such diagrams, the y-axis represents the enthalpy of the reactants and products, and the x-axis represents the reaction pathway. The y-axis shows the enthalpy of the reactants on the left at the start of the reaction and the enthalpy of the products after reaction on the right. If the enthalpy of the products is less than that of the reactants, the reaction is exothermic, as shown in Figure 6.3a. If the enthalpy of the products is greater than the enthalpy of the reactants, then the reaction is endothermic, as shown in Figure 6.3b. In

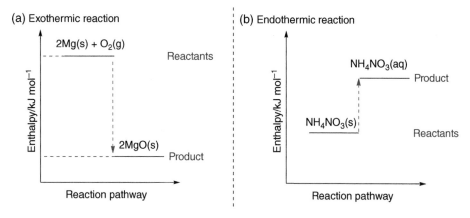

Figure 6.3 Reaction pathways for (a) an exothermic reaction; (b) an endothermic reaction.

each case, the enthalpy change is given by $\Delta H = H_{\text{products}} - H_{\text{reactants}}$. The value of ΔH is negative in an exothermic reaction and positive in an endothermic reaction.

Bond breaking and bond making

To understand the enthalpy changes in a chemical reaction, it helps to be aware of the processes taking place. When a compound such as methane, CH_4, burns in oxygen, O_2, the C–H bonds in CH_4 and the O=O bonds in O_2 first have to be broken. Bond breaking is an endothermic process and requires energy, which is usually provided in the form of heat. As the carbon atoms and hydrogen atoms combine with oxygen atoms in the products (carbon dioxide, CO_2, and water, H_2O) new bonds are formed. Bond formation is an exothermic process, so energy is released. In the combustion of methane, more energy is released in bond formation than is required in bond breaking, so overall the reaction is exothermic. The equation for the reaction showing the bonds that are broken in the products and the new bonds that are formed in the reactants is as follows. We will look at this in more detail in Section 6.1.7.

6.1.4 Measuring enthalpy changes

Different types of reactions have specific named enthalpy changes that you will meet in this chapter. Some enthalpy changes can be measured directly by experiments, and others have to be calculated indirectly by using other known enthalpy changes. In every case, the value of the standard enthalpy change is defined as the energy change for the reaction quoted under standard conditions.

Standard enthalpy of reaction, $\Delta_r H^\ominus$

The **standard enthalpy of reaction, $\Delta_r H^\ominus$**, is a general term referring to the enthalpy change for any type of chemical reaction when carried out under standard conditions. It is the difference between the enthalpy of the products and that of the reactants.

The standard enthalpy of reaction, $\Delta_r H^\ominus$, is the enthalpy change that occurs when reactants are converted to products in a chemical reaction under standard conditions.

> In this book, we will use the convention $\Delta_r H^\ominus$ to represent an enthalpy change for a reaction under standard conditions and carried out at 298 K. In some textbooks, you may see this symbol written as $\Delta H^\ominus{}_r$.

Standard enthalpy of formation, $\Delta_f H^\ominus$

The standard enthalpy of formation, $\Delta_f H^\ominus$, is the enthalpy change for the formation of one mole of the substance under standard conditions from its elements in their standard states.

There are a few points to note about this definition:

- The temperature is assumed to be 298 K unless otherwise stated.

- The enthalpy change quoted is for one mole of the substance.

- All substances are in **their standard states.** For example, oxygen is a gas, sodium is a solid, and water is a liquid. The standard state of carbon is graphite, C(graphite).

- Values of standard enthalpies of formation can be exothermic or endothermic, and the sign of ΔH indicates this.

The chemical equation that represents the standard enthalpy of formation of ethanol would be given by:

$$2C(graphite) + \tfrac{1}{2} O_2(g) + 3H_2(g) \rightarrow C_2H_5OH(l), \quad \Delta_f H^\ominus(C_2H_5OH(l)) = -278 \text{ kJ mol}^{-1}.$$

The enthalpy change for the formation of 1 mole of ethanol is 278 kJ, and the reaction is exothermic. Although we can define $\Delta_f H^\ominus(C_2H_5OH(l))$ and write an equation to represent the reaction, this process could not be used to obtain ethanol or to derive a value for the enthalpy change as it would lead to a mixture of products. In a case such as this, we would have to use an indirect method, which will be explained in Section 6.1.5.

6.1.5   Measuring enthalpy changes using calorimetry

Many values for enthalpy changes are obtained by **calorimetry.** Calorimetry is the name given to the general technique of carrying out a reaction in an insulated container and measuring the resultant temperature change. Calculations are then made to determine the enthalpy change.

Measuring the enthalpy of reaction in a calorimeter

The simplest type of calorimeter is a polystyrene cup with a lid, as shown in Figure 6.4. The lid has two holes to allow the insertion of a stirrer paddle and a thermometer. This type of apparatus is suitable for measuring enthalpy changes that take place in aqueous solution.

The polystyrene cup containing the reactants can be placed inside another cup to improve the insulation. The liquid reactant is measured and added to the inner cup. The temperature of the liquid is measured at intervals over a

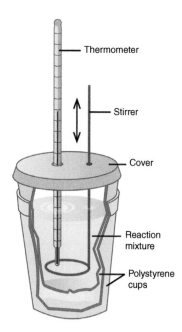

Figure 6.4 A polystyrene cup and lid used as a calorimeter. *Source:* Calorimetry, https://opentextbc.ca/chemistry/chapter/5-2-calorimetry/.

five-minute period to ensure that it is stable, and then the second reactant added. The initial temperature is recorded. The mixture is stirred by the paddle inserted through the lid, and the temperature is measured at regular intervals and recorded until no further temperature rise is observed and natural cooling begins. A graph is plotted of temperature against time, as shown in Figure 6.5. Because not all the heat is transferred instantaneously at the start of the experiment, and to compensate for the heat lost from the calorimeter in an exothermic reaction, the maximum temperature rise is obtained by extrapolating the cooling curve on the graph back to the time of mixing. The temperature rise is labelled ΔT in Figure 6.5.

Calculating the enthalpy change

The enthalpy change in the experiment is the same as the energy transferred as heat because the experiment is carried out at constant pressure. The heat change is given by:

$$q = m\,c\,\Delta T$$

where:

q is the heat lost or gained in J

m is the mass of the water or reacting liquid in g

c is the specific heat capacity in $J\,g^{-1}\,°C^{-1}$ or $J\,g^{-1}\,K^{-1}$

ΔT is the temperature rise in °C or K

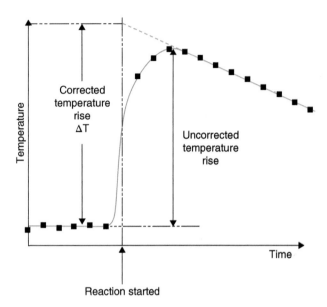

Figure 6.5 Plot of temperature against time for a simple calorimeter, showing correction for cooling losses by extrapolating back to the time of mixing. *Source:* Redrawn from Energetics, www.scienceskool.co.uk/energetics.html.

The specific heat capacity of any substance is the amount of heat energy required to raise the temperature of one gram of the substance by one degree Celsius or one Kelvin. Specific heat capacity has the symbol c and units $J\,g^{-1}\,°C^{-1}$ or $J\,g^{-1}\,K^{-1}$.

The specific heat capacity, c, is the amount of energy required to raise the temperature of one gram of water by one degree Celsius. The specific heat capacity of water is $4.18\,J\,g^{-1}\,°C^{-1}$, which means that it takes $4.18\,J$ of energy to raise the temperature of one gram of water by one degree Celsius. If the volume of water is known, it can be converted to a mass in grams using the density, which is $1\,g\,cm^{-3}$. The assumption is made that the calorimeter is perfectly insulated from the surroundings and all the heat energy lost in the reaction is used to raise the temperature of the water. As we know the mass of water and the temperature rise, we can calculate the total heat energy lost. This quantity is equivalent to the enthalpy change, ΔH, for the reaction as the reaction is carried out at constant pressure.

Worked Example 6.1 An experiment was carried out to determine the enthalpy of reaction between hydrochloric acid and sodium hydroxide. A $25\,cm^3$ portion of $1\,M$ hydrochloric acid was added to $25\,cm^3$ of $1\,M$ sodium hydroxide in a calorimeter. The maximum temperature recorded was $6.7\,°C$. Calculate the enthalpy of reaction, assuming there are no heat losses. The specific heat capacity of water is $4.18\,J\,g^{-1}\,°C^{-1}$.

Solution
The heat change on mixing the acid and base is obtained from the equation:

$$q = m\,c\,\Delta T$$

Identify the values of m, c, and ΔT from the information in the question. The mass of water, m, is obtained from the total volume of liquid used. The total volume $= (25 + 25)$ cm^3. As the density of water is 1 g cm^{-3}, the total mass of water, $m = 50$ g.
The heat capacity of water, c, is a constant whose value $= 4.18$ J g^{-1} °C^{-1}.
The temperature rise, $\Delta T = 6.7$ °C.
Substituting the values in the equation, we obtain:

$$q = 50 \text{ g} \times 4.18 \text{ J g}^{-1}\text{ °C}^{-1} \times 6.7 \text{ °C} = 1400 \text{ J} = 1.4 \text{ kJ}.$$

The process is exothermic so the accompanying enthalpy change is negative, i.e.

$$\Delta H = -1.4 \text{ kJ}.$$

> The density of a substance relates the mass of a substance to its volume.
> The equation for density is $d = \dfrac{m}{V}$.
> The density of water is 1 g cm^{-3}, so 50 cm^3 of water has a mass of 1 g cm$^{-3} \times 50$ cm$^3 = 50$ g.

Measuring the standard enthalpy of solution in a calorimeter

Box 6.2 Worked example 6.1 gives a method for measuring the standard enthalpy of neutralisation, which is defined as **the enthalpy change when one mole of water is formed from the reaction of an acid with an alkali under standard conditions**. The equation for this neutralisation reaction is:

$$HCl(aq) + NaOH(aq) \rightarrow NaCl(aq) + H_2O(l).$$

In the example, the amount in moles of hydrochloric acid ($n(HCl)$) used

$$= c \times V = 1 \text{ mol dm}^{-3} \times 25 \times 10^{-3} \text{ dm}^3 = 25 \times 10^{-3} \text{ mol}.$$

The amount in moles of water formed is therefore also 25×10^{-3} mol. Thus the enthalpy change per mole:

$$= -\frac{1400 \text{ J}}{25 \times 10^{-3} \text{ mol}} = -56 \times 10^3 \text{ J mol}^{-1} = -56 \text{ kJ mol}^{-1}.$$

The standard enthalpy of solution, $\Delta_{sol}H^{\ominus}$, is the enthalpy change when one mole of a solute is dissolved in a solvent to form an infinitely dilute solution under standard conditions.

When sodium chloride is dissolved in water, the lattice first breaks up into sodium ions and chloride ions. This is an endothermic process requiring energy. The ions then become surrounded by water molecules. This is an exothermic process, as new interactions are formed between the ions and water molecules. Water molecules are polar as they contain a permanent dipole: the slightly negative oxygen atoms of the water molecules surround the positive sodium ions, and the slightly positive hydrogen atoms surround the negative chloride ions forming hydrated ions, as shown in Figure 6.6. The difference between the energy required to break the lattice and the energy released as the ions become hydrated is the enthalpy of solution, $\Delta_{sol}H$.

A calorimeter provides a method for determining the enthalpy of solution of a solid, as shown in the following example.

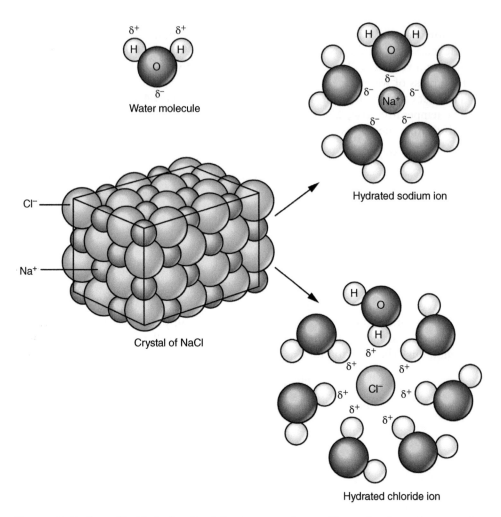

Figure 6.6 Sodium chloride lattice dissolving in water. *Source:* Illustration from 'Anatomy & Physiology', Connexions website, http://cnx.org/contents/e4e45509-bfc0-4aee-b73e-17b7582bf7e1@4, Jun 19, 2013. Licensed under CC-BY-4.0.

Worked Example 6.2   In an experiment to determine the enthalpy of solution of magnesium sulfate, 6.0 g of the anhydrous salt was weighed and dissolved with stirring in a polystyrene cup calorimeter containing 250 cm^3 of water. The maximum temperature rise was recorded to be 4.2 °C. Calculate the enthalpy change of solution of magnesium sulfate. The specific heat capacity of water is 4.18 J g^{-1} °C^{-1}.

Solution
The term *anhydrous* means that the magnesium sulfate is completely dry and contains no water of crystallisation. We can therefore assume the formula of the salt is $MgSO_4$.
The steps in the solution of this problem are:

1. Calculate the heat change when the solid is dissolved in water using $q = m c \, \Delta T$.

2. Calculate the enthalpy change that would be obtained using one mole of $MgSO_4$.

First identify the values of m, c, and ΔT from the information in the question.

The mass of water, m, is obtained from the volume of liquid, which is 250 cm³. As the density of water is 1 g cm⁻³, the total mass of water, m = 250 g.

The heat capacity of water, c, is a constant with value = 4.18 J g⁻¹ °C⁻¹.

The temperature rise = 4.2 °C.

Heat change on dissolving $MgSO_4$:

$$q = m\,c\,\Delta T = 250\,\text{g} \times 4.18\,\text{J g}^{-1}\text{°C}^{-1} \times 4.2\text{°C} = 4389\,\text{J} = 4.4\,\text{kJ}$$

This amount of heat is released when 6.0 g $MgSO_4$ is dissolved.

Molar mass of $MgSO_4$ = 120 g mol⁻¹.

Amount in moles of $MgSO_4$ used $= \dfrac{6\,\text{g}}{120\,\text{g mol}^{-1}} = 0.05\,\text{mol}$

Therefore the heat released by dissolving 1 mole $MgSO_4 = \dfrac{4.4\,\text{kJ}}{0.05} = 88\,\text{kJ}$

As the reaction is exothermic and conducted at constant pressure the enthalpy of solution of magnesium sulfate is therefore –88 kJ mol⁻¹.

Box 6.3 Maths help! An alternative way to do this calculation is the stepwise method shown here:

6.0 g will release 4.4 kJ heat energy

1.0 g will release $\dfrac{4.4\,\text{kJ}}{6.0}$ heat energy

120 g will release $\dfrac{4.4\,\text{kJ}}{6.0} \times 120 = 88\,\text{kJ}$ heat energy

Clearly both methods give the same result.

Measuring the enthalpy of combustion in a calorimeter

The amount of heat energy given out when a substance burns is a measure of the enthalpy of combustion of the material. This is defined in the following way:

The standard enthalpy of combustion, $\Delta_c H^{\ominus}$, is the enthalpy change when one mole of a substance is completely burnt in excess oxygen under standard conditions.

There are a few points to note about this definition:

- The temperature is assumed to be 298 K unless otherwise stated.

- The enthalpy change quoted is for one mole of the substance burnt.

- All substances are in their **standard states**.

- Values of standard enthalpies of combustion are always exothermic.

The standard enthalpy of combustion of methane, for example, is the enthalpy change for the reaction:

$$CH_4(g) + 2O_2(g) \rightarrow CO_2(g) + 2H_2O(l)$$

The value for this is given by $\Delta_c H^{\ominus}(CH_4(g)) = -890.4 \, kJ \, mol^{-1}$.

The sign of the enthalpy change is negative even though you may have to provide heat to start the reaction. The value refers to the *complete* combustion of methane, so that the products are carbon dioxide, and water, and not carbon monoxide, CO. The value of $-890.4 \, kJ \, mol^{-1}$ is the heat energy released when one mole of methane is completely burnt. If only 0.5 mol methane was used, the enthalpy change would be $-0.5 \times 890.4 \, kJ$.

To act as an efficient fuel, a substance must have a high enthalpy of combustion per mass of fuel. The **fuel value** of a substance is the amount of heat energy generated by one gram of the substance. The enthalpy of combustion, or fuel value, can be measured in a simple **flame calorimeter** or a more sophisticated **bomb calorimeter**.

The Flame Calorimeter

A flame calorimeter can be used to determine the enthalpy of combustion of fuels such as alcohols or liquid alkanes. A simple form of this apparatus consists of a small spirit burner that contains the liquid whose enthalpy of combustion is to be measured. The burner is fitted with a wick that dips through the lid of the burner and into the liquid. When the wick is lit, the energy released by burning the fuel is used to heat a known volume of water in an insulated container placed above the flame, as shown in Figure 6.7.

The mass of the spirit burner, its cap, and the liquid being studied are measured at the start of the experiment. A known mass or volume of water is added to the container above, and a lid is fitted to reduce heat loss by evaporation. The temperature of the water at the start of the experiment is measured. The wick is lit so the fuel burns, and the water is stirred until a rise in temperature of about 10 °C is obtained. The cap is then replaced on the spirit burner to extinguish the flame, and the final temperature of the water is recorded. The spirit burner and contents are reweighed. The difference in mass is equivalent to the mass of fuel burnt.

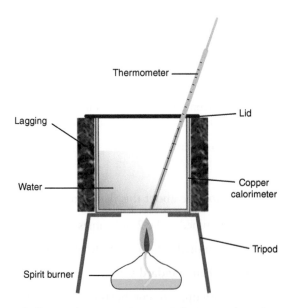

Figure 6.7 A simple flame calorimeter. *Source:* Calorimetry, http://www.cyberphysics. co.uk/topics/heat/calorimetry.htm © Cyberphysics.

The heat energy lost by the burning fuel increases the temperature of the water and that of the calorimeter. The amount of heat lost to the water and the calorimeter can be calculated using $q = m\,c\,\Delta T$.

The following worked example shows how to calculate the enthalpy of combustion of biodiesel in such an experiment.

Worked Example 6.3   In an experiment to determine the enthalpy of combustion of biodiesel, $5.00\,cm^3$ of biodiesel was added to a flame calorimeter. The calorimeter was filled with $1000\,cm^3$ of water at $20.0\,°C$. The biodiesel was lit and burnt until completely oxidised. After correcting for heat losses due to the calorimeter, a maximum temperature of $60.0\,°C$ was recorded. Calculate the enthalpy of combustion of biodiesel in units of $kJ\,kg^{-1}$. (Specific heat capacity of water = $4.18\,J\,g^{-1}\,°C^{-1}$; density of biodiesel = $0.880\,g\,cm^{-3}$; density of water = $1\,g\,cm^{-3}$.)

Solution
The steps involved in solving this question are:

1. Determine the total heat energy lost using the temperature increase and the mass of the water.

2. Calculate the heat energy lost per kg fuel, knowing the original mass of fuel used.

The total heat lost is obtained from the equation $q = m\,c\,\Delta T$. We are told in the question that heat losses have been corrected for, so we can assume all the heat energy lost from burning the fuel is used in raising the temperature of the water.

In this type of problem, it is useful to clearly identify the variables required for the equation:

q = heat energy lost from fuel

Mass of water in the calorimeter, $m = 1000\,cm^3 \times 1\,g\,cm^{-3} = 1000\,g$

$c = 4.18\,J\,g^{-1}\,°C^{-1}$

$\Delta T = 60.0 - 20.0\,°C = 40.0\,°C$

Inserting the values in the equation for ΔH:

$$q = m\,c\,\Delta T = 1000\,g \times 4.18\,J\,g^{-1}°C^{-1} \times 40.0°C$$
$$= 167\,200\,J = 167\,kJ$$

This temperature rise is produced by $5.00\,cm^3$ of biodiesel. To obtain the heat lost per kilogram first convert the volume of biodiesel to mass using the density:

Mass of biodiesel = $d \times V = 0.880\,g\,cm^{-3} \times 5.00\,cm^3 = 4.40\,g$
From the calculation, $4.40\,g$ biodiesel releases $167\,kJ$ heat energy
$1.0\,g$ biodiesel releases $\dfrac{167\,kJ}{4.40} = 37.95\,kJ$
$1000\,g = 1\,kg$ releases $37.95 \times 1000\,kJ$ energy = $37\,950\,kJ$
    = $38\,000\,kJ$

As the process is exothermic, the enthalpy of combustion is $-38\,000$ $kJ\,kg^{-1}$.
$1000\,kJ = 1\,MJ$ (1 megajoule), so the value in MJ is $-38.0\,MJ\,kg^{-1}$.

The Bomb Calorimeter

The flame calorimeter is a simple apparatus and can easily be constructed and used in the laboratory for approximate measurements of enthalpies of combustion. A more sophisticated piece of apparatus used for accurate work is the **bomb calorimeter** (Figure 6.8).

A bomb calorimeter consists of a central chamber in which a known mass of the sample is placed. The chamber is filled with oxygen gas under pressure. The central chamber is surrounded by a reservoir containing a fixed volume of water. This reservoir is thermally insulated to minimise heat losses to the surroundings. A mechanical stirrer and a thermocouple can be inserted into the water reservoir. The heat capacity of the bomb calorimeter is very high and is determined initially by electrical measurements. This constant is specific to a particular calorimeter. The sample is ignited by an electrical spark passed through conducting wires. Because the sample is surrounded by oxygen under pressure, it burns efficiently and completely. The heat lost is absorbed by the calorimeter and the water in the reservoir. The temperature rise is measured accurately by the thermocouple. The enthalpy of combustion of the fuel is obtained by calculation as for the flame calorimeter.

Enthalpy changes are defined at constant pressure. In the bomb calorimeter, the pressure is not constant, but the volume is. Thus, in practice, a small correction must be made to compensate for this.

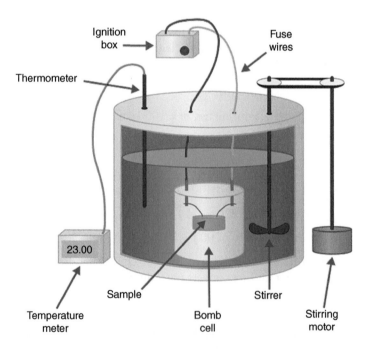

Figure 6.8 A bomb calorimeter. *Source:* From https://www.learner.org/courses/chemistry/text/text.html?dis=U&num=Ym5WdElUQS9PQ289&sec=YzJWaklU QS9OeW89.

Worked Example 6.4 In an experiment to determine the enthalpy of combustion of methanol, a weighed amount of methanol was burnt completely in a calorimeter. The temperature rise in the calorimeter, which contained 100 cm^3 of water, was measured. Given the following information, calculate the enthalpy of combustion of methanol.

Results:

Mass of methanol used = 3.20 g

Initial temperature of water = 20.5 °C

Final temperature of water = 32.5 °C

(Density of water = 1 g cm^{-3}; heat capacity of calorimeter, C_{cal} = 5.30 kJ °C^{-1}; specific heat capacity of water, c = 4.18 J g^{-1} °C^{-1})

Solution

In this problem, the heat energy lost by the fuel raises the temperature of the water in the calorimeter *and* the calorimeter itself. The equation to calculate the heat loss must therefore be adjusted to include the heat energy gained by the calorimeter. This is obtained by multiplying the heat capacity of the calorimeter (C_{cal}) by the temperature rise (ΔT). The total enthalpy change is therefore:

$$q = m_w \times c_w \times \Delta T + C_{cal} \times \Delta T$$

In a problem such as this, it is helpful to clearly lay out the information given and identify the values required to calculate the enthalpy change:

q = heat energy lost from fuel
Temperature rise, $\Delta T = (32.5-20.5) = 12.0$ °C
Mass of water in the calorimeter, $m_w = 100$ cm$^3 \times 1$ g cm$^{-3} = 100$ g
c_w = specific heat capacity of water = 4.18 J g^{-1} °C^{-1}
C_{cal} = heat capacity of calorimeter = 5.30 kJ °C^{-1}
Mass of methanol burnt = 3.20 g

The heat capacity of the calorimeter (C_{cal}) is the amount of heat required to raise the temperature of the calorimeter by one degree Celsius. It does not depend upon the mass of the calorimeter and is a fixed constant for an individual calorimeter.

Inserting the above values in the equation we obtain:

$$q = m_w \times c_w \times \Delta T + C_{cal} \times \Delta T$$

$$= 100 \text{ g} \times 4.18 \text{ J g}^{-1}\text{°C}^{-1} \times 12.0\text{°C} + 5.30 \text{ kJ°C}^{-1} \times 12.0\text{°C}$$

$$= 5\,016 \text{ J} + 63.6 \text{ kJ}$$

A common error in this type of problem is to mistake the mass of fuel burnt for the mass of water in the calorimeter. This is easy to do when the amount of water is given as a volume in cm^3.

Note that the units of energy have different dimensions here so the value in J should be converted to kJ by dividing by 1000 (i.e. multiplying by 10^{-3}).

$q = 5\,016 \times 10^{-3}$ kJ + 63.6 kJ = 68.616 kJ = 68.6 kJ (to 3 significant figures).

Next calculate the heat loss from one mole of methanol:

Molar mass of methanol = 32.0 g mol^{-1}.

$$\text{Amount in moles of methanol used} = \frac{3.20 \text{ g}}{32.0 \text{ g mol}^{-1}} = 0.1 \text{ mol}$$

$$\text{The heat loss from one mole methanol} = \frac{68.6 \text{ kJ}}{0.1 \text{ mol}} = 686 \text{ kJ mol}^{-1}$$

As the reaction is exothermic, the enthalpy of combustion of methanol, $\Delta_c H = -686$ kJ mol^{-1}.
This can be compared with the actual value of -715 kJ mol^{-1}.

Why do the values differ? One reason is that some of the heat will be lost to the surroundings as the apparatus is not completely insulated. Thus the measured temperature change will be smaller than if there were no heat losses. A second reason is that the methanol may not be completely burnt into carbon dioxide and water. Some carbon monoxide and possibly carbon, in the form of soot, may be formed. Again, this will cause the temperature rise to be smaller than if all the methanol was completely burnt in oxygen as stated in the definition.

6.1.6   Hess's law

In many cases, it isn't possible to determine the enthalpy change for a reaction directly by experiment. For example, the equation that represents the standard enthalpy of formation of ethanol is:

$$2C(\text{graphite}) + \tfrac{1}{2}O_2(g) + 3H_2(g) \rightarrow C_2H_5OH(l) \; \Delta_f H^{\ominus}(C_2H_5OH(l))$$

However, in reality ethanol cannot be made directly by this reaction, so the enthalpy change for the formation of ethanol must be determined indirectly.

Indirect measurements of enthalpy changes rely on the **first law of thermodynamics**. This is also known as the **law of conservation of energy**.

The law of conservation of energy states that the total energy of an isolated system is constant; energy can be neither created nor destroyed but can be transformed from one form to another.

This law also applies to chemical reactions. One implication of the law is that when we carry out any reaction with the same starting materials and obtain the same products in the same physical states, the energy change must be the same no matter what route is used to carry out the reaction. The total energy of the reactants and products and their surroundings must be the same, as we cannot create or destroy energy. This is summarised in Hess's law:

Hess's law states that the total enthalpy change in a chemical reaction is independent of the route by which the reaction is carried out, provided the initial and final conditions are the same.

Hess's law is represented graphically in Figure 6.9 in the hypothetical reaction between A and B forming products C and D. Hess's law states that the enthalpy change on going from A and B to C and D is the same whether the reaction is carried out directly in a single step, for which the enthalpy change is ΔH_1, or indirectly in two steps, where the total enthalpy change is $(\Delta H_2 + \Delta H_3)$. In the indirect reaction, the intermediates X and Y are formed after the initial step, and the final products are obtained in the second step.

We can apply this law to find the enthalpy of formation of ethanol. The reaction is represented by the equation:

$$2C(\text{graphite}) + 3H_2(g) + \tfrac{1}{2}O_2(g) \rightarrow C_2H_5OH(l) \; \Delta_f H^{\ominus}(C_2H_5OH(l))$$

As it isn't possible to measure the enthalpy change for this reaction directly, we use known values of enthalpies of combustion to set up an energy triangle or Hess's law cycle. The enthalpy of formation of ethanol forms one side of the

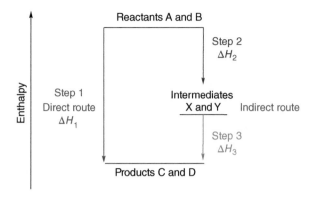

Figure 6.9 Graphical representation of Hess's law. The enthalpy change for direct route Step 1 = enthalpy change for Step 2 + enthalpy change for Step 3. $\Delta H_1 = \Delta H_2 + \Delta H_3$.

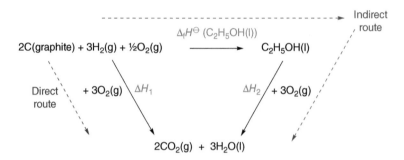

Figure 6.10 Hess's law cycle to determine the enthalpy of formation of $C_2H_5OH(l)$.

triangle, as shown in Figure 6.10. The combustion reactions of the reactants (carbon and hydrogen) and the product (ethanol) to carbon dioxide and water form the other two sides of the triangle. Complete combustion of the reactants, two moles of carbon and three moles of hydrogen, produces two moles of carbon dioxide and three moles of water according to the following equations:

$$2C(graphite) + 2O_2(g) \rightarrow 2CO_2(g), \Delta H = 2 \times \Delta_c H^{\ominus}(C(graphite))$$

$$3H_2(g) + \tfrac{3}{2} O_2(g) \rightarrow 3H_2O(l), \Delta H = 3 \times \Delta_c H^{\ominus}(H_2(g))$$

These reactions form the second side of the energy triangle, and represent the direct route from carbon and hydrogen to combustion products in Figure 6.10. The sum of these enthalpy changes is given by ΔH_1:

$$\Delta H_1 = 2 \times \Delta_c H^{\ominus}(C(graphite)) + 3 \times \Delta_c H^{\ominus}(H_2(g))$$

The same products (i.e. CO_2 and H_2O) are formed when one mole of ethanol is burnt completely in oxygen. This reaction forms the third side of the energy triangle, labelled ΔH_2 in Figure 6.10, and is equivalent to the enthalpy of combustion of ethanol, $\Delta_c H^{\ominus}$ ($C_2H_5OH(l)$), which can be determined directly by calorimetry.

$$C_2H_5OH(l) + 3O_2(g) \rightarrow 2CO_2(g) + 3H_2O(l), \Delta_c H^{\ominus}(C_2H_5OH(l)) = \Delta H_2$$

The energy triangle gives two different routes from 2 moles of carbon and 3 moles of hydrogen to 2 moles of carbon dioxide and 3 moles of water. One is the direct route labelled ΔH_1 in Figure 6.10. The other is the indirect route, which is equivalent to $\Delta_f H^\ominus(C_2H_5OH(l)) + \Delta H_2$. Following the dashed arrows from the reactants (carbon and hydrogen) to the products (carbon dioxide and water) by the two different routes and applying Hess's law, we can write:

$$\Delta H_1 = \Delta H_2 + \Delta_f H^\ominus(C_2H_5OH(l))$$

Rearranging the equation to make $\Delta_f H^\ominus(C_2H_5OH(l))$, the subject gives:

$$\Delta_f H^\ominus(C_2H_5OH(l)) = \Delta H_1 - \Delta H_2$$

Substituting for ΔH_1 and ΔH_2, the equation becomes:

$$\Delta_f H^\ominus(C_2H_5OH(l)) = 2 \times \Delta_c H^\ominus(C(\text{graphite})) + 3 \times \Delta_c H^\ominus(H_2(g)) - \Delta_c H^\ominus(C_2H_5OH(l))$$

This equation contains quantities that are known or measurable, so a value for the enthalpy of formation of ethanol can be calculated as in the following worked example.

Worked Example 6.5 Calculate the standard enthalpy of formation of ethanol given the following information:

$\Delta_c H^\ominus(C(\text{graphite})) = -393.5\,\text{kJ}\,\text{mol}^{-1}$
$\Delta_c H^\ominus(H_2(g)) = -285.8\,\text{kJ}\,\text{mol}^{-1}$
$\Delta_c H^\ominus(C_2H_5OH(l)) = -1367.3\,\text{kJ}\,\text{mol}^{-1}$

Solution
In a problem such as this, an energy triangle must first be drawn, as shown in Figure 6.10. Hess's law is then applied to relate the known enthalpy changes to the standard enthalpy of formation of ethanol, which is the value to be determined:

$$\Delta_f H^\ominus(C_2H_5OH(l)) = 2 \times \Delta_c H^\ominus(C(\text{graphite}))$$
$$+ 3 \times \Delta_c H^\ominus(H_2(g)) - \Delta_c H^\ominus(C_2H_5OH(l))$$

$$\Delta_f H^\ominus(C_2H_5OH(l)) = 2 \times -393.5\,\text{kJ}\,\text{mol}^{-1} + 3 \times -285.8\,\text{kJ}\,\text{mol}^{-1}$$

$$-\left(-1367.3\,\text{kJ}\,\text{mol}^{-1}\right) = -277.1\,\text{kJ}\,\text{mol}^{-1}$$

Worked Example 6.6 Calculate the enthalpy change for the reduction of ethene, C_2H_4, to ethane, C_2H_6, by hydrogen using the following information:

$\Delta_c H(C_2H_4(g)) = -1409\,\text{kJ}\,\text{mol}^{-1}$
$\Delta_c H(H_2(g)) = -285.8\,\text{kJ}\,\text{mol}^{-1}$
$\Delta_c H(C_2H_6(g)) = -1560\,\text{kJ}\,\text{mol}^{-1}$

Solution
The first step in a question of this type is to write the equation for the reaction whose enthalpy change is to be determined. This forms one side of the Hess's law triangle. We will call this $\Delta_r H$.

$$C_2H_4(g) + H_2(g) \xrightarrow{\Delta_r H} C_2H_6(g)$$

In the question, we are given the value for the enthalpy of combustion of ethane, $\Delta_c H(C_2H_6(g))$. The products from this reaction are 2 moles of CO_2 and 3 moles of water. This reaction can form a second side of the enthalpy triangle.

$$C_2H_4(g) + H_2(g) \xrightarrow{\Delta_r H} C_2H_6(g)$$

$+3.5O_2(g)$

$\Delta_c H (C_2H_6(g))$

$$2CO_2(g) + 3H_2O(l)$$

The question also gives us information about the enthalpy of combustion of ethene and hydrogen. Ethene and hydrogen are the reactants on the left-hand side of the equation. When 1 mole of ethene and 1 mole of hydrogen are burnt completely in oxygen, we obtain 2 moles of CO_2 and 3 moles of water. These are the same combustion products as obtained from 1 mole of ethane. The enthalpy triangle can be completed, as shown here.

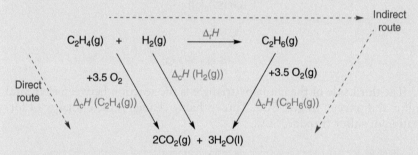

Both routes require the same amount of oxygen for the combustion reaction. This can be seen to be 3.5 moles and has been included on the enthalpy triangle. To simplify enthalpy diagrams, the oxygen required for combustion is often omitted.

The enthalpy triangle shows that there are two routes to produce 2 moles of CO_2 and 3 moles of H_2O from 1 mole of ethene. The route depicted by the blue dashed arrow on the left is the direct route. The second route is shown by the red dashed arrows and involves first reducing ethene to ethane and then combusting the ethane to obtain the same products as from the direct route. The total enthalpy change for each route is the same, and the only unknown enthalpy change is that for the formation of ethane from ethene, $\Delta_r H$.

Equating the enthalpy changes for the two routes gives:

$$\Delta_c H(C_2H_4(g)) + \Delta_c H(H_2(g)) = \Delta_r H + \Delta_c H(C_2H_6(g))$$

The equation can be rearranged to make $\Delta_r H$ the subject:

$$\Delta_r H = \Delta_c H(C_2H_4(g)) + \Delta_c H(H_2(g)) - \Delta_c H(C_2H_6(g))$$

Inserting the values given in the question:

$$\Delta_r H = -1409 \, kJ \, mol^{-1} + \left(-285.8 \, kJ \, mol^{-1}\right) - \left(-1560 \, kJ \, mol^{-1}\right)$$

$$= -134.8 \, kJ \, mol^{-1}$$

Worked Example 6.7   Calculate the enthalpy change for the reaction of sulfur dioxide, SO_2, with oxygen to give sulfur trioxide, SO_3, given the following information:

$$S(s) + O_2(g) \rightarrow SO_2(g) \quad \Delta_f H = -297 \, kJ \, mol^{-1}$$
$$S(s) + \tfrac{3}{2} O_2(g) \rightarrow SO_3(g) \quad \Delta_f H = -396 \, kJ \, mol^{-1}$$

Solution
This question uses values for the enthalpy of formation of the reactants and products to determine the enthalpy change for the reaction. Start by writing the equation for the reaction whose enthalpy change is required. This forms the first side of the enthalpy triangle:

$$SO_2(g) + \tfrac{1}{2} O_2(g) \xrightarrow{\Delta_r H} SO_3(g)$$

SO_2 is formed from sulfur and oxygen, and the enthalpy change for this reaction is the enthalpy of formation of SO_2, $\Delta_f H(SO_2(g))$. This equation forms the second side of the enthalpy triangle.

The third side of the enthalpy triangle is the reaction between elemental sulfur and oxygen to give SO_3. This enthalpy change is the enthalpy of formation of sulfur trioxide, SO_3.

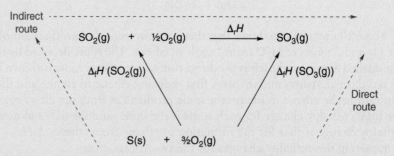

There are therefore two routes to form SO_3. The direct route (blue dashed arrow) reacts sulfur and oxygen to give SO_3. The indirect route (red dashed arrow) reacts sulfur and oxygen to give SO_2, which then reacts further with oxygen to give SO_3. Next, apply Hess's law to relate the enthalpy change for the direct route to the enthalpy changes for the indirect route:

$$\Delta_f H(SO_2(g)) + \Delta_r H = \Delta_f H(SO_3(g))$$

Rearrange to make $\Delta_r H$ the unknown:

$$\Delta_r H = \Delta_f H(SO_3(g)) - \Delta_f H(SO_2(g))$$

And insert values for the enthalpy changes from the question:

$$\Delta_r H = -396 \, kJ \, mol^{-1} - \left(-297 \, kJ \, mol^{-1} \right) = -99 \, kJ \, mol^{-1}$$

6.1.7 Bond energies and enthalpy changes

Enthalpy changes in chemical reactions result from bonds being broken and bonds being formed. The overall enthalpy change for a reaction is the difference between the enthalpy changes for the bond-breaking and bond-making processes.

Bond breaking is an endothermic process and requires energy to occur to break the attractive forces holding atoms or ions together. Bond formation is an exothermic process that releases energy when new interactions are made. Each chemical bond has a certain amount of energy associated with it. The stronger the bond, the more energy it takes to break that bond and the more energy is released when that bond is formed. The **bond energy**, or **bond dissociation enthalpy**, is the amount of energy required to break a certain type of bond, and the symbol for bond energy is E. The type of bond referred to is written in brackets after the symbol. For example, the bond energy of the O—H bond in H_2O is written as $E(O—H)$, and its value is $+463 \, kJ \, mol^{-1}$. The value is positive because energy must be applied to break the bond.

The bond energy refers to the energy required to break one mole of bonds of the specified type in a molecule in the gaseous state. For a Cl—Cl bond, it is defined as:

$$Cl_2(g) \rightarrow 2Cl(g) \quad E(Cl—Cl) = +242 \, kJ \, mol^{-1}$$

When a new bond is formed, the amount of energy released is the same as the energy required to break the bond. This is a consequence of the first law of thermodynamics. The enthalpy change for the formation of a molecule of chlorine from two chlorine atoms is represented by:

$$2Cl(g) \rightarrow Cl_2(g) \quad E(Cl—Cl) = -242 \, kJ \, mol^{-1}$$

The magnitude of the enthalpy change is the same, but the sign is opposite, as bond formation is exothermic.

The values quoted are **mean bond energies**. This is because the strength of any bond depends upon the chemical environment around that bond. For example, the O—H bond energy in the water molecule, H_2O, will be slightly different to the value for the O—H bond in ethanol, C_2H_5OH, and different again from the value in hydrogen peroxide, H_2O_2.

The mean bond energy is obtained by taking the average of bond energies from a number of the same types of bond in different molecules or chemical environments.

Water   Ethanol   Hydrogen peroxide

Table 6.1 gives some typical values for mean bond dissociation energies. It can be seen from the table that the values for bond energies vary quite a lot depending upon the nature of the bond. For bonds between atoms of the same element, the mean bond energy increases with bond order. It takes more energy to break a double bond than a single bond and more energy again to break a triple bond. This can be seen for the bond energies for different types of C—C bonds, as shown:

The *bond order* refers to whether the bond is a single, double or triple bond.

$$E(C—C) = 348 \, kJ \, mol^{-1}$$

$$E(C=C) = 612 \, kJ \, mol^{-1}$$

$$E(C\equiv C) = 837 \, kJ \, mol^{-1}$$

Table 6.1 Values of mean bond energies.

Bond	$E/\text{kJ mol}^{-1}$	Bond	$E/\text{kJ mol}^{-1}$
H—H	436	C—H	412
C—C	348	N—H	388
C=C	612	O—H	463
C≡C	837	F—H	562
N—N	163	Si—H	318
N=N	409	P—H	322
N≡N	944	S—H	338
O—O	146	Cl—H	431
O=O	496	Br—H	366
F—F	158	I—H	299
Si—Si	176	C—O	360
P—P	172	C=O	743
S—S	264	C—N	305
Cl—Cl	242	C—F	484
Br—Br	193	C—Cl	338
I—I	151	C—Br	276
		C—I	238

Source: Data from JG Stark and HG Wallace, *Chemistry Data Book*, John Murray, London, 1989.

Notice the values for the bond energies of the C—X bonds, where X is a halogen element. As the halogen atom gets larger, the bond enthalpy becomes smaller, meaning that the bond is becoming weaker. This is because when the two atoms in a bond are of a similar size, the atomic orbitals overlap more efficiently, and the bonds require more energy to break apart. The C—F bond is the strongest at $484\,\text{kJ mol}^{-1}$, and the C—I bond is the weakest at $238\,\text{kJ mol}^{-1}$. This order of bond strengths has consequences in organic chemistry when substituting the halogen atom for another functional group.

Using bond energies to calculate enthalpy changes

Mean bond energies provide us with another way of calculating enthalpy changes for a chemical reaction. This is useful for reactions where the enthalpy change cannot be measured directly.

For a chemical reaction to occur, the bonds in the reactants must be broken and new bonds formed in the products. Let's take the reaction between oxygen and hydrogen to form water as an example. This reaction occurs explosively once it starts! The equation for the reaction is:

$$O_2(g) + 2H_2(g) \rightarrow 2H_2O(g)$$

The reaction is represented in Figure 6.11.

The initial step involves the breaking of the O=O and H—H bonds in the reactants O_2 and H_2. This is an endothermic process, and the enthalpy change can be calculated from the mean bond enthalpies of the O=O double bond and the H—H single bonds.

The intermediate phase is represented by free atoms of oxygen and hydrogen in Figure 6.11. The intermediate phase is very short-lived and is known as the **transition state** as the nature of the species here is difficult to determine. The intermediate complexes have been represented as free atoms here for simplicity. Finally, new bonds are formed between the oxygen and hydrogen atoms to produce water molecules, and energy is released. The enthalpy change for this step can be calculated from the bond-formation energies of the four new O—H bonds.

Note that quoted values of mean bond energies assume the molecule in which the bonds are being broken or formed to be in the gaseous state. In this example water vapour is formed, not liquid water.

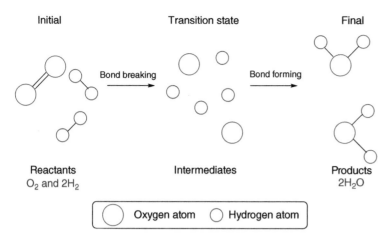

Figure 6.11 Representation of the reaction between oxygen (1 mole) and hydrogen (2 moles) to produce water (2 moles), showing bond breaking and bond formation.

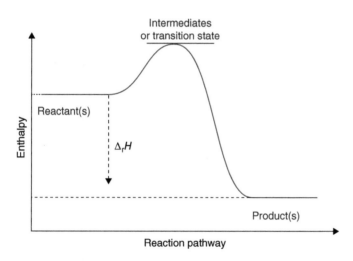

Figure 6.12 Reaction pathway showing the energy of intermediates in the transition state in a reaction.

The enthalpy changes occurring can be represented by a reaction pathway diagram, such as in Figure 6.12. This shows the relative enthalpies of the reactants, intermediates, and products. The diagram here is similar to the reaction pathway shown in Figure 6.3, except in Figure 6.12, the transition state is included to show that it is at a higher energy than that of reactants or products.

The difference in enthalpy between the reactants and products gives us the overall enthalpy of reaction, $\Delta_r H$. This can be calculated using the mean bond energies of the reactants and products. A calculation of this type is best carried out using a table to keep track of bonds broken and formed, as shown in the next example.

Worked Example 6.8   Use the values of mean bond energies given in Table 6.1 to calculate the enthalpy change that occurs on the formation of two moles of water vapour.

Solution

For a problem of this type, first write a balanced stoichiometric equation:

$$O_2(g) + 2H_2(g) \rightarrow 2H_2O(g)$$

Next, draw up a table showing the numbers and types of each bond that must be broken and formed (Table 6.2).

Table 6.2 Calculation of enthalpy of reaction using mean bond energies.

	Bond breaking				Bond formation		
Bond	How many bonds	Bond dissociation enthalpy/kJ	Total/kJ	Bond	How many bonds	Bond formation enthalpy/kJ	Total/kJ
O=O	1	+496	+496	O–H	4	–463	–1852
H–H	2	+436	+872				
Sum			+1368				–1852
$\Delta_r H$	+1368–1852 = –484 kJ						

Look up the mean bond energies for each bond. Add together the total mean bond energies in the reactants (i.e. $E(O=O)$ and $2 \times E(H–H)$). This gives the total enthalpy change required for bond breaking and is a positive value (+1368 kJ). Next, add the total mean bond energies in the products (i.e. $4 \times E(O–H)$). These quantities are all negative, and summing gives an overall enthalpy change for bond formation (–1852 kJ). By adding together the enthalpy change for bond formation to the enthalpy change for bond dissociation, we can obtain the overall enthalpy change for the reaction:

$$\Delta_r H = \text{enthalpy change for bond breaking}$$
$$+ \text{enthalpy change for bond formation}$$
$$= + 1368 - 1852 \text{ kJ} = -484 \text{ kJ}$$

The value for the enthalpy change of reaction of one mole of oxygen with two moles of hydrogen to form two moles of gaseous water is therefore –484 kJ.

This can be written as: $O_2(g) + 2H_2(g) \rightarrow 2H_2O(g)$ $\Delta_r H = -484$ kJ.

The enthalpy of formation of one mole of water vapour is half this value. The equation for this reaction is:

$$\tfrac{1}{2} O_2(g) + H_2(g) \rightarrow H_2O(g) \; \Delta_f H(H_2O(g)) = -242 \text{ kJ mol}^{-1}$$

Worked Example 6.9   Calculate the enthalpy change for the complete combustion of methane, CH_4, using mean bond energies as given in Table 6.1. The experimentally determined value for the enthalpy of combustion of methane is –890 kJ mol^{-1}. Explain why your calculated value differs from this.

Solution
First write the balanced chemical equation for the reaction:

$$CH_4(g) + 2O_2(g) \rightarrow CO_2(g) + 2H_2O(l)$$

Next, draw up a table showing bonds to be broken and bonds formed, and insert the number of each type of bond and the mean bond energy.

Calculate the total enthalpy change for bond breaking and bond formation, and insert the values in the table.

	Bond breaking				Bond formation		
Bond	How many bonds	Bond dissociation enthalpy/kJ	Total/kJ	Bond	How many bonds	Bond formation enthalpy/kJ	Total/kJ
C—H	4	+412	+1648	C=O	2	−743	−1486
O=O	2	+496	+992	O—H	4	−463	−1852
Sum/kJ			+2640				−3338
$\Delta_r H$	+2640−3338 = −698 kJ						

Add the enthalpy change for bond breaking to the enthalpy change for bond formation:

$$\Delta_r H = +2640 - 3338 = -698 \text{ kJ}$$

The overall enthalpy change for the reaction is therefore calculated to be −698 kJ. This value is quite a lot smaller in size than the experimentally determined value. There are two reasons for this. The first is that the calculation uses mean bond energies, and these are not specific for a particular molecule. The second reason in this example is that the definition of enthalpy of combustion assumes that the product, water, is in the liquid state. However, the definition of bond energies requires the water molecules to be in the gaseous state. As water vapour is converted to liquid water, more energy is released (enthalpy of condensation), and the calculation has not taken this enthalpy change into consideration.

Worked Example 6.10   Estimate the standard enthalpy of reaction when one mole of ethene, $C_2H_4(g)$, reacts with gaseous fluorine, $F_2(g)$, to form gaseous 1,2-difluoroethane, $C_2H_4F_2(g)$.

Solution
The procedure for solving this problem is as in the previous example. First write the chemical equation for the reaction. In this case, it may help to draw the bonds involved on each side of the equation, as we have both double and single C—C bonds.

Next, draw up the table showing bonds broken and bonds formed, and insert the number and type of each bond in the table along with its mean bond energy.

Note that in this example the C-H bonds are retained in the product so there is no need to include them in the calculation. However you may wish to include bond energies for all the bonds in the reactants and the products to ensure completeness.

	Bond breaking			Bond formation			
Bond	How many bonds	Bond dissociation enthalpy/kJ	Total/kJ	Bond	How many bonds	Bond formation enthalpy/kJ	Total/kJ
C=C	1	+612	+612	C—C	1	−348	−348
C—H	4	+412	+1648	C—H	4	−412	−1648
F—F	1	+158	+158	C—F	2	−484	−968
Sum/kJ			+2418				−2964
$\Delta_r H$	+2418 − 2964 = −546 kJ						

Calculate the total bond enthalpy for bond dissociation (+2418 kJ) and the total bond enthalpy for formation (−2964 kJ). Add the two values together, as in the final row of the table, to obtain the overall enthalpy of reaction: $\Delta_r H = +2418 − 2964 = −546$ kJ.

6.1.8   Born–Haber cycles

Lattice enthalpy

Section 6.1.5 showed how Hess's law can be applied using energy cycles to work out enthalpy changes that are difficult, or impossible, to measure in the laboratory. These are useful for calculating enthalpy changes involving covalent molecules. However, a different approach is used for ionic materials. A **Born–Haber cycle** is an application of Hess's law used to calculate the **lattice enthalpy** of a solid. The lattice enthalpy of an ionic solid is the energy that holds the oppositely charged cations and anions together to create the three-dimensional crystal lattice. When an ionic solid is formed, oppositely charged ions attract each other and energy is released. Strong electrostatic forces are formed that hold the crystal lattice together; the energy released is called the *lattice enthalpy of formation* and has the symbol $\Delta_{latt}H^{\ominus}$. The lattice enthalpy of formation is defined in the following way:

The lattice enthalpy of formation, $\Delta_{latt}H^{\ominus}$, of an ionic substance is the enthalpy change when one mole of the substance is formed from its constituent ions in the gas phase under standard conditions.

The following equations represent the lattice enthalpy of formation for sodium chloride and calcium chloride:

$$Na^+(g) + Cl^-(g) \rightarrow NaCl(s) \quad \Delta_{latt}H^{\ominus}(NaCl(s)) = -771 \text{ kJ mol}^{-1}$$

$$Ca^{2+}(g) + 2Cl^-(g) \rightarrow CaCl_2(s) \quad \Delta_{latt}H^{\ominus}(CaCl_2(s)) = -2237 \text{ kJ mol}^{-1}$$

New bonds are being formed in the ionic lattice so the enthalpy change is highly exothermic in this direction.

When the lattice is broken down, for example by heat, electrical energy, or by dissolving in a solvent, these strong electrostatic forces are broken, and energy must be supplied. The enthalpy change in this direction is called the *lattice enthalpy of dissociation*, and it is defined as follows.

The lattice enthalpy of dissociation is the enthalpy change when one mole of an ionic compound dissociates into its ions in the gaseous state under standard conditions.

The lattice enthalpy of dissociation of sodium chloride can be represented by:

$$NaCl(s) \rightarrow Na^+(g) + Cl^-(g)\ \Delta_{latt}H^\ominus(NaCl(s)) = +771\,kJ\,mol^{-1}$$

The lattice enthalpy of dissociation of calcium chloride can be represented by:

$$CaCl_2(s) \rightarrow Ca^{2+}(g) + 2Cl^-(g)\ \Delta_{latt}H^\ominus(CaCl_2(s)) = +2237\,kJ\,mol^{-1}$$

Note that the magnitude of these lattice enthalpies are identical for the same ionic solid; it is just the direction of the energy change that differs. We will work with lattice enthalpies of formation in this course, so values will be exothermic.

Calculating the lattice enthalpy of a compound

An energy cycle such as the one shown in Figure 6.13 can be used to determine the lattice enthalpy of an ionic compound.

Applying Hess's law to the energy triangle in Figure 6.13 allows us to write the equation:

$$\Delta_i H^\ominus + \Delta_{latt} H^\ominus = \Delta_f H^\ominus$$

The term $\Delta_i H^\ominus$ is shown in Figure 6.13 and explained below.

Rearranging the terms, we can obtain the following expression for the lattice enthalpy:

$$\Delta_{latt} H^\ominus = \Delta_f H^\ominus - \Delta_i H^\ominus$$

The lattice enthalpy is therefore dependent upon the enthalpy of formation of the solid, $\Delta_f H^\ominus$, and the enthalpy changes involved in converting the elements in their standard states to gaseous ions. This term is represented by $\Delta_i H^\ominus$ and is composed of the enthalpy of atomisation of the elements and their ionisation enthalpies. In Figure 6.14, endothermic reactions are represented by upward-pointing arrows and exothermic enthalpy changes represented by downward-pointing arrows. This type of diagram is known as a Born–Haber diagram and is a way of depicting Hess's law for the formation of an ionic compound from its elements.

Different textbooks define lattice enthalpy in different directions. Here we are defining *lattice enthalpy* as the enthalpy change associated with forming the lattice, so the sign for the lattice enthalpy in this directions is negative (–ve), as energy is released and the process is exothermic. To clarify the direction of the enthalpy change, this is called the *lattice enthalpy of formation*. The enthalpy change required to break down the lattice into its constituent ions is termed the *lattice enthalpy of dissociation* and has a positive value as it is endothermic.

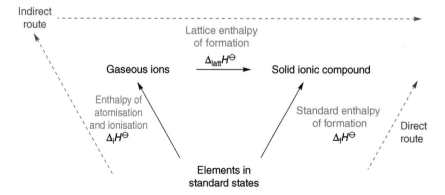

Figure 6.13 General energy triangle used to calculate lattice enthalpy for an ionic compound.

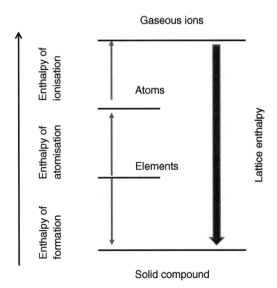

Figure 6.14 General Born–Haber diagram for the calculation of lattice enthalpies.

Before we go any further, there are some enthalpy changes involved that we need to define.

Standard enthalpy of atomisation, $\Delta_{at}H^{\ominus}$

The standard enthalpy of atomisation, $\Delta_{at}H^{\ominus}$ is the enthalpy change accompanying the production of separate gaseous atoms of the element under standard conditions.

So, for example, the standard enthalpy of atomisation of chlorine, Cl(g), is expressed by:

$$\tfrac{1}{2}Cl_2(g) \rightarrow Cl(g) \quad \Delta_{at}H^{\ominus} = +121\,kJ\,mol^{-1}$$

The standard enthalpy of atomisation of iron, Fe(g), is expressed in exactly the same way:

$$Fe(s) \rightarrow Fe(g) \quad \Delta_{at}H^{\ominus} = +418\,kJ\,mol^{-1}$$

Box 6.4 Note that although the standard enthalpy of atomisation involves forming gaseous atoms of the element, it does not have the same value as the bond energy. The bond energy relates to the energy required to break **one mole of the bond,** whereas the enthalpy of atomisation relates to the formation of **one mole of gaseous atoms.** Thus the bond energy for diatomic molecules is twice the enthalpy of atomisation.

$$O_2(g) \rightarrow 2O(g) \quad E(O{=}O) = +496\,kJ\,mol^{-1}$$

Energy required to break one mole of O=O bonds

$$\tfrac{1}{2}O_2(g) \rightarrow O(g) \quad \Delta_{at}H^{\ominus} = +248\,kJ\,mol^{-1}$$

Energy required to form one mole of gaseous O atoms

Values of standard enthalpies of atomisation are always endothermic.

Ionisation energy, IE

To form positive ions of elements from neutral atoms, electrons must be removed. This requires energy, known as the **ionisation energy** of the element.

The ionisation energy (IE) is defined as the amount of energy required to remove one mole of electrons from one mole of gaseous atoms of the element. The first ionisation energy, IE_1, is the energy required to remove the first electron in an atom from its outer (valence) shell.

The general equation that represents the first ionisation energy of an element M is:

$$M(g) \rightarrow M^+(g) + e^-$$

There are a few points to note about ionisation energies:

- Values of ionisation energies are always endothermic i.e. positive.

- Electrons are removed from atoms of the element in the gaseous state.

- Electrons are removed in a step-wise process, one at a time.

- Successive ionisation energies increase (i.e. $IE_1 < IE_2 < IE_3$, etc.) because the positive ion formed exerts a greater pull on the electron than a neutral atom.

- The energy required to remove a specific electron depends upon the orbital that the electron occupies.

Successive ionisation energies

It is possible to remove more than one electron from an atom; however, the amount of energy for successive ionisations will be greater than that for removing the first electron. On removal of the first electron, an ion with a positive charge is left. The amount of energy required to remove an electron from this positive ion is larger than that for the first ionisation energy. This is because there are now more protons in the nucleus than there are electrons orbiting the nucleus, so each electron has a greater electrostatic interaction with the nucleus. The first ionisation energy of sodium is represented by:

$$Na(g) \rightarrow Na^+(g) + e^- \quad IE_1 = +494 \, kJ \, mol^{-1}$$

The electron removed is from the outer 3s orbital:

$$(Na) \, 1s^2 2s^2 2p^6 3s^1 \rightarrow (Na^+) \, 1s^2 2s^2 2p^6$$

The second ionisation energy of sodium is represented by:

$$Na^+(g) \rightarrow Na^{2+}(g) + e^- \quad IE_2 = +4560 \, kJ \, mol^{-1}$$

The change in electron configuration is represented by:

$$(Na^+) \, 1s^2 2s^2 2p^6 \rightarrow (Na^{2+}) \, 1s^2 2s^2 2p^5$$

Removing the electron from the filled second shell of the $Na^+(g)$ ion clearly requires a far greater amount of energy than removing the first outer electron from the neutral $Na(g)$ atom.

The variation of ionisation energy across the periodic table is covered in Chapter 10.

Electron affinity, $\Delta_{EA}H^{\ominus}$

Elements on the right-hand side of the periodic table tend to form negative ions by gaining electrons. The energy associated with this process is called the **electron affinity, $\Delta_{EA}H^{\ominus}$**.

The electron affinity of an element, $\Delta_{EA}H^{\ominus}$ is the standard enthalpy change that occurs when one mole of gaseous atoms of the element accepts one mole of electrons to form one mole of gaseous ions. The first electron affinity, $\Delta_{EA1}H^{\ominus}$, is the energy change when a gaseous atom accepts an electron to form an ion with a –1 charge.

This process is the opposite of ionisation, where the atom loses an electron to form a positive ion.

The electron affinity of chlorine is represented by:

$$Cl(g) + e^- \rightarrow Cl^-(g) \quad \Delta_{EA}H^{\ominus} -364 \, kJ \, mol^{-1}$$

There are a few points to note about the definition of electron affinities:

- Values of first electron affinities are exothermic, but successive ones are endothermic.

- Values of electron affinities relate to adding an electron to a gaseous atom or ion.

- Electrons are added in a step-wise manner.

Electron affinity differs from electronegativity in that it represents the process of gaining an electron and forming a negative ion rather than simply attracting electrons from within a bond. Electron affinities are covered in more detail in Chapter 10.

Born–Haber cycles

The general Born–Haber diagram shown in Figure 6.14 has been redrawn in the form of the more familiar Hess's law cycle in Figure 6.15 to show the enthalpy changes used to calculate the lattice enthalpy of solid sodium chloride, NaCl.

Applying Hess's law to this enthalpy cycle and starting from the elements, we obtain the equation:

$$\Delta_i H^{\ominus}(NaCl(s)) + \Delta_{latt}H^{\ominus}(NaCl(s)) = \Delta_f H^{\ominus}(NaCl(s))$$

The enthalpy term for the ionisation of sodium chloride can be broken down into two parts, i.e. the enthalpy change for the formation of gaseous sodium ions from sodium metal and the enthalpy change for the formation of gaseous chloride ions from chlorine molecules.

The enthalpy change for the formation of sodium ions involves the enthalpy of atomization of sodium, $\Delta_{at}H^{\ominus}(Na(g))$ and the first ionisation energy of sodium, $IE_1(Na(g))$.

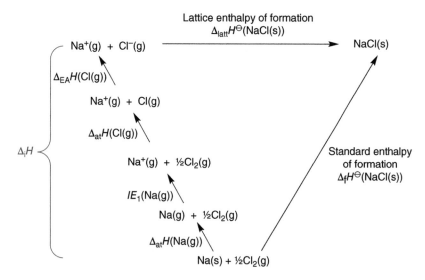

Figure 6.15 A Hess's law cycle to calculate the lattice enthalpy of formation of sodium chloride.

$$Na(s) \rightarrow Na(g) \quad \Delta_{at}H^{\ominus}(Na(g))$$

The first ionisation energy for sodium is represented by:

$$Na(g) \rightarrow Na(g)^+ + e^- \quad IE_1(Na(g))$$

> The ionisation energy requires sodium to be present as gaseous atoms, so the metal must first be vaporised.

Together, these terms equal the enthalpy required to form gaseous sodium ions from sodium metal:

$$\Delta_i H^{\ominus}(Na(g)) = \Delta_{at}H^{\ominus}(Na(g)) + IE_1(Na(g))$$

To form gaseous chloride ions from chlorine gas, the molecular chlorine must first be atomised and then the gaseous atoms each gain one electron to form chloride ions. The enthalpy changes involved are the enthalpy of atomisation, $\Delta_{at}H^{\ominus}(Cl(g))$, and the first electron affinity, $\Delta_{EA}H^{\ominus}(Cl(g))$.

$$\tfrac{1}{2}Cl_2(g) \rightarrow Cl(g) \quad \Delta_{at}H^{\ominus}(Cl(g))$$

$$Cl(g) + e^- \rightarrow Cl^-(g) \quad \Delta_{EA}H^{\ominus}(Cl(g))$$

Note that only half a mole of chlorine gas is required to form one mole of chlorine atoms.

Together, these two terms equal the enthalpy change for the formation of one mole of gaseous chloride ions from half a mole of chlorine molecules in their standard state:

$$\Delta_i H^{\ominus}(\tfrac{1}{2}Cl_2(g)) = \Delta_{at}H^{\ominus}(Cl(g)) + \Delta_{EA}H^{\ominus}(Cl(g))$$

The Born–Haber cycle gives us the following equation to obtain the lattice enthalpy:

$$\Delta_i H^{\ominus}(NaCl(s)) + \Delta_{latt}H^{\ominus}(NaCl(s)) = \Delta_f H^{\ominus}(NaCl(s))$$

We can substitute the enthalpy changes for the ionisation of sodium and chlorine into the equation to obtain:

$$\{\Delta_{at}H^{\ominus}(Na(g)) + IE_1(Na(g))\} + \{\Delta_{at}H^{\ominus}(Cl(g)) + \Delta_{EA}H^{\ominus}(Cl(g))\}$$

$$+ \Delta_{latt}H^{\ominus}(NaCl(s)) = \Delta_f H^{\ominus}(NaCl(s))$$

This equation can be rearranged to obtain an expression for the lattice enthalpy of sodium chloride:

$$\Delta_{latt}H^{\ominus}(NaCl(s)) = \Delta_f H^{\ominus}(NaCl(s)) - [\{\Delta_{at}H^{\ominus}(Na(g)) + IE_1(Na(g))\}$$

$$+ \{\Delta_{at}H^{\ominus}(Cl(g)) + \Delta_{EA}H^{\ominus}(Cl(g))\}]$$

As values for the enthalpy terms on the right-hand side of the expression are known and can be found in tables, a value can be calculated for the lattice enthalpy of sodium chloride.

Worked Example 6.11   Write equations, including state symbols, and give the correct symbols for the following standard enthalpy changes:

  a. The enthalpy of atomisation of oxygen, O_2

  b. The enthalpy of atomisation of magnesium, Mg

  c. The first ionisation energy of potassium, K

  d. The second ionisation energy of calcium, Ca

  e. The electron affinity of oxygen, O

  f. The electron affinity of O^-

Solution

  a. $\frac{1}{2}O_2(g) \rightarrow O(g)$ $\Delta_{at}H^{\ominus}(O(g))$

  b. $Mg(s) \rightarrow Mg(g)$ $\Delta_{at}H^{\ominus}(Mg(g))$

  c. $K(g) \rightarrow K^+(g) + e^-$ $IE_1(K(g))$

  d. $Ca^+(g) \rightarrow Ca^{2+}(g) + e^-$ $IE_2(Ca(g))$

  e. $O(g) + e^- \rightarrow O^-(g)$ $\Delta_{EA}H^{\ominus}(O(g))$

  f. $O^-(g) + e^- \rightarrow O^{2-}(g)$ $\Delta_{EA}H^{\ominus}(O^-(g))$

Constructing a Born–Haber diagram for calculating lattice enthalpy

A Born–Haber diagram provides a clear way of representing the energy changes occurring during the formation of an ionic lattice. A Born–Haber cycle shows each step in an energy level diagram and relates the enthalpy of formation of an ionic solid from its elements to the enthalpy changes involved in ionisation of the elements and formation of the lattice from its constituent ions. The main difference between a diagram showing a Born–Haber cycle and a Hess's law triangle is that enthalpy changes are shown in a vertical manner in a Born-Haber diagram. A Born–Haber diagram for calculating the lattice enthalpy of NaCl is shown in Figure 6.16.

The procedure for drawing a Born–Haber cycle such as that for NaCl is as follows:

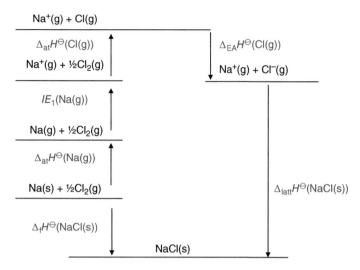

Figure 6.16 Born–Haber diagram for calculating the lattice enthalpy of formation of NaCl.

1. The energies of the elements in their standard states are drawn on the left-hand side of the diagram.

2. The direct route from the elements to the ionic solid is the enthalpy of formation, $\Delta_f H^\ominus$, and is represented by a vertical downwards arrow on the left of the cycle. The energy of the ionic solid is represented by a horizontal line across the lowest part of the diagram.

3. Vertical upwards steps from the elements represent the enthalpy of atomisation of sodium, the first ionisation energy of sodium, and the enthalpy of atomisation of chlorine. These steps are all endothermic and therefore associated with an increase in energy. The next step, the electron affinity of chlorine, is exothermic and is shown by a downward arrow.

4. A final downwards arrow then connects the gaseous ions of the elements to the ionic sold. This enthalpy change is the lattice enthalpy of formation of sodium chloride, $\Delta_{latt} H^\ominus$.

5. On the Born–Haber diagram, endothermic enthalpy changes are shown as upward arrows and exothermic enthalpy changes as downward arrows.

Worked Example 6.12   Calculate the lattice enthalpy of formation of sodium chloride given the following data:

$\Delta_{at} H^\ominus(Na(g)) = +109\,\text{kJ mol}^{-1}$

$\Delta_{at} H^\ominus(Cl(g)) = +121\,\text{kJ mol}^{-1}$

$IE_1(Na(g)) = +494\,\text{kJ mol}^{-1}$

$\Delta_{EA} H^\ominus(Cl(g)) = -364\,\text{kJ mol}^{-1}$

$\Delta_f H^\ominus(NaCl(s)) = -411\,\text{kJ mol}^{-1}$

Solution

An expression was derived earlier for the lattice enthalpy of sodium chloride by applying Hess's law to the enthalpy cycle linking the enthalpy of formation of sodium chloride to the enthalpy of formation of gaseous ions of sodium and chlorine (Figure 6.16):

$$\Delta_{latt}H^{\ominus}(NaCl(s)) = \Delta_f H^{\ominus}(NaCl(s)) - \left[\{\Delta_{at}H^{\ominus}(Na(g)) \right.$$
$$\left. + IE_1(Na(g))\} + \{\Delta_{at}H^{\ominus}(Cl(g)) + \Delta_{EA}H^{\ominus}(Cl(g))\}\right]$$

The values for each of the terms can be inserted into the expression to calculate a value for the lattice enthalpy of sodium chloride:

$$\Delta_{latt}H^{\ominus}(NaCl(s)) = \left(-411 \ kJ \ mol^{-1}\right) - \left[\left(+109 \ kJ \ mol^{-1}\right)\right.$$
$$\left. + \left(+494 \ kJ \ mol^{-1}\right) + \left(+121 \ kJ \ mol^{-1}\right) + \left(-364 \ kJ \ mol^{-1}\right)\right]$$
$$= -411 \ kJ \ mol^{-1} - \left[360 \ kJ \ mol^{-1}\right] = -771 \ kJ \ mol^{-1}$$

Check your answer – this is the value for the lattice enthalpy of formation of NaCl, so we would expect an exothermic figure due to the formation of electrostatic forces of attraction between the oppositely charged ions.

The Born–Haber diagram and calculation are slightly different for a $1:2$ ionic solid such as calcium chloride, $CaCl_2$, as can be seen in Figure 6.17.

The Ca^{2+} ion has a double positive charge and therefore must lose two electrons upon ionisation. The total enthalpy for ionisation of calcium therefore includes both the first and second ionisation enthalpies. In addition, as there are

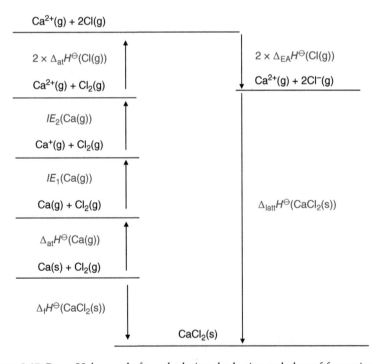

Figure 6.17 Born–Haber cycle for calculating the lattice enthalpy of formation of $CaCl_2$.

two atoms of Cl in each formula unit of $CaCl_2$, we need to multiply values of the enthalpy of atomisation and the electron affinity for chlorine by two. Applying Hess's law to the Born–Haber cycle for calcium chloride, we obtain the expression:

$$\Delta_f H^{\ominus}(CaCl_2(s)) = \Delta_{at}H^{\ominus}(Ca(g)) + IE_1(Ca(g)) + IE_2(Ca(g))$$

$$+ 2 \times \Delta_{at}H^{\ominus}(Cl(g)) + 2 \times \Delta_{EA}H^{\ominus}(Cl(g)) + \Delta_{latt}H^{\ominus}(CaCl_2(s))$$

Therefore:

$$\Delta_{latt}H^{\ominus}(CaCl_2(s)) = \Delta_f H^{\ominus}(CaCl_2(s)) - \{\Delta_{at}H^{\ominus}(Ca(g)) + IE_1(Ca(g))$$

$$+ IE_2(Ca(g)) + 2 \times \Delta_{at}H^{\ominus}(Cl(g)) + 2 \times \Delta_{EA}H^{\ominus}(Cl(g))\}$$

In the following worked example, both ions in the formula unit have a double charge, so the electron affinity value must be adjusted to account for this.

Worked Example 6.13   Draw a Born–Haber diagram for calcium oxide, and calculate the lattice enthalpy of formation of the solid given the following data:

$\Delta_{at}H^{\ominus}(Ca(g)) = +193 \text{ kJ mol}^{-1}$

$\Delta_{at}H^{\ominus}(O(g)) = +248 \text{ kJ mol}^{-1}$

$IE_1(Ca(g)) = +590 \text{ kJ mol}^{-1}$

$IE_2(Ca(g)) = +1150 \text{ kJ mol}^{-1}$

$\Delta_{EA}H^{\ominus}(O(g)) = -142 \text{ kJ mol}^{-1}$

$\Delta_{EA}H^{\ominus}(O^-(g)) = +844 \text{ kJ mol}^{-1}$

$\Delta_f H^{\ominus}(CaO(s)) = -635 \text{ kJ mol}^{-1}$

Solution
The Born–Haber diagram is first drawn using the rules given earlier.

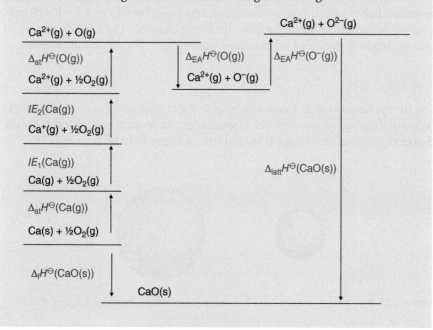

Applying Hess's law to the Born–Haber cycle, we obtain:

$$\Delta_f H^\ominus(CaO(s)) = \Delta_{at} H^\ominus(Ca(g)) + \Delta_{at} H^\ominus(O(g)) + IE_1(Ca(g)) + IE_2(Ca(g))$$
$$+ \Delta_{EA} H^\ominus(O(g)) + \Delta_{EA} H^\ominus(O^-(g)) + \Delta_{latt} H^\ominus(CaO(s))$$

Rearranging to obtain an expression for the lattice enthalpy of CaO:

$$\Delta_{latt} H^\ominus(CaO(s)) = \Delta_f H^\ominus(CaO(s)) - \{\Delta_{at} H^\ominus(Ca(g)) + \Delta_{at} H^\ominus(O(g))$$
$$+ IE_1(Ca(g)) + IE_2(Ca(g)) + \Delta_{EA} H^\ominus(O(g)) + \Delta_{EA} H^\ominus(O^-(g))\}$$

Substituting the values given in the question:

$$\Delta_{latt} H^\ominus(CaO(s)) = -635\,kJ\,mol^{-1} - \{193\,kJ\,mol^{-1} + 248\,kJ\,mol^{-1}$$
$$+ 590\,kJ\,mol^{-1} + 1150\,kJ\,mol^{-1} + \left(-142\,kJ\,mol^{-1}\right) + 844\,kJ\,mol^{-1}\}$$

$$= -3518\,kJ\,mol^{-1}$$

6.1.9 Factors affecting the size of the lattice enthalpy

The calculation in Worked Example 6.13 shows that the lattice enthalpy of formation of CaO is very high with a value of $-3518\,kJ\,mol^{-1}$. This is much higher than that calculated earlier for NaCl, ($-771\,kJ\,mol^{-1}$). The following section explains how the size and charge on the ions in a solid affect the value of the lattice enthalpy.

Charge on the ions

For compounds with the same arrangement of ions, the higher the charge on the ions, the stronger the electrostatic attraction between them, and so the stronger the lattice. Both sodium chloride and calcium oxide have the same solid-state structure. But in calcium oxide, the cation and anion both have a double charge, whereas in sodium chloride the ions are singly charged. The lattice enthalpy of calcium oxide is therefore substantially higher than that of sodium chloride.

Size of the ions

The charge density on an ion depends upon the charge on the ion divided by the radius of the ion. For ions with the same charge, the larger the ion, the smaller the charge density, as the charge is spread over a larger volume (Figure 6.18).

Figure 6.18 The charge density on a Li^+ ion is greater than that on a larger Cs^+ ion.

If the charge density is low, the electrostatic attraction between ions will be smaller and the lattice enthalpy weaker. So for compounds with the same anion, the lattice enthalpy gets smaller as the cation gets bigger. This can be seen in Figure 6.19a for the Group 1 chlorides. For compounds with the same cation, the larger the anion, the weaker the lattice. This is shown in Figure 6.19b for the halides of sodium.

Polarisation

Pure ionic bonding assumes ions are spherical in shape, so their electric field is spread evenly in every direction. However, in practice, this is not always the case, and some ions become *polarised* when in proximity to other ions. This polarisation leads to a distortion of the shape of the ions and the resultant electric field. When ions become polarised, the bonding between them has more covalent character. Polarisation occurs when a cation attracts the electron cloud of a neighbouring anion, which becomes distorted in shape, as shown in Figure 6.20.

Certain cations are more strongly *polarising* than others. Cations with a high charge density have greater polarising power. Small, highly charged cations such as Fe^{3+} and Al^{3+} are strongly polarising.

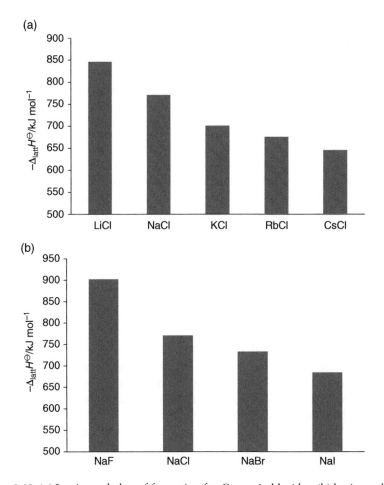

Figure 6.19 (a) Lattice enthalpy of formation for Group 1 chlorides; (b) lattice enthalpy of formation for sodium halides.

Figure 6.20 Polarisation of an anion (–ve) by a small, highly charged cation (+ve).

Some anions are more *polarisable* than others. Anions that are more easily polarised are large anions with a high charge as their electron cloud can be distorted more readily. Anions such as carbonate, CO_3^{2-}, sulfate, SO_4^{2-} nitride, N^{3-}, and nitrate, NO_3^-, are polarisable.

Li^+ is especially strongly polarising despite having only a single positive charge. Salts of lithium such as LiF and Li_2CO_3 have a high degree of covalent nature because of the strong polarising power of the small Li^+ ion, which increases the electron density between the cation and anion. The Born–Haber calculation assumes solids are purely ionic but, in fact, many have a degree of covalent nature just as many covalent molecules are polar and have a degree of ionic character. We will see how polarisation affects the lattice enthalpies and therefore the properties of solids in Chapter 11.

Worked Example 6.14

  a. Arrange the following cations in order of polarising power:
$$Sr^{2+}, Be^{2+}, Ba^{2+}, Mg^{2+}, Ca^{2+}$$

  b. Arrange the following anions in order of polarisability:
$$Br^-, Cl^-, F^-, I^-$$

Solution

  a. All cations have the same charge, so the smaller the cation, the more strongly polarising it is. The order of polarising power is therefore: $Be^{2+} > Mg^{2+} > Ca^{2+} > Sr^{2+} > Ba^{2+}$.

  b. All anions have the same charge, so the larger the anion, the more polarisable it is: $I^- > Br^- > Cl^- > F^-$.

6.2 Entropy and Gibbs free energy

So far, we have looked at enthalpy changes in chemical reactions and seen that these can be either exothermic (when energy is released) or endothermic (when energy is absorbed). In an exothermic reaction, the enthalpy of the products is less than the enthalpy of the reactants. From this, we can state that the products are more stable than the reactants. It might therefore be logical to assume that chemical reactions will proceed only if they are exothermic and the products are in a lower energy state than the reactants. However, this is not the case, because

Figure 6.21 Sherbet Fountain is composed of citric acid and sodium hydrogen carbonate that react endothermically in water. *Source:* Barrattsweets.

many chemical reactions are endothermic: for example, dissolving ammonium nitrate in water or mixing citric acid with sodium hydrogen carbonate. The confectionary Sherbet Fountain contains a mixture of citric acid and sodium hydrogen carbonate. These react together endothermically with saliva in the mouth to produce carbon dioxide. The carbon dioxide causes the fizzing sensation, and the endothermic reaction makes the tongue feel cold (Figure 6.21).

It is therefore clear that we cannot use energy changes alone to predict whether a chemical reaction will take place. To make a more reliable prediction, we need to consider the **entropy change** that occurs in a reaction.

6.2.1 Entropy

The entropy, S, of a system can be defined in several ways. Possibly the most straightforward way of thinking about the entropy of a system is the number of ways the energy in a system can be distributed at a specific temperature. This is also a measure of the disorder of a system. The higher the disorder, the higher the entropy. If the disorder of a system increases then the entropy increases. When the system is a chemical reaction, the overall energy of the system can increase or decrease. If the energy of the system increases, then there will be more energy to distribute over the particles (these can be molecules, ions, etc.) in the system and therefore more ways of distributing the energy. The entropy is therefore greater.

Another factor that determines the disorder of a system is the number of particles, such as molecules or ions, in the system. If the system has more particles, then there are more ways of distributing the energy over those particles, so the entropy is higher. To compare values of entropy, we must use standard conditions for measurements. Standard conditions refer to a pressure of 1.00×10^5 Pa with reactants and products in their standard states. Values are normally quoted at a specific temperature, usually 298 K.

Table 6.3 Standard molar entropy values for selected elements and compounds.

Substance	$S^{\ominus}/\text{J K}^{-1}\text{ mol}^{-1}$	Substance	$S^{\ominus}/\text{J K}^{-1}\text{ mol}^{-1}$
$H_2(g)$	131	$CO(g)$	198
$He(g)$	126	$CO_2(g)$	214
$Li(s)$	28.0	$CH_4(g)$	186
$C(s)$ (diamond)	2.4	$CH_3OH(l)$	127
$C(s)$ (graphite)	5.7	$C_2H_5OH(l)$	161
$N_2(g)$	192	$C_2H_4(g)$	219
$O_2(g)$	205	$C_6H_6(l)$	173
$F_2(g)$	203	$H_2O(l)$	69.9
$Ca(s)$	41.6	$H_2O(g)$	189
$Fe(s)$	27.2	$NH_3(g)$	193

The standard molar entropy, $S^{\ominus}$, is the entropy of one mole of a substance in its standard state. The units of entropy are $\text{J K}^{-1}\text{ mol}^{-1}$.

Table 6.3 lists values of standard molar entropies for various elements and compounds.

There are several points to note from Table 6.3 regarding standard molar entropies:

- Values of standard molar entropies are always positive.

- Unlike standard molar *enthalpies* of elements, which are defined as being equal to zero, elements in their standard states have positive values of standard molar *entropies*.

- Gases generally have higher standard molar entropies than liquids, which have higher standard molar entropies than solids.

- Molecules with fewer atoms tend to have lower entropies than more complex molecules. Carbon monoxide has $S^{\ominus} = 198\,\text{J K}^{-1}\text{ mol}^{-1}$, whereas carbon dioxide has $S^{\ominus} = 214\,\text{J K}^{-1}\text{ mol}^{-1}$.

- The unit of entropy is $\text{J K}^{-1}\text{ mol}^{-1}$, whereas enthalpy changes are normally quoted in kJ mol^{-1}.

For the same substance, the entropy increases as the substance changes from solid to liquid to gas: i.e. on melting and boiling, there is an increase in entropy. For example, water vapour has $S^{\ominus} = 189\,\text{J K}^{-1}\text{ mol}^{-1}$, whereas liquid water has $S^{\ominus} = 69.9\,\text{J K}^{-1}\text{ mol}^{-1}$. The molecules in a liquid move randomly within the space occupied by the liquid. Contrast this to the situation in a solid where the molecules are in fixed positions. In a gas, molecules move even faster and more freely than in the liquid state, so the entropy is even higher. This is associated with the greater amount of energy contained in the gas and liquid phases compared to the solid phase.

Entropy changes in reactions

Most processes occur with a change in entropy. If we consider entropy as a measure of the disorder in a system, it is possible to make predictions as to the direction of entropy change in a chemical reaction.

Because gas molecules have higher entropies than liquids and solids, a chemical reaction that produces gas molecules will proceed with an increase in entropy. For example, in the thermal decomposition of ammonium nitrate, one mole of solid ammonium nitrate produces one mole of gaseous nitrous oxide (N_2O) and two moles of water vapour according to the following equation:

$$NH_4NO_3(s) \rightarrow N_2O(g) + 2H_2O(g)$$

There is an increase in entropy of the system because three moles of gas molecules with high entropy have been produced from one mole of solid, which has a lower entropy.

We have seen that a system with a greater number of particles has a higher entropy, as there are more ways of distributing the energy over those particles. So if we heat a metal salt and it decomposes to release a gas, we would expect an increase in entropy, as the number of product molecules is higher than the number of reactant molecules. An example of this is the thermal decomposition of calcium carbonate, which produces calcium oxide and carbon dioxide on heating:

$$CaCO_3(s) \rightarrow CaO(s) + CO_2(g)$$

The reaction proceeds with an increase in the number of 'particles' from 1 to 2, so there is an increase in entropy. One of the products is a gas, which further increases the entropy.

The overall entropy change of the system is given by the difference between the final and initial entropies of the system: $\Delta S^\ominus = S^\ominus_{final} - S^\ominus_{initial}$.

Worked Example 6.14 Predict the direction of entropy change in the following reactions:

  a. $NaCl(s) \rightarrow Na^+(aq) + Cl^-(aq)$

  b. $H_2O(l) \rightarrow H_2O(g)$

  c. $NaNO_3(s) \rightarrow NaNO_2(s) + \frac{1}{2}O_2(g)$

  d. $2Mg(s) + O_2(g) \rightarrow 2MgO(s)$

  e. $CO(g) + H_2O(g) \rightleftharpoons CO_2(g) + H_2(g)$

Solution

  a. Increase. Sodium chloride solid is dissolving in water and produces a solution of sodium ions and chloride ions. There is an increase in entropy as we go from an ordered crystalline system to free ions in an aqueous solution.

  b. Increase. As molecules of water escape from the liquid phase to the vapour phase, there is an increase in energy and disorder, and therefore the entropy increases.

  c. Increase. The thermal decomposition of solid sodium nitrate involves the release of oxygen gas and the formation of more moles of species in the products compared to reactants. Both these changes result in an increase in entropy.

d. Decrease. Heating metallic magnesium in oxygen results in the formation of solid magnesium oxide. The products are more ordered than the reactants, and there are fewer moles of products compared to reactants. Both these effects result in a decrease in entropy.

e. Very little change. In both the reactants and products, we have two moles of gaseous molecules, so there should be little change in entropy in this reaction.

6.2.2 Spontaneous processes and the second law of thermodynamics

When a chemical reaction takes place, there is usually a change in entropy of the *system*. The *system* is considered to be the chemical reaction itself. This entropy change is associated with redistribution of energy between the reactants and products. As many reactions proceed with an exchange in energy between the system and the surroundings, we can conclude that the surroundings must also experience a change in entropy. The overall entropy change is therefore due to the change in entropy of the system plus the change in entropy of the surroundings:

$$\Delta S^{\ominus}{}_{total} = \Delta S^{\ominus}{}_{system} + \Delta S^{\ominus}{}_{surroundings}$$

Unlike energy changes, the total entropy of the system and surroundings can increase or decrease in a chemical reaction. If there is an overall increase in total entropy, then the entropy change is positive. Reactions in which there is an overall increase in entropy are said to be **spontaneous.**

A spontaneous process is one that occurs naturally without the addition of external energy.

Spontaneous processes aren't restricted to chemical reactions. A ball rolling down a hill, gas escaping from a leaking pipe, and a dab of perfume evaporating off the skin are all physical processes that occur naturally and result in a thermodynamically more stable state. Perhaps one of the most familiar examples of entropy is a student room that becomes more disorganised and untidy as time goes on (Figure 6.22). It requires effort in the form of energy to correct this.

These changes are examples of the **second law of thermodynamics** in action. The second law of thermodynamics states that when any change takes place, the entropy of the universe increases. In this case, the *universe* means the system and the surroundings.

The second law of thermodynamics states that the total entropy of the universe tends to increase with time.

The second law of thermodynamics can be written as $\Delta S_{total} > 0$ for a spontaneous change.

$$\Delta S_{total} = \Delta S_{system} + \Delta S_{surroundings} > 0.$$

To calculate the total entropy change, we must determine the entropy change of the system plus the entropy change of the surroundings. In the case of a chemical reaction, the system is the reaction itself, so from this point, we will use the symbol $\Delta S_{reaction}$ to represent the entropy change of the system, ΔS_{system}.

Figure 6.22 A student bedroom tends to a maximum state of disorder and maximum entropy without intervention. *Source:* Zoonar RF/Getty Images.

Entropy change of the system

For a chemical reaction, the entropy change of the system can be expressed by:

$$\Delta S_{\text{reaction}} = S_{\text{products}} - S_{\text{reactants}}$$

In calculating the entropy change in a chemical reaction, we need to ensure that entropy values are measured under the same conditions of temperature and pressure, so we use standard entropy values, $\Delta S^{\ominus}$:

$$\Delta S^{\ominus}_{\text{reaction}} = S^{\ominus}_{\text{products}} - S^{\ominus}_{\text{reactants}}$$

The entropy of the products and reactants is calculated by using tables of known entropy values and correcting for the number of moles of each component in the chemical equation. Worked Example 6.15 shows how this is carried out.

Worked Example 6.15 Use values in Table 6.3 to calculate the standard entropy change for the following reaction, and explain the sign of the change:

$$2NH_3(g) \rightleftharpoons N_2(g) + 3H_2(g)$$

Solution
To calculate the entropy change in a reaction, we use the equation: $\Delta S^{\ominus}_{\text{reaction}} = S^{\ominus}_{\text{products}} - S^{\ominus}_{\text{reactants}}$ and tabulated entropy values.

Applying the equation to the reaction where ammonia is the reactant and nitrogen and hydrogen gas the products, we obtain:

$$\Delta S^{\ominus}_{\text{reaction}} = \left[S^{\ominus}(N_2(g)) + 3 \times S^{\ominus}(H_2(g)) \right] - 2 \times \left[S^{\ominus}(NH_3(g)) \right]$$

Note that we multiply the entropy values for each component by the stoichiometric coefficient. Use Table 6.3 to determine the entropy values required, and substitute the values in the equation:

$$\Delta S^{\ominus}{}_{reaction} = \left[S^{\ominus}(N_2(g)) + 3 \times S^{\ominus}(H_2(g)) \right] - 2 \times \left[S^{\ominus}(NH_3(g)) \right]$$

$$= \left[192\,J\,K^{-1}\,mol^{-1} + 3 \times 131\,J\,K^{-1}\,mol^{-1} \right]$$

$$- \left[2 \times 193\,J\,K^{-1}\,mol^{-1} \right] = 199\,J\,K^{-1}\,mol^{-1}$$

The value for the entropy change of the system is positive, which indicates an increase in entropy. This is predicted as the number of gaseous molecules is increasing from two on the reactants side of the equation to four in the products.

Worked Example 6.16   Use data in Table 6.3 to calculate the standard entropy change for this reaction:

$$2H_2(g) + O_2(g) \rightarrow 2H_2O(g)$$

Solution
The problem is solved in the same manner as Worked Example 6.15. The main point to note here is that the value for the standard entropy of water depends upon whether water is formed as a liquid or a gas. In the equation given, water is shown in the gas phase, so it is important to use the correct value. Substitute the entropy values for the components into the equation for the entropy change of the reaction:

$$\Delta S^{\ominus}{}_{reaction} = S^{\ominus}{}_{products} - S^{\ominus}{}_{reactants} = \left[2 \times S^{\ominus}(H_2O(g)) \right]$$

$$- \left[2 \times S^{\ominus}(H_2(g)) + S^{\ominus}(O_2(g)) \right] = \left[2 \times 69.9\,J\,K^{-1}\,mol^{-1} \right]$$

$$- \left[2 \times 131\,J\,K^{-1}\,mol^{-1} + 205\,J\,K^{-1}\,mol^{-1} \right] = -327.2\,J\,K^{-1}\,mol^{-1}$$

The overall entropy change for the reaction shows a large decrease, as we convert 3 moles of gaseous molecules to 2 moles of gaseous water molecules.

Entropy change of the surroundings

Most chemical reactions are accompanied by an enthalpy change, which involves rearranging the distribution of energy between the system and surroundings. This means that there must be a change in entropy in the surroundings. The entropy change of the surroundings depends upon the size of the enthalpy change and the temperature. It is expressed by:

$$\Delta S^{\ominus}{}_{surroundings} = -\frac{\left(\Delta H^{\ominus}{}_{reaction} \right)}{T}$$

The temperature, T, is the temperature in Kelvin at which the reaction is carried out. If the reaction is carried out under standard conditions, the temperature is assumed to be 298 K.

The total entropy change can therefore be written as:

$$\Delta S^{\ominus}{}_{total} = \Delta S^{\ominus}{}_{reaction} + \Delta S^{\ominus}{}_{surroundings} = \Delta S^{\ominus}{}_{reaction} - \frac{\left(\Delta H^{\ominus}{}_{reaction}\right)}{T}$$

The total entropy change is expressed in terms of properties of the reaction only, i.e. in terms of the entropy and the enthalpy change in the reaction.

Worked Example 6.17   Calculate the total entropy change that occurs in the universe when one mole of methane, CH_4, is burnt in oxygen at 298 K according to the following equation. Use entropy values from Table 6.3.

$$CH_4(g) + 2O_2(g) \rightarrow CO_2(g) + 2H_2O(g) \quad \Delta H^{\ominus}{}_{reaction} = -890 \, kJ \, mol^{-1}$$

Solution
To calculate the total entropy change in the universe, the following equation should be used:

$$\Delta S^{\ominus}{}_{total} = \Delta S^{\ominus}{}_{reaction} - \frac{\left(\Delta H^{\ominus}{}_{reaction}\right)}{T}$$

This involves calculating the entropy change of the reaction, $\Delta S^{\ominus}{}_{reaction}$, using the equation $\Delta S^{\ominus}{}_{reaction} = S^{\ominus}{}_{products} - S^{\ominus}{}_{reactants}$. We will do that first:

$$\Delta S^{\ominus}{}_{surroundings} = - \frac{\left(\Delta H^{\ominus} reaction\right)}{T} = - \frac{890 \, kJ \, mol^{-1}}{298 \, K} = -2.99 \, kJ \, K^{-1} \, mol^{-1}.$$

Calculate the entropy change of the universe, $\Delta S^{\ominus}{}_{total}$, by inserting the calculated values into the equation:

$$\Delta S^{\ominus}{}_{total} = \Delta S^{\ominus}{}_{reaction} - \frac{\left(\Delta H^{\ominus} reaction\right)}{T}$$

$$= -3 \, J \, K^{-1} \, mol^{-1} - \left(-2.99 \, kJ \, K^{-1} \, mol^{-1}\right)$$

$$= -3 \, J \, K^{-1} \, mol^{-1} + 2.99 \, J \, K^{-1} \, mol^{-1}$$

The units of entropy values are quoted in $J \, K^{-1} \, mol^{-1}$ and enthalpy values in $kJ \, mol^{-1}$. Therefore, the units must be rationalised before we can complete the calculation. In this case multiply kJ by 1000 to convert to J.

$$\Delta S^{\ominus}{}_{total} = -3 \, J \, K^{-1} \, mol^{-1} + 2\,990 \, J \, K^{-1} \, mol^{-1} = +2987 \, J \, K^{-1} \, mol^{-1}$$

$$= +2.99 \, kJ \, mol^{-1}$$

The large positive value for the entropy increase suggests the combustion reaction is highly feasible and spontaneous once started.

6.2.3   Gibbs free energy and spontaneous reactions

From the previous section, we have seen that spontaneous changes are associated with an increase in entropy of the universe. This is a consequence of the second law of thermodynamics. The relationship that determines whether there is an increase in total entropy of the universe is:

$$\Delta S^{\ominus}_{total} = \Delta S^{\ominus}_{reaction} - \frac{\left(\Delta H^{\ominus}_{reaction}\right)}{T}.$$

By multiplying both sides of the equation by T, we obtain the expression:

$$T\Delta S^{\ominus}_{total} = T\Delta S^{\ominus}_{reaction} - \Delta H^{\ominus}_{reaction} \quad \text{Equation 1}$$

From the second law of thermodynamics, for a spontaneous reaction, the total entropy of the universe must increase, i.e.:

$$T\Delta S^{\ominus}_{total} > 0.$$

Therefore, for a spontaneous reaction:

$$T\Delta S^{\ominus}_{reaction} - \Delta H^{\ominus}_{reaction} > 0$$

Assuming the temperature is constant, two factors are important here. The first is the change in entropy of the chemical reaction. If this is large and positive so that $T\Delta S^{\ominus}_{reaction}$ is greater than the value of $\Delta H^{\ominus}_{reaction}$, there will be an overall increase in total entropy of the universe. This can be expressed as:

$$T\Delta S^{\ominus}_{reaction} > \Delta H^{\ominus}_{reaction}$$

The second factor is the enthalpy change, $\Delta H^{\ominus}$, of the reaction. In an exothermic reaction, the enthalpy change is negative. If a large negative value for $\Delta H^{\ominus}$ for is inserted in Equation 1, then the total entropy change is likely to be positive, and the reaction will be spontaneous. This can be expressed as:

$$\Delta H^{\ominus}_{reaction} > T\Delta S^{\ominus}_{reaction}$$

To summarise, the conditions for spontaneous reactions are favoured by:

- A large increase in entropy of the reaction:

$$T\Delta S^{\ominus}_{reaction} > \Delta H^{\ominus}_{reaction}$$

and/or

- A highly exothermic reaction with a large negative enthalpy change:

$$\Delta H^{\ominus}_{reaction} > T\Delta S^{\ominus}_{reaction}$$

The conditions for a spontaneous chemical reaction are summarised by the standard Gibbs energy change, $\Delta G^{\ominus}$. The Gibbs energy of reaction is defined as the difference between $\Delta H^{\ominus}_{reaction}$ and $T\Delta S^{\ominus}_{reaction}$, or:

$$\Delta G^{\ominus}_{reaction} = \Delta H^{\ominus}_{reaction} - T\Delta S^{\ominus}_{reaction}$$

The condition for a spontaneous reaction is that $\Delta G^{\ominus}_{reaction} < 0$, i.e. $\Delta G^{\ominus}$ is negative.

There are various ways of defining the standard Gibbs energy change. One way of thinking about the Gibbs energy change of a reaction is that it represents the useful energy remaining in a reaction after entropy changes have taken place. For this reason, it is sometimes simply called the *free energy* of a reaction. Its importance in chemistry lies in the fact that it can be used to predict whether chemical reactions will occur spontaneously, and it allows us to calculate the useful energy that will be produced in a chemical reaction.

To make comparisons of Gibbs energy changes for different reactions, standard conditions must be used for values of enthalpy and entropy.

Table 6.4 illustrates different combinations of entropy and enthalpy for chemical reactions and shows how the sign of the Gibbs energy change is related and therefore whether the reaction is expected to be spontaneous.

Table 6.4 Combinations of possible enthalpy and entropy changes for a reaction and resultant Gibbs energy change.

Enthalpy ΔH	Entropy ΔS	Sign of Gibbs energy $\Delta G = \Delta H - T\Delta S$	Spontaneous or non-spontaneous reaction
Exothermic ΔH negative (–ve)	Increases (+ve)	$(-T\Delta S)$ = –ve, so ΔG always –ve	Spontaneous
Endothermic ΔH positive (+ve)	Decreases (–ve)	$(-T\Delta S)$ = +ve so ΔG always +ve	Non-spontaneous

Exothermic reactions where the entropy decreases and endothermic reactions where the entropy increases can be either spontaneous or non-spontaneous. The Gibbs energy depends upon the relative magnitude of ΔH to $T\Delta S$. If ΔH is more negative than $T\Delta S$ the reaction will be spontaneous. If ΔH is more positive than $T\Delta S$ the reaction will be non-spontaneous.

Worked Example 6.18   Calculate the Gibbs free energy change at 298 K for the following reaction using the data in the table given:

$$CH_4(g) + N_2(g) \rightarrow HCN(g) + NH_3(g)$$

	$CH_4(g)$	$N_2(g)$	$HCN(g)$	$NH_3(g)$
$\Delta_f H^\ominus$ kJ mol^{-1}	–74.8	0	135	–46.1
$S^\ominus$/J K^{-1} mol^{-1}	186	191	202	192

Solution

To calculate the Gibbs energy change, we use the relationship:

$$\Delta G^\ominus = \Delta H^\ominus - T\Delta S^\ominus$$

This requires finding the enthalpy change, $\Delta H^\ominus$, and the entropy change, $\Delta S^\ominus$, for the reaction from the information given in the table.

The enthalpy change is given by:

$$\Delta H^\ominus_{reaction} = \Delta_f H^\ominus_{(products)} - \Delta_f H^\ominus_{(reactants)}$$

To find the overall enthalpy change for the reaction, insert the enthalpies of formation of the products (HCN and NH$_3$) and subtract those of the reactants (CH$_4$ and N$_2$):

$$\Delta_r H^\ominus = \Sigma\Delta_f H^\ominus_{(products)} - \Sigma\Delta_f H^\ominus_{(reactants)}$$

$$= \Sigma\Delta_f H^\ominus(HCN + NH_3) - \Sigma\Delta_f H^\ominus(CH_4 + N_2)$$

$$\Delta_r H^\ominus = (135 - 46.1)\,kJ\,mol^{-1} - (-74.8 + 0)\,kJ\,mol^{-1} = (88.9 + 74.8)\,kJ\,mol^{-1}$$

$$= 163.7\,kJ\,mol^{-1}$$

The product, N$_2$(g), is a pure element, and the enthalpy of formation of a pure element in its standard state is zero.

To find the overall entropy change in the reaction, we insert the absolute entropy values of the products at 298 K and subtract those of the reactants at 298 K:

$$\Delta S^{\ominus} = \Sigma S^{\ominus}_{(products)} - \Sigma S^{\ominus}_{(reactants)}$$

$$= (202 + 192) \text{ J K}^{-1}\text{mol}^{-1} - (186 + 192) \text{ J K}^{-1}\text{mol}^{-1}$$

$$= (394 - 378) \text{ J K}^{-1}\text{mol}^{-1}$$

$$= 16 \text{ J K}^{-1}\text{mol}^{-1} = 16 \times 10^{-3} \text{ kJ K}^{-1}\text{mol}^{-1} = 0.016 \text{ kJ K}^{-1}\text{mol}^{-1}$$

Because the enthalpy values are in kJ and entropy values in J, the entropy value must be multiplied by 10^{-3} to ensure that the units are equivalent.

We now have the two values required for the Gibbs equation. We are given the temperature, 298 K, so all the parameters can be inserted into the equation:

$$\Delta_r G^{\ominus} = \Delta_r H^{\ominus} - T\Delta S^{\ominus} = \left(163.7 \text{ kJ mol}^{-1}\right) - (298 \text{ K}) \times \left(0.016 \text{ kJ K}^{-1}\text{mol}^{-1}\right)$$

$$= 158.932 \text{ kJ mol}^{-1} = 159 \text{ kJ mol}^{-1}$$

The answer is given to three significant figures, and rounding is carried out in the final step.

The importance of Gibbs free energy and work done in a reaction

The previous sections explain the importance of the Gibbs energy in determining whether a reaction will be spontaneous at a certain temperature. However, the value of the Gibbs energy is an important property of the reaction itself. We have seen that the expression for Gibbs energy is: $\Delta G = \Delta H - T\Delta S$, which shows the Gibbs energy is equal to the enthalpy of the reaction less $T\Delta S$. The term $T\Delta S$ also has units of energy, and one way of thinking about the Gibbs energy is that it is the energy left over after entropy changes have taken place in the reaction. The Gibbs energy is the energy in a chemical reaction that is available to do useful work. When any chemical reaction occurs, such as the combustion of a fuel or electrolysis in an electrochemical cell, a certain amount of work is done in redistributing the energy. The remaining energy, known as the Gibbs energy, is available to do useful work, such as moving a piston, turning an engine, or driving charge as current.

Gibbs energy and temperature changes

Standard conditions assume reactants and products are in their standard states.

Standard conditions assume that chemical reactions are carried out at 1 bar (1.00×10^5 Pa) pressure and a specific temperature, usually 298 K. The equation that allows us to calculate the Gibbs energy, and hence determine if a reaction is spontaneous, includes temperature as a variable. For the purposes of the calculations in this book, we will assume there is no variation in ΔH and ΔS with temperature.

The equation for Gibbs energy depends upon the temperature at which the reaction is carried out: $\Delta G = \Delta H - T\Delta S$.

For the reaction to be spontaneous, the value of ΔG must be negative. Therefore the temperature of the reaction is one of the factors that determines spontaneity.

The temperature at which the reaction will become spontaneous, at fixed values of enthalpy change and entropy change, can be predicted. ΔG changes sign when:

$$T \Delta S = \Delta H$$

$$T = \frac{\Delta H}{\Delta S}$$

The reaction will therefore be spontaneous at any temperatures above the value calculated for $\frac{\Delta H}{\Delta S}$.

Worked Example 6.19 Hydrogen can be produced by heating methane in steam in the presence of a nickel catalyst at a temperature of around 1000 °C. Calculate the ΔG value for the reaction, and state whether the reaction is feasible at this temperature. Determine the minimum temperature at which the reaction is feasible.

	$CH_4(g)$	$H_2O(g)$	$CO(g)$	$H_2(g)$
$\Delta_f H^{\ominus}/kJ\ mol^{-1}$	–74.9	–242	–111	0
$S^{\ominus}/J\ K^{-1}\ mol^{-1}$	186	189	198	131

Solution
First write a balanced chemical equation for the reaction, with state symbols:

$$CH_4(g) + H_2O(g) \rightarrow CO(g) + H_2(g)$$

The equation that allows us to calculate values of the Gibbs energy is: $\Delta G^{\ominus} = \Delta H^{\ominus} - T\Delta S^{\ominus}$.

To use the equation, we need to determine the enthalpy change during reaction and the entropy change during reaction.

The enthalpy change is given by:

$$\Delta H^{\ominus}_{reaction} = \Sigma \Delta_f H^{\ominus}_{(products)} - \Sigma \Delta_f H^{\ominus}_{(reactants)}$$

Inserting the values from the table:

$$\Delta H_{reaction} = (-111 + 0) - (-74.9 - 242)\ kJ\ mol^{-1} = +205\ kJ\ mol^{-1}$$

The entropy change is given by:

$$\Delta S^{\ominus} = \Sigma S^{\ominus}_{(products)} - \Sigma S^{\ominus}_{(reactants)}$$

Inserting the values from the table:

$$\Delta S = (198 + 3 \times 131) - (186 + 189)\ J\ K^{-1}\ mol^{-1} = 216\ J\ K^{-1}\ mol^{-1}$$

ΔG can be found by inserting the values in the equation for the Gibbs energy, remembering to convert the units of entropy to $kJ\ K^{-1}\ mol^{-1}$. The temperature is given as 1000 °C and so must be converted to K by adding 273 to give a value for T of 1273 K:

$$\Delta G = \Delta H - T\Delta S$$

$$= +205\ kJ\ mol^{-1} - 1273\ K \times 0.216\ kJ\ K^{-1}\ mol^{-1}$$

$$= -69.97 = -70.0\ kJ\ mol^{-1}$$

The value of ΔG is negative, which suggests the reaction is feasible at this temperature.

To find the minimum temperature at which the reaction is feasible, we set ΔG to be equal to zero:

$$0 = \Delta H - T\Delta S$$

Therefore, $T\Delta S = \Delta H$.

Note that as the reaction is carried out at 1000 °C, we can only calculate ΔG at this temperature and not $\Delta G^{\ominus}$. ΔH values and S values vary with temperature, but we will assume that they are close to the values of $\Delta G^{\ominus}$ and $S^{\ominus}$ given in the table.

$$T = \frac{\Delta H}{\Delta S} = \frac{205 \text{ kJ mol}^{-1}}{0.216 \text{ kJ K}^{-1}\text{mol}^{-1}} = 949 \text{ K}.$$

To convert the temperature to Celsius, subtract 273:

$$T = 949 - 273 = 676°C$$

This is the temperature below which the value of ΔG will be positive, and the reaction no longer spontaneous.

Quick-check summary

- Energy is transferred in a chemical reaction to or from the surroundings.

- If heat energy is lost from the reaction, it is said to be exothermic, and the products have less energy than the reactants. If heat energy is gained by the reaction, it is said to be endothermic, and the products have more energy than the reactants.

- The energy lost or gained in a chemical reaction when the pressure remains constant is called the enthalpy change and has the symbol $\Delta_r H$.

- The enthalpy changes in a chemical reaction can be represented in an enthalpy profile diagram.

- If the reaction is exothermic, the enthalpy change is negative; and if the reaction in endothermic, the enthalpy change is positive.

- To be able to compare enthalpy changes, values are quoted under standard conditions of pressure (10^5 Pa) and usually measured at 25 °C (298 K). The symbol used in this case is $\Delta H^{\ominus}$.

- There are several different types of enthalpy changes, depending upon the type of reaction. The enthalpy changes you should be familiar with are:

  ○ Standard enthalpy of formation, $\Delta_f H^{\ominus}$

  ○ Standard enthalpy of combustion, $\Delta_c H^{\ominus}$

  ○ Standard enthalpy of atomisation, $\Delta_{at} H^{\ominus}$

  ○ Ionisation energy, IE

  ○ Electron affinity, $\Delta_{EA} H^{\ominus}$

- Heat energy changes can be measured in a calorimeter by using the relationship: $q = m \times c \times \Delta T$

- The first law of thermodynamics states that energy can be neither created nor destroyed in a chemical reaction. Hess's law is another way of expressing this. Hess's law states that the enthalpy change on going from one set of reactants to the same set of products in the same physical states is independent of the route by which the reaction takes place.

- When substances react, the bonds between the atoms must be broken. This process is endothermic.

- When products are formed, new bonds are made, and this process is exothermic.

- The bond energy, $E(X-X)$, is a measure of the energy required to break a specific type of chemical bond, i.e. $X-X$.

- The Born–Haber cycle is an application of Hess's law to the formation of an ionic lattice and allows us to calculate the lattice enthalpy for an ionic solid.

- Entropy (S) is a measure of disorder or randomness in a system. It is related to the distribution of energy in a system. The greater the amount of energy and the larger the number of species, the greater the entropy.

- When a system becomes more disordered, it becomes more energetically stable.

- The second law of thermodynamics states that the total entropy of the universe increases in a spontaneous process.

- The total entropy change in a reaction is given by:
$\Delta S^{\ominus}_{total} = \Delta S^{\ominus}_{system} + \Delta S^{\ominus}_{surroundings}$.

- The entropy change in a system is given by:
$\Delta S^{\ominus}_{system} = S^{\ominus}_{products} - S^{\ominus}_{reactants}$.

- The entropy change of the surroundings is given by:
$\Delta S^{\ominus}_{surroundings} = -\Delta H^{\ominus}/T$

- The Gibbs free energy change for a reaction is related to the enthalpy change by the equation $\Delta G^{\ominus} = \Delta H^{\ominus} - T\Delta S^{\ominus}$. A spontaneous reaction is one that is likely to happen and has a negative value of ΔG.

End-of-chapter questions

1 Calculate the standard enthalpy of hydrogenation of cyclohexene ($C_6H_{10}(l)$) to cyclohexane(l) ($C_6H_{12}(l)$) given that the standard enthalpies of combustion of the two compounds are -3752 kJ mol^{-1} (cyclohexene) and -3953 kJ mol^{-1} (cyclohexane) and the standard enthalpy of combustion of hydrogen is -286 kJ mol^{-1}.

2 Copper(I) oxide and copper(II) oxide can both be used in the ceramics industry to give blue, green, or red tints to glasses and enamels. The table lists some values for enthalpies of formation of some copper compounds.

Compound	$\Delta_f H^{\ominus}$/kJ mol^{-1}
$Cu_2O(s)$	-168.6
$CuO(s)$	-157.3
$Cu(NO_3)_2(s)$	-302.9
$NO_2(g)$	$+33.2$

a. Copper(II) oxide can be produced by heating copper(II) nitrate. Use suitable values from the table to calculate $\Delta_r H^{\ominus}$ for the following reaction:

$$Cu(NO_3)_2(s) \rightarrow CuO(s) + 2NO_2(g) + \tfrac{1}{2}O_2(g)$$

b. Copper(I) oxide can be produced by heating copper(II) oxide. Use suitable values from the table to calculate $\Delta_r H^{\ominus}$ for this reaction:

$$2CuO(s) \rightarrow Cu_2O(s) + \tfrac{1}{2}O_2(g)$$

3   Ethane burns in the presence of excess oxygen as follows:

$$C_2H_6(g) + \tfrac{7}{2}O_2(g) \rightarrow 2CO_2(g) + 3H_2O(l)$$

Given the mean bond enthalpies in the table, calculate the enthalpy change for the reaction.

Bond	Bond enthalpy (kJ mol^{-1})	Bond	Bond enthalpy (kJ mol^{-1})
C—C	347	C—O	360
C=O	740	O—H	460
C—H	412	O=O	496

4   In an experiment to determine the enthalpy of combustion of pentane, C_5H_{12}, using a flame calorimeter, a spirit burner was half-filled with pentane, the cap put in place, and its mass recorded as 60.7 g. The calorimeter was filled with 500 cm^3 of water and the initial temperature measured as 20.2 °C. After lighting the flame and allowing the pentane to burn for several minutes, the temperature of the water was found to have increased to 28.0 °C. At the end of the experiment, the mass of the spirit burner and cap was found to be 57.1 g. Calculate the enthalpy of combustion of pentane and its fuel value in kJ g^{-1}.
The heat capacity of the calorimeter is 22.1 kJ °C^{-1}.
(Specific heat capacity of water = 4.18 J g^{-1} °C^{-1}; density of water = 1 g cm^{-3}.)

5   Give the correct names and thermodynamic symbols for the enthalpy changes represented by the following equations:

a. $Br_2(l) \rightarrow 2Br(g)$

b. $\tfrac{1}{2}Br_2(l) \rightarrow Br(g)$

c. $Br(g) + e^- \rightarrow Br^-(g)$

d. $\tfrac{1}{2}Br_2(l) + K(s) \rightarrow KBr(s)$

e. $K^+(g) + Br^-(g) \rightarrow KBr(s)$

f. $K(s) \rightarrow K(g)$

6   In an experiment to determine the standard enthalpy of combustion of propan-1-ol, a 0.60 g sample of the liquid was completely burnt in a calorimeter containing 0.5 dm^3 of water at 25.0 °C. The temperature of the calorimeter and its contents increased to 33.4 °C. The heat capacity of the calorimeter was determined independently to be $C_{cal} = 0.30$ kJ °C^{-1}. Calculate the standard molar enthalpy of combustion of propan-1-ol.
(Heat capacity of water = 4.18 J K^{-1} g^{-1}.)

7   Rocket engines can use methylhydrazine, CH_3NHNH_2, and dinitrogen tetroxide, N_2O_4, as a fuel. Given the following standard heats of formation, calculate the enthalpy change for the reaction:

$$4CH_3NHNH_2(l) + 5N_2O_4(l) \rightarrow 4CO_2(g) + 12H_2O(l) + 9N_2(g)$$

$$\Delta_f H^\ominus (CH_3NHNH_2\ (l)) = +53.0\,kJ\,mol^{-1}$$

$$\Delta_f H^\ominus (CO_2\ (g)) = -393\,kJ\,mol^{-1}$$

$$\Delta_f H^\ominus (N_2O_4\ (l)) = 20.0\,kJ\,mol^{-1} \quad \Delta_f H^\ominus (H_2O\ (l)) = -286\,kJ\,mol^{-1}$$

8   Use the data in the following table to calculate the lattice enthalpy for CsBr by constructing a suitable Born–Haber cycle.

Quantity	Value / $kJ\,mol^{-1}$
Enthalpy of atomisation of caesium	+78
First ionisation energy for Cs	+375.5
Enthalpy of atomisation of Br	+112
First electron affinity for Br	−324
Enthalpy of formation of CsBr	−405.9

9   Comment on the direction of entropy change, ΔS, in the following transformations:

a. Evaporation of water from a boiling kettle

b. $2H_2(g) + O_2(g) \rightarrow 2H_2O(g)$

c. Crystallisation of solid sodium chloride from molten salt

d. $2NaNO_3(s) \rightarrow 2NaNO_2(s) + O_2(g)$

10 Thermochemical data for the decomposition of zinc carbonate to zinc oxide and carbon dioxide is as follows:

$$ZnCO_3(s) \rightarrow ZnO(s) + CO_2(g) \quad \Delta H^\ominus = +71\,kJ\,mol^{-1}$$
$$S^\ominus(ZnO(s)) = +43.6\,J\,mol^{-1}\,K^{-1}$$
$$S^\ominus(ZnCO_3(s)) = +82.4\,J\,mol^{-1}\,K^{-1}$$
$$S^\ominus(CO_2(g)) = +213\,J\,mol^{-1}\,K^{-1}$$

a. Calculate the entropy change for the system, $\Delta S^\ominus_{reaction}$, for the reaction. Include the sign and units in your answer.

b. Calculate the entropy change for the surroundings, $\Delta S^\ominus_{surroundings}$, at 298 K.

c. Calculate the total entropy change for the reaction, $\Delta S^\ominus_{total}$, at 298 K.

d. State whether the reaction is likely to occur.

11 Carbon monoxide can be used as a reducing agent to obtain metal ores from their oxides. The reaction between zinc oxide and carbon monoxide can be represented by:

$$ZnO(s) + CO(g) \rightarrow Zn(s) + CO_2(g)$$

Using the thermodynamic data in the table, determine the following quantities (a) to (d) at 1000 K, assuming there is no change in entropy values at this temperature.

	ZnO(s)	CO(g)	Zn(s)	CO$_2$(g)
$\Delta_f H^{\ominus}$ (kJ mol^{-1})	–348	–111	0	–394
$S^{\ominus}$ (J K^{-1} mol^{-1})	43.9	198	41.6	214

a. The enthalpy change for the reaction, $\Delta_r H$

b. The entropy change of the reaction, $\Delta S_{reaction}$

c. The entropy change of the surroundings, $\Delta S_{surroundings}$

d. The total change in entropy, ΔS_{total}.

12 The decomposition of calcium carbonate, $CaCO_3$, is a process that takes place in a blast furnace in the reduction of iron ore. Use the data in the following table to calculate the standard enthalpy change and standard entropy change for the decomposition of calcium carbonate at 298 K. Using these values, determine a value for the decomposition temperature of calcium carbonate.

	CaCO$_3$	CaO	CO$_2$
$\Delta_f H^{\ominus}$ (kJ mol^{-1})	–1207	–635	–394
$S^{\ominus}$ (J K^{-1} mol^{-1})	92.9	40	214

7

Chemical equilibrium and acid-base equilibrium

At the end of this chapter, students should be able to:

- Understand the characteristics of a chemical equilibrium

- Write expressions for the following equilibrium constants given appropriate chemical equations: K_c, K_p, K_a, K_w

- Carry out calculations involving the expressions for the equilibrium constants K_c, K_p, K_a, and K_w to obtain values for the equilibrium constant and/or amounts of reactants and products present at equilibrium, and determine appropriate units for such values

- Use Le Chatelier's principle to predict the change in equilibrium position and equilibrium constant when a reaction at equilibrium is subjected to a change in concentration or partial pressure of one of the reactants, or to a change in reaction conditions

- Understand the difference between homogeneous and heterogeneous reactions

- Define an acid and a base according to the Brønsted–Lowry theory

- Distinguish between strong and weak acids and bases

- Relate conjugate acid/base pairs

- Understand the pH scale and its use in determining the acidity or alkalinity of a solution

- Explain the use of indicators in acid-base titrations and be able to select suitable indicators depending upon the acid-base combination used and the relevant pH titration curve

Foundations of Chemistry: An Introductory Course for Science Students, First Edition.
Philippa B. Cranwell and Elizabeth M. Page.
© 2021 John Wiley & Sons Ltd. Published 2021 by John Wiley & Sons Ltd.
Companion website: www.wiley.com/go/Cranwell/Foundations

- Define a chemical buffer, and explain how acidic and basic buffers are made and used to maintain constant pH values of substances

- Calculate the pH of a buffer solution

7.1   Introduction

In many chemical reactions, the reactants are totally converted to products. Such reactions are said to *go to completion*. For example, if magnesium wire is burnt in air, the metal is oxidised to a mixture of magnesium oxide and magnesium nitride. The reaction is said to be complete when all the metal has been oxidised. The products formed are not easily converted back to metallic magnesium, oxygen, and nitrogen.

However, many chemical reactions don't go to completion and the reactants are never all converted to products. In fact, in these reactions, once a certain amount of product is formed it is converted back to the original reactants. This type of reaction is called a **reversible reaction**. When the rate of the forward reaction is equal to the rate of the backward reaction, the reaction is said to be at **equilibrium**. In this chapter we will consider what is meant by equilibrium reactions and how they can be tailored to give useful products.

7.2   Equilibrium and reversible reactions

A reversible reaction is a reaction that does not go to completion. Reversible reactions may be either physical or chemical processes. For example, evaporation and condensation are reversible physical processes. If perfume is stored in a stoppered bottle at room temperature some of the liquid evaporates. Eventually the rate at which the perfume evaporates is the same as the rate at which the perfume vapour condenses back to the liquid phase. This is an example of a physical process that is at equilibrium.

Chemical processes that reach equilibrium are ones in which the products react back together to give the original reactants. Consider the reaction between gaseous hydrogen and gaseous iodine in a sealed glass tube at 400 °C. Hydrogen and iodine react together to give gaseous hydrogen iodide, HI:

$$H_2(g) + I_2(g) \rightleftharpoons 2HI(g)$$

At the start of the reaction the hydrogen and iodine react together quickly to give hydrogen iodide. However, as soon as hydrogen iodide is formed, it begins to decompose to give back hydrogen and iodine gas. The symbol $\rightleftharpoons$ means that the reaction proceeds in both directions and does not go to completion but an equilibrium is established. At some point, the concentrations of hydrogen, iodine, and hydrogen iodide remain constant. When this happens the rate of the forward reaction to form hydrogen iodide is the same as the rate of the backward reaction in which hydrogen iodide decomposes. The equilibrium is said to be a **dynamic equilibrium** as the forward and backward reactions still take place, but the concentrations of reactants and products remain constant. The rates of the forward and backward reactions in such a reversible reaction are plotted against time of reaction and are shown in Figure 7.1.

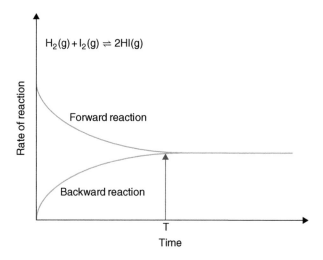

Figure 7.1 The rates of the forward and backward reactions are equal once the reaction reaches equilibrium at time, T.

7.2.1 Characteristics of an equilibrium

- A reversible reaction is one that can proceed spontaneously in either direction.

- A chemical equilibrium is dynamic and is established when the rates of the forward and backward reactions are the same.

- The concentrations of reactants and products in a reversible reaction remain constant when the reaction has reached equilibrium.

- A chemical reaction can only reach equilibrium if none of the reactants or products can escape from the reaction mixture. To reach equilibrium a reaction involving gases should be carried out in a **closed system** in which matter cannot escape to the surroundings. A reaction involving liquids can be carried out in an open container provided none of the reactants or products can escape.

- A reversible reaction that can reach equilibrium is denoted by the symbol $\rightleftharpoons$.

A spontaneous process is one that occurs naturally without the addition of external energy. For a reminder about spontaneous processes see Chapter 6.

7.2.2 The equilibrium mixture and the equilibrium constant, K_c

Consider a hypothetical reaction in which reactants A and B are converted to products C and D:

$$A + B \rightleftharpoons C + D$$

The reaction is said to reach equilibrium when the concentrations (or pressures, in the case of gases) of A, B, C, and D are constant and the rate of formation of C and D from A and B is the same as the rate of formation of A and B from C and D.

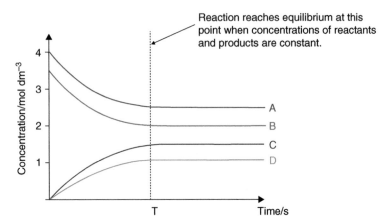

Figure 7.2 Plot of concentrations of reactants (A and B) and products (C and D) against time in reaction aA + bB ⇌ cC + dD. The reaction reaches equilibrium at time, T.

Consider the same reaction where the stoichiometric amounts of A, B, C, and D in the equation are a, b, c, and d, respectively:

$$aA + bB \rightleftharpoons cC + dD$$

At the beginning of the reaction, there is a mixture of A and B only as C and D have not yet been formed. As time proceeds the amount of C and D increases, whereas the amount of A and B decreases, as shown in Figure 7.2. When the reaction reaches equilibrium, the concentrations of A, B, C, and D remain constant.

The ratio of the concentrations of reactants to products at equilibrium gives chemists a useful way of predicting how far the reaction will proceed and therefore how much product will be formed. This ratio is called the *equilibrium constant* and has the symbol K_c. For this hypothetical reaction, the equilibrium constant is given by the general equilibrium expression:

$$K_c = \frac{[C]^c \, [D]^d}{[A]^a \, [B]^b}$$

where:

K_c is the equilibrium constant in terms of concentrations of reactants and products.

$[A]^a$ is the equilibrium concentration of A (usually in mol dm^{-3}) raised to the power 'a', where 'a' is the stoichiometric coefficient or number of moles of A in the equation for the reaction. The terms $[B]^b$, $[C]^c$ and $[D]^d$ are defined in the same way.

The expression for the equilibrium constant indicates that the value of K_c depends upon the ratio of the concentrations of C and D to A and B in the equilibrium mixture. A large value of K_c indicates that the reaction lies over to the right-hand side of the equation towards the products. The larger the value of K_c, the more efficient the reaction is at producing products. The value of K_c is constant for the reaction in the direction written, and does not depend upon the starting concentrations of reactants or products. The value of K_c varies only with temperature.

Let's consider one of the equilibrium reactions we have met already – the reaction of hydrogen and iodine to give hydrogen iodide. The equation for the reaction is:

$$H_2(g) + I_2(g) \rightleftharpoons 2HI(g)$$

K_c is the equilibrium constant and is obtained when concentration terms are used in the expression. The general symbol for an equilibrium constant is K_{eq}.

The equilibrium expression is given by:

$$K_c = \frac{[HI]^2}{[H_2][I_2]}$$

The value of the equilibrium constant at 490 °C (763 K) is 45.6. This value is obtained by mixing different initial concentrations of hydrogen, iodine, and hydrogen iodide at a temperature of 490 °C and allowing the gases to react until they reach equilibrium. When at equilibrium, the concentrations of each gas are measured.

The equilibrium constant for the reaction is determined by first writing the equilibrium expression: $K_c = \frac{[HI]^2}{[H_2][I_2]}$.

For a particular reaction carried out at a temperature of 490 °C, the equilibrium concentrations of the gases were found to be [HI] = 1.54×10^{-3} mol dm^{-3} and, $[H_2]$ = $[I_2]$ = 0.228×10^{-3} mol dm^{-3}. These values are inserted into the expression and a value for K_c calculated. An example calculation is given here using the results from one experiment:

$$K_c = \frac{[HI]^2}{[H_2][I_2]} = \frac{\left(1.54 \times 10^{-3} \text{ mol dm}^{-3}\right)^2}{\left(0.228 \times 10^{-3} \text{mol dm}^{-3}\right)\left(0.228 \times 10^{-3}\text{mol dm}^{-3}\right)} = 45.6$$

When the results from other experiments conducted at the same temperature are calculated, the value of K_c is always found to be the same and does not depend upon the initial concentrations of hydrogen, iodine, or hydrogen iodide. Note that the value of K_c for this equilibrium expression has no units as the units on the top and bottom of the equation cancel.

The reaction is reversible so an equilibrium constant can be written for the reverse reaction:

$$2HI(g) \rightleftharpoons H_2(g) + I_2(g)$$

The equilibrium expression for the reverse reaction is given by: $K_c' = \frac{[H_2][I_2]}{[HI]^2}$.

K_c' is the equilibrium constant for the reaction in the reverse direction.

Mathematically, this expression is the inverse of the expression for K_c:

$$K_c' = \frac{1}{K_c} = \frac{1}{45.6} = 0.022.$$

The results from the experiment can also be used to calculate the value of K_c':

$$K_c' = \frac{\left(0.228 \times 10^{-3}\text{mol dm}^{-3}\right)\left(0.228 \times 10^{-3}\text{mol dm}^{-3}\right)}{\left(1.54 \times 10^{-3} \text{ mol dm}^{-3}\right)^2} = 0.022$$

Note that the equilibrium constant for the reverse reaction also has no units.

Worked Example 7.1   Write the equilibrium constant expressions for the following equilibrium reactions, and give the unit of the constant, K_c, in each case, if the concentrations are measured in mol dm^{-3}.

  a. $N_2(g) + 3H_2(g) \rightleftharpoons 2NH_3(g)$

  b. $2SO_2(g) + O_2(g) \rightleftharpoons 2SO_3(g)$

  c. $N_2O_4(g) \rightleftharpoons 2NO_2(g)$

Solution

  a. $K_c = \dfrac{[NH_3]^2}{[N_2][H_2]^3}$

  Units: $K_c = \dfrac{[\text{mol dm}^{-3}]^2}{[\text{mol dm}^{-3}][\text{mol dm}^{-3}]^3} = \dfrac{[\text{mol dm}^{-3}]^2}{[\text{mol dm}^{-3}]^4} = \dfrac{1}{[\text{mol dm}^{-3}]^2}$

  $= \text{mol}^{-2}\, \text{dm}^6$

  b. $K_c = \dfrac{[SO_3]^2}{[SO_2]^2[O_2]}$

  Units: $K_c = \dfrac{[\text{mol dm}^{-3}]^2}{[\text{mol dm}^{-3}]^2[\text{mol dm}^{-3}]} = \dfrac{[\text{mol dm}^{-3}]^2}{[\text{mol dm}^{-3}]^3} = \dfrac{1}{[\text{mol dm}^{-3}]}$

  $= \text{mol}^{-1}\, \text{dm}^3$

  c. $K_c = \dfrac{[NO_2]^2}{[N_2O_4]}$

  Units: $K_c = \dfrac{[\text{mol dm}^{-3}]^2}{[\text{mol dm}^{-3}]} = \text{mol dm}^{-3}$

Worked Example 7.2   The reaction between gaseous sulfur dioxide and oxygen reaches an equilibrium with the formation of gaseous sulfur trioxide, which is an essential feedstock for the production of sulfuric acid. The equation for the reaction is:

$$2SO_2(g) + O_2(g) \rightleftharpoons 2SO_3(g)$$

A mixture of sulfur dioxide and oxygen was heated at 800 K until equilibrium was reached. The concentrations of each gas at equilibrium were:

$$SO_2 = 2.5 \times 10^{-3}\ \text{mol dm}^{-3}; O_2 = 3.0 \times 10^{-3}\ \text{mol dm}^{-3};$$
$$SO_3 = 3.8 \times 10^{-2}\ \text{mol dm}^{-3}$$

  a. Calculate the equilibrium constant for the reaction at this temperature.

  b. Write the equation for the reverse reaction and give an expression for the equilibrium constant K_c' in this direction. Calculate the value of K_c'.

  c. Calculate the equilibrium constant for the following reaction:

$$SO_2(g) + \tfrac{1}{2}O_2(g) \rightleftharpoons SO_3(g)$$

Solution

a. **Step 1:** Write the equilibrium constant expression using the balanced equation for the reaction given: $K_c = \dfrac{[SO_3]^2}{[SO_2]^2[O_2]}$.

Step 2: Identify the equilibrium concentrations of the three gases and insert them into the expression for the equilibrium constant:

$$K_c = \frac{\left(3.8 \times 10^{-2} \text{mol dm}^{-3}\right)^2}{\left(2.5 \times 10^{-3} \text{mol dm}^{-3}\right)^2 \left(3.0 \times 10^{-3} \text{mol dm}^{-3}\right)}$$

$$= \frac{14.44 \times 10^{-4} \text{mol}^2 \, \text{dm}^{-6}}{18.75 \times 10^{-9} \text{mol}^3 \, \text{dm}^{-9}} = 7.7 \times 10^4 \, \text{mol}^{-1} \, \text{dm}^3$$

b. The equation for the reverse reaction is $2SO_3(g) \rightleftharpoons 2SO_2(g) + O_2(g)$.

The equilibrium expression for the reverse reaction is $K_c' = \dfrac{[SO_2]^2[O_2]}{[SO_3]^2}$.

The equilibrium constant for the reverse reaction, K_c', is given by:

$$K_c' = \frac{1}{K_c} = \frac{1}{7.7 \times 10^4 \, \text{mol}^{-1} \, \text{dm}^3} = 1.3 \times 10^{-5} \, \text{mol dm}^{-3}.$$

c. The equilibrium expression for the reaction $SO_2(g) + \frac{1}{2}O_2(g) \rightleftharpoons SO_3(g)$

is given by: $K_c'' = \dfrac{[SO_3]}{[SO_2][O_2]^{\frac{1}{2}}}$.

This expression for $K_c'' = \sqrt{\dfrac{[SO_3]^2}{[SO_2]^2[O_2]}}$. So $K_c'' = \sqrt{K_c}$.

$K_c'' = \sqrt{7.7 \times 10^4 \, \text{mol}^{-1} \, \text{dm}^3} = 2.8 \times 10^2 \, \text{mol}^{-\frac{1}{2}} \, \text{dm}^{\frac{3}{2}}.$

In some examples of equilibrium calculations the equilibrium concentrations are not specifically stated but must be worked out from the balanced equation for the reaction and other information. Worked Example 7.3 explains how to go about this. It is always safer to draw up an equilibrium table to track concentrations.

Worked Example 7.3 Nitrosyl chloride, NOCl, undergoes thermal decomposition to yield nitrous oxide (NO) and chlorine gas. An initial sample of 1.00 mole of NOCl was heated to 500 K in a 2 dm^3 flask and an equilibrium established. Analysis of the equilibrium mixture showed it contained 0.054 moles of chlorine. Determine the equilibrium constant for the reaction at this temperature. The equation for the decomposition reaction is:

$$2NOCl(g) \rightleftharpoons 2NO(g) + Cl_2(g)$$

Solution

The key to equilibrium calculations of this type lies in the balanced equation for the reaction:

$$2NOCl(g) \rightleftharpoons 2NO(g) + Cl_2(g)$$

This shows that two moles of NOCl decompose to give two moles of NO and one mole of Cl_2. We need to find the equilibrium concentrations of these gases in order to insert them into the expression for the equilibrium constant.

Step 1: Start by writing the expression for the equilibrium constant, K_c:

$$K_c = \frac{[NO]^2[Cl_2]}{[NOCl]^2}$$

Step 2: Next, set up the equilibrium table. An equilibrium table sets out the known information about the reaction and its chemical equation. The rows in the table can be labelled: (i) **Ratio of number of moles** from the balanced equation for the reaction, (ii) **Initial number of moles** from the experimental data, (iii) **Change in number of moles**, and (iv) **Equilibrium number of moles** (**RICE**). Let the number of moles of Cl_2 gas formed at equilibrium be equal to x.

Row		NOCl	NO	Cl_2
1	Ratio of number of moles from equation	2	2	1
2	Initial number of moles/mol	1	0	0
3	Change in number of moles/mol	–2x	+2x	+x
4	Equilibrium number of moles/mol	1 – 2x	2x	x
5	Numerical values of mol at equilibrium/mol	0.892	0.108	0.054
6	Concentration at equilibrium/ mol dm^{-3}	$\frac{0.892}{2} = 0.446$	$\frac{0.108}{2} = 0.054$	$\frac{0.054}{2} = 0.0127$

Row 1 gives the ratio of the number of moles from the balanced equation. These are the same as the stoichiometric coefficients.

Row 2 gives the initial number of moles stated in the question. We are told there is one mole of NOCl at the start.

Row 3 gives the change in number of moles. We assume that x moles of Cl_2 are formed at equilibrium. From row 1 the number of moles of NOCl that dissociate is 2x, and the number of moles of NO formed is also 2x.

Row 4 gives the number of moles of each gas at equilibrium in terms of x. To calculate this we add the change in number of moles of each gas to the initial number of moles.

Row 5 gives the actual numerical value for the number of moles of each gas at equilibrium which is the information required for the equilibrium expression. These values are obtained using the information that the number of moles of Cl_2 at equilibrium is 0.054. This is equal to x, so 2x = 0.108 moles. The number of moles of NO at equilibrium is 0.108.

The number of moles of NOCl at equilibrium is equal to (1 – 2x) = (1 – 0.108) = 0.892.

Row 6 gives the values of concentrations that are required for the calculation. As the experiment is conducted in a vessel of 2 dm^3 volume, the concentration of each gas is obtained by dividing the number of moles by 2 dm^3.

Thus $[NOCl] = \dfrac{0.892}{2} = 0.446 \text{ mol dm}^{-3}$.

$$[NO] = \dfrac{0.108}{2} = 0.054 \text{ mol dm}^{-3}$$

$$[Cl_2] = \dfrac{0.054}{2} = 0.0127 \text{ mol dm}^{-3}$$

Step 3: Insert the values of the equilibrium concentrations into the equilibrium expression to obtain the value of K_c:

$$K_c = \dfrac{[NO]^2[Cl_2]}{[NOCl]^2} = \dfrac{\left(0.054 \text{ mol dm}^{-3}\right)^2\left(0.0127 \text{ mol dm}^{-3}\right)}{\left(0.446 \text{ mol dm}^{-3}\right)^2}$$

$$= 1.86 \times 10^{-4} \text{ mol dm}^{-3} = 2 \times 10^{-4} \text{ mol dm}^{-3} \text{ (to 1 significant figure)}$$

An important use of equilibrium constants is to enable the determination of concentrations of products in the equilibrium mixture. If the equilibrium constant for a reaction at a specific temperature is known, along with the concentrations of one or more of the reactants at equilibrium, the concentration of the other species at equilibrium can be determined.

Calculations such as this are straightforward if you follow the procedure in the following example.

Worked Example 7.4 n-butane is converted to *iso*-butane in a reaction that reaches equilibrium and can be represented by the equation below. The equilibrium constant for the reaction is 2.50 at 300 K. If the starting concentration of n-butane is 1.00 mol dm^{-3} calculate the equilibrium concentration of each gas at this temperature.

n-butane   iso-butane

Solution

The equation for the reaction is n-butane(g) $\rightleftharpoons$ *iso*-butane(g) in which one reactant, n-butane, is converted directly to the product, *iso*-butane.

Step 1: Write an expression for the equilibrium constant for the reaction:

$$K_c = \dfrac{[iso\text{-butane}]}{[n\text{-butane}]}$$

Step 2: Draw up an equilibrium table and insert information given in the question. Note that we are given concentrations here rather than number of moles of reactant and product. Let the decrease in concentration of n-butane equal x mol dm^{-3} at equilibrium.

	n-butane	iso-butane
Ratio of moles	1	1
Initial concentration/mol dm^{-3}	1.00	0
Change in concentration/mol dm^{-3}	−x	+x
Equilibrium concentration/mol dm^{-3}	1.00 − x	x

Step 3: Insert the equilibrium concentrations in terms of x into the expression for the equilibrium constant, K_c. This can be equated to the value of K_c of 2.50 given in the question:

$$K_c = \frac{x}{(1.00 - x)} = 2.50$$

Multiply out the expression to obtain an equation for x.

$$x = 2.50(1.00 - x)$$

$$x = 2.50 - 2.50x$$

$$3.50x = 2.50$$

$$x = 0.714$$

The equilibrium concentration of *iso*-butane is $0.714 \, \text{mol dm}^{-3}$.
The equilibrium concentration of *n*-butane is:
$(1.00 - x) = (1.00 - 0.714) = 0.286 \, \text{mol dm}^{-3} = 0.29 \, \text{mol dm}^{-3}$.

Worked Example 7.5   4.00 moles of hydrogen and 4.00 moles of iodine were mixed in a $2.0 \, \text{dm}^3$ vessel at 700 K until equilibrium was reached. If K_c at this temperature is 54, calculate the equilibrium concentrations of each gas in the mixture. The equation for the reaction is:

$$H_2(g) + I_2(g) \rightleftharpoons 2HI(g)$$

Solution
Step 1: Write the expression for the equilibrium constant, K_c:

$$K_c = \frac{[HI]^2}{[H_2][I_2]}$$

Step 2: Draw up the equilibrium table to include the information given in the question.

		H_2	I_2	HI
1	Ratio of moles	1	1	2
2	Initial no. moles/mol	4.00	4.00	0
3	Change in no. moles/mol	−x	−x	+2x
4	Equilibrium no. moles/mol	(4.00 − x)	(4.00 − x)	2x
5	Equilibrium concentration/mol dm^{-3}	$\dfrac{(4.00 - x)}{2}$	$\dfrac{(4.00 - x)}{2}$	$\dfrac{2x}{2} = x$

In row 1 determine the mole ratio of each gas from the chemical equation.

In row 2 insert the information about the initial number of moles of each gas.

In row 3 determine the change in number of moles of each gas at equilibrium by assuming that x moles of H_2 and I_2 react, so 2x moles of HI are formed.

In row 4 give the equilibrium number of moles of each gas using the initial number of moles and the change.

In row 5 give the equilibrium concentration of each gas in terms of x by dividing through by the volume of the container ($2.0\,dm^3$).

Step 3: We now have a set of values for the equilibrium concentrations in terms of x. These can be inserted into the expression for K_c and a value calculated for x:

$$54 = \frac{x^2}{\{(4.00-x)/2.0\}\{(4.00-x)/2.0\}}$$

$$= \frac{x^2}{\{(4.00-x)/2.0\}^2}$$

As we have two squared terms on the right-hand side of this equation the easiest way to solve it is to take square roots of both sides:

$$\sqrt{54} = \sqrt{\frac{x^2}{\{(4.00-x)/2.0\}^2}}$$

$$7.35 = \frac{x}{\{(4.00-x)/2.0\}}$$

$$7.35(4.00-x) = 2x$$

$$29.4 - 7.35x = 2x$$

$$29.4 = 9.35x$$

$$x = \frac{29.4}{9.35} = 3.14$$

Step 4: Now that we know the value of x we can find the equilibrium concentrations of each gas:

At equilibrium, $[HI] = (2x\,mol\,dm^{-3})/2 = x\,mol\,dm^{-3} = 3.14\,mol\,dm^{-3}$

$$= 3.1\,mol\,dm^{-3}$$

At equilibrium, $[H_2] = \{(4.00-x)/2\}\,mol\,dm^{-3} = \{(4.00-3.14)/2\}\,mol\,dm^{-3}$

$$= 0.43\,mol\,dm^{-3}$$

At equilibrium, $[I_2] = \{(4.00-x)/2\}\,mol\,dm^{-3} = (4.00-3.14)/2\,mol\,dm^{-3}$

$$= 0.43\,mol\,dm^{-3}$$

Worked Example 7.6 Dinitrogen tetroxide decomposes to nitrogen dioxide and an equilibrium is reached:

$$N_2O_4(g) \rightleftharpoons 2NO_2(g)$$

1.0 mole of dinitrogen tetroxide was added to a container of volume $10\,dm^3$ at a temperature of 70 °C. At equilibrium, 50% of the N_2O_4 had dissociated. Calculate the value of K_c for the reaction at this temperature.

Solution

This question is slightly different in that we are told the proportion of N_2O_4 that dissociates rather than the actual equilibrium concentration. It can be solved in a similar manner by following the earlier steps.

Step 1: Write an expression for the equilibrium constant, K_c:

$$K_c = \frac{[NO_2]^2}{[N_2O_4]}$$

Step 2: Set up the equilibrium table.

		N_2O_4	NO_2
1	Ratio of moles	1	2
2	Initial no. moles/mol	1.0	0
4	Change in moles/mol	−0.5	+1
5	Equilibrium no. moles/mol	1−0.5 = 0.5	0 + 1 = 1
6	Equilibrium concentrations/mol dm^{-3}	$\frac{0.5}{10} = 0.05$	$\frac{1}{10} = 0.1$

To obtain the change in number of moles in row 4 we use the information that 50% of the initial number of moles of N_2O_4 has dissociated. This means the change is −0.5 mol.

The mole ratio shows us that each 1 mole N_2O_4 gives 2 moles of NO_2, so the change in number of moles of $NO_2 = +2 \times 0.5 = +1.0$ mol.

In this example it is simpler to determine the number of moles at equilibrium (row 5) before calculating the concentration of each gas in row 6 by dividing by the volume of the container.

Step 3: Insert the equilibrium concentrations in the equation for the equilibrium constant to obtain a value for K_c:

$$K_c = \frac{[NO_2]^2}{[N_2O_4]} = \frac{(0.1 \text{ mol dm}^{-3})^2}{(0.05 \text{ mol dm}^{-3})} = \frac{0.01 \text{ mol}^2 \text{dm}^{-6}}{0.05 \text{ mol dm}^{-3}} = 0.2 \text{ mol dm}^{-3}$$

7.2.3 The effects of changing the reaction conditions on the position of equilibrium

Many important industrial processes rely on chemical reactions that reach equilibrium. Such reactions are not especially profitable as there is incomplete conversion of reactants to products. However, in chemical reactions that reach equilibrium, it is possible to optimise the amount of products obtained by changing the conditions under which the reaction is carried out. Such reaction conditions normally include temperature, pressure, and the amount of reactant used. In this way chemists and chemical engineers can increase the proportion of products to reactants and make the reaction more profitable.

Le Châtelier's principle

Le Châtelier's principle allows us to predict the direction in which a reaction at equilibrium will shift if we change the conditions. When the equilibrium position shifts to give more products, we say the reaction shifts to the right. If the reverse occurs and more reactants are formed, the equilibrium position has shifted to the left. In this way, the direction of the shift allows us to determine whether we will obtain more reactants or products when the conditions are changed.

Le Châtelier's principle states that if a reaction at equilibrium is disturbed by a change in one of the reaction conditions, the position of equilibrium will shift so as to minimise the change.

We will now look at the effect of making changes in reaction conditions on the position of equilibrium and the value of the equilibrium constant, K_c.

Effect of changing concentration

In the reaction $A(aq) + B(aq) \rightleftharpoons C(aq) + D(aq)$, all the reactants and products are in the same physical phase, i.e. aqueous solution. At equilibrium their concentrations are constant.

Consider adding more of reactant A. Le Châtelier's principle states the reaction will move in the direction to minimise that increase in concentration. Therefore, the excess A will react with reactant B to give more of products C and D. The equilibrium shifts to the right. A new equilibrium position is quickly re-established, and the new concentrations of A, B, C, and D will be constant once the rates of the forward and backward reactions are again equal. Adding more of reactant B to the equilibrium mixture would have the same effect. Consider adding more of product C. Again, Le Châtelier's principle states that the reaction will shift in such a direction as to reduce this increase in concentration of C. This is achieved by C reacting with D, so more of A and B are produced. The reaction is said to shift to the left. The same effect would be observed if excess D were added.

The equilibrium expression for the reaction is $K_c = \dfrac{[C][D]}{[A][B]}$.

If the concentration of any of the reactants or products is changed, a new equilibrium position is established. However, the value of the equilibrium constant remains unchanged.

Remember: A change in concentration of a reactant or product at equilibrium shifts the position of equilibrium, but the equilibrium constant is unchanged.

Effect of changing the pressure of a gaseous reactant or product

In a reaction involving gases, the number of moles of a certain gas present is equivalent to the pressure of that gas. We saw in Chapter 4 that in a fixed volume of gas at a constant temperature, the pressure of each gas in a mixture is proportional to the number of moles present. Consider the reaction between gaseous hydrogen and iodine to give gaseous hydrogen iodide:

$$H_2(g) + I_2(g) \rightleftharpoons 2HI(g).$$

The amount of each gas is measured by its partial pressure, p. This is the pressure that each gas in the mixture exerts. If the partial pressure of hydrogen, $p(H_2(g))$, is increased, Le Châtelier's principle states that the reaction will shift to decrease the partial pressure of hydrogen. The excess hydrogen reacts with the iodine to produce more hydrogen iodide. The equilibrium shifts to the right and more products are formed.

What if the partial pressure of one of the gases is decreased? Again, Le Châtelier's principle can be used and allows us to predict that the reaction will shift to minimise that decrease in pressure. If the partial pressure of hydrogen iodide is decreased by removing some of the gas, the hydrogen and iodine react together to produce more hydrogen iodide until a new equilibrium is established. The equilibrium shifts to the right. Clearly, if the hydrogen iodide was continually removed, hydrogen and iodine would continue to react together until one or both was used up. In this case the reaction could not establish a new equilibrium as it would no longer be a closed system. Removing one of the products is a strategy that can be used to increase the amount of desired product in an equilibrium reaction.

Again there is no change in the value of the equilibrium constant if the partial pressure of a gas is changed once a new equilibrium has been established.

Changing the partial pressure of a gas in an equilibrium mixture has the same effect on the equilibrium position as changing the concentration of the gas.

Effect of changing the overall pressure of a reaction

If a reaction does not involve gases then changing the overall or external pressure will have no effect on the equilibrium position. However, for a reaction where gases are involved, if there is a different number of moles of gas on each side of the equation, then Le Châtelier's principle can be applied to determine the effect of changing the external pressure.

Consider the reaction in which the colourless gas dinitrogen tetroxide decomposes to brown nitrogen dioxide:

$$N_2O_4(g) \rightleftharpoons 2NO_2(g)$$

colourless   brown

An equilibrium is established when the gases are held in a sealed container at a constant temperature and pressure. If the external pressure on the gases is increased – for example, by decreasing the volume, as in Figure 7.3 – the

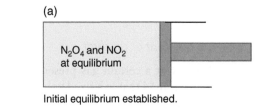

(a)

N$_2$O$_4$ and NO$_2$
at equilibrium

Initial equilibrium established.

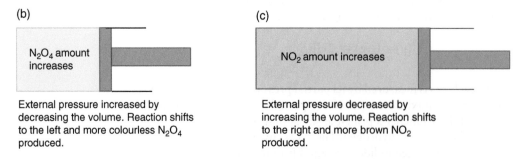

(b)

N$_2$O$_4$ amount increases

External pressure increased by decreasing the volume. Reaction shifts to the left and more colourless N$_2$O$_4$ produced.

(c)

NO$_2$ amount increases

External pressure decreased by increasing the volume. Reaction shifts to the right and more brown NO$_2$ produced.

Figure 7.3 Effect of changing the external pressure on the dinitrogen tetroxide/nitrogen dioxide equilibrium.

equilibrium shifts so as to minimise the increase in pressure. The way it does this is by decreasing the number of moles of gas present at equilibrium, as the overall pressure of the reaction mixture is proportional to the number of moles of gas present. The side of this equation with fewer moles is the left-hand side – the N_2O_4 side. Increasing the overall pressure will cause the equilibrium to shift to the left. NO_2 molecules are forced to react together to produce N_2O_4 and the overall pressure is reduced.

Changing the overall pressure only affects the equilibrium position in a reaction where there is a change in the number of moles of gas on either side of the equation. If the number of moles of gas is the same on both sides – for example, in the $H_2(g)$ + $I_2(g) \rightleftharpoons 2HI(g)$ equilibrium – changing the external pressure will have no effect on the equilibrium position because the reaction cannot shift to increase or reduce the total pressure.

Changing the overall pressure of a reaction at equilibrium has no effect on the value of the equilibrium constant.

Effect of changing the temperature of reaction

Most chemical reactions are either exothermic or endothermic processes. The energy change is measured by the enthalpy of reaction, $\Delta_r H^\ominus$.

If the temperature of a chemical reaction at equilibrium is changed, Le Châtelier's principle can be used to predict the effect of the change in temperature on the equilibrium position.

Consider again the decomposition of dinitrogen tetroxide into nitrogen dioxide. As with most chemical reactions that involve the decomposition of one molecule into smaller ones, the reaction is endothermic in the forward direction and the enthalpy change is positive:

$$N_2O_4(g) \rightleftharpoons 2NO_2(g) \quad \Delta_r H^\ominus = +58\,\text{kJ}\,\text{mol}^{-1}$$
colourless brown

An exothermic reaction releases heat energy, and the enthalpy of reaction is negative. An endothermic reaction absorbs heat energy, and the enthalpy of reaction is positive.

If the temperature of the reaction mixture is increased, Le Châtelier's principle predicts that the reaction will proceed so as to reduce the effect of the increased temperature. Increasing the temperature will cause the equilibrium to shift in the endothermic direction to absorb the heat. In this case the equilibrium shifts to the right and more NO_2 product is formed. The reaction mixture becomes darker. Figure 7.4 shows the colour and composition of the reaction mixture at three different temperatures. At low temperatures we have only colourless N_2O_4 (Figure 7.4a). As the temperature is increased to $0\,^\circ$C, N_2O_4 dissociates, and an equilibrium mixture of N_2O_4 and NO_2 is formed (Figure 7.4b). When the temperature is increased further the reaction shifts to produce more NO_2 and the mixture becomes even darker (Figure 7.4c). A new equilibrium position is established at the new temperature. If the temperature is decreased the equilibrium shifts in the opposite direction to form more N_2O_4.

If the temperature of a reaction at equilibrium is changed, the value of the equilibrium constant also changes. Therefore, equilibrium constants are quoted at specific temperatures.

A thermal decomposition reaction requires heat to take place and is endothermic. An increase in temperature provides heat energy so more reactants decompose and more products are formed. For an equilibrium to be established, the reaction must be carried out in a sealed container.

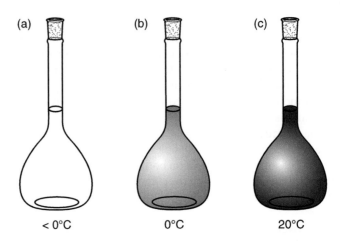

Figure 7.4 The $N_2O_4 \rightleftharpoons 2NO_2$ equilibrium at different temperatures. An increase in temperature causes the equilibrium to shift to the right to produce more brown NO_2.

Effect of introducing a catalyst

A *catalyst* is defined as a substance that speeds up a reaction but remains chemically unchanged itself. Catalysts and their action will be explored more fully in the next chapter.

Catalysts are used in many industrial processes as they speed up reactions so that more product can be obtained in a shorter time. Catalysts have no effect on the equilibrium position but simply decrease the time taken to reach equilibrium.

Table 7.1 illustrates the effect of changing various conditions upon the equilibrium position and the equilibrium constant.

Table 7.1 Summary of the effect of changes in reaction conditions upon the equilibrium position and equilibrium constant.

Change	Effect on equilibrium position	Effect on equilibrium constant
Increase in concentration or partial pressure of reactant	Equilibrium shifts to the right	No change
Increase in concentration or partial pressure of product	Equilibrium shifts to the left	No change
Increase in overall pressure when number of moles increases, e.g. $A \rightleftharpoons 2B$	Equilibrium shifts to the left	No change
Increase in overall pressure when number of moles decreases, e.g. $2A \rightleftharpoons B$	Equilibrium shifts to the right	No change
Increase in temperature in endothermic reaction, $\Delta_r H = +ve$	Equilibrium shifts to the right	Increases
Increase in temperature in exothermic reaction, $\Delta_r H = -ve$	Equilibrium shifts to the left	Decreases
Decrease in temperature in endothermic reaction, $\Delta_r H = +ve$	Equilibrium shifts to the left	Decreases
Decrease in temperature in endothermic reaction, $\Delta_r H = -ve$	Equilibrium shifts to the right	Increases
Introduce a catalyst	No change	No change

Box 7.1 The importance of equilibrium conditions on a reaction used industrially – the production of ammonia, NH_3.

Ammonia is a critically important chemical used to make fertilisers, textiles, dyes, explosives, and other chemicals. Before it was manufactured industrially, it was obtained from nitrates in guano (seabird droppings) shipped from South America. During the First World War, sea routes to South America were blocked by the British, and Germany looked for alternative methods to obtain ammonia for ammunition and explosives. A process that uses nitrogen from the air to produce ammonia was developed in Germany by Fritz Haber and Carl Bosch in the early twentieth century. They optimised the conditions of the equilibrium reaction between nitrogen and hydrogen to ensure maximum yields of ammonia. The overall reaction is:

$$2N_2(g) + 3H_2(g) \rightleftharpoons 2NH_3(g) \quad \Delta_r H^\ominus = -92 \, kJ \, mol^{-1}$$

This is an equilibrium reaction and the challenge was to tweak the equilibrium conditions to shift the reaction to the right without using extreme conditions of temperature and pressure.

The forward reaction involves a reduction in overall number of moles of gas. A high pressure is favoured as this causes the equilibrium to shift to the right and increases the formation of ammonia. The reaction uses an external pressure of around 200 atm (20 000 kPa). It would be preferable to use even higher pressures for even higher yields but gases under pressure require expensive plants to ensure safety, so a compromise must be reached.

The forward reaction is exothermic. This means that the formation of ammonia is favoured by low temperatures. However, running the reaction at a low temperature reduces the rate of reaction and hence the time taken to reach equilibrium. Again, a compromise was found whereby a temperature of around 400 °C (670 K) is used with a finely divided iron catalyst.

One further method for increasing the production of ammonia would be to remove the product, thus prompting nitrogen and hydrogen to react to produce more. Removing ammonia from the reaction vessel is difficult to achieve, but it can be condensed out from reaction mixture and the nitrogen and hydrogen recycled.

The Haber–Bosch process played a significant role in Germany's war effort. It is still deployed today to produce over 450 million tonnes of nitrogen fertiliser a year. For his part, Fritz Haber was awarded the Nobel Prize in 1918.

Worked Example 7.7   Predict the effect of the following changes in reaction conditions on the equilibrium position and the equilibrium constant in the following reaction:

$$H_2(g) + CO_2(g) \rightleftharpoons H_2O(g) + CO(g) \quad \Delta_r H^\ominus = +41.2 \, kJ \, mol^{-1}$$

a. Increase the partial pressure of CO_2.

b. Increase the temperature.

c. Increase the overall pressure.

d. Decrease the concentration of CO.

Solution

a. Increasing the partial pressure of a gas in a reaction mixture is equivalent to increasing its concentration. If the partial pressure of CO_2 is increased the reaction will shift to reduce the extra CO_2 and the equilibrium position will move to the right, so more products are formed. There is no change in the value of the equilibrium constant.

b. The reaction is endothermic in the forward direction. If the temperature is increased, Le Châtelier's principle states that the equilibrium will shift to minimise the increase in temperature. It does this by shifting in the endothermic direction, to the right, so more products are formed. The equilibrium constant increases.

c. If the overall pressure is increased the reaction will shift to reduce the overall pressure by moving to the side with fewer moles of gas. In this reaction there is no change in number of moles of gas so an increase in overall pressure has no effect on the equilibrium position or the equilibrium constant.

d. Decreasing the concentration of CO by removing the gas causes the equilibrium to shift to the right and more CO and H_2O are formed. There is no change in the value of the equilibrium constant.

7.2.4   Heterogeneous and homogeneous equilibria

So far, in all the examples of equilibrium reactions we have looked at, the reactants and products have been in the same physical state. In fact, they have all been gases. Reactions in which all the reactants are in the same phase are called **homogeneous reactions**.

Many equilibria involve solids, liquids, or substances dissolved in a solvent such as water. An example of an equilibrium reaction where the reactants are all liquids is the reaction between ethanol and ethanoic acid to produce the ester ethyl ethanoate. You will meet this reaction again when you study organic chemistry. The equation for the reaction is:

$$C_2H_5OH(l) + CH_3COOH(l) \rightleftharpoons C_2H_5COOCH_3(l) + H_2O(l)$$

ethanol   ethanoic acid   ethyl ethanoate   water

As you can see from the symbols in the equation all the liquids are in the same physical phase and water is one of the liquids present. The concentrations of each component can be measured in $mol\,dm^{-3}$ and the equilibrium expression written:

$$K_c = \frac{[C_2H_5COOCH_3][H_2O]}{[C_2H_5OH][CH_3COOH]}$$

Some equilibrium reactions have products or reactants that are in a different physical phase from each other. These reactions are termed **heterogeneous** equilibria. If one of the components is a solid, the concentration of this species can be ignored in the expression for the equilibrium constant. This is because the concentration of a solid material is a constant. Thus, the amount of solid reactant or product has no effect on the value of the equilibrium constant.

An example of a reaction that reaches equilibrium in a closed system where one of the components is a gas and the others are solids is the thermal decomposition of calcium carbonate:

$$CaCO_3(s) \rightleftharpoons CaO(s) + CO_2(g)$$

The expression for the equilibrium constant would be written as $K_c = \frac{[CaO][CO_2]}{[CaCO_3]}$. However, as both $CaCO_3$ and CaO are solids, their concentration values are constant, so the equilibrium constant expression becomes $K_c = [CO_2]$.

The terms *heterogeneous* and *homogeneous* are derived from Greek. The prefix *hetero* means *different*, and the prefix *homo* means the *same*.

7.2.5   The equilibrium constant, K_p

When chemical reactions involve mixtures of gases the concentration of a gas is normally expressed by its partial pressure, p, with a unit of Pa or atm. The partial pressure, p, of a gas is the pressure exerted by that gas in a mixture. So in a mixture of three different gases, A, B, and C, the total pressure of the gas, p_{total}, is obtained by adding the partial pressures of each gas in the mixture: $p_{total} = p_A + p_B + p_C$.

p_A, p_B, and p_C are the partial pressures of the gases A, B, and C, respectively.

An equilibrium expression can be written for a reaction in terms of partial pressures of the gases. The equilibrium constant is given by the symbol K_p when the amounts of gas present at equilibrium are measured in terms of partial pressures.

As an example, take the following reaction between sulfur dioxide and oxygen to give sulfur trioxide:

$2SO_2(g) + O_2(g) \rightleftharpoons 2SO_3(g)$.

The equilibrium constant, K_p, is given by:

$$K_p = \frac{p^2{}_{SO_3}}{p^2{}_{SO_2} \times p_{O_2}}$$

Note the following:

Concentration is defined as $\frac{amount\ in\ moles}{volume}$. As the amount in moles of a pure material is proportional to its mass, the expression for the concentration of a solid is effectively a measure of its density, and so is constant; i.e. density $= \frac{mass}{volume}$.

- The partial pressures of the gases are represented by lowercase 'p' with the formula for the gas given as a subscript.

- The value for the partial pressure is raised to a power equal to the number of moles of the gas in the balanced equation. The power is written as a superscript after the 'p'.

- Units of partial pressure are units of pressure. The units of K_p are determined by the expression for K_p.

The units of K_p for the previous reaction can be determined by inserting the units for partial pressure into the expression. When the partial pressures are measured in Pa the units can be found to be Pa^{-1}, as shown:

$$K_p = \frac{Pa^2}{Pa^2 \times Pa} = Pa^{-1}$$

Worked Example 7.8   Determine the units of K_p for the following equilibrium reactions if the partial pressures are measured in Pa:

a. $PCl_5(g) \rightleftharpoons PCl_3(g) + Cl_2(g)$

b. $H_2(g) + I_2(g) \rightleftharpoons 2HI(g)$

c. $2NOCl(g) \rightleftharpoons 2NO(g) + Cl_2(g)$

Solution

a. First write the expression for K_p: $\dfrac{p_{PCl_3} \times p_{Cl_2}}{p_{PCl_5}}$

Next insert units for each gas: $\dfrac{Pa \times Pa}{Pa}$

Next rationalise the units: $\dfrac{\cancel{Pa} \times Pa}{\cancel{Pa}} = Pa$

b. First write the expression for K_p: $\dfrac{p^2_{HI}}{p_{H_2} \times p_{I2}}$

Insert units for each gas: $\dfrac{Pa^2}{Pa \times Pa}$

Rationalise the units, $\dfrac{\cancel{Pa^2}}{\cancel{Pa} \times \cancel{Pa}}$, so there are no units.

c. First write the expression for K_p: $\dfrac{p^2_{NO} \times p_{Cl_2}}{p^2_{NOCl}}$

Insert the units: $\dfrac{Pa^2 \times Pa}{Pa^2}$

Rationalise the units: $\dfrac{Pa^2 \times Pa}{Pa^2} = Pa$

Worked Example 7.9   The equation for the formation of ammonia used by the Haber process is:

$$N_2(g) + 3H_2(g) \rightleftharpoons 2NH_3(g) \quad \Delta H^\ominus = -92\,kJ\,mol^{-1}$$

At equilibrium 6.8 moles of hydrogen, 2.2 moles of nitrogen, and 1.4 moles of ammonia were found to be present in a reaction carried out at 700 K. The total pressure of the system was 10 atm.

a. Use a copy of the following table to calculate the mole fraction of each gas present at equilibrium.

b. Calculate the partial pressure of each gas at equilibrium.

c. Calculate a value for K_p.

d. Predict the effect of increasing the temperature to 900 K on the amount of ammonia produced.

	Number of moles	Mole fraction	Partial pressure/atm
$N_2(g)$	2.2		
$H_2(g)$	6.8		
$NH_3(g)$	1.4		
Total			

Solution

	Number of moles	Mole fraction	Partial pressure/atm
$N_2(g)$	2.2	2.2/10.4 = 0.21	0.21 × 10 = 2.1
$H_2(g)$	6.8	6.8/10.4 = 0.65	0.65 × 10 = 6.5
$NH_3(g)$	1.4	1.4/10.4 = 0.14	0.14 × 10 = 1.4
Total	10.4	1	10

a. The mole fraction of each gas is calculated by dividing the amount in moles of each gas by the total number of moles of gas. The total number of moles of gas is 10.4.

b. The partial pressure of each gas is obtained by multiplying the mole fraction of each gas by the total pressure (10 atm), as given in column 4 of the table.

c. $K_p = \dfrac{p^2_{NH_3}}{p_{N_2} \times p^3_{H_2}} = \dfrac{(1.4 \text{ atm})^2}{(2.1 \text{ atm}) \times (6.5 \text{ atm})^3} = 3.4 \times 10^{-3} \text{ atm}^{-2}$

d. The enthalpy change for the reaction is exothermic in the forward reaction. This means that an increase in temperature to 900 K would favour the backward reaction, and more N_2 and H_2 would be formed.

Worked Example 7.10 Nitrogen and oxygen react together to form nitric oxide, NO, at temperatures of around 2500 K. An equilibrium is established, and the value of the equilibrium constant, K_p, is 0.050. The equation for the reaction is $N_2(g) + O_2(g) \rightleftharpoons 2NO(g)$. Determine the equilibrium concentrations of each gas if the initial pressures of N_2 and O_2 are 0.2 atm.

Solution

The expression for $K_p = \dfrac{p^2_{NO}}{p_{N2} \times p_{O2}} = 0.050$.

To determine the equilibrium concentrations of each gas, a RICE table must be constructed. Let the pressures of N_2 and O_2 that dissociate at equilibrium be equal to x. The new partial pressures of each gas are given in row 4.

		N_2	O_2	NO
1	Ratio of moles	1	1	2
2	Initial partial pressure/atm	0.2	0.2	0
3	Change in partial pressure/atm	−x	−x	2x
4	Equilibrium partial pressure/atm	0.2 − x	0.2 − x	2x

Insert the equilibrium partial pressures into the expression for K_p:

$$K_p = \frac{(2x)^2}{(0.2-x) \times (0.2-x)} = 0.050$$

Multiplying out the denominator gives: $0.050 = \dfrac{(2x)^2}{(0.2-x)^2}$

To solve this take square roots of both sides of the equation:

$$\sqrt{0.050} = \sqrt{\frac{(2x)^2}{(0.2-x)^2}}; \text{therefore } 0.22 = \frac{2x}{0.2-x}$$

Multiplying out: $0.22(0.2 - x) = 2x$

$$0.044 - 0.22x = 2x$$

$$0.044 = 2.22x,$$

$$x = 0.02$$

$$p_{N2} = p_{O2} = 0.20 - 0.020 \text{ atm} = 0.18 \text{ atm}$$

$$p_{NO} = 2 \times 0.02 \text{ atm} = 0.04 \text{ atm}$$

7.3   Acid-base equilibria

7.3.1   The Brønsted–Lowry theory of acids and bases

Acids

There are three theories used to identify acids and bases. The Brønsted-Lowry theory, the Arrhenius theory and the Lewis theory. The Arrhenius theory defines an acid as a species that dissociates in water to produce hydrogen ions and a base as a species that dissociates in water to produce hydroxide ions. The Arrhenius theory and the Brønsted-Lowry theory are similar. The Lewis theory defines acids and bases in terms of transfer of a pair of electrons. Each of the theories has advantages in different applications.

The Brønsted–Lowry theory defines acids as substances that act as proton donors and bases as proton acceptors. Protons are positively charged hydrogen ions, H^+, and are found in all substances that can act as acids.

A Brønsted–Lowry acid is a proton (H^+) donor.

Free H^+ ions are protons and cannot exist as such in aqueous solutions. They very quickly combine with water molecules to give the H_3O^+ (hydroxonium) ion with the formation of a dative covalent bond between one lone pair on oxygen and the H^+ ion:

$$H^+(aq) + H_2O(l) \rightarrow H_3O^+(aq)$$

hydroxonium
ion

The acid, hydrochloric acid, is obtained when the gas, hydrogen chloride, (HCl) dissolves in water.

$$HCl(g) \xrightarrow{H_2O} HCl(aq) \rightarrow H^+(aq) + Cl^-(aq)$$

One molecule of HCl splits up or **dissociates** to form a hydrogen ion (proton) and a chloride ion. Both ions exist in aqueous solution surrounded by water molecules.

The equation for the dissolution of hydrogen chloride in water can also be written as:

$$HCl(g) + H_2O(l) \rightarrow H_3O^+(aq) + Cl^-(aq)$$

The proton formed by dissociation of HCl quickly reacts with a water molecule to form the hydroxonium ion, H_3O^+. This second equation more correctly represents the process occurring. Textbooks and other materials frequently use the symbol $H^+(aq)$ as shorthand for the hydrated proton instead of using the more correct symbol $H_3O^+(aq)$.

Table 7.2 lists several relatively well-known acids and the ions that are produced when they dissolve in water.

You will notice that all the species listed as acids in Table 7.2 contain hydrogen atoms; when they dissolve in water, hydrogen ions or hydroxonium ions are formed. These species are all acids according to the Brønsted–Lowry definition.

Table 7.2 Some common acids and the ions they form when dissolved in water.

Acid	Formula	Ions formed in water
Hydrochloric acid	HCl	H^+, Cl^-
Sulfuric acid	H_2SO_4	$2H^+$, SO_4^{2-}
Nitric acid	HNO_3	H^+, NO_3^-
Ethanoic acid	CH_3COOH	H^+, CH_3COO^-
Citric acid	$C_3H_5O(COOH)_3$	$3H^+$, $C_3H_5O(COO)_3^{3-}$
Carbonic acid	H_2CO_3	H^+, HCO_3^-

Bases

A Brønsted–Lowry base is a proton acceptor.

Table 7.3 lists some common bases. Many of them are derived from metal oxides or hydroxides. A base that is soluble in water is called an **alkali**. Alkalis that you may be familiar with are sodium and potassium hydroxide and ammonia.

Table 7.3 Some common bases and the ions they form when dissolved in water.

Base	Common name	Formula	Ions formed in water
Sodium hydroxide	Caustic soda	$NaOH$	Na^+, OH^-
Potassium hydroxide	Caustic potash	KOH	K^+, OH^-
Calcium hydroxide	Slaked lime	$Ca(OH)_2$	Ca^{2+}, $2OH^-$
Ammonia		NH_3	NH_4^+, OH^-

Group 1 metal hydroxides dissolve in water to produce aqueous hydroxide ions, OH$^-$:

$$NaOH(s) \xrightarrow{H_2O} Na^+(aq) + OH^-(aq)$$

The hydroxide ion acts as a base by accepting a proton according to this equation:

$$OH^-(aq) + H^+(aq) \rightleftharpoons H_2O(l)$$

The equation can also be written as:

$$OH^-(aq) + H_3O^+(aq) \rightleftharpoons H_2O(l) + H_2O(l)$$

The product of the reaction is water.

Note that when a reaction is reversible an equilibrium sign ($\rightleftharpoons$) is used to show the reaction can go in either direction.

Water as an acid and a base

You may have noticed from the previous sections that water is frequently involved when acids and bases dissociate. This is not always the case, but water is a very extraordinary molecule and can behave as either an acid, by losing a proton, or a base, by accepting a proton. The donation of a proton from the acid, HCl, to water can be seen in the following equation (state symbols have been removed):

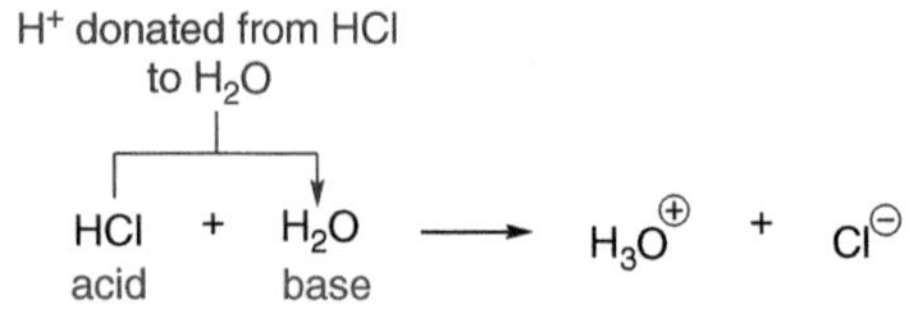

The proton is accepted by the water molecule and so, in this case, water is acting as a base by accepting a proton.

The hydroxide ion acts as a base by accepting a proton from the solvent, water. In this case, the water molecule is acting as an acid by donating a proton:

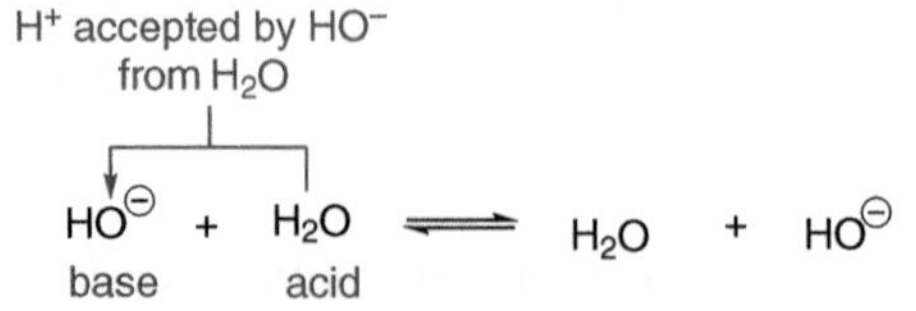

Ammonia, NH_3, is also a base. It can accept a proton from a water molecule to form an ammonium (NH_4^+) ion and a hydroxide ion. Here again the water molecule is acting as an acid:

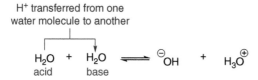

In some cases water loses a proton, acting as an acid; and in other cases water accepts a proton, acting as a base. A substance, such as water, that can behave as both an acid and a base is said to be **amphoteric**.

The way in which the water molecule can simultaneously act as an acid and a base is shown in the following equation for the self-ionisation of water:

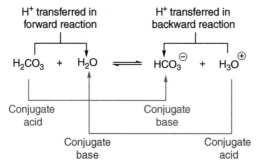

Conjugate acids and bases

When an acid dissociates and loses a proton, the species that remains is called a **conjugate base**. This is because the species remaining can accept a proton and so behave as a base. This can be seen for the dissociation of H_2CO_3, carbonic acid, the acid present in acid rain:

H^+ transferred in forward reaction H^+ transferred in backward reaction

$$H_2CO_3 + H_2O \rightleftharpoons HCO_3^{\ominus} + H_3O^{\oplus}$$

Conjugate acid   Conjugate base

Conjugate base   Conjugate acid

H_2CO_3 acts as an acid and loses a proton to the water molecule. The ion formed is the HCO_3^- ion (hydrogen carbonate). The HCO_3^- ion is called the conjugate base of H_2CO_3 as it can accept a proton and re-form H_2CO_3. The H_2CO_3 and HCO_3^- form a conjugate acid-base pair. Similarly, H_3O^+ and H_2O are a conjugate acid-base pair.

A conjugate acid-base pair are linked by the transfer of a proton.

Worked Example 7.11   Write equations for the following processes, showing the ions formed in solution:

a. The reaction of nitric acid, HNO_3, with water

b. The reaction of potassium hydroxide, KOH, with water

c. The reaction of the base ethylamine, $C_2H_5NH_2$, with water. (Amines behave in a similar manner to ammonia.)

d. The reaction of ethanoic acid, CH_3COOH, with water

Solution

a. $HNO_3(aq) + H_2O(l) \rightleftharpoons NO_3^-(aq) + H_3O^+(aq)$

b. $KOH(aq) + H_2O(l) \rightleftharpoons K^+(aq) + H_2O(l) + OH^-(aq)$

c. $C_2H_5NH_2(aq) + H_2O(l) \rightleftharpoons C_2H_5NH_3^+(aq) + OH^-(aq)$

d. $CH_3COOH(aq) + H_2O(l) \rightleftharpoons CH_3COO^-(aq) + H_3O^+(aq)$

Worked Example 7.12   Give the conjugate bases of the following acids:

a. H_2SO_4

b. H_3PO_4

c. CH_3COOH

Solution

a. Loss of a proton from H_2SO_4 leaves the HSO_4^- ion as the conjugate base: $H_2SO_4(aq) \rightleftharpoons HSO_4^-(aq) + H^+(aq)$.

b. Loss of a proton from H_3PO_4 gives the $H_2PO_4^-$ ion as the conjugate base: $H_3PO_4(aq) \rightleftharpoons H_2PO_4^-(aq) + H^+(aq)$.

c. Loss of a proton from CH_3COOH leaves the CH_3COO^- ion as the conjugate base:

$$CH_3COOH(aq) \rightleftharpoons CH_3COO^-(aq) + H^+(aq)$$

Worked Example 7.13   Give the conjugate acids of the following bases:

a. NH_3

b. H_2O

c. $HCOO^-$

Solution

a. After accepting a proton, the NH_3 molecule forms NH_4^+ as its conjugate acid:

$$NH_3(g) + H^+(aq) \rightleftharpoons NH_4^+(aq)$$

b. When H_2O accepts a proton, the H_3O^+ ion is formed:

$$H_2O(l) + H^+(aq) \rightleftharpoons H_3O^+(aq).$$

c. When $HCOO^-$ accepts a proton, the HCOOH molecule is formed as the conjugate acid:

$$HCOO^-(aq) + H^+(aq) \rightleftharpoons HCOOH(aq).$$

7.3.2 The pH scale

When an acid dissolves in water, hydrogen ions are formed in solution. The number of hydrogen ions formed depends on the number of molecules of acid that dissociate. The pH of the solution is a measure of the concentration of hydrogen ions in the solution. pH is defined in the following way:

$$pH = -\log_{10}[H^+(aq)]$$

Expressed in words we say that: 'pH is the negative log to the base 10 of the hydrogen ion concentration'.

This mathematical expression has two important consequences. The first is that the pH is not directly a measure of the hydrogen ion concentration. The expression uses the log to the base 10 of the hydrogen ion concentration. The log term converts very small numbers (such as 1×10^{-13}) or relatively large numbers (such as 0.1) to a simple scale.

The symbol pH is made up from two parts. The lowercase p is the first letter of the German word 'potenz', which means 'power' in English. The uppercase H represents the hydrogen ion.

Box 7.2 The *logarithm to the base 10* ($\log_{10}$) of a number is the power that the base 10 must be raised to in order to get that number. For example, the number 100 can be written as 10^2 (or 10×10). 10 is called the **base**, and 2 is called the **power**. Therefore *log to the base 10* of 100 is 2, as this is the power that 10 must be raised to in order to get the number 100.

For numbers smaller than 1, such as 0.1, $\log_{10}$ is obtained in the same way. The number 0.1 can be expressed as 10^{-1}, where 10 is the base and −1 is the power. So $\log_{10}$ of 0.1 is −1. On a calculator, the $\log_{10}$ key is normally labelled **log**.

$$100 = 10 \times 10 = 10^2$$

$$\log_{10}100 = \log_{10}10^2 = 2$$

number = 100   base = 10   power = 2

Taking the log of a number to the base 10

The second point to note is that the pH is the negative log of the hydrogen ion concentration. A large hydrogen ion concentration gives a small pH and vice versa.

The pH scale is a set of numbers typically between the values of 0 and 14. The following examples will demonstrate how pH is calculated.

An example of a strong acid with a high concentration is 0.5 M HCl ($0.5\ mol\ dm^{-3}$ HCl). The hydrogen ion concentration in this solution is $0.5\ mol\ dm^{-3}$. The pH of this acid is equal to 0.3 and is calculated as shown:

$$pH = -\log_{10}[H^+] = -\log_{10}\left(0.5\ mol\ dm^{-3}\right) = -(-0.3) = 0.3$$

If the acid is diluted to $0.1\ mol\ dm^{-3}$, the new pH can be calculated:

$$pH = -\log_{10}[H^+] = -\log_{10}\left(0.1\ mol\ dm^{-3}\right) = -(-1) = 1$$

Note that the concentration of the acid has decreased, but the pH has increased.

Because pH uses $\log_{10}$ of the hydrogen ion concentration, a solution that is 10 times more dilute has a pH that is just one pH point higher.

High concentrations of hydrogen ions lead to low pH values, indicating acidic solutions. Low hydrogen ion concentrations lead to high pH values, indicating basic solutions.

Solutions of bases have extremely low hydrogen ion concentrations. A solution of $0.1 \, mol \, dm^{-3}$ NaOH has a hydrogen ion concentration of $1 \times 10^{-13} \, mol \, dm^{-3}$. The pH can be determined in the same way:

$$pH = -\log_{10}[H^+] = -\log_{10}\left(1 \times 10^{-13} \, mol \, dm^{-3}\right) = -(-13) = 13$$

Measuring pH

The approximate pH of a solution can be obtained using an indicator. An **indicator** is a substance that changes colour in solutions of different pH. Universal indicator (Figure 7.5) is a useful indicator as it gives a slightly different colour for every pH value and so can be used as a rough indicator of pH. The colours shown on the pH scale in Figure 7.5 are the colours obtained when universal indicator is added to a solution of that pH. A low pH between 0 and 6 means the solution is acidic, a pH of 7 means the solution is neutral, and a high pH between 8 and 14 means the solution is basic. The term **neutral** means that the solution is neither acidic nor basic. Universal indicator, shown in Figure 7.5, gives approximate values of pH to the nearest whole number.

For more precise measurement a pH meter is used (Figure 7.6). A pH meter consists of a glass rod containing an electrode sensitive to hydrogen ions and a reference electrode. These are connected to a digital voltmeter. The glass rod is dipped into a solution of unknown pH, and a potential difference is set up between the glass electrode and the reference electrode. The potential difference is registered as a pH on the meter and can be read directly to one or two decimal places, depending upon the accuracy of the meter.

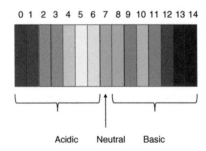

Figure 7.5 The pH scale showing the typical colour of universal indicator in solutions of each pH. *Source:* Elizabeth Page.

Figure 7.6 Using a pH meter to measure hydrogen ion concentration. *Source:* hannainst.com.

7.3.3   Strong and weak acids and bases

Strong acids

Strong acids are acids that dissociate completely when dissolved in water. Hydrochloric acid is an example of a strong acid. When the gas hydrogen chloride is passed into water it dissolves and splits into hydroxonium ions, H_3O^+, and chloride ions, Cl^-.

$$HCl(g) + H_2O(l) \rightarrow H_3O^+(aq) + Cl^-(aq)$$

Because the acid dissociates completely, the concentration of hydroxonium ions formed in solution will be the same as the original concentration of the HCl used. Starting with a 0.1 mol dm^{-3} hydrochloric acid solution we obtain the concentrations of hydroxonium ions and chloride ions shown in the table.

	$HCl(aq)$	$H_3O^+(aq)$	$Cl^-(aq)$
Original concentration/mol dm^{-3}	0.1	0	0
Concentration after dissociation/ mol dm^{-3}	0	0.1	0.1

Strong acids dissociate completely in water to give solutions with low pH. The next example shows how to calculate the pH of a solution of a strong acid.

Worked Example 7.14   Calculate the pH of nitric acid, HNO_3, of concentration 0.0125 mol dm^{-3}.

Solution
If the concentration of the undissociated nitric acid is 0.0125 mol dm^{-3}, then the concentration of hydrogen ions in the aqueous solution is also 0.0125 mol dm^{-3}. The pH is obtained from this equation:

$$pH = -\log_{10}[H^+] = -\log_{10}\left(0.0125 \text{ mol dm}^{-3}\right) = -(-1.90) = 1.90$$

Weak acids

A *weak* acid is an acid that dissociates only partially in aqueous solution to give hydrogen ions. An equilibrium is established between the undissociated acid and the ions in solution. For a general acid of formula HA, the following equation can be written:

$$HA(aq) + H_2O(l) \rightleftharpoons H_3O^+(aq) + A^-(aq)$$

The symbol that represents an equilibrium, $\rightleftharpoons$, is used here to show only partial dissociation of the acid in water. The extent of dissociation is indicated by the size of the equilibrium constant. The equilibrium constant that represents the equilibrium between a weak acid and its dissociated ions is K_a and is called the **acid dissociation constant**. This is directly analogous to the equilibrium constant, K_c, and is defined in the same way. K_a is related to the concentrations of the species in solution at equilibrium and is given by the following expression for the weak acid HA:

$$HA(aq) + H_2O(l) \rightleftharpoons H_3O^+(aq) + A^-(aq)$$

$$K_a = \frac{[H_3O^+(aq)][A^-(aq)]}{[HA(aq)][H_2O(l)]}$$

Strong acids dissociate completely in water to give high concentrations of hydroxonium ions. Weak acids dissociate only partially in water to give low concentrations of hydroxonium ions.

Because the concentration of water is hugely in excess it is not included in the expression for the acid dissociation constant as its concentration is essentially constant. So the expression becomes

$$K_a = \frac{[H_3O^+(aq)][A^-(aq)]}{[HA(aq)]}$$

A high value of K_a suggests the equilibrium lies to the right. This results in high concentrations of hydrogen ions and low pH values. Values of K_a are generally very small numbers.

To make it easier to handle these very small values, **pK_a** is usually quoted for an acid rather than K_a. The pK_a is defined as:

$$pK_a = -\log_{10} K_a$$

Remember: Another name for a hydrogen ion is a proton, H^+. A hydrated proton is known as a hydroxonium ion, and we use the fomula H_3O^+ to represent this.

This expression has the same form as the expression for pH. An acid that is more highly dissociated has a high K_a value, a low pK_a value, and a low pH. Table 7.4 lists some common weak acids along with their K_a and pK_a values.

Many weak acids are *dibasic* acids, which means that they have two hydrogen ions (or protons) that can dissociate; *tribasic* acids possess three hydrogen ions. For example, phosphoric acid has the formula H_3PO_4. It can lose a maximum of three hydrogen ions and it does this in a step-wise fashion. There is an associated K_a value for each hydrogen ion that is lost. To distinguish between the different acid dissociation constants they are given the symbols K_1, K_2, K_3:

$$H_3PO_4(l) \rightleftharpoons H_2PO_4^-(aq) + H^+(aq) \quad K_1 = 7.1 \times 10^{-3} \text{ mol dm}^{-3}$$

$$H_2PO_4^-(aq) \rightleftharpoons HPO_4^{2-}(aq) + H^+(aq) \quad K_2 = 6.17 \times 10^{-8} \text{ mol dm}^{-3}$$

$$HPO_4^{2-}(aq) \rightleftharpoons PO_4^{3-}(aq) + H^+(aq) \quad K_3 = 4.37 \times 10^{-13} \text{ mol dm}^{-3}$$

Calculating the pH of a solution of a weak acid is not as straightforward as the calculation for a strong acid. A reaction table must be set up in the same way as for the equilibrium calculations carried out earlier in this chapter. An example of this type of calculation is given here for ethanoic acid, CH_3COOH.

Table 7.4 Some weak acids and their K_a and pK_a values.

Acid	Formula	Ions formed in aqueous solution	K_a/mol dm^{-3}	pK_a
Ethanoic acid	CH_3COOH	CH_3COO^- and H^+	1.74×10^{-5}	4.76
Methanoic acid	$HCOOH$	$HCOO^-$ and H^+	1.78×10^{-4}	3.75
Benzoic acid	C_6H_5COOH	$C_6H_5COO^-$ and H^+	6.31×10^{-5}	4.20
Hydrofluoric acid	HF	F^- and H^+	5.62×10^{-4}	3.25

Worked Example 7.15 Calculate the pH of a solution of ethanoic acid of concentration 0.05 mol dm^{-3} if the K_a value for ethanoic acid is 1.74×10^{-5} mol dm^{-3}.

Solution
Step 1: Write a balanced equation for the dissociation of ethanoic acid in water:

$$CH_3COOH(aq) + H_2O(l) \rightleftharpoons CH_3COO^-(aq) + H_3O^+(aq)$$

Step 2: Write an expression for the equilibrium constant, K_a.

$$K_a = \frac{[CH_3COO^-(aq)][H_3O^+(aq)]}{[CH_3COOH(aq)]} = 1.74 \times 10^{-5} \text{ mol dm}^{-3}$$

This is effectively the equilibrium constant for the reaction:

$$CH_3COOH(aq) \rightleftharpoons CH_3COO^-(aq) + H^+(aq)$$

Notice that $[H_2O(l)]$ has not been included as the water is in excess and its concentration constant.

Step 3: Draw up a reaction table showing initial and equilibrium concentrations of each species.

	CH_3COOH	CH_3COO^-	H^+
Ratio	1	1	1
Initial concentration/mol dm^{-3}	0.05	0	0
Change in concentration/mol dm^{-3}	$-x$	$+x$	$+x$
Equilibrium concentration/mol dm^{-3}	$0.05 - x$	x	x

Step 4: Insert the equilibrium values into the expression for K_a:

$$K_a = \frac{[CH_3COO^-(aq)][H^+(aq)]}{[CH_3COOH(aq)]} = \frac{(x) \times (x)}{(0.05 - x) \text{ mol dm}^{-3}}$$

The value of x is going to be very small compared to 0.05 as this is a weak acid and only a very small amount dissociates. Therefore we can approximate $(0.05 - x) \approx 0.05$. The equation becomes:

$$K_a = \frac{x^2}{0.05 \text{ mol dm}^{-3}} = 1.74 \times 10^{-5} \text{ mol dm}^{-3}$$

We can now check if we were justified in ignoring x in the expression $(0.05 - x)$. Insert $x = 9.33 \times 10^{-4}$ into the expression for the equilibrium concentration of CH_3COOH.

$$(0.05 - 9.33 \times 10^{-4}) = 0.0490$$

If $[CH_3COOH] = 0.0490 \, mol \, dm^{-3}$, the new value of

$[H^+] = \sqrt{0.0490 \times 1.74 \times 10^{-5}} \, mol \, dm^{-3}$
$= 9.233 \times 10^{-4} \, mol \, dm^{-3}$.

So

$$pH = -\log(9.233 \times 10^{-4}) = 3.03.$$

The value is identical to that obtained earlier at this level of accuracy.

Step 5: Solve the equation by multiplying out both sides:

$$x^2 = (1.74 \times 10^{-5} \times 0.05) \, mol^2 \, dm^{-6} = 8.7 \times 10^{-7} \, mol^2 \, dm^{-6}$$

Take square roots of both sides: $x = 9.33 \times 10^{-4} \, mol \, dm^{-3}$.

This is the concentration of hydrogen ions in the solution of ethanoic acid. The final step is to use this concentration to determine the pH of the solution.

Step 6: Calculate the pH of the solution using:

$$pH = -\log_{10}[H^+] = -\log_{10}\left(9.33 \times 10^{-4} \, mol \, dm^{-3}\right) = 3.03$$

The pH of the ethanoic acid is 3.03.

Weak bases

A weak base is a base that only partially dissociates in aqueous solution. A base is a substance that accepts a proton from water. This gives a solution containing hydroxide ions (OH^-). Ammonia, NH_3, is a gas that dissolves readily in water. It is a weak base because the equilibrium

$$NH_3(g) + H_2O(l) \rightleftharpoons NH_4^+(aq) + OH^-(aq)$$

lies to the left. The equilibrium constant for this reaction is the base dissociation constant, K_b, and is expressed in a similar way to the acid dissociation constant, K_a.

$$K_b = \frac{[NH_4^+(aq)][OH^-](aq)}{[NH_3(aq)]}$$

The pK_b value of a weak base is defined in the same way as pK_a:

$$pK_b = -\log_{10}K_b$$

A small value for pK_b indicates greater ionisation of the base in water and a higher pH. Organic amines such as methylamine (CH_3NH_2) and phenylamine ($C_6H_5NH_2$) are examples of weak bases that you will meet later in the course.

7.3.4   The ionisation of water

Water is amphoteric and can act as an acid or a base. Water undergoes self-ionisation forming both a hydroxonium ion (a hydrated proton) and a hydroxide ion:

$$H_2O(l) + H_2O(l) \rightleftharpoons H_3O^+(aq) + OH^-(aq)$$
$$\text{acid} \qquad \text{base} \qquad \text{acid} \qquad \text{base}$$

For a single molecule of water, the equation can be simplified to:

$$H_2O(l) \rightleftharpoons H^+(aq) + OH^-(aq).$$

An equilibrium expression can be written for the dissociation:

$$K_c = \frac{[H^+(aq)][OH^-(aq)]}{[H_2O(l)]}$$

Because $[H_2O(l)]$ is a constant, a new equilibrium constant, K_w, can be written where $K_w = K_c \times [H_2O(l)]$.

$$K_w = [H^+(aq)][OH^-(aq)].$$

The equilibrium constant K_w is known as the **ionic product** for water and represents the dissociation constant for the self-ionisation of water. The value is a constant at constant temperature and is equal to $1.0 \times 10^{-14}\ mol^2\ dm^{-6}$ at 298 K.

Because one molecule of H_2O dissociates to give one H^+ ion and one OH^- ion, the concentration of the $H^+(aq)$ ion and the $OH^-(aq)$ ion are equal at 298 K in pure water, i.e. $[H^+(aq)] = [OH^-(aq)]$.

The expression for K_w can be written as:

$$K_w = [H^+(aq)]^2.$$

At 298 K, $K_w = 1.0 \times 10^{-14}\ mol^2\ dm^{-6}$
Therefore $[H^+] = 1.0 \times 10^{-7}\ mol\ dm^{-3}$.
Because $[H^+] = [OH^-]$, $[OH^-] = 1.0 \times 10^{-7}\ mol\ dm^{-3}$

pOH

The pOH of a solution is used to express the basicity of an alkaline solution. The pOH is defined in the same way as pH, but substituting OH^- for H^+:

$$pOH = -\log_{10}[OH^-]$$

In aqueous solution at 298 K, the product of the H^+ ion concentration and the OH^- ion concentration is a constant, K_w:

$$K_w = [H^+(aq)][OH^-(aq)] = 1.0 \times 10^{-14}\ mol^2\ dm^{-6}$$

A solution with a high concentration of hydroxide ions, such as a strong base, has a low pOH but a high pH. This can be seen in the following example.

An important relationship to remember is:

pH + pOH = 14

This is obtained from the expression:

$[H^+(aq)][OH^-(aq)] = 1.0 \times 10^{-14}\ mol^2\ dm^{-3}$.

Taking $\log_{10}$ of both sides of this expression we obtain:

$\log_{10}[H^+(aq)] + \log_{10}[OH^-(aq)] = -14$

Substituting for $\log_{10}[H^+(aq)] = -pH$

and $\log_{10}[OH^-(aq)] = -pOH$:

$-pH - pOH = -14$

Therefore:

$pH + pOH = 14$

Worked Example 7.16 Calculate the pH and pOH of a solution of sodium hydroxide of concentration $0.005\ mol\ dm^{-3}$.

Solution
Sodium hydroxide is a strong base and so fully dissociated in aqueous solution. The pOH of the solution is calculated from:

$$pOH = -\log_{10}[OH^-] = -\log_{10}(0.005\ mol\ dm^{-3}).$$

$$pOH = 2.3$$

$$pH + pOH = 14$$

$$pH = 14 - 2.3 = 11.7$$

Worked Example 7.17 Determine the pH and pOH of an aqueous solution of $0.001\ mol\ dm^{-3}$ potassium hydroxide at 298 K.

Solution
Potassium hydroxide is a strong base of formula KOH. It dissociates completely in aqueous solution to give K^+ and OH^- ions:

$$KOH(s) \rightarrow K^+(aq) + OH^-(aq).$$

A 0.001 mol dm^{-3} solution of KOH has [K$^+$(aq)] = 0.001 mol dm^{-3} and because it is a strong base, [OH$^-$(aq)] = 0.001 mol dm^{-3}:

$$pOH = - \log_{10}[OH^-(aq)] = - \log_{10}\left(0.001 \text{ mol dm}^{-3}\right)$$

$$= - \log_{10}\left(1.00 \times 10^{-3} \text{ mol dm}^{-3}\right) = -(-3) = +3$$

Using pH + pOH = 14.

$$pH = 14 - pOH = 14 - 3 = 11$$

The solution has a pH of 11.

Always think about the answer. You know this is a strong base and will therefore have a **high** pH and a **low** pOH.

Variation of the pH of water with temperature

The value of K_w is a constant and is equal to 1.0×10^{-14} mol^2 dm^{-6} in any aqueous solution at 298 K.

However, the ionisation of water is endothermic in the forward direction:

$$H_2O(l) \rightleftharpoons H^+(aq) + OH^-(aq) \quad \Delta_r H^\ominus = + 57.3 \text{ kJ mol}^{-1}$$

When water is heated the equilibrium shifts to the right and more water molecules dissociate. The value of K_w increases and the concentration of hydrogen ions also increases. Therefore, when water is heated, the pH decreases. This does not mean that water gets more acidic at high temperatures. Water remains neutral because the concentration of hydroxide ions also increases because [H$_3$O$^+$] must equal [OH$^-$].

7.3.5   Acid-base reactions

An early definition of an acid stated that it is a substance that reacts with a base to give a salt and water. An example of this might be the reaction of hydrochloric acid with sodium hydroxide to produce sodium chloride and water:

$$HCl(aq) + NaOH(aq) \rightarrow NaCl(aq) + H_2O(l)$$

acid   base   salt   water

For any monoprotic acid, the overall ionic equation for reaction with a base is the same. If the aqueous species are written in their dissociated states, we have the following equation, and the ions that are present on both sides can be cancelled out:

$$H^+(aq) + \cancel{Cl^-(aq)} + \cancel{Na^+(aq)} + OH^-(aq) \rightarrow \cancel{Na^+(aq)} + \cancel{Cl^-(aq)} + H_2O(l)$$

This leaves the **ionic equation** for the acid-base reaction

$$H^+(aq) + OH^-(aq) \rightarrow H_2O(l)$$

that represents the chemical process that takes place when an acid and base react together.

In an acid-base titration, an acid reacts with a base under controlled conditions. A titration is used to determine the concentration or amount in moles of either the acid or the base, knowing the amount in moles of the other reactant.

7.3.6   Carrying out a titration

The apparatus for carrying out a titration is shown in Figure 7.7. An acid of known concentration is added to the burette. An accurately measured volume of base, whose concentration is to be determined, is added to the conical flask. The acid in the burette is run into the flask and the solution swirled to mix. When just enough acid has been added to the base to completely neutralise it, the titration is stopped. The volume of acid added is noted and used to calculate the concentration of the base. Alternatively, the burette is filled with base which is added to the acid in the flask until just neutralised.

Accurate values of pH changes occurring during an acid-base titration can be monitored using a pH meter (Figure 7.6). The results from the pH titration can be plotted, as shown in Figure 7.8, where the pH of the solution in the flask is plotted against the volume of hydrochloric acid added. If 0.1 M sodium hydroxide is in the flask the initial pH will be 13. Addition of HCl causes a very slight decrease in pH. Close to the neutralisation point the pH changes very rapidly and a very small increase in the volume of acid causes a sharp decrease in pH. This is called the **equivalence point** of the titration. It is represented by the vertical region on the plot in Figure 7.8. The pH at the equivalence point for the reaction between a strong acid and strong base is pH 7. If 0.1 M HCl is used in the burette, the volume required to completely neutralise 25 cm^3 of 0.1 M NaOH is 25 cm^3. When a pH meter is used to monitor the reaction, the equivalence point is determined from the mid point of the vertical section on the plot of volume against pH.

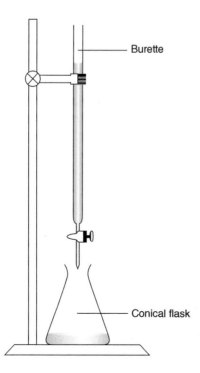

Burette

Conical flask

Figure 7.7 Titration apparatus.

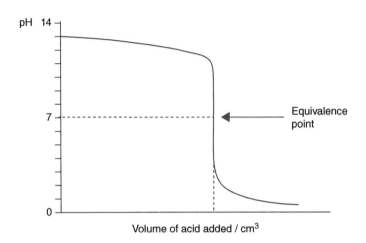

Figure 7.8 Plot of pH change in the titration of a strong acid against a strong base, e.g. HCl and NaOH.

Indicator	pH	1	2	3	4	5	6	7	8	9	10	11	12	13	14
Universal indicator															
Methyl orange					change										
Methyl red						change									
Bromothymol blue							change								
Phenolphthalein										change					

Figure 7.9 Colour changes of some common indicators.

7.3.7   Indicators

In practice, when carrying out a titration, the equivalence point is generally determined using an **indicator**. An **acid-base indicator** is a chemical that shows a different colour at different pH values. The colour changes of some commonly used indicators are shown in Figure 7.9. Indicators may also be derived from natural plant materials such as lichens and red cabbage, as shown in Box 7.3.

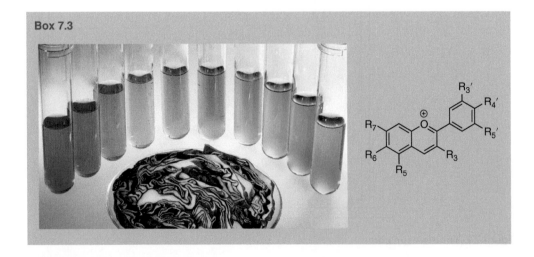

Box 7.3

Many plants, including fruits, vegetables, and flowers, can be used as acid-base indicators. Most contain anthocyanins based on the structure shown. A simple indicator can be made by boiling red cabbage in water. The solution formed shows a range of colours depending upon the pH of the solution it is added to. Source: NutritionFacts.

The indicator chosen to determine the neutralisation point of an acid and base titration should change colour when one additional drop (0.05 cm^3) of acid or base has been added beyond the equivalence point. This is known as the **end point** of a titration.

Box 7.4 Most acid-base indicators can be considered as weak acids in which the unionised form (HIn) and its conjugate base (In$^-$) have different colours:

$$HIn(aq) \rightleftharpoons H^+(aq) + In^-(aq)$$

conjugate acid   conjugate base

colour 1   colour 2

In acidic solutions the equilibrium is shifted to the left by excess H$^+$ ions and the indicator shows colour 1. When base is added it removes H$^+$ ions to form water and therefore pulls the equilibrium over to the right so the indicator shows colour 2. The exact colour shown by the indicator depends upon the relative amounts of HIn and In$^-$. If there are roughly equal amounts of HIn and In$^-$, the colour of the indicator will be midway between colour 1 and colour 2. So, for example, bromothymol blue is yellow in acids (HIn) and blue in bases (In$^-$). At the half-way point when [HIn] equals [In$^-$], the colour is green. This change in colour takes place over a pH range of approximately two units for most indicators. For bromothymol blue the colour change takes place between pH 6.0 and pH 7.6. The half-way point, which is green, occurs at pH 7.0.

Different indicators change colour over different pH ranges. The colour changes of some common indicators and pH values at which the changes occur are shown in Figure 7.9.

Universal indicator has a different colour in solutions of different pH values and is used primarily to give an approximate pH value rather than to determine the end point in titrations. Most other indicators have different colours in acid and base solutions and change colour over a fairly narrow pH range. The choice of indicator in a titration depends upon the pH of the equivalence point, and this depends upon the type of acid and base used in the titration.

7.3.8   Acid-base titrations

Titration of a strong acid and strong base

Figure 7.10 shows the pH change during the titration of a strong acid (in the burette) and strong base. The equivalence point is around pH 7, which is the mid-point of the vertical section of the pH curve. Figure 7.10 shows the pH curve overlaid with bands showing the pH ranges at which the indicators phenol-phthalein and methyl orange change colour. Both change colour over the vertical portion of the pH curve and so both would be suitable indicators. However, as the colour change from pink to colourless is easier to detect than yellow to pink, phenolphthalein is preferred over methyl orange. Both methyl red and bromothymol blue could also be used for this titration but are not included on the plot for clarity.

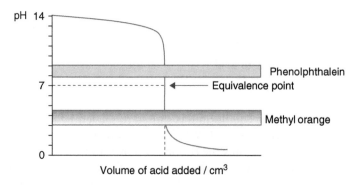

Figure 7.10 Plot of pH against volume of acid in a titration of a strong acid and strong base showing the pH range of phenolphthalein and methyl orange indicators.

Titration of a weak acid and strong base

Figure 7.11 shows the pH curve obtained for the titration of a strong base (in the flask) against a weak acid. This pH curve has a much shorter vertical section and the pH of the equivalence point (about 8.5) is higher than that with

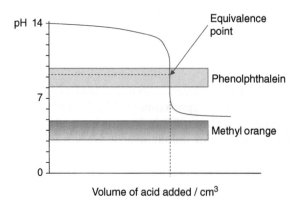

Figure 7.11 Plot of pH against volume of acid in a titration of a weak acid and strong base showing the pH range of phenolphthalein and methyl orange indicators.

a strong acid and strong base. Methyl orange does not change colour over the vertical section of the curve and so would not be a suitable indicator in this case. Phenolphthalein changes colour from pink to colourless on the vertical section of the curve and so is a suitable indicator with a clear colour change.

Titration of a strong acid and weak base

Figure 7.12 shows the pH curve for the titration of a strong acid (in the flask) and weak base. The shape of the curve shows that the vertical section is again shorter than with a strong acid and strong base and the equivalence point occurs at a lower pH, around 5.5 in this example. Overlaid on the curve are the pH ranges for the colour changes of methyl orange and phenolphthalein. Phenolphthalein changes colour over the pH range 8.2 to 10, which is higher than the vertical section of the titration curve. Methyl orange changes colour over the pH range 3.2 to 4.4, which is on the vertical section of the curve. Therefore methyl orange would be a suitable indicator, although its colour change is difficult to detect precisely.

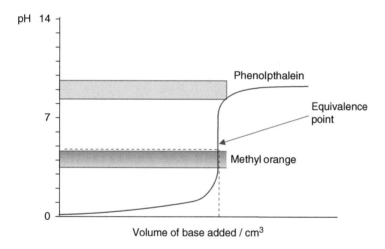

Figure 7.12 Plot of pH against volume of base in a titration of a strong acid (flask) and weak base showing the pH range of phenolphthalein and methyl orange indicators.

Titration of a weak acid and weak base

The pH curve for the titration of a weak acid and weak base is shown in Figure 7.13. This curve is different to the three previous curves in that there is no clear vertical section where the pH changes rapidly for a small increase in volume of acid. The equivalence point actually occurs at the mid-point of the steepest section of the curve, but the gradient is very shallow. It is therefore difficult

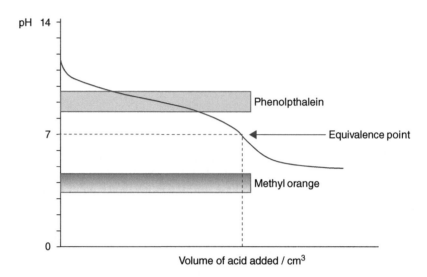

Figure 7.13 Plot of pH against volume of acid in a titration of a weak acid and weak base showing the pH range of phenolphthalein and methyl orange indicators.

to judge the end point precisely. Neither phenolphthalein nor methyl orange would be suitable indicators as their pH change ranges fall outside the equivalence point region. In fact, because of the very gradual pH change around the equivalence point, there is no suitable acid-base indicator for this combination. The best estimate of the equivalence point is obtained by using a pH meter.

Worked Example 7.18   Write a balanced chemical equation and determine a suitable indicator for each of the following acid-base titrations:

    a. 0.10 M nitric acid (HNO_3) against 0.10 M potassium hydroxide (KOH)

    b. 0.10 M benzoic acid (C_6H_5COOH) against 0.10 M sodium hydroxide (NaOH)

    c. 0.10 M ammonium hydroxide (NH_4OH) against 0.10 M hydrochloric acid (HCl)

Solution

    a. Nitric acid is a strong acid and potassium hydroxide a strong base. The pH curve for this titration is similar to that in Figure 7.10, and the pH of the equivalence point is around 7. The equation for the reaction is:

$$HNO_3 + KOH \rightarrow KNO_3 + H_2O$$

       The salt formed is that of a strong acid and strong base so the pH at the equivalence point is around pH 7. Any acid-base indicator such as methyl orange, methyl red, bromothymol blue, or phenolphthalein would be a suitable indicator.

b. Benzoic acid is a weak acid and sodium hydroxide a strong base. The pH curve for the titration is similar to that shown in Figure 7.11. The equivalence point of the titration is around pH 9. The equation for the reaction is:

$$C_6H_5COOH + NaOH \rightarrow C_6H_5COONa + H_2O$$

The salt formed is sodium benzoate (C_6H_5COONa). This is the salt of a weak acid and strong base, so C_6H_5COONa is a **basic salt**. This means that solutions of sodium benzoate are basic. A suitable indicator would be phenolphthalein.

The benzoate ion is the conjugate base of benzoic acid:

$$C_6H_5COOH(s) + H_2O(l) \rightleftharpoons C_6H_5COO^-(aq) + H_3O^+(aq)$$

c. Ammonium hydroxide is a weak base and hydrochloric acid a strong acid. The pH curve for the titration is similar to that in Figure 7.12. The equivalence point of the titration is around pH 5. The equation for the reaction is:

$$NH_4OH + HCl \rightarrow NH_4Cl + H_2O$$

The salt formed is ammonium chloride (NH_4Cl) and is an acidic salt as it is the salt of a strong acid and weak base. A suitable indicator for the titration would be methyl orange. Methyl red and bromothymol blue could also be used.

Ammonium chloride is an acidic salt formed in the reaction between ammonium hydroxide and hydrochloric acid. The ammonium ion (NH_4^+) is the conjugate acid of the base ammonia (NH_3):

$$NH_4^+(aq) + H_2O(l) \rightleftharpoons NH_3(aq) + H_3O^+(aq)$$

Therefore, it can donate a proton to water, increasing the concentration of hydrogen ions in solution and lowering the pH. This is why it is called an acidic salt.

7.3.9   Buffers

A **buffer** is a solution that can resist changes in pH upon the addition of small amounts of acid or base.

Buffers are used widely industrially to ensure constant pH during many processes such as dye production, pharmaceutical manufacture, fermentation, and electroplating. Most personal products such as shampoos and soaps are buffered so they are slightly alkaline and not harmful when in contact with skin. Many foodstuffs contain buffers necessary for maintaining the pH at which enzymes operate and proteins are soluble. When manufacturing foodstuffs, buffers are added to ensure that the properties of the food remain stable under different conditions.

Equally important are physiological buffers such the hydrogen carbonate ion, HCO_3^-, and dihydrogen phosphate and hydrogen phosphate ($H_2PO_4^-$ and HPO_4^-) in blood, which maintain a pH between 7.35 and 7.45; essential for bodily functions.

Box 7.5 The pH of human blood must be maintained at around 7.35–7.45 to allow the many biochemical processes in the body to take place. Blood is a complex mixture that contains various buffers, including:

hydrogen carbonate, HCO_3^-, and carbonic acid, H_2CO_3, and

dihydrogen phosphate, $H_2PO_4^-$, and hydrogen phosphate, HPO_4.

Human cells produce carbon dioxide through aerobic respiration. This combines with water in the blood to form carbonic acid, H_2CO_3, which itself dissociates to produce a solution containing hydrogen carbonate ions and hydrogen ions:

$$CO_2(aq) + H_2O(l) \rightleftharpoons H_2CO_3(aq) \rightleftharpoons HCO_3^-(aq) + H^+(aq)$$

The mixture of the weak acid, carbonic acid, and its salt, hydrogen carbonate, HCO_3^-, creates a buffer that controls the pH of blood at around 7.40. The pK_a for carbonic acid is 6.1, and the ratio of the concentration of HCO_3^- to H_2CO_3 is normally around 20 : 1, which ensures that the pH of blood can be maintained in the required range.

Buffers work by keeping the concentration of hydrogen ions and hydroxide ions in a solution almost constant by shifting the position of an equilibrium reaction. There are two main types of buffers: **acidic buffers** and **basic buffers**. The main difference between the two types is the pH range that the particular buffer can control.

Acidic buffers

Acidic buffers typically are composed of a weak acid and a soluble salt of the acid. They are used to maintain the pH of acidic solutions.

An example of an acidic buffer is a mixture of ethanoic acid and sodium ethanoate. Ethanoic acid is a weak acid and, in aqueous solution, it dissociates to produce ethanoate ions and hydrogen ions:

$$CH_3COOH(aq) \rightleftharpoons CH_3COO^-(aq) + H^+(aq) \tag{7.1}$$

High concentration   Low concentration

As this is a weak acid it partially dissociates, as shown by the equilibrium symbol, $\rightleftharpoons$.

Salts of sodium and potassium are always fully ionised in aqueous solution, so sodium ethanoate dissociates completely to give ethanoate ions and sodium ions in solution:

$$CH_3COONa(s) \xrightarrow{H_2O} CH_3COO^-(aq) + Na^+(aq) \tag{7.2}$$

High concentration

In an aqueous solution made from a mixture of ethanoic acid and sodium ethanoate there is a high concentration of the ethanoate ion, $C_2H_5COO^-$ (from the CH_3COONa in Eq. (7.2)) and a low concentration of hydrogen ions (from the ethanoic acid in Eq. (7.1)). This is because the high concentration of the CH_3COO^- ion drives the equilibrium in Eq. (7.1) to the left. As the equilibrium in Eq. (7.1) lies to the left, there is a high concentration of undissociated CH_3COOH.

Behaviour on the addition of base
If a little base is added to the acidic buffer the OH^- ions from the base react with the H^+ ions, forming water:

$$H^+(aq) + OH^-(aq) \rightleftharpoons H_2O(l)$$

As H^+ ions are used up their concentration decreases, so the equilibrium in Eq. (7.1) shifts to the right to produce more H^+ ions. Because the concentration of undissociated CH_3COOH is high, a large reservoir of H^+ ions is available. The pH therefore remains constant until all the CH_3COOH is used up.

Behaviour on the addition of acid
If a little acid is added to the buffer solution the concentration of H^+ ions increases. This shifts the equilibrium in Eq. (7.1) to the left, and the additional H^+ reacts with the CH_3COO^- ions:

$$CH_3COO^-(aq) + H^+(aq) \rightleftharpoons CH_3COOH(aq) \quad \text{(reverse of Eq.(7.1))}$$

There is a high reservoir of CH_3COO^- ions from the CH_3COONa, and the pH will remain constant until all the CH_3COO^- is used up.

Basic buffers

A basic buffer is used when a constant pH above 7 is required. A basic buffer is composed of a weak base and the soluble salt of the base. For example, the weak base could be ammonium hydroxide and the soluble salt ammonium chloride.

Ammonium hydroxide dissociates partially in aqueous solution to produce ammonium ions and hydroxide ions:

$$NH_4OH(aq) \rightleftharpoons NH_4^+(aq) + OH^-(aq)$$

High concentration   Low concentration

(7.3)

Ammonium chloride is a soluble salt that is fully ionised in aqueous solution:

$$NH_4Cl(aq) \rightarrow NH_4^+(aq) + Cl^-(aq)$$

High concentration

(7.4)

Because ammonium chloride is fully ionised according to Eq. (7.4), there is a high concentration of NH_4^+ ions in the buffer solution. This drives the equilibrium in Eq. (7.3) back to the left-hand side, producing high concentrations of undissociated NH_4OH.

Behaviour on the addition of base
If a small quantity of base is added the OH^- ion concentration increases. The excess OH^- ions react with the NH_4^+ ions, causing the equilibrium in Eq. (7.3) to shift to the left. There is a high concentration of NH_4^+ ions from Eq. (7.4), so the pH remains stable until all the NH_4Cl is used up.

Behaviour on the addition of acid
Addition of a small quantity of acid causes the H^+ concentration to increase. The OH^- ions mop up the additional H^+ ions to form water:

$$H^+(aq) + OH^-(aq) \rightleftharpoons H_2O(l)$$

As the OH^- ions are removed more of the ammonium hydroxide dissociates according to Eq. (7.3) to replace them. The amount of undissociated ammonium hydroxide present is high, as the equilibrium in Eq. (7.3) lies over to the left. Thus the pH remains constant until all the NH_4OH is used up, when a small amount of additional acid will lower the pH.

7.3.10   Calculating the pH of a buffer solution

The pH of a buffer solution is determined by the concentration of hydrogen ions in the solution. In an acidic buffer, the hydrogen ion concentration can be found directly from the equilibrium expression for the weak acid present in the buffer:

$$HA(aq) \rightleftharpoons H^+(aq) + A^-(aq)$$

$$K_a = \frac{[H^+(aq)][A^-(aq)]}{[HA(aq)]}$$

$$[H^+(aq)] = \frac{K_a \times [HA(aq)]}{[A^-(aq)]}$$

In the buffer solution the concentration of HA is equivalent to the concentration of weak acid added, [HA(aq)]. The concentration of A^- is mainly due to the salt and can be equated to the concentration of the salt of the weak acid, [A^-(aq)]. Substituting:

$$[HA(aq)] = [acid] \quad and \quad [A^-(aq)] = [salt]$$

the equation for [H^+(aq)] becomes:

$$[H^+(aq)] = \frac{K_a \times [acid]}{[salt]}$$

Taking the log to the base 10 of both sides of this equation:

$$\log_{10}[H^+(aq)] = \log_{10}(K_a) + \log_{10}\left(\frac{[acid]}{[salt]}\right)$$

Multiply through by –1:

$$-\log[H^+(aq)] = -\log(K_a) - \log\left(\frac{[acid]}{[salt]}\right)$$

As $pH = -\log[H^+(aq)]$ and $pK_a = -\log(K_a)$, the equation becomes:

$$pH = pK_a + \log\left(\frac{[salt]}{[acid]}\right)$$

Maths alert! The value of
$-\log_{10}\left(\frac{[A]}{[B]}\right) = +\log_{10}\left(\frac{[B]}{[A]}\right)$

This equation is known as the Henderson–Hasselbalch equation and allows calculation of the pH of a buffer solution made up from a known concentration of weak acid and its salt. Notice that the pH of the buffer is determined by the ratio of the concentration of the conjugate base (salt) to conjugate acid (acid).

Worked Example 7.19   Calculate the pH of a buffer solution made up from benzoic acid of concentration $0.050 \, mol \, dm^{-3}$ and sodium benzoate of concentration $0.100 \, mol \, dm^{-3}$ (pK_a benzoic acid = 4.20).

Solution
The pH of a buffer solution can be calculated using the Henderson–Hasselbalch equation, $pH = pK_a + \log \left(\frac{[salt]}{[acid]} \right)$.

Substituting in the values given in the equation we obtain:

$$pH = 4.20 + \log \left(\frac{(0.100 \, mol \, dm^{-3})}{(0.500 \, mol \, dm^{-3})} \right) = 4.20 + \log (0.2) = 4.20 + (-0.70)$$

$$= 3.50$$

In some cases a weak acid and strong base are mixed to produce a buffer solution. For example, the reaction of benzoic acid with sodium hydroxide leads to the formation of the salt sodium benzoate. To calculate the pH of the resulting buffer solution, the amount of sodium benzoate formed must first be calculated.

Worked Example 7.20   Calculate the pH of the buffer solution formed when 500 cm^3 of 0.100 $mol \, dm^{-3}$ NaOH is added to 500 cm^3 of 0.500 $mol \, dm^{-3}$ benzoic acid (pK_a benzoic acid = 4.20).

Solution
The strong base NaOH reacts with the weak acid benzoic acid, to give the salt sodium benzoate. The concentration of the benzoate ion, $C_6H_5COO^-$, must first be calculated from this equation:

$$C_6H_5COOH(aq) + NaOH(aq) \rightleftharpoons C_6H_5COONa(aq) + H_2O(l)$$

This equation shows that benzoic acid and sodium hydroxide react together in a 1:1 ratio:

Number of moles of NaOH added

$$= c \times V = 0.100 \, mol \, dm^{-3} \times 500 \times 10^{-3} \, dm^3 = 0.05 \, mol$$

Therefore, the number of moles of sodium benzoate formed is 0.05 mol in a solution of total volume 1000 cm^3.
Concentration of sodium benzoate

$$= \frac{n(\text{sodium benzoate})}{\text{total volume}} = \frac{0.05 \, mol}{1000 \times 10^{-3} \, dm^3} = 0.05 \, mol \, dm^{-3}.$$

This is the [salt] in the equation to calculate pH.
Next we must calculate the concentration of weak acid, [C_6H_5COOH(aq)]. We know the total number of moles of benzoic acid used initially, but some of this has reacted with the sodium hydroxide, so we need to calculate the remaining number of moles of benzoic acid:

Initial number of moles of benzoic acid

$$= c \times V = 0.500 \, mol \, dm^{-3} \times 500 \times 10^{-3} \, dm^3 = 0.250 \, mol$$

Number of moles of benzoic acid remaining

= initial number of moles of benzoic acid – number of moles of benzoic acid reacted

= initial number of moles of benzoic acid – number of moles

of NaOH = 0.250 mol – 0.05 mol = 0.20 mol

This amount of benzoic acid is present in a solution of total volume 1000 cm^3. The concentration of benzoic acid is therefore

$$= \frac{n(\text{benzoic acid})}{V_{total}} = \frac{0.20\,\text{mol}}{1000 \times 10^{-3}\,\text{dm}^3} = 0.20\,\text{mol dm}^{-3}.$$

This value is therefore equal to [acid] in the equation for pH.

The equation to be used is $\text{pH} = pK_a + \log\left(\frac{[\text{salt}]}{[\text{acid}]}\right)$.

Inserting the calculated values for [salt] = 0.05 mol dm^{-3} and [acid] = 0.20 mol dm^{-3}, we obtain

$$\text{pH} = 4.20 + \log\left(\frac{0.05\,\text{mol dm}^{-3}}{0.20\,\text{mol dm}^{-3}}\right)$$
$$= 4.20 + \log(0.25) = 4.20 + (-0.60) = 3.60.$$

Calculating the pH change upon the addition of acid

Buffer solutions can withstand the addition of small quantities of acid and base without showing a significant pH change. The pH change obtained can be calculated again using the Henderson–Hasselbalch equation.

Worked Example 7.21   Calculate the pH change observed in 1.0 dm^3 of an acidic buffer made up of 500 cm^3 0.10 M ethanoic acid and 500 cm^3 0.10 M sodium ethanoate when 20 cm^3 of HCl of concentration 1.0 M is added (pK_a ethanoic acid = 4.76).

Solution

First calculate the initial pH of the buffer using $\text{pH} = pK_a + \log\left(\frac{[\text{salt}]}{[\text{acid}]}\right)$

$$[\text{acid}] = \frac{n(\text{benzoic acid})}{V\,(\text{solution})} = \frac{0.10 \times 500 \times 10^{-3}}{1000 \times 10^{-3}} = 0.05\,\text{mol dm}^{-3}$$

$$[\text{salt}] = \frac{n(\text{sodium ethanoate})}{V(\text{solution})} = \frac{0.10 \times 500 \times 10^{-3}}{1000 \times 10^{-3}} = 0.05\,\text{mol dm}^{-3}$$

Substituting the values into the equation for pH gives:

$$\text{pH} = 4.76 + \log\left(\frac{0.05}{0.05}\right) = 4.76 + \log(1.0) = 4.76 + 0 = 4.76$$

The hydrochloric acid acts to increase the total acid concentration and decrease the salt concentration.

The final volume of the buffer solution, V_f is (1000 + 20) cm^3 and the final total acid concentration is:

$$[acid]_f = \frac{n(\text{ethanoic acid}) + n(\text{HCl})}{V_f}$$

$$= \frac{(0.05 + 1.00 \times 20 \times 10^{-3})\ \text{mol}}{(1000 + 20) \times 10^{-3}\ \text{dm}^3}$$

$$= 0.0686\ \text{mol dm}^{-3}$$

The final salt concentration is:

$$[salt]_f = \frac{n(\text{sodium ethanoate}) - n(\text{HCl})}{V_f}$$

$$= \frac{(0.05 - 1.00 \times 20 \times 10^{-3})\ \text{mol}}{(1000 + 20) \times 10^{-3}\ \text{dm}^3}$$

$$= 0.0294\ \text{mol dm}^{-3}$$

We now have the new concentrations of acid and salt to insert into the equation for pH:

$$pH = pK_a + \log\left(\frac{[salt]_f}{[acid]_f}\right) = 4.76 + \log\left(\frac{0.0294\ \text{mol dm}^{-3}}{0.0686\ \text{mol dm}^{-3}}\right)$$

$$= 4.76 + \log 0.429 = 4.76 + (-0.37) = 4.39$$

So the pH has changed from 4.76 to 4.39, a total of 0.37 pH units, which is a very small difference in pH and only detectable by a pH meter despite the addition of 20 cm^3 of a strong acid of concentration 1 M.

7.3.11   Lewis acids and bases

In this chapter we have been referring to the Brønsted–Lowry definition of an acid as a proton donor and a base as a proton acceptor. This is useful when discussing pH and acidity, as pH is dependent upon hydrogen ion (proton) concentration.

However, there is an alternative theory relating to acids and bases called the *Lewis theory*. This theory defines an acid as an electron-pair acceptor and a base as an electron-pair donor. Using this convention, the H$^+$ ion is an electron-pair acceptor as it can accept a lone pair of electrons from a water molecule, for example, to produce the H$_3$O$^+$ ion.

The H$^+$ ion acts as both a Lewis acid and a Brønsted–Lowry acid and the water molecule acts as a Lewis base and a Brønsted–Lowry base.

In the following reaction boron trifluoride, BF_3, acts as a Lewis acid by accepting a pair of electrons from ammonia, NH_3, which acts as a Lewis base. However, BF_3 cannot be an acid under the Brønsted–Lowry theory as it has no protons. But all Brønsted–Lowry acids are Lewis acids. So the Lewis acid-base theory is more widely applicable than the Brønsted–Lowry definition.

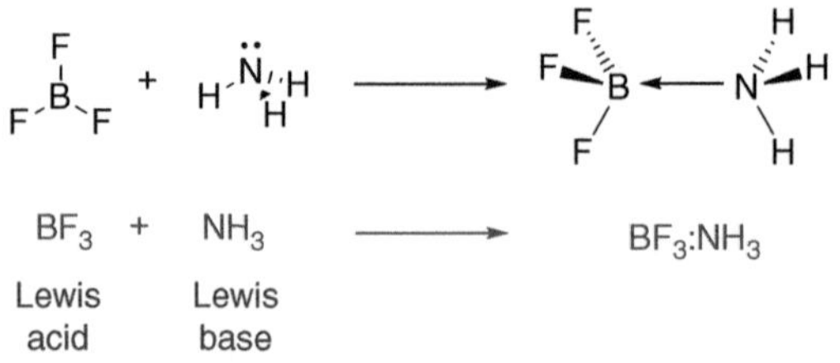

$$BF_3 \quad + \quad NH_3 \quad \longrightarrow \quad BF_3{:}NH_3$$

Lewis acid   Lewis base

Quick-check summary

- A reversible chemical reaction is one that does not go to completion and the products can be converted back to the reactants.

- A reversible chemical reaction reaches equilibrium when the rates of the forward and backward reactions become equal.

- A chemical equilibrium is a dynamic equilibrium because reaction continues to occur in the forward and backward directions.

- The equilibrium constant for a reaction, K_c, relates the concentrations of products at equilibrium to the concentrations of reactants.

- Le Châtelier's principle states that when a reaction at equilibrium is subjected to a change, the equilibrium shifts to minimise the change.

- Changes in temperature, pressure, and concentration affect the position of equilibrium, but only a change in temperature affects the equilibrium constant.

- The equilibrium constant, K_c, can be written in terms of concentrations of reactants and products and K_p can be written for gaseous reactions in terms of partial pressures of reactants and products.

- A homogeneous equilibrium is one in which the reactants and products are in the same physical state whereas a heterogeneous reaction has reactants and products in more than one physical state.

- The Brønsted–Lowry theory of acids and bases defines an acid as a proton donor and a base as a proton acceptor.

- A strong acid is totally ionised (or dissociated) in aqueous solution, whereas a weak acid is only partially ionised (or dissociated).

- The pH of a solution is a measure of the acidity of the solution and is expressed on a scale of 0 to 14. The expression for pH $= -\log_{10}[H^+]$.

- The acid dissociation constant, K_a, represents the equilibrium constant for the dissociation of a weak acid.

- Acid strengths are measured by pK_a values, where $pK_a = -\log_{10}[K_a]$.

- The base dissociation constant, K_b, represents the equilibrium constant for the dissociation of a weak base.

- Base strengths are measured by pK_b values, where $pK_b = -\log_{10}[K_b]$.

- A conjugate acid and base are linked by the gain and loss of a proton.

- Indicators are weak acids that change colour at different pH values and can therefore show approximate pH values of solutions.

- A titration is a procedure that can be used to determine the concentration of an acid or base, and the end point or equivalence point is shown by the use of an indicator or a pH meter.

- The pH of the end point of a titration depends upon the combination of acid and base used. A chemical indicator cannot be used to determine the end point in the titration of a weak acid against a weak base; a pH meter must be used.

- A buffer is a solution that can resist changes in pH caused by the addition of small amounts of acids or bases. An acidic buffer is made up from a weak acid and the salt of the weak acid, and a basic buffer is made up from a weak base and the salt of the weak base.

- The Henderson–Hasselbalch equation allows the calculation of the pH of a buffer solution.

- The Lewis theory of acids and bases defines an acid as an electron-pair acceptor and a base as an electron-pair donor.

End-of-chapter questions

1 When N_2O_4 is kept in a flask at 1.0 atm and 298 K, it dissociates according to the following equation:

$$N_2O_4\,(g) \rightleftharpoons 2NO_2\,(g)$$

   If the final equilibrium partial pressure of N_2O_4 is 0.82 atm, what is the equilibrium partial pressure of NO_2 (g)?

2 For the reaction

$$N_2O_4\,(g) \rightleftharpoons 2NO_2\,(g)$$

   K_p is 1.78×10^4 Pa at 600 °C and 2.82×10^4 Pa at 1000 °C. Deduce whether the following statements are true or false:

   a. The formation of NO_2(g) is favoured by a rise in temperature.

   b. ΔH is negative.

   c. The units of K_p are atm^{-1}.

   d. The formation of NO_2(g) is favoured by a decrease in pressure.

3 Five moles of nitrogen gas and ten moles of hydrogen gas were mixed in a vessel of volume 10 dm^3 at 400 K. At equilibrium 1.8 moles of NH_3 gas were formed. The equilibrium can be represented by this equation:

$$N_2(g) + 3H_2(g) \rightleftharpoons 2NH_3(g)\ \Delta H = -92.4\,kJ\,mol^{-1}$$

a. Use the information provided to calculate the equilibrium constant for the reaction at 400 K.

b. Explain the effect of the following changes in reaction conditions upon the equilibrium position and the equilibrium constant:

    i. An increase in temperature

    ii. An increase in the external pressure

    iii. Addition of more nitrogen gas

4 Esters are important organic compounds widely used as flavourings. Esters such as ethyl ethanoate can be formed by the reaction of carboxylic acids with alcohols. The reaction reaches equilibrium:

$$CH_3COOH(aq) + CH_3CH_2OH(aq) \rightleftharpoons CH_3COOC_2H_5(aq) + H_2O(l)$$

At 298 K, the mixture at equilibrium was found to contain 0.5 moles of ethanoic acid, 0.5 moles of ethanol, 0.1 moles of ethyl ethanoate, and 0.1 mole of water. The final volume of the solution was 500 cm^3.

Calculate the value of K_c for the equilibrium.

5 The following equation represents the reaction used to produce sulfur trioxide, a starting material required for the Contact process for making sulfuric acid:

$$2SO_2(g) + O_2(g) \rightleftharpoons 2SO_3(g) \; \Delta H = -197 \, kJ \, mol^{-1}$$

a. 3.00 moles of sulfur dioxide were mixed with 1.50 moles of oxygen gas and allowed to reach equilibrium in a vessel of volume 5.00 dm^3 at a fixed temperature. At equilibrium there were found to be 1.20 moles of sulfur trioxide present. Calculate the value of K_c for the equilibrium at this temperature.

b. If the total pressure of the system is 200 kPa, calculate a value for K_p.

6* If K_p for the following reaction is 0.54 atm and the initial partial pressure of PCl_5 was 1.20 atm, what is the partial pressure of PCl_5 at equilibrium?

$$PCl_5(g) \rightleftharpoons PCl_3(g) + Cl_2(g)$$

7 a. Describe the difference between heterogeneous and homogeneous equilibria.

b. The equation for the thermal decomposition of ammonium carbamate, $NH_4(NH_2CO_2)$ is:

$$NH_4(NH_2CO_2)(s) \rightleftharpoons 2NH_3(g) + CO_2(g)$$

In this dissociation, 25.00 g of $NH_4(NH_2CO_2)$ (s) were placed in an evacuated 250 cm^3 flask and held at 25 °C. After some hours equilibrium was reached and 0.022 g of CO_2 were present. How many moles of CO_2 and NH_3 are present at equilibrium?

c. Calculate a value for the equilibrium constant, K_c, at this temperature.

d. What mass of CO_2 would be formed at equilibrium if 50.00 g of ammonium carbamate had been present initially?

e. Explain whether the amount of CO_2 and NH_3 is likely to increase or decrease when the temperature is raised to $50\,°C$.

8   Calculate the pH of the following:

a. 1.0×10^{-4} M HCl

b. 1.0×10^{-4} M NaOH

c. 0.01 M propanoic acid $(K_a = 1.45 \times 10^{-5}\ \text{mol dm}^{-3})$

d. 1.0 M NH_4OH $(K_b = 1.7 \times 10^{-5}\ \text{mol dm}^{-3})$

e. A solution that is 0.015 M HF $(K_a = 3.5 \times 10^{-4}\ \text{mol dm}^{-3})$ and 0.011 M NaF

9   Butanoic acid, $CH_3(CH_2)_2COOH$, is the acid formed when butter turns rancid and tastes sour. A solution of butanoic acid was prepared with a concentration of $0.250\ \text{mol dm}^{-3}$. The K_a of butanoic acid is $1.51 \times 10^{-5}\ \text{mol dm}^{-3}$.

a. Calculate the pH of the solution of butanoic acid.

b. $100\ \text{cm}^3$ of sodium butanoate of concentration $0.100\ \text{mol dm}^{-3}$ was added to $100\ \text{cm}^3$ of the butanoic acid solution. Calculate the pH of the resultant buffer formed.

c. $20\ \text{cm}^3$ of HCl of concentration $0.10\ \text{mol dm}^3$ was added to the buffer solution. Calculate the new pH.

10   The following plot shows the titration curves (a and b) for two different acids with 0.1 M sodium hydroxide. For each of the curves, state:

a. Whether the acid is a strong or weak acid

b. A suitable indicator (if any) for the titration

c. The approximate pH of the end point

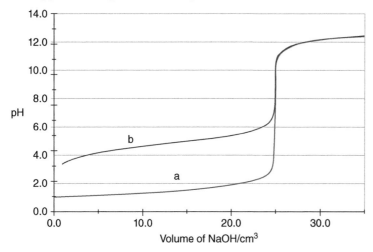

*This question requires you to use the quadratic formula: $x = -b \pm \sqrt{(b^2 - 4ac)}/2a$ which gives the roots of x for the quadratic equation $ax^2 + bx + c = 0$. You may not be expected to cover this in your course.

8

Chemical kinetics – the rates of chemical reactions

At the end of this chapter, students should be able to:

- Define and understand the following terms:

  - Rate of reaction

  - Reaction mechanism

  - Rate expression

  - Rate constant

  - Overall order of a reaction

  - Order with respect to a reactant

  - Half-life of a reaction

  - Rate-determining step

- Use collision theory to explain how changes in reaction conditions of temperature, concentration, and the presence of a catalyst affect the rate of a chemical reaction

- Outline the difference in behaviour between heterogeneous and homogeneous catalysts

- Define the average rate of reaction and the instantaneous rate of reaction, and understand how the latter changes throughout a reaction

- Describe how experiments can be used to determine the rate expression for a chemical reaction

Foundations of Chemistry: An Introductory Course for Science Students, First Edition.
Philippa B. Cranwell and Elizabeth M. Page.
© 2021 John Wiley & Sons Ltd. Published 2021 by John Wiley & Sons Ltd.
Companion website: www.wiley.com/go/Cranwell/Foundations

- Be familiar with plots of concentration against time for zero-, first-, and second-order reactions

- Be familiar with the appropriate plots used to obtain the rate constant from the integrated rate expressions for zero-, first-, and second-order reactions

- Understand how the half-life of a reaction can be obtained from concentration against time plots, and how half-life depends upon the order of the reaction

- Predict a rate expression consistent with the reaction mechanism for a simple multi-step reaction

- Use the Arrhenius equation to obtain the activation energy for a reaction from rate constant and temperature data

8.1 Introduction

We saw in the previous chapter that the economic viability of industrial chemical processes depends upon a number of factors. Chemists and chemical engineers must be able to control reaction conditions in order to obtain maximum yields by optimising factors such as temperature and pressure. Equally important is the time required to produce products. Fast reaction times reduce costs and energy requirements. Reaction times are critical when designing drugs and in drug delivery. To be effective a drug must be sufficiently inert that it reaches its target before breaking down but must then react quickly and produce waste products that are readily removed from the body. The area of chemistry concerned with studying and controlling the rates of chemical reactions is known as **chemical kinetics**. Chemists study the rates at which chemical reactions occur so they can control them. The series of molecular processes that occur when a chemical reaction takes place is called the **mechanism** of the reaction, and understanding the mechanism helps chemists control the **rate of reaction**.

8.2 The rate of reaction

8.2.1 Defining the rate of a chemical reaction

The rate of a chemical reaction is defined as the increase in concentration of one of the products of reaction divided by the time taken. Alternatively, it can be defined as the decrease in concentration of one of the reactants divided by the time:

$$\text{Rate of reaction} = \frac{\text{change in concentration of reactant (or product)}}{\text{time taken for the change}}$$

A plot of concentration against time is given in Figure 8.1 for the hypothetical reaction of reactant A being converted to product B, as represented by the equation A $\rightarrow$ B. The rate can be expressed as:

$$\text{Rate of reaction} = \frac{\text{change in concentration of B}}{\text{time}} \text{ or } \frac{\Delta[B]}{\Delta t}$$

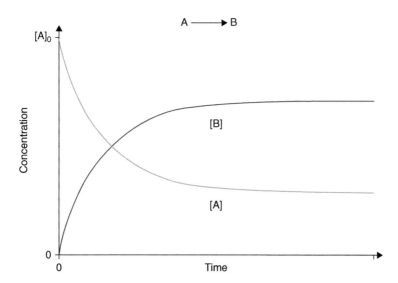

Figure 8.1 Plot of the concentration of A and B against time for the reaction $A \rightarrow B$. $[A]_0$ is the concentration of A at time t=0.

The symbol Δ (Greek letter delta) means 'a change', so $\Delta[B]$ represents a change in concentration of B and Δt is the time taken for this change to occur. The units for reaction rate are therefore units of concentration divided by time: typically, $mol\,dm^{-3}\,s^{-1}$. If the reactant is a gas, the unit may be $Pa\,s^{-1}$ or $atm\,min^{-1}$, depending upon the units in which pressure and time are measured.

Rates of reaction must always be positive. So, if the rate is expressed as a change in concentration of the reactant A, a negative sign must be introduced into the expression. This is because the quantity $\Delta[A]$ will itself be a negative number as the amount of A is decreasing through the reaction. The rate of reaction with respect to reactant A is given by:

$$\text{Rate of reaction} = -\frac{\Delta[A]}{\Delta t}$$

In the reaction $A \rightarrow B$, the rate at which B is being produced must be the same as the rate at which A is being used, so we can state:

$$\text{Rate of reaction} = \frac{\Delta[B]}{\Delta t} = -\frac{\Delta[A]}{\Delta t}$$

Most reactions involve more than one product or reactant. For example, in the reaction between hydrogen and iodine to form hydrogen iodide, two different molecules combine to produce a single product:

$$H_2(g) + I_2(g) \rightarrow 2HI(g).$$

The rate of reaction can be measured by the rate of formation of the product HI with time and we will use the symbol R' to represent this rate: $R' = \frac{\Delta[HI]}{\Delta t}$. This is the rate of reaction with respect to HI.

Another way of expressing the rate of reaction is as the rate of disappearance of iodine with time, and we will use the symbol R'' to represent this rate:

$$R'' = -\frac{\Delta[I_2]}{\Delta t}.$$

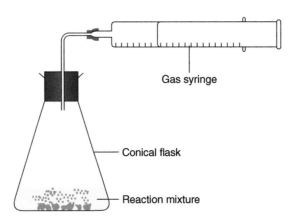

Figure 8.2 Apparatus to measure the amount of gas produced in a reaction over time. Source: www.markedbyteachers.com/gcse/science/chemistry-rate-of-reaction-coursework-for-calcium-carbonate-and-hydrochloric-acid.html.

In this reaction the two rates are not the same because two moles of HI are formed for every one mole of I_2 used. Therefore, HI is being formed faster than I_2 is being used; in fact, twice as fast. The relationship between the two rates of reaction is:

$$\frac{\Delta[HI]}{\Delta t} = 2 \times \left(-\frac{\Delta[I_2]}{\Delta t}\right).$$

This equation can be rearranged to obtain an expression for the rate of reaction with respect to iodine:

$$-\frac{\Delta[I_2]}{\Delta t} = \frac{1}{2}\frac{\Delta[HI]}{\Delta t}.$$

The rate at which hydrogen gas is being used up is the same as the rate at which iodine is disappearing, so we can also write:

$$R'' = -\frac{\Delta[I_2]}{\Delta t} = -\frac{\Delta[H_2]}{\Delta t} = \frac{1}{2}\left(\frac{\Delta[HI]}{\Delta t}\right)$$

The actual rate at which a chemical reaction occurs can only be determined by carrying out a suitable experiment to measure how the amount of a product or reactant formed changes with time. For example, if the reaction is one in which a gas is produced, the volume of gas obtained is measured at successive time intervals using apparatus such as that shown in Figure 8.2.

Worked Example 8.1   The equation for the combustion of methane is

$$CH_4(g) + 2O_2(g) \rightarrow CO_2(g) + 2H_2O(g)$$

The rate of consumption of oxygen, $\dfrac{-\Delta[O_2]}{\Delta t}$, was found to be = 0.2 mol s^{-1}.

a. What is the average reaction rate with respect to the combustion of methane?

b. What is the rate of production of gaseous carbon dioxide?

c. If initially there were 10.00 moles of methane, how many moles of methane would be present after 40 seconds?

Solution

a. The equation for the reaction shows that oxygen is used up twice as fast as methane. This can be expressed as:

$$\frac{-\Delta[O_2]}{\Delta t} = 2 \times \frac{-\Delta[CH_4]}{\Delta t}.$$

If the rate of disappearance of oxygen is $0.2\ mol\ s^{-1}$, then the rate of disappearance of methane is $0.1\ mol\ s^{-1}$.

b. The equation for the reaction shows that carbon dioxide is produced at the same rate as methane is used up. So, if the rate of disappearance of methane is $0.1\ mol\ s^{-1}$, the rate of formation of carbon dioxide is the same: i.e. $0.1\ mol\ s^{-1}$.

c. From part (a), methane is being used up at $0.1\ mol\ s^{-1}$. In 40 s, the amount in moles of methane that is used up $= 40\ s \times 0.1\ mol\ s^{-1}$ $= 4\ mol$.
The amount of methane left is $(10-4)\ mol = 6\ mol$.

8.2.2 Collision theory

In order to understand how chemical reactions take place we use a model of molecular behaviour called **collision theory**. This is based on experiment and states that in order for particles to react chemically they must first collide with each other in the correct orientation and they must have sufficient energy. By *particles* we mean atoms, molecules, or ions.

In Chapter 4, kinetic theory explained that particles are in constant motion in a liquid and a gas. There are therefore many collisions between particles in a reaction involving liquids and gases. Some of these interactions may lead to

(a)

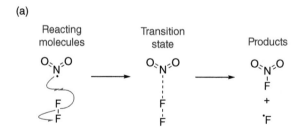

Reacting molecules in correct orientation so reaction occurs

(b)

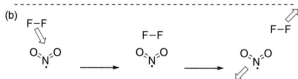

Reacting molecules in incorrect orientation so no reaction occurs

Figure 8.3 The reaction of NO_2 with F_2: (a) successful collision leading to the formation of the transition state and product NO_2F; (b) unsuccessful collision where the orientation of molecules is unsuitable for reaction to take place.

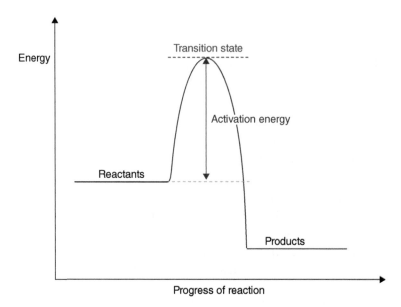

Figure 8.4 Energy profile for an exothermic reaction.

the formation of new products. However, in many interactions the particles simply collide and bounce away like two balls on a snooker table. If the colliding particles have enough energy a successful reaction may take place, provided the particles meet in the correct orientation. Figure 8.3 shows two possible outcomes of collisions between a molecule of nitrogen dioxide, NO_2 and fluorine, F_2 to produce nitryl fluoride, NO_2F. In pathway (a) the F_2 molecule collides with the nitrogen atom of the NO_2 molecule. Providing the two molecules have sufficient energy, a successful collision takes place with the formation of products. In pathway (b) the F_2 molecule collides with an oxygen atom of the NO_2 molecule and no reaction occurs. The molecules bounce away.

The minimum energy required for a successful collision is called the **activation energy**. Activation energy has the symbol E_a. A successful reaction proceeds through a **transition state** in which bonds in the reactants are being broken and new bonds in the products are being formed. The transition state for the successful reaction between NO_2 and F_2 is depicted in Figure 8.3a. The rate of the reaction depends upon the number of successful collisions that take place. Figure 8.4 shows the energy profile for an exothermic reaction with the activation energy labelled. The activation energy is the difference in energy between the reactants and the transition state. The activation energy is required to initiate bond breaking. Once new bonds are formed, energy is released that propagates the reaction.

8.2.3   Factors that affect the rate of a reaction

Collision theory allows us to predict the effect of various changes in conditions upon the rate of reaction. The main variables that can be changed are the concentration or partial pressure of reactants, the external pressure, the temperature, and the presence of a catalyst.

Concentration or pressure of reactants

The rate of a reaction depends upon the number of successful collisions that take place between the reacting particles in a certain period of time. The greater the number of particles in a fixed volume, the greater the frequency of collisions.

Therefore, the rate of reaction depends upon the concentration of the particles involved in the reaction. If the reaction takes place in solution the rate will depend upon the concentration of the reactant measured in mol dm^{-3}. If the reaction is between gas molecules, the rate of reaction will depend upon the partial pressures of the gases measured in atmospheres or pascals or some other unit of pressure. The partial pressure is equivalent to the concentration of the gas.

We will study how the concentration of reactants affects the rate of reaction in more detail later in this chapter.

The rate of reaction also depends upon the external pressure of the gas mixture. Increasing the external pressure (or decreasing the volume) results in more molecules of gas per unit volume, and the molecules are forced closer together, so again the frequency of collisions and the rate of reaction increase.

Temperature

Increasing the temperature of a reacting mixture increases the average kinetic energy of the particles. This increases the likelihood of molecular collisions in a sample and will therefore increase the rate of reaction. However, the effect on the reaction rate is not simply due to the increased number of collisions; it also depends upon the energy of the particles. Not all particles in a mixture have the same energy. A few will have a relatively low amount of energy, some will have much higher energy, and the rest will have an amount of energy that lies somewhere in between. The distribution of the energies possessed by particles in a sample at a specific temperature is said to follow a **Boltzmann distribution**, which is shown in the graph in Figure 8.5. The activation energy, E_a, for a specific reaction has been marked on the x-axis. The area under the curve to the right of the activation energy represents the number of molecules possessing the activation energy or higher. These molecules will have sufficient energy to go on and react in this particular reaction at this temperature.

If the temperature is increased, the average energy of the particles increases, so the peak in the curve shifts to a higher energy. Figure 8.6 shows the

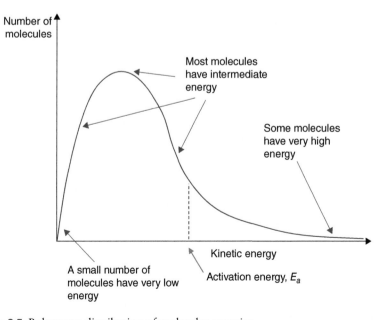

Figure 8.5 Boltzmann distribution of molecular energies.

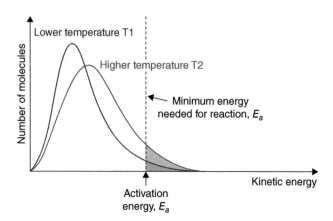

Figure 8.6 Boltzmann distribution of molecular energies at two different temperatures.

distribution of molecular energies at two different temperatures, T1 and T2, where T1 (shown in blue) is lower than T2 (shown in red). The area under each curve represents the total number of molecules, which is the same in each case. As the temperature increases and the average kinetic energy of the molecules increases, the curve becomes flatter and the maximum shifts to a higher energy.

When the temperature is increased from T1 to T2, the number of molecules with energy greater than the activation energy (i.e. molecules under the appropriate curve to the right of the line representing E_a) increases. Thus the proportion of successful collisions increases and the rate of reaction will increase. For every ten degree rise in temperature the number of molecules possessing the activation energy roughly doubles, as does the rate of reaction. We will study how temperature affects the rate of a chemical reaction in more detail later in the chapter.

In summary, an increase in temperature increases the rate of reaction because:

- The average kinetic energy of the particles increases, which increases the frequency of collisions.

- The number of particles possessing the activation energy increases, which leads to more successful collisions.

Catalysts

A **catalyst** is a substance that speeds up the rate of a chemical reaction but remains chemically unchanged itself after reaction. Catalysts are therefore critical in increasing the rate of reactions and are widely used industrially in many chemical processes as they allow shorter reaction times and the use of lower temperatures and/or pressures to obtain the maximum yield of products. Biological catalysts known as **enzymes** are used widely in physiological processes. Enzymes are large protein molecules that enable biochemical reactions to take place quickly.

Both catalysts and enzymes work by providing an alternative reaction pathway that has a lower activation energy. Figure 8.7 shows the energy profile for a catalysed and uncatalysed reaction. The catalysed reaction represented by the red line has a lower activation energy than the uncatalysed reaction (blue line). Because a catalyst provides a pathway with a lower activation energy, there will

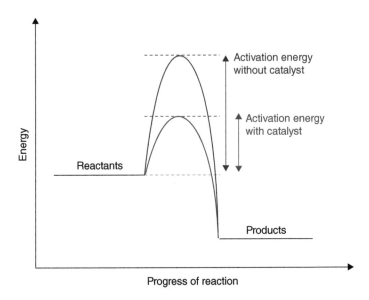

Figure 8.7 Energy profile for a catalysed (red) and uncatalysed (blue) reaction.

be many more molecules in the reaction mixture that possess the activation energy and therefore more successful collisions.

Homogeneous and heterogeneous catalysts
Catalysts can be defined as either homogeneous or heterogeneous. A **homogeneous** catalyst is one that is in the same physical state as the reactants. This type of catalysis usually occurs in the liquid or gas phase. An enzymatic reaction is an example of homogeneous catalysis in a solution.

Box 8.1 An undesirable example of homogeneous catalysis occurs in the gas phase in the atmosphere. Chlorofluorocarbons (CFCs) in the atmosphere break down to form chlorine radicals. This occurs when a single C-Cl bond in the chlorofluorocarbon molecule is broken evenly so that the atoms forming the bond each accept one electron from the bond. A chlorine radical, represented by Cl•, is formed; and this goes on to react with ozone, O_3. The chlorine radicals convert ozone to oxygen and then are regenerated as chlorine radicals that can react with more ozone molecules, as illustrated in the following equations:

$$Cl• + O_3 \rightarrow ClO• + O_2$$

$$ClO• + O_3 \rightarrow Cl• + 2O_2$$

A small quantity of CFCs can destroy a large quantity of ozone. The ozone in the atmosphere provides the beneficial effect of absorbing ultraviolet radiation from sunlight. Now, decades after CFCs and other ozone-depleting chemicals were banned in the late 1980s, we are seeing that the hole in the ozone layer is the smallest it has been since measurements began. This is hopefully the start of a long-term recovery.

False-colour view of total ozone over the Antarctic pole (September 2019). The purple and blue colours are where there is the least ozone, and the yellows and reds are where there is more ozone. Source: From NASA/Ozone Watch.

Heterogeneous catalysis occurs when the catalyst is in a different physical state than the reacting molecules. This type of catalysis is more common in industrial processes such as the hydrogenation (reduction with hydrogen) of alkenes to alkanes (Box 8.2). Catalytic converters made from platinum or

Box 8.2

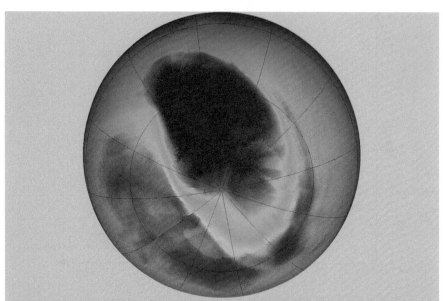

Possible mechanism for reduction (hydrogenation) of ethene to ethane on a heterogeneous metal catalyst.

In a gas-phase reaction, such as the hydrogenation of an alkene, a nickel catalyst on a solid carbon or silicon support provides a surface that can hold the gaseous reactants in place and bring about an effective collision that would be less likely to occur successfully if the gaseous molecules simply collided. An example of this is the reduction of ethene, C_2H_4, by H_2. Molecules of hydrogen gas are adsorbed onto the surface of the catalyst and activated. The approaching ethene molecule then collides with the activated hydrogen, and the double bond is reduced to form ethane, C_2H_6.

rhodium metals installed in vehicle exhausts are examples of heterogeneous catalysts. The catalysts convert gases such as nitrogen oxide to nitrogen, and oxygen and carbon monoxide and unburnt hydrocarbons to carbon dioxide, which are slightly less harmful than the original gases.

8.3   Determining the rate of a chemical reaction

8.3.1   Methods for monitoring the rate of a chemical reaction

We have seen that the rate of a chemical reaction is expressed as the rate of formation of a product or the rate of disappearance of a reactant over time. Therefore, to determine the rate of a particular reaction, a method for measuring the amount of a product or a reactant present at any time is required. This can only be achieved by carrying out an experiment. The actual experimental technique used to determine the rate depends upon the particular reaction.

There are two commonly used approaches to monitoring the amount of product or reactant at any one time. The first is called *sampling* and involves taking a sample of the reaction mixture at various times throughout the reaction and analysing the content. The reaction mixture must be 'quenched' to stop it or slow it down whilst the sample is removed. This method could be used to investigate the hydrolysis of an ester, for example, where one of the products is a carboxylic acid:

$$CH_3COOC_2H_5(aq) + H_2O(l) \rightarrow CH_3COOH(aq) + C_2H_5OH(aq)$$
$$\text{ester} \qquad \text{water} \qquad \text{carboxylic acid} \qquad \text{alcohol}$$

Samples are removed at regular intervals and titrated with a base of known concentration to determine the amount of acid formed at each time during the reaction.

The second method is a continuous method where a physical property of the reaction is monitored over time. The property is related to the amount of one of the reactants or products. Possible monitoring techniques include measuring gas pressure or volume (as in Figure 8.2), colorimetry, or conductivity.

For example, in the reaction of H_2 with I_2, the amount of iodine could be tracked using a colorimeter, as iodine has a characteristic purple colour:

$$H_2(g) + I_2(g) \rightleftharpoons 2HI(g)$$

Iodine absorbs a specific wavelength of light and the intensity of light absorbed is directly related to the amount of iodine present. The amount of light absorbed is measured at various time intervals during the reaction and is plotted directly onto a graph of intensity against time.

8.3.2   The instantaneous rate of reaction

The change in concentration of a reactant or product over the duration of the whole reaction gives us the **average** rate of reaction. This is the quantity obtained from the expression: Rate $= -\dfrac{\Delta[A]}{\Delta t}$,

where the symbol $\Delta[A]$ indicates a large change in concentration of A and t represents the time of the whole reaction. Of more interest to chemists when studying reaction kinetics is the rate of reaction at any instant of time during the reaction. This is equivalent to the change in concentration of A over a very

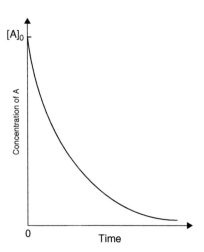

Figure 8.8 Plot of concentration of reactant A against time.

short period of time and can be expressed as: Rate $= -\dfrac{d[A]}{dt}$.

The symbol 'd' represents a small change in the quantity. A typical plot of concentration of a reactant A against time is shown in Figure 8.8.

At the start of the reaction the line is very steep and the reaction very fast. The concentration of A at the start of the reaction ($t = 0$) is represented by $[A]_0$. As time goes on the curve becomes less steep as the reaction slows down. There is less reactant remaining and so the rate decreases. Towards the end of the reaction the concentration of A is decreasing very slowly. At any individual time, t, the rate of reaction is expressed by the small change in concentration of A divided by the time taken. When the time period is very small this can be expressed by the quantity $\dfrac{-d[A]}{dt}$. The quantity $\dfrac{-d[A]}{dt}$ is equivalent to the gradient of a tangent to the curve at any point, as shown in Box 8.3. By drawing tangents to the reaction curve at several points, the rate of reaction can be obtained

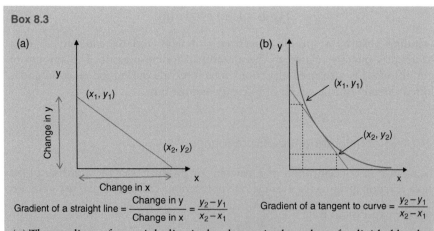

Box 8.3

(a) The gradient of a straight line is the change in the value of y divided by the change in the value of x. (b) The gradient of a tangent (orange line) to a curve (blue line) at any point is the change in the value of y divided by the change in the value of x. In both (a) and (b) the gradient will be negative.

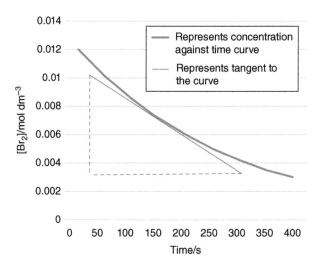

Figure 8.9 Plot of [Br$_2$] against time with a tangent drawn at $t = 150$ seconds.

at each time during reaction. An example is given in the next section for the reaction of bromine with methanoic acid, as shown in Figure 8.9.

8.3.3   An example of measuring rate of reaction at any time

The reaction of bromine with methanoic acid can be monitored by using a colorimeter to measure the concentration of bromine at regular time intervals. Bromine is a brown colour and the reaction takes place in solution. The brown colour of the solution decreases as the reaction proceeds, so the absorbance measured by the colorimeter also decreases and is directly proportional to the concentration of bromine. The equation for the reaction is:

$$HCOOH(aq) + Br_2(aq) \rightarrow 2Br^-(aq) + 2H^+(aq) + CO_2(g)$$

The reaction takes place reasonably slowly. Some typical results are given in Table 8.1, and a plot of bromine concentration against time is given by the blue line in Figure 8.9. The rate of reaction at each time can be obtained by calculating the gradient of the tangent to the curve at each time, t. The results have been added to Table 8.1 for each time point. On Figure 8.9, a tangent has been drawn

Table 8.1 Concentration, rate, and time data for the reaction of bromine with methanoic acid.

Time/s	[Br$_2$]/mol dm^{-3}	Rate/mol dm^{-3} s^{-1}
0	0.0120	4.20×10^{-5}
50	0.0101	3.52×10^{-5}
100	0.0085	2.96×10^{-5}
150	0.0071	2.49×10^{-5}
200	0.0060	2.09×10^{-5}
250	0.0050	1.75×10^{-5}
300	0.0042	1.48×10^{-5}
350	0.0035	1.23×10^{-5}
400	0.0030	1.04×10^{-5}

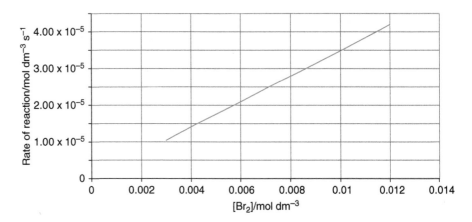

Figure 8.10 Plot of rate against concentration of Br_2.

The values for the gradients obtained in this way ($d[Br_2]/dt$) will be negative because the concentration of bromine is decreasing as the reaction proceeds. The rate of the reaction is equal to

rate = $-d[Br_2]/dt$

and so results in a positive value for the rate.

to the curve at time $t = 150\,s$ (orange line). The gradient of the tangent is calculated as shown in Box 8.3.

A graph of rate of reaction against bromine concentration can be plotted (Figure 8.10). As the graph is a straight line, it shows that the rate of this reaction is directly proportional to the concentration of bromine. As expected, at the start of the reaction when the concentration of bromine is high, the rate is also high. As reaction proceeds and the bromine concentration drops, the rate of reaction becomes slower.

Note that in Figure 8.10 the start of the reaction is when the concentration of bromine is highest - this is to the right of the straight line. When we plot time on the x axis, as in Figure 8.9, the start of the reaction is to the left of the plot.

8.4   The rate expression

The relationship between the rate of a chemical reaction and the concentration of the reactants is given by the **rate expression**. The rate expression can only be obtained by carrying out an experiment such as the one described previously in which the rate of reaction of bromine with methanoic acid was investigated. In that experiment, it was shown that the rate of reaction is directly proportional to the concentration of bromine. We can therefore write:

$$\text{Rate} \propto [Br_2].$$

For a general reaction, $aA + bB \rightarrow$ products, the rate expression is in the form:

$$\text{Rate} = k[A]^x[B]^y$$

The parameters x and y are the powers that the concentrations of the reactants A and B are raised to in the rate expression. x and y are small whole integers, usually 0, 1, or 2. The integer x is called the **order of the reaction** with respect to A, and the integer y is the **order of the reaction** with respect to B. The overall order of the reaction is equal to x + y.

The quantity k is called the **rate constant** for the reaction. It is a constant for a specific reaction at a fixed temperature.

Box 8.4

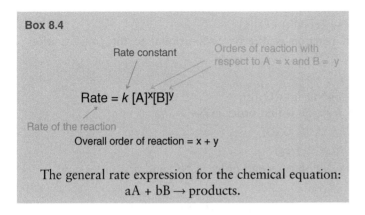

Rate constant

Orders of reaction with respect to A = x and B = y

$$\text{Rate} = k\,[A]^x[B]^y$$

Rate of the reaction

Overall order of reaction = x + y

The general rate expression for the chemical equation:
aA + bB → products.

Table 8.2 Some chemical reactions and their related rate expressions and overall reaction orders.

Chemical equation	Rate expression	Overall order of reaction
$NO_2(g) + CO(g) \rightarrow NO(g) + CO_2(g)$	Rate = $k[NO_2]^2$	2
$CH_3CHO(g) \rightarrow CH_4(g) + CO(g)$	Rate = $k[CH_3CHO]^2$	2
$H_2(g) + I_2(g) \rightarrow 2HI(g)$	Rate = $k[H_2][I_2]$	2
$2H_2(g) + 2NO(g) \rightarrow 2H_2O(g) + N_2(g)$	Rate = $k[H_2][NO]^2$	3

The parameters in the rate expression are summarised in Box 8.4.

The rate expression can only be obtained by carrying out experiments. This has been stated before in this chapter, but it is really important to understand that the rate expression cannot be obtained from the chemical equation in the way that the equilibrium constant can. The orders of the reaction, x and y, are obtained from physically investigating how the rate depends upon the concentration of each reactant, as we will see in the following section. They are not related to the stoichiometric coefficients in any way. Table 8.2 lists some balanced chemical equations and their related rate expressions. From this you can see there is no connection between the orders of reaction and the stoichiometry. (Note that k is used to represent the rate constant in each rate expression, but it is numerically different for every reaction.)

The following sections outline various experimental methods that are used to determine the rate expression. The chosen method depends upon the actual reaction being studied.

In some textbooks you may find the terms *rate law* and *rate equation* in place of *rate expression*. As several laws and equations are used in chemical kinetics here we will use the term **rate expression** to refer to the equation that relates the rate of a reaction to the concentration of reactants and the rate constant.

Worked Example 8.2 Write the rate expression for the following chemical reactions, given the information in each question that is derived from experiments. Determine the overall order of the reaction in each example. Use k to represent the rate constant in each case.

a. $2H_2O_2(aq) \rightarrow 2H_2O(l) + O_2(g)$
The rate of reaction is directly proportional to the concentration of H_2O_2.

b. $2N_2O_5(g) \rightarrow 4NO_2(g) + O_2(g)$
The rate of reaction is first-order in N_2O_5.

c. $2NO(g) + O_2(g) \rightarrow 2NO_2(g)$
The rate of reaction is second-order in NO and first-order in O_2.

d. $NO(g) + CO(g) + O_2(g) \rightarrow NO_2(g) + CO_2(g)$
The rate of reaction is second-order in NO and zero-order in CO and O_2.

Solution

a. Rate = $k[H_2O_2]$. Overall order = 1.

b. Rate = $k[N_2O_5]$. Overall order = 1.

c. Rate = $k[NO]^2[O_2]$. Overall order = 3.

d. Rate = $k[NO]^2$. Any reactant that is zero-order is not involved in the rate expression. This is because if we write $[CO]^0$, the value is always = 1. So Rate = $k[NO]^2[CO]^0[O_2]^0 = k[NO]^2 \times 1 \times 1 = k[NO]^2$. Overall order = 2.

8.4.1   Determining the rate expression using instantaneous rates

This method involves measuring the concentration of a reactant during the course of the reaction. We have seen that in the reaction between bromine and methanoic acid a plot of concentration of bromine against time can be made (Figure 8.9). By drawing tangents to the curve in Figure 8.9, instantaneous rates can be obtained throughout the reaction. These values can be plotted against the concentration of bromine, as in Figure 8.10. If a straight line is obtained, then the rate of reaction is directly proportional to the concentration of bromine, and we can write a rate expression in the form Rate = $k_1[Br_2]$. In this case k_1 is the rate constant, and the reaction is first-order in bromine. If a straight line is not obtained from a plot of rate against concentration, then we cannot say that the reaction is first-order in that reactant. This method therefore has limited use.

8.4.2   Determining the rate expression using the initial rates method

An alternative approach is to use the **initial rates** method. In this method the reaction under study is carried out several times. In each experimental run the initial concentration of just one reactant is varied and the concentrations of the other reactants are kept constant. This is repeated for all the reactants. The initial reaction rate is then measured and the dependence of the rate upon the concentration of each starting material is calculated.

Table 8.3 Initial rates data for the reaction $2NO(g) + O_2(g) \rightarrow 2NO_2(g)$.

Run number	Initial [NO]/ mol dm^{-3}	Initial [O$_2$]/ mol dm^{-3}	Initial rate/ mol dm^{-3} s^{-1}
1	1.00×10^{-3}	1.00×10^{-3}	7.00×10^{-4}
2	1.00×10^{-3}	4.00×10^{-3}	28.0×10^{-4}
3	2.00×10^{-3}	1.00×10^{-3}	28.0×10^{-4}
4	2.00×10^{-3}	2.00×10^{-3}	56.0×10^{-4}
5	3.00×10^{-3}	4.00×10^{-3}	255.6×10^{-4}

An example is given in Table 8.3 for the reaction of nitrogen monoxide, NO, with oxygen. The equation for the reaction is $2NO(g) + O_2(g) \rightarrow 2NO_2(g)$.

Some experimental data for the reaction are given in which a series of runs has been carried out and the initial rates of reaction measured.

In this procedure we start with the assumption that the rate expression has this general form:

$$\text{Rate} = k[NO]^x[O_2]^y$$

To determine the rate expression we must find the values of x and y. To find x, which is the order with respect to NO, we first look for two runs in which the concentration of NO is varied and the concentration of O_2 is kept constant. Two such runs are Run 1 and Run 3. Alternatively, we could use Run 2 and Run 5.

In going from Run 1 to Run 3 the concentration of NO doubles. However, the rate of reaction increases by a factor of 4. The rate of reaction with respect to NO is therefore linked to the square of the concentration of NO, so we can write:

$$\text{Rate} \propto [NO]^2.$$

This means that the reaction is second-order in NO.

To obtain the value of y, the order with respect to O_2, we must choose two runs in which the concentration of NO is kept constant and the concentration of O_2 is varied. Such combinations of runs are Run 1 and Run 2, or Run 3 and Run 4. The same procedure is carried out as for NO.

If we choose Runs 3 and 4, the concentration of O_2 doubles whilst the concentration of NO remains constant. The rate of reaction increases from 28.0×10^{-4} to 56.0×10^{-4} mol dm^{-3} s^{-1}, so the rate also doubles. The rate of reaction is therefore directly proportional to the concentration of O_2, and we can write:

$$\text{Rate} \propto [O_2].$$

The reaction is therefore first-order in O_2.

An overall rate expression can be written that combines the two relationships:

$$\text{Rate} = k[NO]^2[O_2]$$

Worked Example 8.3 The equation for the reaction between nitrogen(II) oxide and hydrogen at 1000 °C is

$$2NO(g) + 2H_2(g) \rightarrow N_2(g) + 2H_2O(g)$$

Use the data in the table to determine the order of the reaction with respect to NO and H$_2$. Calculate the overall order of the reaction, and write the rate expression if the rate constant is k.

Run number	[NO]/ mol dm^{-3}	[H$_2$]/mol dm^{-3}	Initial rate/mol dm^{-3} s^{-1}
1	2.0×10^{-3}	2.0×10^{-3}	6.0×10^{-6}
2	4.0×10^{-3}	2.0×10^{-3}	2.4×10^{-5}
3	2.0×10^{-3}	4.0×10^{-3}	1.2×10^{-5}

Solution

To determine the order with respect to (wrt) NO, choose two runs in which the NO concentration changes but H$_2$ is constant. These are Runs 1 and 2.

On going from Run 1 to Run 2, the concentration of NO doubles, but the rate quadruples or increases by a factor of 4 (i.e. 2^2). The reaction is therefore second-order in NO.

To determine the order wrt H$_2$, choose two runs in which the H$_2$ concentration changes but NO is constant. These are Runs 1 and 3. On going from Run 1 to Run 3, the concentration of H$_2$ doubles, and the rate also doubles ($1.2 \times 10^{-5} = 2 \times 6.0 \times 10^{-6}$). The rate is directly proportional to the concentration of H$_2$, so the reaction is first-order in H$_2$.

The rate expression is: Rate = k[NO]^2[H$_2$].

Worked Example 8.4   The equation for the oxidation of bromide ions in acidic solution by bromate ions is:

$$5Br^-(aq) + BrO_3^-(aq) + 6H^+(aq) \rightarrow 3Br_2(aq) + 3H_2O(l)$$

Four different experimental runs were conducted to determine the rate expression for the reaction, and the results are given in the table.

Run number	[Br$^-$]/ mol dm^{-3}	[BrO$_3^-$]/ mol dm^{-3}	[H$^+$]/ mol dm^{-3}	Initial rate/ mol dm^{-3} s^{-1}
1	0.2	0.2	0.2	1.6×10^{-3}
2	0.2	0.4	0.2	3.2×10^{-3}
3	0.4	0.4	0.2	6.4×10^{-3}
4	0.2	0.2	0.4	6.4×10^{-3}

a. Find the orders of the reaction with respect to each of the reactants.

b. Write the rate expression for the reaction if the rate constant is k, and give the overall order.

c. Calculate the value of the rate constant, and give its units.

Solution

a. To determine the order wrt Br$^-$, choose two runs in which the Br$^-$ concentration changes and the concentrations of the other reactants remain constant, i.e. Runs 2 and 3. On going from Run 2 to Run 3, [Br$^-$] doubles and the initial rate also doubles. The rate is directly proportional to the Br$^-$ concentration, and the reaction is first-order in Br$^-$.

To determine the order wrt BrO$_3^-$, choose two runs in which the BrO$_3^-$ concentration changes and the concentrations of the other reactants remain constant, i.e. Runs 1 and 2. On going from Run 1 to

Run 2, $[BrO_3^-]$ doubles and the initial rate also doubles. The rate is directly proportional to the BrO_3^- concentration, and the reaction is first-order in BrO_3^-.

To determine the order wrt H^+, choose two runs in which the H^+ concentration changes and the concentrations of the other reactants remain constant, i.e. Runs 1 and 4. On going from Run 1 to Run 4, $[H^+]$ doubles, and the initial rate increases by 4 times. When the concentration of H^+ doubles, the rate is squared (i.e. increases by a factor of 4), so the reaction is second-order in H^+.

b. The rate expression is: Rate $= k[Br^-][BrO_3^-][H^+]^2$, and the reaction is overall fourth-order. This is very unusual as it suggests a successful collision is required between four particles simultaneously.

c. The rate constant is found by substituting the results for one of the runs into the rate expression and rearranging to get a value for k:

$$\text{Rate} = k[Br^-][BrO_3^-][H^+]^2$$

Select the results for Run 1, and insert it into the rate expression:

$$1.6 \times 10^{-3} \text{ mol dm}^{-3} \text{ s}^{-1} = k \times 0.2 \text{ mol dm}^{-3}$$
$$\times 0.2 \text{ mol dm}^{-3} \times \left(0.2 \text{ mol dm}^{-3}\right)^2$$

$$k = \frac{1.6 \times 10^{-3} \text{ mol dm}^{-3} \text{ s}^{-1}}{0.2 \text{ mol dm}^{-3} \times 0.2 \text{ mol dm}^{-3} \times (0.2 \text{ mol dm}^{-3})^2}$$

$$= \frac{1.6 \times 10^{-3} \text{ mol dm}^{-3} \text{ s}^{-1}}{1.6 \times 10^{-3} \left(\text{mol}^4 \text{ dm}^{-12}\right)} = 1.0 \text{ mol}^{-3} \text{ dm}^9 \text{ s}^{-1}$$

8.4.3   Determining the rate expression by inspection

The profile of plots of concentration of a reactant against time can give an indication of the order of the reaction with respect to that reactant for zero-, first-, and second-order reactions, although the results are not always definitive. Figure 8.11 shows how the **concentration** of reactant changes with **time** for a zero-, first-, and second-order reaction.

Zero-order reactions

In a zero-order reaction the plot of concentration of reactant against time is a straight line with a negative gradient, shown by the black line in Figure 8.11. The rate of reaction is obtained from the gradient of the concentration against time plot. Because this is a straight line, the gradient – and therefore the rate of reaction – is constant throughout the reaction.

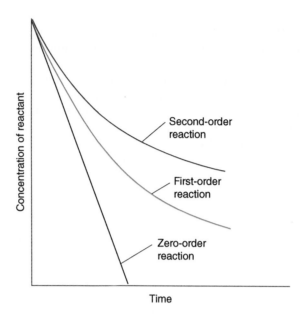

Figure 8.11 Plots of concentration against time for a zero-, first-, and second-order reaction.

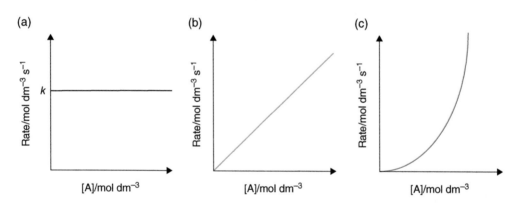

Figure 8.12 (a) Plot of rate against concentration for a zero-order reaction; (b) plot of rate against concentration for a first-order reaction; (c) plot of rate against concentration for a second-order reaction.

The plot of **rate** against **concentration** is a straight horizontal line (Figure 8.12a). The rate of reaction is equal to the rate constant throughout the reaction. Enzyme reactions typically show this behaviour.

For a zero-order reaction where $A \rightarrow B$, the general rate expression is given by:

$$\text{Rate} = k\,[A]^0 = k$$

Maths Alert! Any number raised to the power zero has a value equal to 1. For example, $10^0 = 1$ and $[A]^0 = 1$.

First-order reactions

For a first-order reaction the plot of **concentration** of reactant A against **time** is shown by the green line in Figure 8.11 and is an exponential decay curve. The rate of reaction at any time (instantaneous rate) is obtained from the gradients of tangents to this curve at different concentrations. When the **rate** is plotted against **concentration** a straight line is obtained with a gradient k, as shown in Figure 8.12b. This shows that Rate = $k[A]^1$ for a first-order reaction. We have already seen an example of this with the bromine/methanoic acid reaction.

Second-order reactions

For a second-order reaction the plot of **concentration** of reactant A against **time** is shown by the red curve in Figure 8.11. This curve drops more steeply at the start of the reaction than the curve for the first-order reaction, and then levels off. This indicates the reaction is very fast at the start and then slows down as reactant is used up. If instantaneous **rates** are measured and plotted against **concentration**, the curve in Figure 8.12c is obtained. This is an indication that the reaction is second-order but doesn't prove it conclusively.

8.4.4 Determining the rate expression using the integrated rate expression

An **integrated rate expression** (or integrated rate equation) tells us about how the concentration of a reactant varies with time throughout the reaction. Such an equation must be derived by carrying out experiments in which the concentration of a reactant is measured at short time intervals throughout the reaction.

The way in which the concentration of a reactant varies throughout the experiment depends upon the **order** of that reactant in the rate expression.

Zero-order reactions

The rate of a reaction is defined as the rate of decrease in concentration of a reactant with time and represented by the expression $-\dfrac{d[A]}{dt}$, where A is a reactant. For a zero-order reaction the rate is directly proportional to the concentration of a reactant raised to the power zero:

$$\text{Rate of a zero-order reaction} = -\frac{d[A]}{dt} = k\,[A]^0 = k \times 1 = k$$

Integration of this rate expression gives the integrated rate expression for a zero-order reaction, which is: $[A]_t = [A]_0 - kt$.

The integrated rate expression for a zero-order reaction has the same format as the equation of a straight line, i.e. $y = mx + c$. In this expression m is the gradient of the line, and c is the intercept on the y-axis.

Plotting $[A]_t$ against time, we obtain a straight line as in Figure 8.13 for a zero-order reaction. If a plot of concentration of a reactant against time is a

$[A]_t$ is the concentration of reactant A at any time, t, and $[A]_0$ is the concentration of reactant A at the start of the reaction when $t = 0$.

Integration is a mathematical operation in which small slices or quantities are added together to get the whole. In integrating the rate expression, we are adding together the rates at infinitely small time intervals to get the overall rate expression. See Box 8.5 for an explanation of the mathematical procedure of **integration**.

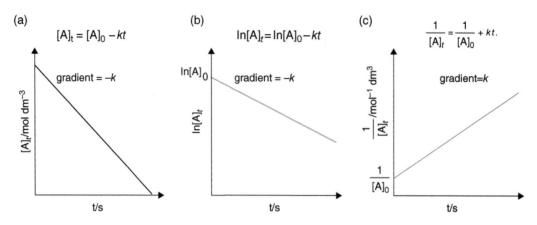

Figure 8.13 Integrated rate expression plots: (a) plot of $[A]_t$ against time for a zero-order reaction; (b) plot of $\ln[A]_t$ against time for a first-order reaction; (c) plot of $1/[A]_t$ against time for a second-order reaction.

straight line with a negative slope, the reaction must be zero-order in that reactant. The gradient of the slope is $-k$ and so allows a determination of the rate constant.

First-order reactions

The rate expression for a first-order reaction is: $\dfrac{-d[A]}{dt} = k\,[A]$. Integration of this rate expression gives the equation called the **integrated rate expression** for a first-order reaction:

$$\ln [A]_t = \ln [A]_0 - kt$$

$[A]_t$ is the concentration of A at any time t during the reaction, and $[A]_0$ is the concentration of A at time $t = 0$. The integrated rate equation has the same format as the equation of a straight line, $y = mx + c$, where m is the gradient of the line and is equivalent to $(-k)$ and c is the intercept on the y-axis, which is equivalent to $\ln[A]_0$. A plot of $\ln[A]_t$ against time is a straight line with gradient $-k$, as in Figure 8.13b.

Second-order reactions

The rate expression for a second-order reaction is: $\dfrac{-d[A]}{dt} = k[A]^2$. Integration of this rate expression gives the integrated rate expression for a second-order reaction:

$$\frac{1}{[A]_t} = \frac{1}{[A]_0} + kt$$

$$y = c + mx$$

If $\dfrac{1}{[A]_t}$ is plotted against time, a straight line is obtained, as in Figure 8.13c. In this case, the gradient of the line is positive and equal to k, the rate constant. The intercept on the y-axis is $\dfrac{1}{[A]_0}$.

Box 8.5 Maths alert!

Integration is a mathematical operation. The following workings show how the integration of the rate expression for a first-order reaction leads to the integrated rate expression. If you have not covered integration in maths, you do not need to worry about how it is carried out, as you will not normally be required to derive the equation. The derivation is only shown here for a first-order reaction, and the procedure for zero- and second-order reactions can be found in other textbooks.

For a first-order reaction $A \rightarrow B$, we can write the following rate expression, which states that the rate of reaction is directly proportional to the concentration of reactant A:

$$\frac{d[A]}{dt} = -k[A]$$

To find the mathematical relationship between the concentration of reactant A and time, we integrate both sides of the previous equation.

The equation is first rearranged and written as

$$\frac{1}{[A]} d[A] = -k dt$$

We then integrate both sides of the equation between the conditions at the start of the reaction, time $t = 0$, and any time throughout the reaction, t:

$$\int_{[A]_t}^{[A]_0} \frac{1}{[A]} d[A] = -k \int_0^t dt \qquad (8.1)$$

This uses the standard integral $\int_{x_0}^{x_t} \frac{1}{x} dx = [\ln x]_{x_1}^{x_2} = \ln x_2 - \ln x_1$

Applying this to Eq. (8.1) gives:

$$[\ln [A]]_{[A]_0}^{[A]_t} = -k [t]_0^t$$

$$\ln [A]_t - \ln [A]_0 = -k (t - 0)$$

And we can rearrange this to:

$$\ln [A]_t = \ln [A]_0 - k t$$

You can think of integration as being a mathematical black box, if you haven't covered it in maths. You should still be able to apply and use the integrated rate expression even if you don't understand the maths used to derive it.

Experimental rate data can be manipulated in this way to obtain the rate constant for the reaction and to determine whether the reaction is zero-, first-, or second-order.

Worked Example 8.5 The decomposition of gaseous nitrogen dioxide, NO_2, was monitored at 298 K:

$$NO_2(g) \rightarrow NO(g) + \tfrac{1}{2} O_2(g)$$

and the following data were collected.

Time/s	$[NO_2]$/mol dm^{-3}
0.0	0.100
5.0	0.017
10.0	0.0090
15.0	0.0062
20.0	0.0047

a. Determine the order of the reaction with respect to nitrogen dioxide.

b. Calculate the value of the rate constant and its unit.

Solution
With a set of experimental results it is possible to first predict the order of reaction by inspecting the data. Once a prediction has been made, the appropriate straight-line plot can be drawn to prove the order and obtain the rate constant.

a. An initial prediction can be made by looking at how the concentration varies with time. A rough sketch can be drawn to determine the approximate shape of the concentration-against-time curve. For the data provided in the question the concentration drops sharply at the start of the reaction and then levels off. This suggests second-order behaviour. The integrated rate expression for a second-order reaction is $\frac{1}{[A]_t} = \frac{1}{[A]_0} + kt$. Thus values of $1/[NO_2]$ should be calculated and plotted on the y-axis against time on the x-axis.

Time/s	$[NO_2]$/mol dm^{-3}	$1/[NO_2]$ mol^{-1} dm^3
0	0.1	10.0
5	0.017	58.8
10	0.009	111.1
15	0.0062	161.3
20	0.0047	212.8

The line obtained from the plot is a straight line, proving that the reaction is second-order in NO_2.

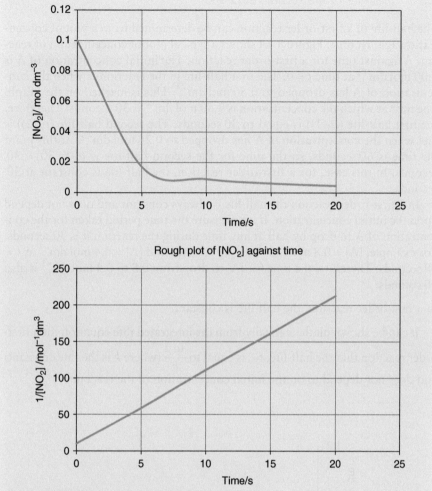

Rough plot of [NO_2] against time

Plot of $1/[NO_2]$ against time to show a straight line.

b. The gradient of the line is equal to the rate constant, k. The gradient can be obtained either using Excel, if the graph is plotted electronically, or by calculation, if the graph is hand-drawn. The gradient is found to be +10.2. The gradient has the units of the y-axis (mol^{-1} dm^3) divided by the unit of the x-axis (s). So the value of the rate constant is 10.2 $mol^{-1}dm^3s^{-1}$.

8.5   The half-life of a reaction

The **half-life** of a reaction is defined as the time taken for the concentration of a reactant to drop by half of its original value. The symbol for half-life is $t_{1/2}$, and the units are units of time. A large value for a half-life indicates that the reaction is proceeding quite slowly, whereas a short half-life means that the reactant concentration is disappearing quickly and the reaction is therefore fast.

8.5.1   Half-life of first-order reactions

The half-life of a first-order reaction can be determined from a plot of concentration against time. Figure 8.14 shows a typical plot of concentration of reactant A against time for a first-order reaction. The initial concentration of A is 1.00 mol dm^{-3} at time $t = 0$. The first half-life in the reaction is when the concentration of A has dropped to 0.50 mol dm^{-3}. This is marked on the graph. The time at which the concentration is 0.50 mol dm^{-3} is 30 seconds. Therefore, the first half-life ($t_{1/2}(1)$) is equal to 30 seconds. The second half-life ($t_{1/2}(2)$) is met when the concentration of A has dropped to 0.25 mol dm^{-3}. At this point the time is 60 seconds, so the time for the second half-life is $(60 - 30) = 30$ seconds. In this case, for a first-order reaction, the half-life is constant at 30 seconds.

In a first-order reaction the half-life is always constant and does not depend upon the initial concentration. If we measure the time period taken for the concentration of A to drop by half at any time during the reaction it is 30 seconds. For example, $[A] = 0.8 \text{ mol dm}^{-3}$ at $t = 9$ seconds and $[A] = 0.4 \text{ mol dm}^{-3}$ at $t = 39$ seconds. Therefore, the time for $[A]$ to drop from 0.8 to 0.4 mol dm^{-3} is also 30 seconds.

In a first-order reaction, the half-life is constant.

It can be shown mathematically from the integrated rate equation for a first-order reaction that the half-life, $t_{1/2}$, is equal to $\dfrac{\ln 2}{k}$ (where k is the rate constant) and does not depend upon the initial concentration of the reactant.

Radioactive decay is when a radioactive isotope (radionuclide) transforms into another radioisotope and emits radiation. The decay is exponential and follows first-order reaction kinetics. A radionuclide with a short half-life is very active when first produced, but the activity drops off quickly with time. A radionuclide such as carbon-14 with a half-life of 5700 years can be used for carbon-dating as its half-life is so long.

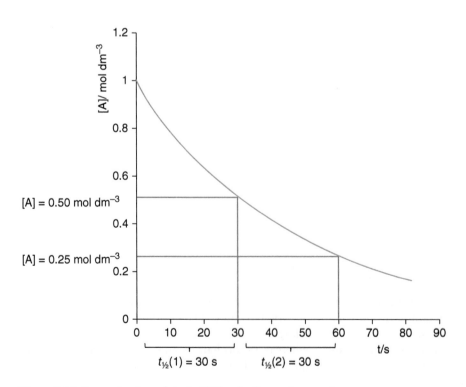

Figure 8.14 Determination of the half-life of a first-order reaction.

Box 8.6 Maths Alert!

The equation that links half-life to the rate constant for a first-order reaction can be derived using the integrated expression for the reaction: $\ln[A]_t = \ln[A]_0 - kt$.

After one half-life, $[A]_t = [A]_0/2$, by definition, and $t = t_{1/2}$. The expression becomes:

$$\ln \frac{[A]_0}{2} = \ln[A]_0 - kt_{1/2}$$

Rearranging the equation: $\ln \dfrac{[A]_0}{2} - \ln[A]_0 = -kt_{1/2}$

$$\ln \frac{[A]_0}{2[A]_0} = -kt_{1/2} \quad \left(\text{Because } \ln[A] - \ln[B] = \ln \frac{[A]}{[B]} \right)$$

Therefore, $-kt_{1/2} = \ln \dfrac{1}{2}$

and $kt_{1/2} = \ln 2 \quad \left(\text{Because if } \ln \dfrac{1}{A} = -x, \text{then } \ln A = +x \right)$

Therefore, $t_{1/2} = \dfrac{\ln 2}{k}$

8.5.2   Half-life of zero-order reactions

The half-life of a zero-order reaction is not constant but decreases with time. Figure 8.15 shows the plot of concentration against time for a zero-order reaction. The first two half-lives are marked on the plot. The first half-life is the time taken for the concentration to drop from 0.5 to 0.25 mol dm^{-3}, and $t_{1/2}(1)$ is roughly 12.5 seconds. The second half-life ($t_{1/2}(2)$) is the time taken for the concentration to drop from 0.25 to 0.125 mol dm^{-3} and is roughly equal to 5.5 seconds. The half-life of a zero-order reaction is not constant but decreases through the reaction.

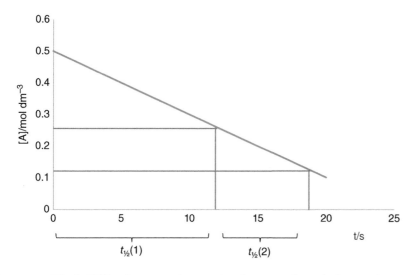

Figure 8.15 The half-life of a zero-order reaction decreases through the reaction.

8.5.3   Half-life of second-order reactions

The half-life of a second-order reaction is not constant but increases with time. A plot of concentration of A against time is given for a second-order reaction in Figure 8.16. The first two half-lives are marked on the plot. The first half-life, $t_{1/2}(1)$, is the time taken for the concentration to drop from 7.2 to 3.6 mol dm^{-3} and is equal to about 190 seconds. The second half-life, $t_{1/2}(2)$, is the time taken for the concentration to drop from 3.6 to 1.8 mol dm^{-3} and is equal to about 380 seconds. The half-life therefore increases through the reaction. The integrated rate expression can be used to show that the half-life for a second-order reaction is: $t_{1/2} = \dfrac{1}{k[A]_0}$ and so depends upon the initial concentration of the reactant. In fact, because k is constant, this equation shows that each successive half-life must be twice the previous one. When $[A]_0$ is halved, $t_{1/2}$ doubles.

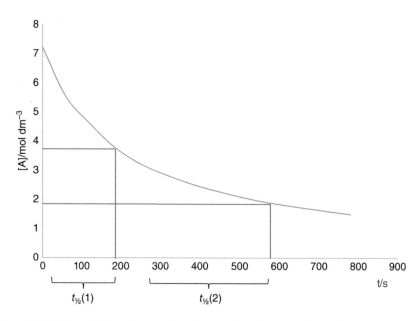

Figure 8.16 The half-life of a second-order reaction increases through the reaction.

Box 8.7 The half-life of a reaction is the time it takes for the initial concentration of a reactant to fall by a half.
In a zero-order reaction the half-life decreases as reaction proceeds:
$$t_{1/2} = \frac{[A]_0}{2k}.$$
In a first-order reaction the half-life is constant:
$$t_{1/2} = \frac{\ln 2}{k}.$$
In a second-order reaction the half-life increases as reaction proceeds:
$$t_{1/2} = \frac{1}{k[A]_0}.$$
Each successive half-life is twice the preceding one.
For all three orders of reaction the rate constant, k, can be determined from the half-life.

Worked Example 8.6   The decay of radionuclides by loss of alpha- or beta-particles follows first-order reaction kinetics. Some data for the radioactive decay of thorium-234 are given in the table. Sketch a plot of percentage of radionuclide remaining against time, and from this determine the half-life of thorium-234.

Years	Percentage remaining
0	100
8	80
18	60
20	55
40	35
60	20
85	10
110	5

Solution
The plot obtained is as shown, and it follows exponential decay. By measuring the time taken for the percentage of thorium-234 to drop from 80 to 40 and then 40 to 20, for example, the half-life can be measured to be around 25 years.

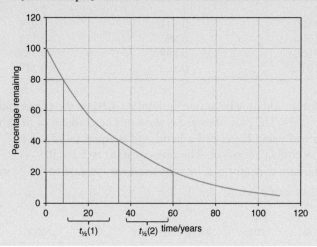

Worked Example 8.7   The dissociation of sulfuryl chloride, SO_2Cl_2, is first-order with respect to SO_2Cl_2:

$$SO_2Cl_2(g) \rightarrow SO_2(g) + Cl_2(g)$$

The reaction was carried out at 298 K, and the following data were obtained.

Time/s	$[SO_2Cl_2]/\text{mol dm}^{-3}$
0	1.000
2500	0.947
5000	0.895
7500	0.848
10 000	0.803

a. State the rate expression for the dissociation.

b. Using a graphical method, calculate the value of the rate constant, k, giving its units.

c. Calculate the half-life for the reaction.

Solution

a. If the reaction is first-order in SO_2Cl_2, the rate expression is: Rate = $k[SO_2Cl_2]$.

b. As we are told that the reaction is first-order in SO_2Cl_2, the plot required to determine the rate constant is the integrated rate expression for a first-order reaction, which is: $\ln[A]_t = \ln[A]_0 - kt$. The rate constant is obtained by measuring the gradient of the straight line that is obtained. Therefore, we first need to calculate values of $\ln[SO_2Cl_2]$. This problem is best solved using Excel. Note that, as for log values, ln values have no units.

Time/s	$[SO_2Cl_2]/\text{mol dm}^{-3}$	$\ln[SO_2Cl_2]$
0	1	0
2500	0.947	−0.054
5000	0.895	−0.111
7500	0.848	−0.165
10 000	0.803	−0.219

A plot should now be drawn with $\ln[SO_2Cl_2]$ on the y-axis and time on the x-axis. The plot obtained is shown. The gradient of the line can be obtained using Excel or by hand and is equal to -2.2×10^{-5} s^{-1}, and this is equivalent to $-k$. The rate constant is 2.2×10^{-5} s^{-1}. (The unit of the rate constant is s^{-1} as the gradient is equal to $\frac{y_2 - y_1}{x_2 - x_1}$, which has units of $\frac{1}{s}$ or s^{-1}.)

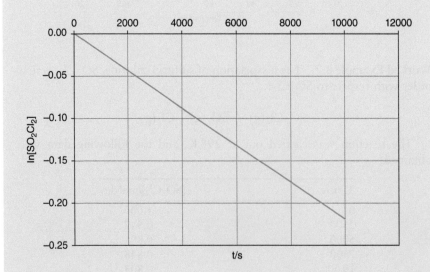

c. As the reaction is first-order, the half-life is constant. There are insufficient data to determine the half-life from a plot of $[SO_2Cl_2]$ against time, so we must use the formula derived from the integrated rate expression:

$$t_{1/2} = \frac{\ln 2}{k}.$$

$$t_{1/2} = \frac{\ln 2}{2.2 \times 10^{-5}\,\text{s}^{-1}} = 3.2 \times 10^4\,\text{s}$$

Worked Example 8.8   The decomposition of nitrogen dioxide to nitric oxide and oxygen takes place by second-order kinetics:

$$2NO_2(g) \rightarrow 2NO(g) + O_2(g)$$

At time $t = 10$ seconds, the pressure of NO_2 is 80 atm; after a further 5 seconds, the pressure has dropped to 40 atm. Determine the time at which the pressure is 20 atm.

Solution
This is a second-order reaction, so the half-life is not constant. The first half-life is the time taken for the pressure to drop from 80 to 40 atm. The time taken for this pressure drop is 5 seconds. So $t_{1/2}(1) = 5$ seconds, and the total reaction time at this point is $10 + 5 = 15$ seconds (because the time when the pressure was 80 atm is 10 seconds, so we need to add another 5 seconds to obtain the time when the pressure is 40 atm). We need to determine $t_{1/2}(2)$, the length of time taken for the pressure to drop from 40 to 20 atm.

The half-life of a second-order reaction increases through the reaction and the time of each successive half-life is twice the preceding one. So $t_{1/2}(2) = 2 \times 5 = 10$s. The time at which a pressure of NO_2 of 20 atm is therefore $15 + 10 = 25$s. (Again, we add the time for the drop from 40 to 20 atm to the time at 40 atm, which was 15s.)

Worked Example 8.9   Consider the first-order reaction $S \rightarrow T$ in which S molecules are converted to T molecules:

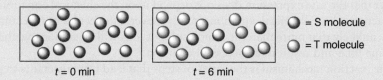

$t = 0$ min $t = 6$ min

Determine the half-life of the reaction and the number of S (blue) molecules and T (brown) molecules present after one half-life.

Solution
In this question we are told that the reaction is first-order, and the concentration of reactants and products are represented by spheres. At the start of the reaction ($t = 0$), there are 16 molecules of S; and at $t = 6$ minutes, 4 molecules of S remain. As this is a first-order reaction the half-life is constant and can be determined from this information, as shown in the following working:

$$[S] = 16 \text{ at } t = 0s$$

$$[S] = 8 \text{ at } t = t_{1/2}(1)$$

$$[S] = 4 \text{ at } t = t_{1/2}(2)$$

We know that the time elapsed after two half-lives is 6 minutes, so $t_{1/2} = 3$ minutes.

After the first half-life, eight molecules of S remain. As one mole of S is converted directly to one mole of T, the number of molecules of T must also be eight.

8.6   Reaction mechanisms

8.6.1   Reaction mechanisms and the rate-determining step

Most chemical reactions have more than one step between reactants and products. The series of steps that takes place makes up the **reaction mechanism**. You may have noticed that in some rate expressions the rate only seems to depend upon the concentration of one of the reactants, even though two or more molecules may be involved in the stoichiometric equation. In fact, some rate expressions contain species that are not present in the stoichiometric equation. For example, the rate expression for the reaction of iodine with propanone in acidic solution is:

$$\text{Rate} = k[CH_3COCH_3][H^+]$$

The stoichiometric equation for the reaction is:

$$CH_3COCH_3 + I_2 \xrightarrow{H^+} CH_3COCH_2I + HI$$

The rate expression indicates that the rate is proportional to the concentration of H^+ even though H^+ is not a reactant but is present as a catalyst. On the other hand, I_2 is a reactant but does not appear in the rate expression. This reaction therefore cannot occur in a single step.

Another indication that reactions don't take place in a single step is given by the fact that the rate expression does not depend upon the chemical equation for the reaction. Some chemical equations involve several molecules as reactants. It is very unlikely that more than three molecules or ions will all collide together at the same time and result in a successful collision.

The reaction mechanism is the series of steps that lead from reactants to products. In some steps very short-lived, unstable intermediate species are formed that go on to react in the following steps. Most of the steps in a reaction mechanism are very fast. Because the mechanism involves a series of steps, the slowest step determines the overall rate of the reaction. The slowest step is like a bottleneck in a journey or a procedure. Imagine driving from Birmingham to Manchester on the M6. If there are no hold-ups, the journey should take around 1 hour and 45 minutes. However, if there is congestion at any one point on the M6, your journey time will be affected by how long you are stuck in the jam. The traffic jam is like the slowest step in a reaction mechanism; it will determine the overall time it takes for you to get to your destination.

Similarly, the slowest step in a reaction mechanism determines the rate of the overall reaction and the rate expression. Any steps in the mechanism after the slowest step have no effect on the rate. However, if an intermediate species is involved in the slowest step, the rate of formation of the intermediate can affect the overall rate.

If we look again at the reaction between propanone and iodine, the overall equation is:

$$CH_3COCH_3 + I_2 \xrightarrow[\text{catalyst}]{H^+} CH_3COCH_2I + HI$$

The rate expression for the reaction has been shown to be:
$$\text{Rate} = k[CH_3COCH_3][H^+].$$
The mechanism for the reaction is as follows:

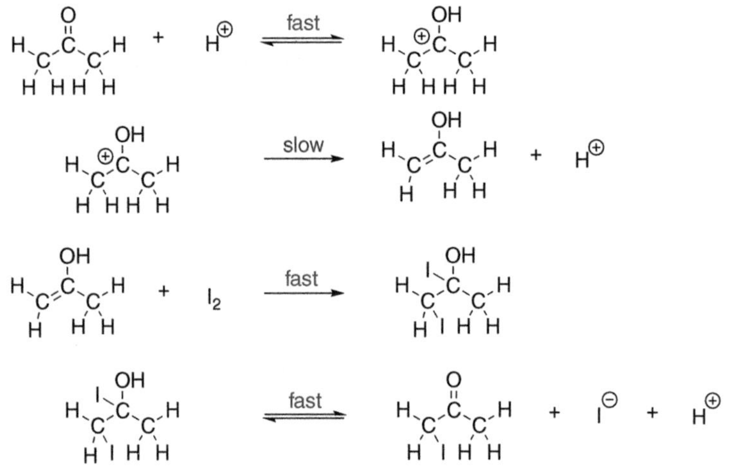

The slowest step in the reaction is the second step and involves the rearrangement of the intermediate $CH_3C^+(OH)CH_3$ formed in the first step of the reaction. Neither propanone nor H^+ is actually involved in the slowest step in the reaction. However, the reaction of propanone and H^+ in the first step forms the intermediate and therefore determines the rate of the overall reaction. The intermediate, $CH_3C^+(OH)CH_3$, is unstable and quickly rearranges to lose H^+ and form a C=C double bond. Iodine can react with the alkene formed, breaking the double bond and ultimately forming the product iodopropanone and losing another H^+ ion. Thus the catalyst H^+ is reformed in the mechanism. The reaction rate depends solely upon the concentration of propanone and H^+, and the steps after the slowest step do not affect the rate.

8.6.2 Using the rate expression to determine the mechanism of a reaction

Earlier in this chapter we saw that there are several different approaches to determining the rate expression for a reaction. One of the reasons that the rate expression is so important is that it can provide information about the mechanism of a reaction. Once chemists know the mechanism of a reaction they can start to plan ways to increase the rate and the yield and so decrease

You will meet nucleophilic substitution in Chapter 16 where it is explained that a nucleophile is a species that is electron rich and seeks areas of low electron density in another molecule.

production costs. Understanding the mechanisms involved in the reactions of drugs in the body helps improve the efficiency of the drug so side effects can be minimised.

In organic chemistry the hydrolysis of halogenoalkanes can take place by two different mechanisms. These are named S_N1 and S_N2. The S_N refers to nucleophilic ($_N$) substitution (S), which is a type of mechanism in organic chemistry. The number, either 1 or 2, refers to the order of the reaction. This can be understood more clearly by looking at examples of each type of reaction.

Hydrolysis of bromobutane by alkali takes place by nucleophilic substitution and leads to the formation of butanol according to the following equation:

$$C_4H_9Br + OH^- \rightarrow C_4H_9OH + Br^-$$

The OH^- ion is the nucleophile and substitutes the Br^- group in the bromobutane. The two possible mechanisms are outlined next.

S_N1 mechanism

In this mechanism the first step is slow and involves the C—Br bond breaking. The rate constant for this slow step is k_1.

$$\text{Step 1: } C_4H_9Br \xrightarrow{k_1} C_4H_9{}^+ + Br^- \text{ (slow)}$$

Step 2 is fast and involves attack of OH^- on the carbocation formed:

$$\text{Step 2: } C_4H_9{}^+ + OH^- \rightarrow C_4H_9OH \text{ (fast)}$$

The rate expression for the S_N1 mechanism is: Rate = $k_1[C_4H_9Br]$. The reaction is first-order and depends upon the concentration of bromobutane alone.

S_N2 mechanism

This mechanism has only one step in which the C—Br bond is broken and the OH^- attacks at the same time. This step must be the rate-determining step of the reaction and the rate constant for this step is k_2.

$$C_4H_9Br + OH^- \xrightarrow{k_2} C_4H_9OH + Br^- \text{ (slow)}$$

The rate expression for the S_N2 mechanism is: Rate = $k_2[C_4H_9Br][OH^-]$. This reaction is second-order overall as the rate depends upon the concentration of both bromobutane and hydroxide ions.

Bromobutane has three structural isomers, all with formula C_4H_9Br:

1-bromobutane   2-bromobutane   2-bromo-2-methylpropane

The rate expression for the hydrolysis of 1-bromobutane is:

$$\text{Rate} = k_2[\text{C}_4\text{H}_9\text{Br}][\text{OH}^-].$$

This isomer reacts by the S_N2 mechanism in which bond breaking and bond forming occur at the same time. The reaction is second-order as the slow step involves the collision of two species.

2-bromo-2-methylpropane is a structural isomer of 1-bromobutane.

The rate expression for the hydrolysis of 2-bromo-2-methylpropane is: Rate = $k_1[\text{C}_4\text{H}_9\text{Br}]$.

This isomer reacts by the S_N1 mechanism in which the slow step is the ionisation of the bromobutane to form the carbocation:

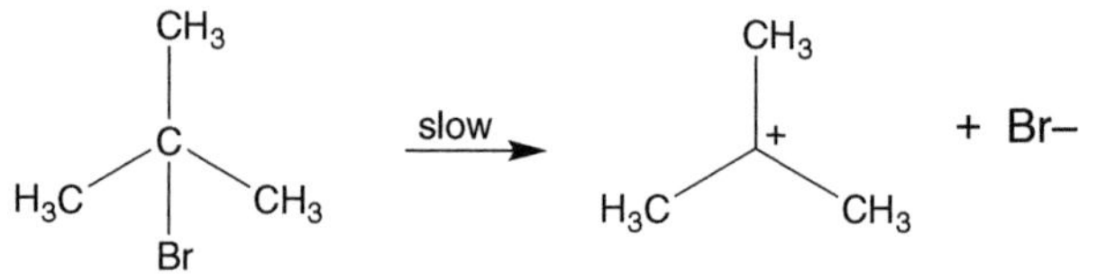

The next step is fast as it involves attack of the OH^- ion at the positively charged carbocation centre. This reaction is therefore first-order, and the rate depends only upon the concentration of bromobutane.

This example shows how an investigation of the kinetics of the reaction gives us the rate expression, which can then be linked to an appropriate mechanism. The rate expression and the order of the reactants in the rate expression do not prove that a specific mechanism is correct, but they can help predict possible mechanisms that can be further explored.

Structural isomers are molecules that have the same number and types of atoms but are bonded together in different ways. You will learn more about isomers in Chapter 12.

Chapter 16 covers nucleophilic substitution in more detail, and you will learn why the different structural isomers react with different mechanisms.

Worked Example 8.10   Gaseous hydrogen iodide is formed according to this equation:

$$\text{H}_2 + \text{I}_2 \rightarrow 2\text{HI}$$

The following is a possible mechanism:

$$\text{I}_2 \xrightarrow[\text{fast}]{k_1} 2\text{I}$$

$$\text{I} + \text{I} + \text{H}_2 \xrightarrow[\text{slow}]{k2} 2\text{HI}$$

Give the rate expression for this reaction if the overall rate constant is k.

Solution
The slow step of the reaction involves two iodine atoms, which are intermediates, and one hydrogen molecule. The rate of the slow step depends upon the concentration of hydrogen molecules and iodine atoms. A rate expression for this step would be: Rate = $k_2[\text{I}]^2[\text{H}_2]$.

However, iodine atoms are intermediates and cannot appear in the rate expression. The concentration of iodine atoms depends upon the concentration of molecular iodine, so a possible rate expression for the overall reaction could be: Rate = $k[\text{I}_2][\text{H}_2]$.

Experiments would be required to prove that this rate expression is correct. Can you suggest a possible way of monitoring the rate of this reaction?

8.7   Effect of temperature on reaction rate

8.7.1   The distribution of the energies of molecules with temperature

In the first section of this chapter, we stated that the rate of a chemical reaction depends upon the following properties of the reaction:

   i. The concentration or pressure of reactants

   ii. The activation energy, E_a

   iii. The temperature, T

   iv. The orientation of the molecules upon collision

In this final section we shall look at how the temperature at which a reaction is carried out quantitatively affects the rate of reaction and the rate constant.

The energies of molecules in a sample have a Boltzmann distribution, as shown in Figure 8.5.

Collision theory states that, for reaction to take place, molecules must possess sufficient energy to overcome the activation energy. The pattern of energies of molecules at two different temperatures, T1 and T2, is shown in Figure 8.6, which shows that an increase in temperature increases the number of molecules possessing the activation energy, and therefore the number of successful collisions. Hence an increase in temperature increases the rate of a chemical reaction.

8.7.2   The Arrhenius equation

The Swedish chemist Svante Arrhenius proposed an empirical equation linking the rate constant for a reaction to the temperature and activation energy. He developed the equation by carrying out experiments, but it was subsequently proved to be consistent with observations and theory from chemical kinetics and thermodynamics.

The Arrhenius equation states that the rate constant for a specific reaction is related to the temperature and activation energy by:

$$\ln k = \ln A - \frac{E_a}{RT}$$

where

   k is the rate constant for the reaction

   E_a is the activation energy

   R is the gas constant

   T is the temperature in Kelvin

   A is known as the Arrhenius factor and is a constant for a specific reaction. It is linked to the orientation and rate of collisions.

The Arrhenius equation has the form of the equation for a straight line, $y = mx + c$. If $\ln k$ is plotted on the y-axis and $1/T$ on the x-axis, a straight-line graph is obtained. The gradient, m, is equal to $-E_a/R$ and so can be used to obtain the activation energy for the reaction. The activation energy is a constant for a given reaction and does not depend upon temperature. It does depend upon the presence of a catalyst, as we have seen, so the equation can be used to find E_a for both the catalysed and uncatalysed reaction.

Although the activation energy for a specific reaction does not vary with the temperature, the rate constant does. The Arrhenius equation shows that the higher the temperature, the higher the value of $\ln k$, and therefore, the higher the rate constant and faster the reaction. Some data are given in the table for the decomposition of nitrogen dioxide, NO_2, into NO and O_2:

$$2NO_2(g) \rightarrow 2NO(g) + O_2(g).$$

$T/°C$	k/s^{-1}
100	1.10×10^{-9}
200	1.80×10^{-8}
300	1.20×10^{-7}
400	4.40×10^{-7}

To obtain the data in the correct format to draw an Arrhenius plot, the temperature must first be converted to Kelvin, by adding 273, and then the reciprocal temperature found in units of K^{-1} which is then plotted on the x-axis. The natural logarithm of the rate constant should be calculated and plotted on the y-axis. The calculated values of $\ln k$ and $1/T$ are given in Table 8.4. A plot of $\ln k$ against $1/T$ for the decomposition of NO_2 is shown in Figure 8.17. The plot is a straight line with gradient = -5505 K. The straight line has the formula $\ln k = \ln A - \frac{E_a}{RT}$, and the gradient is equal to $-E_a/R$. Thus the gradient can be used to find the value of the activation energy, E_a, for the reaction ($R = 8.31 \text{ J K}^{-1} \text{ mol}^{-1}$):

$$-\frac{E_a}{R} = -5505 \text{ K}$$

$$E_a = 5505 \text{ K} \times 8.31 \text{ J K}^{-1} \text{ mol}^{-1} = 45\,746 \text{ J mol}^{-1} = 45.7 \text{ kJ mol}^{-1}$$

Table 8.4 Kinetics data for the decomposition of NO_2 at varying temperatures.

$T/°C$	T/K	k/s^{-1}	$\frac{1}{(T/K)}$	$\ln k$
100	373	1.10×10^{-9}	2.68×10^{-3}	-20.6
200	473	1.80×10^{-8}	2.11×10^{-3}	-17.8
300	573	1.20×10^{-7}	1.75×10^{-3}	-15.9
400	673	4.40×10^{-7}	1.49×10^{-3}	-14.6

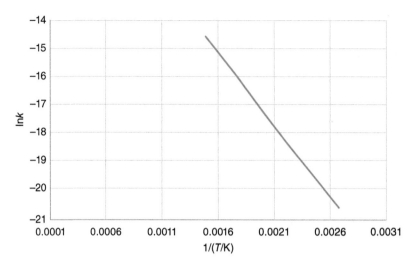

Figure 8.17 An Arrhenius plot for the decomposition of nitrogen dioxide, NO_2: $2NO_2(g) \rightarrow 2NO(g) + O_2(g)$.

CH$_3$· is a methyl radical formed when the C-C single bond in C_2H_6 is broken evenly and each C atom of the bond accepts one unpaired electron.

Worked Example 8.11   The reaction C_2H_6 (g) $\rightarrow$ 2·CH$_3$ (g) was monitored. At 400 K, the rate constant k was found to be $0.052\ s^{-1}$ and at 550 K, the rate constant was $0.540\ s^{-1}$. Calculate the activation energy for the reaction.

Solution
In this problem, we have two data points for the rate constant and temperature. These aren't enough data to plot a graph, but we can use the Arrhenius equation to set up two simultaneous equations to solve for the activation energy.

Taking the Arrhenius equation, $\ln k = \ln A - \dfrac{E_a}{RT}$, we can write two separate equations with the data provided:

Equation (i): $\ln 0.052\ s^{-1} = \ln A - E_a/R \times 400\ K$

Equation (ii): $\ln 0.540\ s^{-1} = \ln A - E_a/R \times 550\ K$

Here we have two equations with two unknowns, so we need to cancel one of the unknowns and then solve for the other.

If we subtract equation (ii) from equation (i), the $\ln A$ values will cancel:

$$\ln 0.052\ s^{-1} - \ln 0.540\ s^{-1} = -\frac{E_a}{R \times 400\ K} - \left(-\frac{E_a}{R \times 550\ K}\right)$$

This becomes:

$$\ln \frac{0.052\ s^{-1}}{0.540\ s^{-1}} = \frac{E_a}{R}\left(-\frac{1}{400\ K} + \frac{1}{550\ K}\right)$$

Working out the numerical terms on both sides gives us:

$$\ln 0.0963 = -\frac{E_a}{R} \times 0.682 \times 10^{-3}\ K^{-1}$$

Rearranging to give an expression for E_a gives:

$$E_a = -\ln 0.0963 \times 8.314 \text{ J mol}^{-1}\text{K}^{-1}/0.682 \times 10^{-3}\text{K}^{-1}$$

$$= -\left(-2.34 \times 8.314 \text{ J mol}^{-1}\text{K}^{-1}/0.682 \times 10^{-3}\text{K}^{-1}\right)$$

$$= 28.53 \times 10^3 \text{ J mol}^{-1} = 28.53 \text{ kJ mol}^{-1}$$

Quick-check summary

- The rate of a reaction describes how quickly reactants are used up and products are formed in a chemical reaction.

- Collision theory states that in order for a chemical reaction to take place, the species reacting must have sufficient energy to overcome the activation energy for the reaction and must collide together in the correct orientation.

- An increase in the concentration of reactants leads to more collisions, which results in an increase in the rate of reaction.

- The energies of molecules in a sample follow a Boltzmann distribution, which plots the number of molecules with a specific energy against the energy of the molecules. As the temperature of the sample is increased, the average energy of the molecules increases so the maximum in the Boltzmann curve shifts to higher energies. The area under the curve that represents the total number of molecules in the sample must remain the same.

- As the temperature is increased, the number of molecules possessing the activation energy increases, so the rate of reaction increases.

- Catalysts are used to increase the rate of chemical reactions by providing an alternative pathway for the reaction with a lower activation energy.

- A heterogeneous catalyst is in a different physical phase from the reactants and works by providing a surface upon which reacting molecules can be absorbed, thus making successful collision more likely.

- A homogeneous catalyst is in the same physical phase as the reactants.

- The rate expression links the rate of a reaction to the concentrations of some, or all, of the chemical species involved in the reaction.

- The reaction mechanism is the series of steps by which the reaction proceeds from reactants to products. Each step has an individual rate expression.

- The general expression for the rate of reaction is given by Rate $= k[A]^x[B]^y$, where k is the rate constant, x is the order of reaction with respect to A, and y is the order of the reaction with respect to B.

- The overall order of a reaction with the rate expression Rate $= k[A]^x[B]^y$ is $x + y$.

- Rate expressions can only be obtained by carrying out experiments that investigate the way in which the concentration of a specific reactant affects the rate of reaction. Such experiments may involve:

○ Measuring the initial rate of the reaction in a series of runs where the concentration of one reactant is varied and the concentrations of the other reactants kept constant. This is called the *initial rates method*.

○ Measuring the concentration of a reactant during an experiment and obtaining the reaction rate from tangents taken at several different points on the concentration against time curve.

○ Using the integrated rate expression and plotting an appropriate graph to prove the order of reaction and determine the rate constant.

- The half-life of a reaction is the time it takes for the concentration of a reactant to drop by a half its initial value.

- The half-life of a first-order reaction is a constant throughout the reaction. It can be obtained from:

○ A plot of concentration against time

○ The relationship $t_{1/2} = \frac{\ln 2}{k}$, if the value of the rate constant is known.

- The half-life of a zero-order reaction decreases throughout the reaction.

- The half-life of a second-order reaction increases throughout the reaction.

- In a reaction mechanism, the slow step is the rate-determining step and therefore determines the overall rate of reaction.

- The rate expression depends upon the reacting species in the slow step of the reaction and the steps leading to the formation of those species.

- The rate expression cannot include an intermediate species even if it is involved in the slow step of the reaction.

- The rate expression gives information about the mechanism of the reaction.

- The Arrhenius equation describes the relationship between the rate constant and temperature. The Arrhenius equation is $\ln k = \ln A - \frac{E_a}{RT}$. A plot of $\ln k$ against $1/T$ is a straight line with a negative slope. The activation energy for the reaction can be obtained from the gradient of the line, which is equal to $-E_a/R$.

- The Arrhenius equation shows that the rate constant for a reaction varies with temperature and determines the rate of reaction.

- The activation energy for a reaction is a constant and does not depend upon temperature. It does depend upon the presence and nature of any catalyst used.

End-of-chapter questions

1 Dinitrogen pentoxide dissociates according to the following chemical equation:

$$2N_2O_5(g) \rightarrow 4NO_2(g) + O_2(g)$$

a. Write three separate equations to express the rate of reaction in terms of the rate of consumption of N_2O_5 and the rates of formation of the products, NO_2 and O_2.

b. Show how the equations for these individual reaction rates are linked.

c. If the average rate of dissociation of N_2O_5 at a certain temperature is 0.02 mol s^{-1}, calculate the amount in moles of NO_2 formed after one minute.

2   The decomposition of hydrogen peroxide into oxygen gas and water proceeds according to the following equation:

$$2H_2O_2(aq) \rightarrow O_2(g) + 2H_2O(l)$$

The reaction can be catalysed by platinum metal.

a. State whether platinum is acting as a homogeneous or heterogeneous catalyst. Explain your answer.

b. Draw a Boltzmann distribution curve to show the distribution of energies of the reacting molecules. Label the axes, and mark the activation energy of the catalysed and uncatalysed reaction on the appropriate axis.

3   For the following reactions and associated rate expressions, give the order of the reaction with respect to each of the reactants and the total reaction order:

a. Rate = $k[S_2O_8{}^{2-}][I^-]$

b. Rate = $k[BrO_3{}^-]^2$

c. Rate = $k[NO][Cl_2]$

4   Ethanal decomposes to methane and carbon monoxide in the vapour phase in the presence of a catalyst. The equation for the reaction is given by:

$$CH_3CHO(g) \rightarrow CH_4(g) + CO(g)$$

Several experiments were conducted with different starting concentrations of ethanal at a fixed temperature to determine the dependence of the rate of reaction on the concentration of ethanal. The following data were obtained.

Experiment	$[CH_3CHO(g)]$/mol dm^{-3}	Initial rate/mol dm^{-3} s^{-1}
1	0.05	9.35×10^{-5}
2	0.10	3.74×10^{-4}
3	0.20	1.50×10^{-3}
4	0.40	5.98×10^{-3}

a. Use the data to determine the order of reaction with respect to ethanal.

b. Write a rate expression for the reaction.

c. Calculate a value for the rate constant and state the units.

d. What possible conclusion could you draw about the mechanism of the reaction from the rate expression?

5   Benzenediazonium chloride, $C_6H_5N_2Cl$, reacts with water in aqueous solution at temperatures above 5 °C according to the following equation:

$$C_6H_5N_2Cl(aq) + H_2O(l) \rightarrow C_6H_5OH(aq) + N_2(g) + HCl(aq)$$

Relative concentrations of benzenediazoniumchloride were determined and plotted against time of reaction and the following curve obtained:

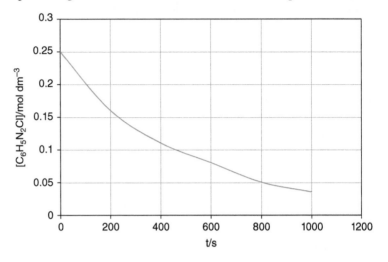

Determine the order of the reaction with respect to benzenediazonium chloride, and obtain an approximate value for the half-life of the reaction.

6   In a hypothetical reaction $X + Y \rightarrow$ Products, the grey spheres in the following figure represent molecules of X and the green spheres represent molecules of Y. The rate of reaction is first-order in both X and Y. The reaction is carried out with different starting concentrations of X and Y represented pictorially in boxes A, B, C, and D. Which of the starting conditions (A, B, C, or D) will result in the highest initial rate of reaction?

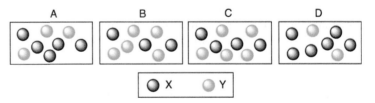

7   The reduction of nitrogen dioxide by carbon monoxide is represented by the following equation:

$$NO_2(g) + CO(g) \rightarrow NO(g) + CO_2(g)$$

The reaction is second-order in NO_2 and zero-order in CO, and the rate constant is k. Two experiments were carried out with different starting concentrations of NO_2 and CO, as shown in the table.

Experiment	[NO_2]/mol dm^{-3}	[CO]/mol dm^{-3}
1	0.2	0.2
2	0.4	0.3

a. Write the rate expression for the reaction.

b. Determine the relative initial rates of reaction in terms of the rate constant k for the experiments.

c. How would the initial rates be affected if the concentration of each starting material was decreased by a half?

d. How would the initial rates be affected if the volume of the container was doubled for each experiment?

8   The following data were obtained for the dimerisation of C_4H_6 to C_8H_{12} in a $1\ dm^3$ vessel. The reaction proceeds according to the following equation:

$$2C_4H_6(g) \rightarrow C_8H_{12}(g)$$

Time/s	$C_4H_6(g)$/mol dm^{-3}
0	0.0100
1000	0.00625
1800	0.00476
2800	0.0037
3600	0.00313
4400	0.0027
5200	0.00241
6200	0.00208

Determine the order of reaction with respect to C_4H_6 by graphical methods. Use an appropriate plot to obtain a value for the rate constant.

9   The reaction between NO_2 and CO, $NO_2(g) + CO(g) \rightarrow NO(g) + CO_2(g)$, may occur according to the following mechanism:

Step 1: $NO_2 + NO_2 \xrightarrow{k_1} NO_3 + NO$ (slow)

Step 2: $NO_3 + CO \xrightarrow{k_2} NO_2 + CO_2$ (fast)

If k is the overall rate constant write a possible rate expression for the reaction.

10   In aqueous solution the reaction between sodium bromide, NaBr, and hydrogen peroxide, H_2O_2, proceeds as follows:

$$2NaBr\ (aq) + H_2O_2\ (aq) \rightarrow Br_2\ (aq) + 2NaOH\ (aq)$$

The proposed rate expression is:

$$Rate = k[NaBr][H_2O_2]$$

a. The experiment was performed with concentrations of

$$[NaBr] = 1.5 \times 10^{-3}\ mol\ dm^{-3}$$

$$[H_2O_2] = 2.5 \times 10^{-3}\ mol\ dm^{-3}$$

and the rate was found to be $8.67 \times 10^{-3} \text{ mol dm}^{-3} \text{ s}^{-1}$.
Calculate the rate constant, k, giving its units.

b. What is the minimum number of steps in the reaction mechanism?

c. Hence, suggest a mechanism for the reaction.

11 The following set of data was obtained for the conversion of cyclopropane to propene at various temperatures. By plotting an appropriate graph determine the activation energy for the reaction.

$T/^{\circ}C$	k/s^{-1}
480	1.8×10^{-4}
530	2.7×10^{-3}
580	3.0×10^{-2}
630	0.26

9

Electrochemistry

At the end of this chapter, students should be able to:

- Identify oxidation and reduction reactions based on the transfer of electrons

- Write balanced equations for redox reactions using half-equations

- Draw and label a diagram of a typical electrochemical cell made up of two half-cells, and write half-equations for the reactions occurring at the electrodes

- Define the standard reduction potential of a half-cell, and explain how this is measured using a standard hydrogen electrode

- Identify the cathode and anode and the direction of flow of electrons based on the standard reduction potential of the half-cells

- Calculate the standard electrode potential (voltage) of an electrochemical cell from the standard reduction potentials of the half-cells

- Use standard reduction potentials to predict the direction of oxidation or reduction when two half-cells are combined

- Use the Nernst equation to calculate the electrode potential of an electro-chemical cell under non-standard conditions

- Write the cell diagram for an electrochemical cell using the half-cell equations

- Understand the relationship between the standard electrode potential of a cell and the Gibbs energy of the reaction, and use this to predict whether the reaction will take place under standard conditions

- Describe some different types of galvanic cells and their uses

Foundations of Chemistry: An Introductory Course for Science Students, First Edition.
Philippa B. Cranwell and Elizabeth M. Page.
© 2021 John Wiley & Sons Ltd. Published 2021 by John Wiley & Sons Ltd.
Companion website: www.wiley.com/go/Cranwell/Foundations

- Describe how electrolysis can be used to decompose ionic substances to obtain elemental species

- Identify the oxidation and reduction reactions occurring at the electrodes in simple electrolytic cells, and write half-equations for these processes

- Calculate the amount of substance deposited in an electrolysis reaction from a knowledge of the current and time used

9.1   Introduction

Earlier in this course, in Chapter 5, we saw that many chemical reactions involve the transfer of electrons in oxidation and reduction reactions, known collectively as *redox processes*. In a redox reaction one chemical species is oxidised and another reduced and electrons are transferred in the process. This movement of electrons can be harnessed within electrochemical cells, and the current generated used to do work in batteries or fuel cells which are used extensively in a wide range of electronic equipment from mobile phones to vehicle engines. The individual half-equations that make up redox reactions are equilibrium processes that can proceed in either direction depending upon the half-reaction that they are coupled with.

The generation of current in this way can be reversed by providing electrons from an external source, such as a power supply. This reverse process is called *electrolysis* and involves using an electrical charge to electrolyse or split a substance into its individual components. Electrolysis is an important industrial process used in metal extraction to obtain pure metals such as aluminium from oxide ore and to produce essential industrial raw materials such as bromine, obtained by the electrolysis of seawater, or chlorine, which can be extracted from brine or concentrated sodium chloride solution.

Electrochemistry and an understanding of redox equilibria underpin all these processes, which have significant commercial applications.

9.2   Using redox reactions

Redox reactions can be broken down into two half-equations: one to represent the oxidation reaction and one to represent the reduction reaction. In the oxidation half-equation the species being oxidised loses electrons and its oxidation number increases. In the reduction half-equation the species being reduced gains electrons and its oxidation number decreases. The important point about redox reactions is that electrons are transferred.

Remember OIL RIG! Oxidation is the loss of electrons. Reduction is the gain of electrons.

Worked Example 9.1   State whether the following half-equations are oxidation or reduction reactions in the direction written:

a. $Sn^{2+}(aq) \rightarrow Sn^{4+}(aq) + 2e^-$

b. $Fe^{3+}(aq) + e^- \rightarrow Fe^{2+}(aq)$

c. $H^+(aq) + e^- \rightarrow \frac{1}{2}H_2(g)$

Solution

    a. Oxidation. The Sn^{2+} ion has lost electrons (OIL) and been oxidised to Sn^{4+}.

    b. Reduction. The Fe^{3+} ion has gained an electron (RIG) and been reduced to Fe^{2+}.

    c. Reduction. The H^{+} ion has gained an electron (RIG) and been reduced to hydrogen gas in its elemental form in which hydrogen has an oxidation state of zero.

Worked Example 9.2   Using the first two half-equations given in Worked Example 9.1, write a balanced equation for the reduction of Fe^{3+} ions to Fe^{2+} by Sn^{2+} ions. State which species is the oxidising agent and which is the reducing agent.

Solution
To write a balanced chemical equation, we must use the two half-equations given. In the direction written the Sn^{4+}/Sn^{2+} reaction represents the oxidation half-equation and the Fe^{3+}/Fe^{2+} reaction represents the reduction half-equation. Adding the two sides of the half-equations will give us the overall equation, but the electrons must cancel. Equation (b) must therefore be multiplied throughout by a factor of 2:

$$2\,Fe^{3+}\,(aq) + 2e^{-} \rightarrow 2Fe^{2+}\,(aq)$$

The equations may now be added and the electrons cancelled:

$$2\,Fe^{3+}\,(aq) + Sn^{2+}\,(aq) + \cancel{2e^{-}} \rightarrow 2Fe^{2+}\,(aq) + Sn^{4+}\,(aq) + \cancel{2e^{-}}$$

$$2\,Fe^{3+}\,(aq) + Sn^{2+}\,(aq) \rightarrow 2Fe^{2+}\,(aq) + Sn^{4+}\,(aq)$$

9.2.1   Redox reactions and electrochemical cells

The electrons transferred in a redox reaction produce energy in the form of an electrical current that can be made to do work. This is the basis of a battery or electrochemical cell. In the following section we will see how knowledge about redox half-equations informs the design of useful batteries and fuel cells.

9.2.2   Electrochemical cells and half-cells

When a strip of metal such as zinc or copper is placed in an aqueous solution of its ions, an equilibrium is set up. The metal atoms from the surface move into the solution in the form of metal ions; the metal atoms have been oxidised. Figure 9.1 shows the arrangement for a length of copper rod in a solution of Cu(II) ions. The half-equation for the oxidation is:

$$Cu(s) \rightarrow Cu^{2+}\,(aq) + 2e^{-}$$

At the same time, aqueous Cu^{2+} ions from the solution are deposited back onto the surface of the metal as copper atoms. The metal ions are reduced:

$$Cu^{2+}\,(aq) + 2e^{-} \rightarrow Cu(s)$$

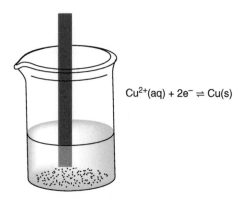

$$Cu^{2+}(aq) + 2e^- \rightleftharpoons Cu(s)$$

Figure 9.1 The Cu^{2+}/Cu half-cell.

For a reminder about equilibrium reactions see Chapter 7.

The rate at which these two reactions occur is the same and an equilibrium is established, such as that in Eq. (9.1):

$$Cu^{2+}(aq) + 2e^- \rightleftharpoons Cu(s) \tag{9.1}$$

Copper is a fairly unreactive metal and so this equilibrium lies over to the right. This means that copper(II) ions are fairly easily reduced to copper metal by gaining electrons but the reverse process is much less likely to occur.

For other metals the position of equilibrium will be different. If the metal is more reactive than copper, for example zinc, the equilibrium will be further over to the left:

$$Zn^{2+}(aq) + 2e^- \rightleftharpoons Zn(s)$$

An arrangement such as this equilibrium process, where electrons are transferred between a metal and a solution of its ions, is called a **half-cell**. An electrical potential (voltage) is set up between the metal and its ions in solution as a result of the movement of electrons. Two such half-cells with different electrical potentials can be connected as in Figure 9.2.

On the left-hand side of the diagram is a half-cell consisting of a strip of zinc suspended in an aqueous solution of zinc sulfate. On the right-hand side of the diagram is a half-cell consisting of a piece of copper suspended in a solution of copper sulfate. Connecting the two beakers is a strip of filter paper soaked in a solution of an inert salt such as potassium nitrate or potassium chloride. This is called a **salt bridge**, and it allows charged species such as ions to flow between the two half-cells. A liquid that contains ions is called an **electrolyte solution**.

Check that your answer is balanced by ensuring that there are equal numbers of each type of atom on both sides of the equation and that the charges on each side are the same.

Wires connect the two strips of metal in each half-cell via a high-resistance voltmeter. This completes the circuit. The arrangement of the two half-cells connected as in Figure 9.2 is a type of electrochemical cell called a **galvanic** or **voltaic cell**. The metal strips through which electrons enter and leave the solution are called **electrodes**. When the two electrodes are connected in a circuit with a voltmeter in series, a **potential difference** between the two electrodes can be measured. For an electrochemical cell as shown in Figure 9.2, when the concentrations of copper(II) sulfate and zinc(II) sulfate are each 1.0 M and the temperature is 298 K, the potential difference measured is 1.1 V (V = volt). This arrangement forms the basis of the **Daniell cell,** which is one of the first batteries developed. A small light bulb connected in series with the voltmeter would be lit by the current generated in the cell.

Electrons are lost at the Zn^{2+}/Zn electrode and move to the Cu^{2+}/Cu electrode. Thus the Cu^{2+}/Cu electrode must be positive relative to the Zn^{2+}/Zn electrode.

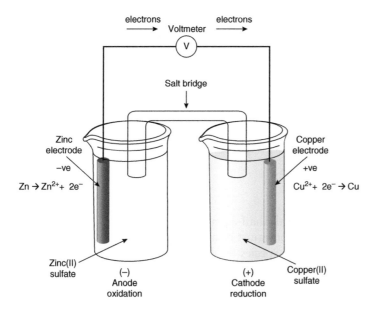

Figure 9.2 Two half-cells connected together to form an electrochemical cell, showing the flow of electrons.

The zinc electrode loses electrons so the equilibrium $Zn^{2+}(aq) + 2e^- \rightleftharpoons Zn(s)$ shifts to the left, and so the concentration of zinc ions in the solution increases.

The copper electrode gains electrons so the equilibrium $Cu^{2+}(aq) + 2e^- \rightleftharpoons Cu(s)$ shifts to the right. The concentration of copper ions in the solution decreases, and metallic copper can be seen to be deposited on the rod. The blue colour of the copper sulfate solution becomes paler. Although electrons flow in the cell from the negative electrode (Zn^{2+}/Zn) to the positive electrode (Cu^{2+}/Cu), the current is said, by convention, to flow in the opposite direction.

Ions move from the salt bridge to the appropriate half-cell to maintain an equal balance of charge.

Box 9.1 Some definitions:

A **half-cell** is one of two electrodes in an electrochemical cell where an oxidised and reduced species are linked in an equilibrium process involving the transfer of electrons.

An **electrochemical cell** comprises two half-cells physically separated but joined by a salt bridge. A current flows when the circuit is complete by joining the half-cells by a wire. This type of cell is also known as a **galvanic** or **voltaic** cell.

An **electrode** is the rod of metal or graphite in the half-cell that conducts electricity.

An **electrolyte** is a solution or liquid that contains ions.

In an electrochemical cell reduction occurs at the **cathode,** which is the more positive electrode or half-cell. Electrons flow to here.

In an electrochemical cell oxidation occurs at the **anode,** which is the more negative electrode or half-cell. Electrons flow from here.

A **salt bridge** is a piece of filter paper that is soaked in an electrolyte such as potassium nitrate and connects two electrical half-cells by allowing charged species to move from one to the other.

The **potential difference** or **electrode potential** is the voltage difference between two half-cells.

Worked Example 9.3   Write half-equations in the direction of the **reduction** process for the reactions taking place in the following half-cells:

  a. Ag^+/Ag

  b. $Cl_2/2Cl^-$

  c. $MnO_4^- + 4H^+/MnO_2$

  d. $O_2 + 2H^+/H_2O_2$

Solution

The question asks for the half-equations to be written as **reduction** reactions. This means the electrons must be added to the left-hand side of the equation and the more **reduced** species (in the lower oxidation state) included on the right-hand side.

  a. Here the more reduced species is the silver metal, Ag, so electrons must be added to the Ag^+ ion: $Ag^+ + e^- \rightleftharpoons Ag$.

  b. Here the more reduced species is the Cl^- ion, so electrons must be added to the Cl_2 on the left-hand side: $Cl_2 + 2e^- \rightleftharpoons 2Cl^-$.
     Each Cl atom in Cl_2 requires one electron for reduction to Cl^-.

  c. Work out the oxidation states of the two Mn species to decide which is the more reduced and determine the number of electrons required to bring about the reduction:

  MnO_4^- contains Mn(VII) or Mn^{7+}.

  MnO_2 contains Mn(IV) or Mn^{4+}.

  Therefore MnO_2 is the more reduced species, and Mn(VII) requires three electrons for reduction to Mn(IV). Write this information in equation form:

$$MnO_4^- + 3e^- \rightleftharpoons MnO_2 \quad \text{UNBALANCED EQUATION}$$

  The question suggests we need to add hydrogen ions to the left-hand side of the equation. These are converted to water molecules by combination with the oxygen from MnO_4^-. There is a total of four O atoms in MnO_4^-. Two O atoms are required for MnO_2, so the two remaining oxygen atoms form two moles of H_2O with the H^+ ions. This requires a total of four H^+ ions to balance the half-equation:

$$MnO_4^- + 4H^+ + 3e^- \rightleftharpoons MnO_2 + 2H_2O$$

Check that your answer is balanced by ensuring that there are equal numbers of each type of atom on both sides of the equation and that the charges on each side are the same.

  d. In this question, oxygen is the species being reduced. Determine the oxidation state of O in the two oxygen-containing species:

O_2 has O in the zero oxidation state.

H_2O_2 contains O_2^{2-} ions and so has O in the -1 oxidation state.

The more reduced species is therefore H_2O_2, and the electron difference is one electron. As there are two O atoms we require two electrons on the left-hand side:

$$O_2 + 2e^- \rightleftharpoons H_2O_2 \qquad \text{UNBALANCED EQUATION}$$

We must balance the atoms and the charge by adding 2 H^+ ions to the left-hand side of the half-equation:

$$O_2 + 2H^+ + 2e^- \rightleftharpoons H_2O_2$$

Worked Example 9.4   When a half-cell consisting of a piece of iron dipped in an iron(II) solution is connected to a half-cell consisting of a piece of copper rod in a copper(II) solution and the circuit is completed an electrochemical cell is formed, and electrons pass from the Fe^{2+}/Fe half-cell to the Cu^{2+}/Cu half-cell.

a. Write equations for the two half-cells in the direction of the reduction reactions.

b. Combine the equations to obtain an overall equation for the electrochemical cell.

c. State which electrode is positive and which is negative, and explain why.

d. Describe what you would observe happening in the Cu^{2+}/Cu half-cell.

Solution

a. $Fe^{2+} + 2e^- \rightleftharpoons Fe$, $Cu^{2+} + 2e^- \rightleftharpoons Cu$

b. We are told that electrons pass from the Fe^{2+}/Fe half-cell to the Cu^{2+}/Cu half-cell. Electrons must therefore be lost at the Fe^{2+}/Fe half-cell, so this half-equation must be reversed to give:

$Fe \rightleftharpoons Fe^{2+} + 2e^-$

This equation must be added to the half-equation for the reduction of Cu^{2+} where the electrons are taken up. The overall equation is therefore:

$Fe + Cu^{2+} \rightleftharpoons Fe^{2+} + Cu$

c. As the electrons are moving from the iron half-cell to the copper half-cell, the iron half-cell must be negative and the copper positive. Oxidation occurs at the Fe^{2+}/Fe electrode, so this is the anode. Reduction occurs at the Cu^{2+}/Cu electrode, so this is the cathode.

d. The overall equation shows Cu^{2+} ions, which are blue, being converted to metallic copper. The colour of the Cu^{2+} solution will therefore become paler, and a reddish-brown deposit of copper metal will form on the copper rod.

9.2.3   Standard electrode potentials, $E^{\ominus}$

So far, we have seen that two half-cells can be joined together to make an electrochemical cell. Electrons flow from the negative electrode (anode) where oxidation takes place to the positive electrode (cathode) where the reduction occurs. When a piece of metal such as zinc is placed in a solution of its ions, a potential difference (voltage) is set up between the metal and its ions in solution. This potential difference is known as the electrode potential for the half-cell, with the symbol E, and is measured in volts. To compare the potentials of half-cells, we need a reference electrode to measure them against. A standard hydrogen electrode is used as the reference electrode for such measurements.

9.2.4   The standard hydrogen electrode

Figure 9.3 shows a diagram of a standard hydrogen electrode. This forms a half-cell or electrode that can be connected to the electrode of a second half-cell whose standard reduction potential is being measured.

The hydrogen electrode uses a platinum electrode connected by a platinum wire to the voltmeter. The platinum electrode is covered with finely divided platinum black, which increases the surface area of the electrode. The electrode is mounted in an inverted glass tube. In a standard hydrogen electrode, hydrogen gas at a pressure of 1 atm (101.325 kPa) is passed over the platinum electrode and bubbled into a solution of hydrogen ions with a concentration of 1 mol dm^{-3}. The solution can be dilute hydrochloric acid (HCl) or sulfuric acid (H_2SO_4). The platinum electrode covered with platinum black is designed to ensure close contact between the hydrogen gas and the hydrogen ions in solution, and an equilibrium is established. Platinum metal is used as it is inert and does not interfere in the reaction. When standard conditions of temperature, pressure, and concentration are used, the potential measured is the standard electrode potential for the hydrogen electrode, $E^{\ominus}$. The electrode potential for the standard hydrogen electrode is arbitrarily set at zero. The electrode reaction and the value of $E^{\ominus}$ are represented by:

$$2H^{+}(aq) + 2e^{-} \rightleftharpoons H_2(g) \quad E^{\ominus} = 0.0\ V$$

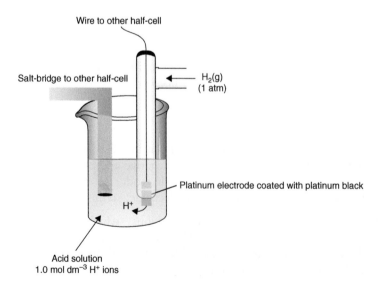

Figure 9.3 Standard hydrogen electrode.

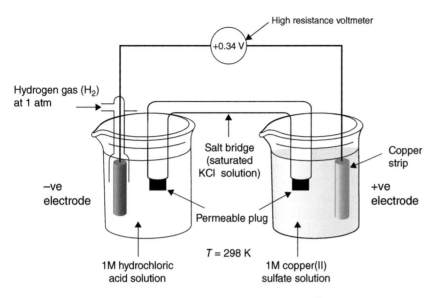

Figure 9.4 Measuring the standard reduction potential of a Cu^{2+}(aq)/Cu(s) half-cell using a standard hydrogen electrode.

The $E^{\ominus}$ value does not depend upon the stoichiometric coefficients in the equation used and has the same value even if the stoichiometric coefficients are halved.

To measure the standard reduction potential of a half-cell such as the Cu^{2+}(aq)/Cu(s) electrode the standard hydrogen electrode half-cell is connected via a salt bridge and voltmeter to the Cu half-cell (Figure 9.4). The value obtained when standard conditions are used is called the *standard electrode potential* and has the symbol $E^{\ominus}$. This is often referred to as the **standard reduction potential** as all values are quoted in the direction of the reduction half-equation.

Standard conditions for measurement of standard reduction potentials are:

- All solutions at a concentration of 1 mol dm^{-3} (1 M)

- All gases at a pressure of 1 atmosphere (1.0 atm)

- All measurements made at 298 K (25 °C)

Standard electrode potentials are always measured as reduction reactions. In quoting the $E^{\ominus}$ value, the oxidised species is always given first followed by a slash and then the reduced species. For example Cu^{2+}/Cu.

The standard reduction potential for a half-cell is the voltage of the half-cell measured under standard conditions when connected to a standard hydrogen electrode.

Values of standard reduction potentials

The value of the standard reduction potential for the copper half-cell is +0.34 V:

$$Cu^{2+}(aq) + 2e^{-} \rightleftharpoons Cu(s) \quad E^{\ominus} = +0.34 \text{ V}$$

The plus sign indicates that the copper electrode is more positive than the hydrogen electrode and electrons will move from the hydrogen electrode to copper. This means that the reduction of $Cu^{2+}(aq)$ to $Cu(s)$ is more likely to occur than the reduction of $H^+(aq)$ to $H_2(g)$ under standard conditions.

The value of the standard reduction potential for the zinc half-cell is -0.76 V:

$$Zn^{2+}(aq) + 2e^- \rightleftharpoons Zn(s) \quad E^\ominus = -0.76 \text{ V}$$

The sign of the electrode potential for the $Zn^{2+}(aq)/Zn(s)$ couple is negative. The minus sign indicates that the zinc electrode is negative with respect to the hydrogen electrode and electrons move from zinc to hydrogen. This means that the reduction of $Zn^{2+}(aq)$ to $Zn(s)$ is less likely to occur than the reduction of $H^+(aq)$ to $H_2(g)$ under standard conditions.

Because the $E^\ominus$ value for the Cu^{2+}/Cu electrode is more positive than that of the Zn^{2+}/Zn electrode, the reduction of $Cu^{2+}(aq)$ to $Cu(s)$ is more likely to occur than the reduction of $Zn^{2+}(aq)$ to $Zn(s)$. In a cell made up of a $Zn^{2+}(aq)/Zn(s)$ electrode and a $Cu^{2+}(aq)/Cu(s)$ electrode, the copper will be the positive electrode and the zinc negative.

> **Box 9.2** A positive value for a standard reduction potential means that the reaction will proceed to the right more readily than the $H^+/\frac{1}{2}H_2$ reaction. This means the species will be more readily reduced. A negative value for a standard reduction potential means that the reaction will proceed in the backward direction and the species will be more readily oxidised. The more positive the standard reduction potential for one of two half-cells, the more readily it will be reduced, and thus it will form the positive electrode in an electrochemical cell made up of the two half-cells.

Table 9.1 gives the standard reduction potentials for some common metal/metal ion half-cells.

Note the following points about the information in Table 9.1:

- The electrode potential refers to the reduction reaction (it is known as the standard reduction potential), so the electrons appear on the left-hand side of the half-equation.

- The more positive the value of the electrode potential, the easier it is to reduce the metal ions (or other species) on the left. In Table 9.1, the Ag^+ ions are most readily reduced to metallic Ag.

- The more negative the electrode potential, the more difficult it is to reduce the metal ions on the left. Look at the metals at the top of Table 9.1. There you find the alkali metals K and Na. We know these metals are very reactive, yet their ions are quite stable in aqueous solution, suggesting that the ions are not easily reduced to the metal.

Table 9.1 Standard reduction potentials for metal/metal ion half-cells.

Electrode reaction		
Oxidised species	Reduced species	$E^{\ominus}/V$
$K^+(aq) + e^- \rightleftharpoons K(s)$		−2.92
$Ca^{2+}(aq) + 2e^- \rightleftharpoons Ca(s)$		−2.87
$Na^+(aq) + e^- \rightleftharpoons Na(s)$		−2.71
$Mg^{2+}(aq) + 2e^- \rightleftharpoons Mg(s)$		−2.38
$Al^{3+}(aq) + 3e^- \rightleftharpoons Al(s)$		−1.66
$Zn^{2+}(aq) + 2e^- \rightleftharpoons Zn(s)$		−0.76
$Fe^{2+}(aq) + 2e^- \rightleftharpoons Fe(s)$		−0.44
$Ni^{2+}(aq) + 2e^- \rightleftharpoons Ni(s)$		−0.25
$2H^+(aq) + 2e^- \rightleftharpoons H_2(g)$		0.0
$Cu^{2+}(aq) + 2e^- \rightleftharpoons Cu(s)$		+0.34
$Ag^+(aq) + e^- \rightleftharpoons Ag(s)$		+0.80

Worked Example 9.5   State in which direction the following half-equations will proceed when connected to a standard hydrogen electrode:

a. Ni^{2+}/Ni

b. Ag^+/Ag

c. Al^{3+}/Al

Solution

a. The Ni^{2+}/Ni half-cell has an $E^{\ominus}$ of −0.25 V. The reduction reaction therefore occurs less readily than the reduction of H^+ to $\frac{1}{2}H_2$. The reaction $Ni^{2+} + 2e^- \rightleftharpoons Ni$ will therefore lie over to the Ni^{2+} ions.

b. The Ag^+/Ag half-cell has an $E^{\ominus}$ of +0.80 V. The reduction reaction therefore occurs more readily than the reduction of H^+ to $\frac{1}{2}H_2$. The reaction $Ag^+ + e^- \rightleftharpoons Ag$ will therefore lie over to the Ag metal.

c. The Al^{3+}/Al half-cell has an $E^{\ominus}$ of −1.66 V. The reduction reaction therefore occurs less readily than the reduction of H^+ to $\frac{1}{2}H_2$. The reaction $Al^{3+} + 3e^- \rightleftharpoons Al$ will therefore lie over to the Al^{3+} ions.

9.2.5   Half-cells involving non-metals and non-metal ions

An electrochemical cell can be made up from half-cells that do not necessarily contain metals and their ions but may involve gases with aqueous solutions of their ions or a combination of ions with different oxidation states. Electrical contact must be made between the two half-cells. When both half-cells include a metal, contact is easily achieved using a metal wire. If a half-cell involves a gaseous component electrical contact can be made using a platinum electrode over which the gas is bubbled, as with the hydrogen electrode. Platinum metal is inert so does not react with the gas or its aqueous ions. The platinum electrode must be in contact with both the gas and the solution of its ions. The standard electrode potential of such a half-cell is again obtained by connecting it to a standard hydrogen electrode.

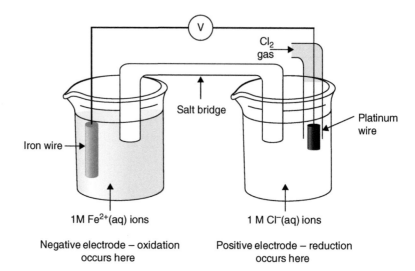

Figure 9.5 A $Cl_2(g)/Cl^-(aq)$ and $Fe^{2+}(aq)/Fe(s)$ electrochemical cell.

Figure 9.5 shows an electrochemical cell made up from a chlorine/chloride electrode in combination with an iron(II)/iron electrode. The right-hand electrode is a Cl_2/Cl^- half-cell. Chlorine gas is passed over a platinum wire in contact with chloride ions at a concentration of $1.0\,mol\,dm^{-3}$. The half-cell reaction is:

$$\tfrac{1}{2}\,Cl_2(g) + e^- \rightleftharpoons Cl^-(aq) \quad E^\ominus = +1.36\,V$$

The left-hand electrode is a $Fe^{2+}(aq)/Fe(s)$ half-cell. A piece of iron wire is in contact with a solution of $1.00\,mol\,dm^{-3}$ Fe^{2+} ions. The half-cell reaction for this electrode is

$$Fe^{2+}(aq) + 2e^- \rightleftharpoons Fe(s) \quad E^\ominus = -0.44\,V$$

The two solutions are connected by a salt bridge and the two electrodes by a wire and a high-resistance voltmeter.

The half-cell with the more positive value of $E^\ominus$ is the positive electrode where reduction occurs. In this case this is the $Cl_2(g)/Cl^-(aq)$ electrode. The chemical reaction occurring represents the reduction of chlorine gas to chloride ions. Knowing that chlorine is far more electronegative than iron, it makes sense that this half-reaction should proceed in the direction shown, i.e. reduction of chlorine to chloride ions. The $Fe^{2+}(aq)/Fe(s)$ electrode is the negative electrode (anode) where electrons are generated as the metal is oxidised to Fe^{2+} ions. The actual direction of reaction at this electrode is therefore:

$$Fe(s) \rightleftharpoons Fe^{2+}(aq) + 2e^- \quad E^\ominus = +0.44\,V$$

9.2.6   The cell diagram

Two half-cells can be combined to form an **electrochemical cell**. We have already seen that a $Zn^{2+}(aq)/Zn(s)$ half-cell and a $Cu^{2+}(aq)/Cu(s)$ half-cell can be combined to give an electrochemical cell that has a measurable voltage of $1.10\,V$.

There is a convention for depicting the components of an electrochemical cell on paper called the **cell diagram**. This avoids having to draw a picture of the two half-cells and the circuit. For the electrochemical cell in Figure 9.2, the cell diagram is written as:

$$Zn(s) \mid Zn^{2+}(aq) \parallel Cu^{2+}(aq) \mid Cu(s)$$

The convention for writing the cell diagram is as follows:

- Write the half-cell undergoing oxidation on the left-hand side of the diagram.

- Write the half-cell undergoing reduction on the right-hand side of the diagram.

- The single lines represent a change in phase of the components, i.e. between the solid phase and aqueous solution in this case.

- The double vertical line represents the salt bridge between the two half-cells.

Worked Example 9.6   Write the cell diagrams for the following electrochemical cells:

    a. $Ni(s) + Cu^{2+}(aq) \rightleftharpoons Ni^{2+}(aq) + Cu(s)$

    b. $Zn^{2+}(aq) + Mg(s) \rightleftharpoons Zn(s) + Mg^{2+}(aq)$

Solution

    a. Use the convention for writing the cell diagrams:

- Which half-cell is undergoing oxidation? Write this on the left. To work this out, check the standard reduction potentials in Table 9.1.
    The Ni^{2+}/Ni half-cell has $E^{\ominus} = -0.25$ V, and $Cu^{2+}/Cu = +0.34$ V. The value for Cu^{2+}/Cu is more positive, so this is where reduction occurs; therefore, oxidation occurs at the Ni/Ni^{2+} electrode. This is therefore written on the left and the Cu^{2+}/Cu half-cell on the right.

- Use a single vertical line to represent a change of phase. There is a change of phase between the oxidised and reduced species in each half-cell.

- Use a double vertical line to represent the salt bridge. This occurs between the two half-cells:

$$Ni(s) \mid Ni^{2+}(aq) \parallel Cu^{2+}(aq) \mid Cu(s)$$

    b. In this cell, Mg^{2+}/Mg has $E^{\ominus} = -2.38$ V and Zn^{2+}/Zn has $E^{\ominus} = -0.76$ V. The Zn^{2+}/Zn is more positive and therefore reduction occurs here. The Mg^{2+}/Mg half-cell is therefore written on the left. The complete cell diagram is therefore

$$Mg(s) \mid Mg^{2+}(aq) \parallel Zn^{2+}(aq) \mid Zn(s)$$

9.2.7   Using $E^\ominus$ values to obtain voltages of electrochemical cells

When two standard half-cells are combined in an electrochemical cell the individual $E^\ominus$ values of the half-cells can be used to determine the voltage of the whole cell. The voltage of the cell is a measure of the **potential difference** between the two electrodes of the half-cells as measured by a voltmeter of high resistance connected in the circuit. It is sometimes referred to as the **electromotive force** or **emf**.

The standard electrode potential of an electrochemical cell is the difference in the standard electrode potentials of the two half-cells. It is calculated by using the expression

$$E^\ominus_{cell} = E^\ominus (\text{half-cell undergoing reduction}) - E^\ominus (\text{half-cell undergoing oxidation})$$

In the Daniell cell reduction occurs at the Cu^{2+}/Cu electrode, so the $E^\ominus$ value for the cell is given by:

$$E^\ominus_{cell} = E^\ominus \left(Cu^{2+}(aq)/Cu(s)\right) - E^\ominus \left(Zn^{2+}(aq)/Zn(s)\right)$$
$$= +0.34\,V - (-0.76\,V) = +1.10\,V$$

We can also use the half-cell equations to obtain the overall cell reaction and $E^\ominus$ value. The two half-cell reactions are:

$$Cu^{2+}(aq) + 2e^- \rightleftharpoons Cu(s) \quad E^\ominus = +0.34\,V \tag{9.2}$$

$$Zn^{2+}(aq) + 2e^- \rightleftharpoons Zn(s) \quad E^\ominus = -0.76\,V \tag{9.3}$$

We know that metallic zinc is oxidised to Zn^{2+} ions, so the direction of this reaction is reversed to obtain the overall redox equation. The sign of the $E^\ominus$ value must also be reversed:

> This method for calculating the $E^\ominus$ value of the cell is exactly the same as the equation given earlier where the $E^\ominus$ value of the half-cell undergoing oxidation is subtracted.

$$Zn(s) \rightleftharpoons Zn^{2+}(aq) + 2e^- \quad E^\ominus = +0.76\,V \tag{9.4}$$

The overall redox equation is obtained by adding the two sides of the half-equations Eqs. (9.2) and (9.4):

$$Cu^{2+}(aq) + Zn(s) \rightleftharpoons Cu(s) + Zn^{2+}(aq)$$

The overall cell potential (emf) is obtained by adding the $E^\ominus$ values for half-Eqs. (9.2) and (9.4):

$$E^\ominus_{cell} = +0.34\,V + 0.76\,V = +1.10\,V$$

> The electrode potential for an electrochemical cell is also known as the **voltage** or **emf** (electromotive force) of the cell or the **cell potential**. These terms all mean the same thing. We will use the term **electrode potential** in this text and **standard electrode potential** when the value is measured under standard conditions. We will refer to standard half-cell potentials as **standard reduction potentials** to ensure that it is clear that the reaction is a reduction process.

Note that if the number of electrons transferred is different in the two half-equations that make up the overall equation, the equations must be adjusted so the electrons cancel. However, the $E^\ominus$ values remain unchanged as the standard reduction potential is a fixed value and does not depend upon the number of moles of electrons in the equation for the half-cell. An example of this can be seen in the following problem where the $Ag^+(aq)/Ag(s)$ half-equation involves the transfer of only one electron, whereas the $Zn^{2+}(aq)/Zn(s)$ requires two.

Worked Example 9.7   Write the cell diagram for an electrochemical cell made up of two standard half-cells, and calculate the cell potential using values in Table 9.1. The two half-cells are $Ag^+(aq)/Ag(s)$ and $Zn^{2+}(aq)/Zn(s)$.

Solution
The half-cell reactions are

$$Ag^+(aq) + e^- \rightleftharpoons Ag(s) \quad E^\ominus = +0.80\ V$$

$$Zn^{2+}(aq) + 2e^- \rightleftharpoons Zn(s) \quad E^\ominus = -0.76\ V$$

The $E^\ominus$ value for the Ag^+/Ag half-cell is more positive than the value for the Zn^{2+}/Zn half-cell. Therefore, the Ag^+/Ag is the positive electrode (cathode) at which reduction occurs. The Zn^{2+}/Zn electrode is negative (anode), and oxidation occurs here.

The cell diagram is given by:

$$Zn(s) \mid Zn^{2+}(aq) \parallel Ag^+(aq) \mid Ag(s)$$

The $E^\ominus$ value for the cell is obtained from this equation:

$$E^\ominus_{cell} = E^\ominus(\text{half-cell undergoing reduction})$$

$$- E^\ominus(\text{half-cell undergoing oxidation})$$

$$E^\ominus_{cell} = E^\ominus\,(Ag^+\,(aq)/Ag(s)) - E^\ominus\,(Zn^{2+}\,(aq)/Zn(s)$$

$$= +0.80\ V - (-0.76\ V) = +1.56\ V$$

Alternatively, the cell potential can be calculated from the overall redox equation for the electrochemical cell. This is obtained by adding the two half-equations in the appropriate directions. As oxidation is occurring at the Zn^{2+}/Zn electrode the half-equation for this reaction, and the sign of the $E^\ominus$ value must be reversed:

$$Zn(s) \rightleftharpoons Zn^{2+}(aq) + 2e^- \quad E^\ominus = +0.76\ V \tag{9.5}$$

$$Ag^+(aq) + e^- \rightleftharpoons Ag(s) \quad E^\ominus = +0.80\ V \tag{9.6}$$

As two electrons are required in Eq. (9.5), we must multiply Eq. (9.6) by two. Note that the $E^\ominus$ value remains unchanged and is the same no matter how many moles are involved:

$$2Ag^+(aq) + 2e^- \rightleftharpoons 2Ag(s) \quad E^\ominus = +0.80\ V \tag{9.7}$$

The left-hand and right-hand sides of Eqs. (9.5) and (9.7) are now added to obtain the overall equation for the reaction. The electrons cancel:

$$Zn(s) + 2Ag^+(aq) \rightleftharpoons Zn^{2+}(aq) + 2Ag(s)$$

The cell potential is obtained by adding the $E^\ominus$ values for each half-cell in the direction shown in the overall equation:

$$+0.76\ V + 0.80\ V = +1.56\ V$$

Worked Example 9.8   Determine the standard reduction potential for electrochemical cells made up from the following combinations of half-cells:

a. $Ag^+(aq) + e^- \rightleftharpoons Ag(s)$ $E^\ominus = +0.80\,V$
    $\frac{1}{2}Br_2(l) + e^- \rightleftharpoons Br^-(aq)$ $E^\ominus = +1.07\,V$

b. $MnO_4^-(aq) + 8H^+(aq) + 5e^- \rightleftharpoons Mn^{2+}(aq) + 4H_2O(l)$ $E^\ominus = +1.52\,V$
    $CO_2(g) + e^- \rightleftharpoons \frac{1}{2}C_2O_4^{2-}(aq)$ $E^\ominus = -0.49\,V$

c. $Cl_2(g) + 2e^- \rightleftharpoons 2Cl^-(aq)$ $E^\ominus = +1.36\,V$
    $Fe^{2+}(aq) + 2e^- \rightleftharpoons Fe(s)$ $E^\ominus = +0.44\,V$

Solution

a. The first step in such problems is to remember that reduction occurs at the electrode with the more positive standard electrode potential. Electrons are consumed at this electrode. In this question, the Br_2/Br^- half-cell has the more positive $E^\ominus$ value, so this will be where reduction occurs. The equation that allows calculation of the cell emf is

$$E^\ominus_{cell} = E^\ominus(\text{half-cell undergoing reduction})$$
$$- E^\ominus(\text{half-cell undergoing oxidation})$$
$$= E^\ominus(Br_2(l)/Br^-(aq)) - E^\ominus(Ag^+(aq)/Ag(s))$$
$$= +1.07 - (+0.80\,V) = +0.27\,V$$

b. Again, first look at the $E^\ominus$ values to decide at which electrode reduction will occur. The higher value of +1.52 V for the MnO_4^-/Mn^{2+} electrode indicates that this reaction is more favourable in the reduction direction. The manganate (MnO_4^-) ion is a strong oxidising agent. Oxidation must occur at the $CO_2/C_2O_4^{2-}$ half-cell.

$$E^\ominus_{cell} = E^\ominus(\text{half-cell undergoing reduction})$$
$$- E^\ominus(\text{half-cell undergoing oxidation})$$
$$= E^\ominus(MnO_4^-(aq)/Mn^{2+}(aq)) - E^\ominus(CO_2(g)/C_2O_4^{2-}(aq))$$
$$= +1.52\,V - (-0.49\,V) = +2.01V$$

The high positive $E^\ominus_{cell}$ value of +2.01 V suggests that the reaction is highly favourable in the direction written, which concurs with MnO_4^- being a strong oxidising agent.

c. The Cl_2/Cl^- electrode has the more positive $E^\ominus$ value, so reduction will occur at this electrode:

$$E^\ominus_{cell} = E^\ominus(\text{half-cell undergoing reduction})$$
$$- E^\ominus(\text{half-cell undergoing oxidation})$$

$$E^\ominus_{cell} = E^\ominus(Cl_2(g)/Cl^-(aq)) - E^\ominus(Fe(s)/Fe^{2+}(aq))$$

$$= +1.36\,V - (+0.44\,V) = +0.92\,V$$

All these examples can also be solved by reversing the equation for the half-cell reaction that involves oxidation and summing the $E^\ominus$ values – which is effectively the same procedure.

9.2.8   Using standard reduction potentials to predict the outcome of redox reactions

If the standard reduction potential for a half-equation is positive, the species is more likely to be reduced than hydrogen. This means that half-cell reactions with positive values of $E^{\ominus}$ will oxidise species whose $E^{\ominus}$ value is less positive. Half-cell reactions with the highest positive $E^{\ominus}$ values are found for half-equations such as the reduction of halogens to halides. Therefore, species with high positive values of $E^{\ominus}$ are strong oxidising agents. Half-cell reactions with the largest negative $E^{\ominus}$ values are for reactions that are unlikely to occur in the direction of reduction. For example, metals at the top of the table (e.g. Li, Na, Ca, Mg) are very reactive, so the reverse reaction in which the metal is oxidised to the positive ion is more favourable. These metals act as strong reducing agents. Values of standard reduction potentials can be listed alphabetically or numerically. A table that contains values listed alphabetically, such as Table 9.1, is easier to use when searching for a specific $E^{\ominus}$ value. However, a table that lists numerical $E^{\ominus}$ values has more chemical use and provides a guide to strong reducing and oxidising agents. A numerically arranged list of some common standard reduction potentials is given in Table 9.2.

When comparing the species on the left of each equation, the strongest oxidising agents are at the bottom of the table and the strongest reducing agents are at the top of the table. Remember that the half-equation with the more positive $E^{\ominus}$ value forms the positive electrode, the cathode, and reduction occurs at this electrode. The half-equation with the lower $E^{\ominus}$ value forms the negative electrode, the anode, and electrons are lost at this electrode, forcing this half-cell reaction to be reversed. The difference between the standard reduction potentials for two half-cells gives the overall cell potential. A positive value of $E^{\ominus}$ suggests the reaction is feasible but does not necessarily mean it will take place.

Table 9.2 List of some common standard reduction potentials in numerical order.

Electrode reaction	$E^{\ominus}$ value/V
$Li^+(aq) + e^- \rightleftharpoons Li(s)$	−3.04
$Ca^{2+}(aq) + 2e^- \rightleftharpoons Ca(s)$	−2.87
$Mg^{2+}(aq) + 2e^- \rightleftharpoons Mg(s)$	−2.38
$Al^{3+}(aq) + 3e^- \rightleftharpoons Al(s)$	−1.66
$Zn^{2+}(aq) + 2e^- \rightleftharpoons Zn(s)$	−0.76
$Fe^{2+}(aq) + 2e^- \rightleftharpoons Fe(s)$	−0.44
$Cr^{3+}(aq) + e^- \rightleftharpoons Cr^{2+}(aq)$	−0.41
$Sn^{2+}(aq) + 2e^- \rightleftharpoons Sn(s)$	−0.14
$2H^+(aq) + 2e^- \rightleftharpoons H_2(g)$	0.00
$Cu^{2+}(aq) + e^- \rightleftharpoons Cu^+(aq)$	+0.15
$Cu^{2+}(aq) + 2e^- \rightleftharpoons Cu(s)$	+0.34
$Cu^+(aq) + e^- \rightleftharpoons Cu(s)$	+0.52
$\frac{1}{2}I_2(s) + e^- \rightleftharpoons I^-(aq)$	+0.54
$O_2(g) + 2H^+(aq) + 2e^- \rightleftharpoons H_2O_2(aq)$	+0.68
$Fe^{3+}(aq) + e^- \rightleftharpoons Fe^{2+}(aq)$	+0.77
$Ag^+(aq) + e^- \rightleftharpoons Ag(s)$	+0.80
$\frac{1}{2}Br_2(l) + e^- \rightleftharpoons Br^-(aq)$	+1.07
$Cr_2O_7^{2-}(aq) + 14H^+(aq) + 6e^- \rightleftharpoons 2Cr^{3+}(aq) + 7H_2O(l)$	+1.33
$\frac{1}{2}Cl_2(g) + e^- \rightleftharpoons Cl^-(aq)$	+1.36
$MnO_4^-(aq) + 8H^+(aq) + 5e^- \rightleftharpoons Mn^{2+}(aq) + 4H_2O(l)$	+1.52
$\frac{1}{2}H_2O_2(aq) + H^+(aq) + e^- \rightleftharpoons H_2O(l)$	+1.77
$\frac{1}{2}F_2(g) + e^- \rightleftharpoons F^-(aq)$	+2.87

Increase in oxidising strength of species on the left of the equation

Worked Example 9.9   Use the data in Table 9.2 to calculate a value of $E^\ominus$ for the reaction in which Fe^{3+} ions oxidise I^- ions in aqueous solution. Give the overall equation for the redox reaction.

Solution
The two standard reduction potentials involved are:

$$Fe^{3+}(aq) + e^- \rightleftharpoons Fe^{2+}(aq) \quad E^\ominus = +0.77\,V$$

$$\tfrac{1}{2}I_2(s) + e^- \rightleftharpoons I^-(aq) \quad\quad E^\ominus = +0.54\,V$$

The Fe^{3+}/Fe^{2+} electrode has the higher $E^\ominus$ value, so reduction occurs at this electrode. We are told in the question that Fe^{3+} ions oxidise I^- ions, so oxidation occurs at the I_2/I^- electrode, which is the anode:

$$E^\ominus_{cell} = E^\ominus(\text{half-cell undergoing reduction})$$
$$- E^\ominus(\text{half-cell undergoing oxidation})$$
$$= E^\ominus(Fe^{3+}(aq)/Fe^{2+}(aq)) - E^\ominus(I_2(s)/I^-(aq))$$
$$= +0.77\,V - (+0.54\,V) = +0.23\,V$$

9.2.9   Relation between $E^\ominus$ and Gibbs energy

We saw in Chapter 6 that the Gibbs energy for a reaction is a measure of its spontaneity – or how likely the reaction is to occur. The Gibbs energy for an electrochemical process is linked to the standard electrode potential for the process by the following equation:

$$\Delta G^\ominus = -nFE^\ominus$$

where

$\Delta G^\ominus$ is the standard Gibbs energy $(kJ\,mol^{-1})$.

n is the number of electrons transferred in the process.

F is a constant known as the Faraday constant $(96\,485\,C\,mol^{-1})$.

$E^\ominus$ is the standard cell potential (V).

There is not space to derive this equation here, but in Chapter 6, we saw that if the Gibbs energy is negative the reaction is said to be thermodynamically feasible. The equation above shows that, if the standard electrode potential for a redox couple is greater than zero, in theory the reaction should be feasible because the Gibbs energy will be negative. If the value of $E^\ominus$ is positive, then the value of $\Delta G^\ominus$ will be negative, so the reaction should proceed under standard conditions according to thermodynamic considerations. However, this does not take into account kinetic feasibility. If the activation energy for the reaction is very high, such that the reaction proceeds very slowly unless heated or in the presence of a catalyst, then it is unlikely that any reaction will be observed even if the standard electrode potential is greater than zero.

In general,

- A high positive value for $E^\ominus$ *usually* means the reaction is thermodynamically feasible.

- A negative value for $E^\ominus$ *usually* means the reaction is not thermodynamically feasible and will not proceed.

9.2.10   The effect of non-standard conditions on cell potential – the Nernst equation

In the examples we have looked at so far all reactants and conditions have been standard: i.e. concentrations of 1 mol dm^{-3}, pressures of 1 atm, and standard temperature. However, the overall reaction occurring in an electrochemical cell is represented by a chemical equation that is an equilibrium reaction. We saw in Chapter 7 that changing the conditions of a reaction shifts the equilibrium position, and this can have an effect on the cell potential. Using concentrations other than 1 mol dm^{-3} can either increase or decrease the cell potential. In turn, this may affect the feasibility of the reaction.

To allow calculation of the reduction potential for half-cells with different concentrations of reacting species, the Nernst equation is applied:

$$E = E^{\ominus} + \frac{0.059\ V}{n} \log_{10} \frac{[\text{oxidised species}]}{[\text{reduced species}]}$$

where n is the number of electrons transferred.

The oxidised and reduced species refer to the concentrations of the oxidised and reduced components in the half-equation. If a solid is involved, such as a metal, its concentration is a constant and can be ignored in the equation, as shown in Worked Example 9.10. The value of 0.059 V is a constant obtained from a combination of other constants used in the derivation of the equation.

Worked Example 9.10   Calculate the value of the electrode potential at 298 K of a Ni^{2+}(aq)/Ni(s) electrode that has a concentration of Ni^{2+} ions of 2.0 mol dm^{-3}. The $E^{\ominus}$ value for the reaction is –0.25 V.

Solution
The half-equation here is Ni^{2+}(aq) $+ 2e^{-} \rightleftharpoons Ni$(s).

The Nernst equation allows us to calculate the half-cell potential under non-standard conditions:

$$E = E^{\ominus} + \frac{0.059\ V}{n} \log_{10} \frac{[\text{oxidised species}]}{[\text{reduced species}]}$$

The oxidised species here is Ni^{2+} with a concentration of 2.0 mol dm^{-3}. The reduced species is the solid Ni metal. The concentration of the solid Ni metal is a constant and does not change, so the equation can be written as:

$$E = E^{\ominus} + \frac{0.059\ V}{n} \log_{10} [\text{oxidised species}]$$

The number of electrons exchanged, n, is 2.

Inserting the values in the equation we obtain:

$$E = -0.25\ V + \frac{0.059\ V}{2} \times \log_{10} 2.0 = -0.25\ V + 0.0295\ V \times 0.301 = -0.24\ V.$$

9.3   Using redox reactions – galvanic cells

9.3.1   Galvanic (voltaic) cells

Electrical energy from redox reactions is harnessed in batteries that rely on the transfer of electrons to produce an electrical current. A *galvanic* cell is simply an electrical cell that produces electrical energy from a spontaneous redox reaction. There are many different types of cells, and research into cheaper, lighter, more sustainable and rechargeable cells that use easily obtainable materials continues to drive developments in this field.

9.3.2   The variety of cells

The first electrical cell to be produced deployed the reaction between zinc and copper(II) sulfate and was the Daniell cell, developed in the nineteenth century.

A cell or battery consists of an anode where oxidation occurs and electrons are lost, and a cathode where reduction occurs and the electrons are used up. The two electrodes are in contact with an electrolyte, which may be a liquid in 'wet' cells or a moist paste in 'dry cells'.

9.3.3   Disposable batteries

Standard alkaline batteries such as AA or AAA are types of galvanic cells (Figure 9.6).

Batteries that are disposable are known as *primary* cells. This means the battery can be used only once. Most common batteries of this type contain a zinc anode and a cathode formed of a carbon rod that is packed around with a mixture of graphite and manganese(IV) oxide. The electrolyte can be a paste of ammonium chloride (Figures 9.6 and 9.7).

Alkaline dry cells have largely replaced carbon–zinc cells as they can maintain a more stable operating voltage at a high current drain. In an alkaline battery, the anode is powdered zinc and the cathode is manganese(IV) oxide in a paste of potassium hydroxide. The electrode reactions are the same in both cases: oxidation of zinc to zinc(II) ions with loss of electrons occurs at the anode. At the cathode manganese(IV) oxide is reduced to manganese(III):

$$Zn(s) \rightleftharpoons Zn^{2+}(aq) + 2e^- \quad \text{oxidation at the anode}$$

$$2MnO_2(s) + 2H_2O(l) + 2e^- \rightleftharpoons 2MnO(OH)(s) + 2OH^-(aq) \quad \text{reduction at the cathode}$$

Figure 9.6 A variety of alkaline batteries that are examples of galvanic cells. *Source:* Elizabeth Page.

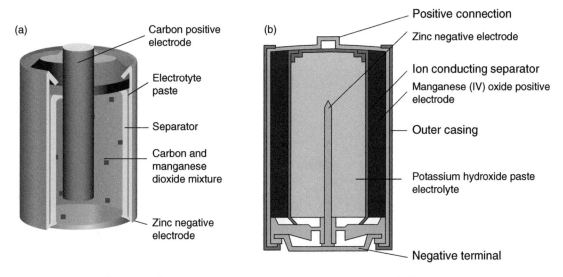

Figure 9.7 Sections through (a) a carbon–zinc battery; (b) a zinc–manganese(IV) oxide battery.

9.3.4 Rechargeable cells

Rechargeable cells are cells that can be recharged by passing a current through in the opposite direction to the normal current flow when in use. These types of cells are known as *secondary* cells. The recharge current reverses the cell reactions so that the original materials are reformed. Nicad (nickel–cadmium) batteries used to be the most prolific type of rechargeable batteries used domestically, but these have been largely replaced by lithium-ion batteries.

9.3.4.1 Lithium-Ion batteries

Lithium-ion batteries are an important type of rechargeable battery widely used in applications where a high energy density is required and weight is critical (see Figure 9.8). Thus they are now increasingly used in mobile devices, digital cameras, power tools, and electric cars. The anode can be lithium or an intercalated lithium–graphite material, and the cathode is a mixed-metal oxide such as lithium cobalt oxide. The electrolyte is a lithium salt in an organic solvent.

Lithium ions, generated at the anode, are able to move through the electrolyte:

$$LiC_6 \rightleftharpoons Li^+ + 6C + e^- \quad \text{oxidation at the anode}$$

Reduction of the metal oxide occurs at the cathode. For example:

$$CoO_2 + Li^+ + e^- \rightleftharpoons LiCoO_2$$

The great advantage of lithium-ion batteries is their light weight and their ability to be charged and discharged many times over without a loss in performance.

Figure 9.8 Lithium-ion batteries. *Sources:* LG; Zoonar/P.Malyshev.

9.3.5   Fuel cells

Fuel cells are galvanic cells that use a fuel to produce electrons at one electrode. The electrons are consumed by oxygen at the other electrode. Fuel cells are increasingly being used to power electric vehicles and manned spacecraft. Fuel cells that use hydrogen as the fuel and oxygen from the air are described as producing 'zero emissions' as, unlike burning traditional fossil fuels, no carbon dioxide is released to the atmosphere. The only gas emitted is water vapour.

In a hydrogen–oxygen fuel cell, gaseous hydrogen and oxygen are bubbled through two porous platinum-coated electrodes where half-cell reactions take place. The two halves of the fuel cell are separated by a proton-exchange membrane that allows hydrogen ions released at the negative electrode to diffuse through to the positive electrode, as shown in Figure 9.9. The electrode reactions are

$$H_2(g) \rightleftharpoons 2H^+(aq) + 2e^- \quad \text{Anode reaction}$$

$$4H^+(aq) + O_2(g) + 4e^- \rightleftharpoons 2H_2O(l) \quad \text{Cathode reaction}$$

The overall reaction in the cell is: $2H_2(g) + O_2(g) \rightleftharpoons 2H_2O(l)$

Electrons flow through the external circuit from the anode to the cathode and can be used to drive the motor of a vehicle. Current will flow until the fuel is exhausted.

The main benefits of hydrogen–oxygen fuel cells over petrol and diesel engines are that no carbon dioxide, CO_2, or nitrogen oxides, NO_x, are emitted. Hydrogen is light and obtained by electrolysis of water, which is abundant, and the energy transfer is very efficient. Clearly, these are very positive factors. The main drawback with hydrogen fuel cells is that extracting hydrogen gas from water requires large quantities of energy because of the highly exothermic enthalpy of formation of water. Unless a renewable energy source or nuclear fuel

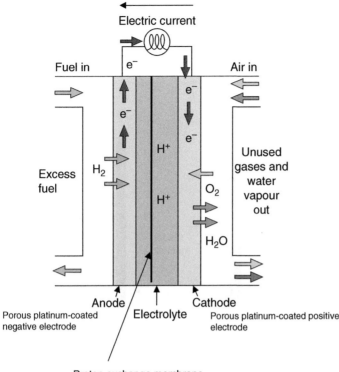

Figure 9.9 Hydrogen–oxygen fuel cell.

is used to achieve this, CO_2 and NO_x gases are released in producing the hydrogen. There is also the hazard associated with carrying large quantities of potentially explosive hydrogen gas under pressure. The cylinders are heavy and the hydrogen extremely unstable in air.

9.4 Using redox reactions – electrolytic cells

The redox reactions described so far in this chapter have been shown to generate a voltage that can be deployed in an electrochemical cell (galvanic or voltaic cell) or a battery. In this type of cell the electric current can be used to do useful work. An electrolytic cell is the chemical opposite of a galvanic cell. In an electrolytic cell, electricity is used from an external source to bring about a chemical reaction. Such cells are frequently used to extract and purify metals from their molten ores and useful chemicals such as halogens from aqueous solutions containing halide ions.

9.4.1 Electrolysis

Electrolysis is a process in which an electric current is used to decompose an ionic substance into its elemental components. An electrolytic cell consists of two electrodes immersed in an **electrolyte**. The electrolyte is the substance being decomposed and may be a molten material or an aqueous solution containing ions. The electrolyte contains ions of opposite charge that are free to move. A diagram of a simple electrolytic cell is shown in Figure 9.10.

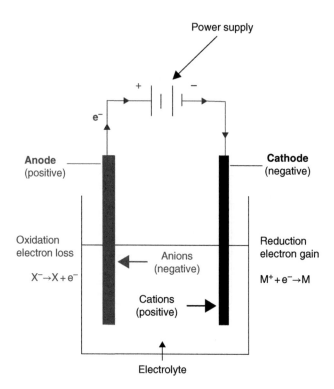

Figure 9.10 Schematic diagram of an electrolytic cell.

The electrodes can be carbon (graphite) rods or a suitable metal and must conduct electricity to the electrolyte. The electrodes are connected via an external circuit to a power supply. One electrode has a positive charge and the other a negative charge, depending upon how the power supply is connected. Negative ions (anions) from the electrolyte move towards the positive electrode, where they lose electrons and are oxidised. The positive ions (cations) are attracted towards the negative electrode, where they combine with electrons and are reduced.

As in a galvanic cell, reduction occurs at the cathode, where the cations gain electrons. But in an electrolytic cell, the cathode has a negative charge, whereas in a galvanic cell, the cathode has a positive charge. Therefore the **sign** or polarity of the electrode where reduction occurs is opposite in galvanic and electrolytic cells. However, reduction always occurs at the cathode.

Oxidation occurs at the anode where the anions lose electrons. In an electrolytic cell the anode has a positive charge, whereas in a galvanic cell the anode has a negative charge (Table 9.3).

> Oxidation occurs at the Anode. Reduction occurs at the Cathode. These processes are the same in both galvanic and electrolytic cells, even though the sign of the electrodes is different. It might help to remember that Oxidation and Anode both begin with a vowel and Reduction and Cathode both begin with a consonant.

9.4.2   Electrolysis of molten substances

When an ionic salt is melted the electrostatic forces between the component ions are broken, and the ions are free to move within the molten liquid. If the molten salt is contained in an electrolytic cell the positive ions (cations) move towards the cathode, and the negative ions (anions) move towards the anode. When a current is applied to molten sodium chloride, for example, sodium ions move towards the cathode where they are reduced to sodium metal. The negatively

Table 9.3 Comparison of galvanic and electrolytic cells.

	Galvanic cell		Electrolytic cell	
Anode	Negative	Oxidation	Positive	Oxidation
Cathode	Positive	Reduction	Negative	Reduction
Process	Current produced – can do useful work		Current required	

charged chloride ions move towards the anode, where they are oxidised to chlorine gas. The electrodes used in the electrolysis of molten sodium chloride are graphite rods. The chlorine gas bubbles off at the anode and is collected. The molten sodium metal floats to the surface of the liquid and is removed. It is important to ensure that the two reactive products are not able to recombine, as an explosive reaction would result.

The half-equations involved are:

$$Na^+ (l) + e^- \rightleftharpoons Na(l) \quad E^\ominus = -2.71 \text{ V} \quad \text{Cathode reaction}$$

$$Cl^- (l) \rightleftharpoons \tfrac{1}{2}Cl_2(g) + e^- \quad E^\ominus = -1.36 \text{ V} \quad \text{Anode reaction}$$

The values of the electrode potentials for each half-equation allow calculation of the voltage required to drive this reaction. Both $E^\ominus$ values are negative in the direction shown, which indicates that neither half-reaction is favourable and electrical energy must be supplied.

9.4.3 Electrolysis of aqueous solutions

Aqueous solutions of ionic salts dissociate in water to give hydrated cations and anions of the salt. Also present in the solution are low concentrations of hydrogen, H^+, and hydroxide, OH^-, ions:

$$H_2O(l) \rightleftharpoons H^+ (aq) + OH^- (aq)$$

As with molten liquids, the cations move to the negative electrode, the cathode. In an aqueous solution of sodium chloride $Na^+(aq)$ and $H^+(aq)$ (or H_3O^+) ions are present. Either of these ions could be reduced at the cathode. The standard reduction potential for H^+ is more positive than that of Na^+, so H^+ is reduced to H_2 gas in preference to Na^+, which remains in solution. The Cl^- ions move to the anode, where they are oxidised to chlorine gas.

9.4.4 Calculating the amount of substance deposited during electrolysis

The amount of charge produced by a power supply during electrolysis is measured in Faradays (symbol F). This is a unit of charge where one Faraday of charge is the amount of charge carried by one mole of electrons. To deposit one mole of silver from one mole of aqueous silver ions would require one Faraday of charge because each mole of silver ions has a single positive charge and requires one mole of electrons to be reduced:

$$Ag^+ (aq) + e^- \rightleftharpoons Ag(s)$$

One Faraday of charge is equal to $96\,500$ C mol^{-1}.

For a divalent metal ion such as copper(II), the equivalent half-equation for the reduction reaction is:

$$Cu^{2+}(aq) + 2e^- \rightleftharpoons Cu(s)$$

In this case, two moles of electrons are required for each mole of copper metal deposited, so the total charge required to deposit one mole of copper is two Faradays or $96\,500$ C mol$^{-1} \times 2$ mol $= 193\,000$ C.

The amount of charge produced by the supply depends upon:

- The time during which the current is passed

- The size of the current

The equation that allows the calculation of charge passed in Coulombs is:

$$Q = I \times t$$

where:

Q = charge (in Coulombs, C)

I = current (in amperes, A)

t = time (in seconds, s)

We can use this equation to determine the amount of charge passed in a fixed length of time and, therefore, the quantity of substance deposited, knowing the half-equation for the process.

Worked Example 9.11   Calculate the mass of tin deposited at the cathode when a current of 2.0 A is passed for 10 minutes through molten tin(II) chloride. (A_r (Sn) = 118.7)

Solution
The equation for the reduction of tin(II) to metallic tin at the cathode is:

$$Sn^{2+}(l) + 2e^- \rightleftharpoons Sn(s)$$

As two moles of electrons are required for each one mole of tin deposited, the total charge required to deposit one mole of tin is two Faradays:
$= 2$ mol $\times 96\,500$ C mol$^{-1} = 193\,000$ C.

The total charge transferred during electrolysis:
$Q = I \times t = 2.0$ A $\times 10 \times 60$ s $= 1\,200$ C.

$193\,000$ C is required to deposit 1 mole or 118.7 g tin.

Therefore, by simple ratios,
mass of tin deposited by $1\,200$ C $= 118.7$ g $\times 1\,200$ C$/193\,000$ C $= 0.74$ g Sn.

Quick-check summary

- An electrochemical cell is composed of two half-cells, each of which forms an electrode in a circuit completed by a salt bridge and external wire.

- The electrode may be a metal rod in a solution of its ions or a gas flowing over an inert material in contact with a solution containing ions formed from the gas.

- A reduction process occurs at the cathode, where a chemical species gains electrons.

- An oxidation process occurs at the anode, where a chemical species loses electrons.

- Electrons flow from the anode to the cathode, and the current is said, by convention, to flow in the opposite direction.

- Each half-cell has a standard reduction potential, $E^\ominus$, which is a measure of the overall voltage of an electrochemical cell when the half-cell is in combination with a standard hydrogen electrode.

- The standard reduction potential of a hydrogen electrode is assigned a value of zero, and standard reduction potentials of other half-cells can be compared with this.

- If the standard reduction potential of a half-cell is greater than zero, the oxidised species will reduce H^+ ions to H_2 under standard conditions.

- If the standard reduction potential of a half-cell is less than zero, the reduced species will reduce H^+ ions to H_2 under standard conditions.

- The overall electrode potential for a cell, $E^\ominus$, can be used to predict whether a reaction will occur, as it is linked to the Gibbs energy by $\Delta G^\ominus = -nF\,E^\ominus$. If $E^\ominus$ is positive the reaction is likely to be spontaneous.

- The Nernst equation can be used to determine the $E^\ominus$ value for a half-cell when the concentrations are not equal to 1 M.

- Electrochemical cells are used in batteries to produce energy.

- Electrochemical cells are sometimes called galvanic or voltaic cells to distinguish them from electrolytic cells.

- Electrolysis occurs in electrolytic cells and is a process in which a current is used to decompose an electrolyte into its constituents, which may be produced as ionic or elemental species.

- In an electrolytic cell the anode is the positive electrode where electrons are lost and oxidation occurs. The cathode is the negative electrode where reduction occurs.

- Electrolysis can occur in either molten liquids or aqueous solutions.

- The nature of the substance liberated at each electrode depends upon whether the electrolyte is molten or in aqueous solution and the standard reduction potential of each species. The more positive the value of $E^\ominus$, the more readily the ion is reduced at the cathode.

- The amount of substance discharged is found by calculating the quantity of charge passed during electrolysis from $Q = I\,t$ and relating this to the number of moles of ions reduced or oxidised. One mole of singly charged ions is said to carry one Faraday of charge.

- One Faraday of charge is equal to the charge on the electron multiplied by the number of electrons in a mole and has the value $96\,500\ \mathrm{C\,mol^{-1}}$.

End-of-chapter questions

1 Consider the following pairs of half-cell equations and standard electrode potential, $E^\ominus$, values. For each pair:

a. At which electrode does reduction occur?

b. Therefore which metal is the better reducing agent?

c. Determine which metal would form the positive electrode in an electro-chemical cell made up of the two half-cells.

d. Write the cell diagram for the overall electrochemical cell.

e. Determine the standard cell potential.

$$
\begin{aligned}
&\text{i.}\quad && Ag^+ (aq) + e^- \rightleftharpoons Ag(s) && E^\ominus = +0.80\ V \\
& && Pb^{2+} (aq) + 2e^- \rightleftharpoons Pb(s) && E^\ominus = -0.13\ V \\[4pt]
&\text{ii.}\quad && Co^{2+} (aq) + 2e^- \rightleftharpoons Co(s) && E^\ominus = -0.28\ V \\
& && Zn^{2+} (aq) + 2e^- \rightleftharpoons Zn(s) && E^\ominus = -0.76\ V \\[4pt]
&\text{iii.}\quad && Fe^{2+} (aq) + 2e^- \rightleftharpoons Fe(s) && E^\ominus = -0.44\ V \\
& && Cu^{2+} (aq) + 2e^- + \rightleftharpoons Cu(s) && E^\ominus = +0.34\ V
\end{aligned}
$$

2 Calculate the value of the electrode potential at 298 K of a $Sn^{2+}(aq)/Sn(s)$ half-cell that has a concentration of $Sn^{2+}(aq)$ ions of 0.0025 M. $E^\ominus$ (cell) = –0.14 V.

3 Determine whether $Fe^{2+}(aq)$ or $Sn^{2+}(aq)$ will reduce $VO_2^+(aq)$ ions to $VO^{2+}(aq)$. Write the overall equation for the redox reaction. The relevant half-equations are:

$$Fe^{3+} (aq) + e^- \rightleftharpoons Fe^{2+} (aq) \qquad\qquad E^\ominus = +0.77\ V$$

$$Sn^{4+} (aq) + 2e^- \rightleftharpoons Sn^{2+} (aq) \qquad\qquad E^\ominus = +0.15\ V$$

$$VO_2^+ (aq) + 2H^+ + e^- \rightleftharpoons VO^{2+} (aq) + H_2O(l) \quad E^\ominus = +1.00\ V$$

4 State in which direction electrons will flow when the following half-cells are connected to a standard hydrogen electrode. Use the $E^\ominus$ values in Table 9.2 to inform your answers.

a. I_2/I^-

b. Cr^{3+}/Cr

5 An electric current of 1.50 A was passed through a solution of dilute sulfuric acid for five minutes. Hydrogen gas is released at the cathode. The electrolysis reaction was carried out at standard temperature and pressure. $(F = 96\ 500\ C\ mol^{-1}, V_m = 22.4\ dm^3$ at STP.)

a. Calculate the total charge passed during the experiment.

b. Write a half-equation for the reduction of hydrogen ions to hydrogen gas.

c. Calculate the volume of hydrogen gas that would be discharged during the electrolysis reaction.

6   When 1 A of electricity was passed through an aqueous solution of copper(II) sulfate for one hour, 1.19 g of copper were deposited according to this equation:

$$Cu^{2+}(aq) + 2e^- \rightarrow Cu(s)$$

1 coulomb is the unit of measurement of electricity passed. The number of coulombs is a product of the current in amperes and the time in seconds, $Q = I\,t$. One electron has a charge of 1.6×10^{-19} coulombs.

a. Calculate the number of moles of copper in 1.19 g.

b. Calculate the number of coulombs of electricity used when 1 A is passed for one hour.

c. Calculate the number of electrons in this number of coulombs.

d. Hence, calculate the number of atoms in one mole of copper.

7   Use the following standard reduction potentials for copper to suggest what would happen if a sample of copper(I) sulfate was added to water:

$$Cu^{2+}(aq) + e^- \rightleftharpoons Cu^+(aq) \quad E^\ominus = +0.15\ V \qquad \text{(i)}$$

$$Cu^+(aq) + e^- \rightleftharpoons Cu(s) \qquad E^\ominus = +0.52\ V \qquad \text{(ii)}$$

8   The standard reduction potentials for a MnO_4^-/Mn^{2+} electrode and Cl_2/Cl^- electrodes are as follows:

$$MnO_4^-(aq) + 8H^+(aq) + 5e^- \rightleftharpoons Mn^{2+}(aq) + 4H_2O(l) \quad E^\ominus = +1.51\ V$$

$$\tfrac{1}{2}Cl_2(g) + e^- \rightleftharpoons Cl^-(aq) \quad E^\ominus = +1.36\ V$$

a. Combine the two half-equations to give an overall equation for the reaction of manganate(VII) ions with chloride ions.

b. State, giving reasons, whether you would expect this reaction to occur under standard conditions.

9   Use the following standard reduction potentials to answer the questions that follow:

Electrode reaction	$E^\ominus$
$Fe^{2+} + 2e^- \rightleftharpoons Fe$	$-0.44\ V$
$Fe^{3+} + 3e^- \rightleftharpoons Fe$	$-0.04\ V$
$H^+ + e^- \rightleftharpoons \tfrac{1}{2}H_2$	$0.00\ V$
$Fe^{3+} + e^- \rightleftharpoons Fe^{2+}$	$+0.77\ V$
$Co^{3+} + e^- \rightleftharpoons Co^{2+}$	$+1.82\ V$

a. Explain why the reaction of iron metal with dilute aqueous hydrochloric acid gives iron(II) chloride and not iron(III) chloride.

b. Construct the cell diagram for an electrolytic cell composed of a platinum electrode in a solution of $Co^{3+}(aq)/Co^{2+}(aq)$ ions connected to a platinum electrode in a solution of $Fe^{3+}(aq)/Fe^{2+}(aq)$ ions. Determine the electrode potential of such a cell.

10 Use the following data to answer the questions below about a lead-acid battery. A lead-acid battery is a type of galvanic cell

$$PbO_2(s) + 4H^+(aq) + SO_4^{2-}(aq) + 2e^- \rightleftharpoons PbSO_4(s) + 2H_2O(l) \quad E^\ominus = +1.69 \text{ V}$$

$$PbSO_4(s) + 2e^- \rightleftharpoons Pb(s) + SO_4^{2-}(aq) \quad E^\ominus = -0.36 \text{ V}$$

a. What substance is used for the negative electrode (anode)?

b. What substance is used for the positive electrode (cathode)?

c. What substance is used for the electrolyte?

d. Give the overall equation for the reaction taking place when the battery discharges.

e. Calculate the electrode potential of the cell.

10

Group trends and periodicity

At the end of this chapter, students should be able to:

- Understand and explain general trends and properties within a period and a group

- Explain the difference between ionic and atomic radii

- Understand and explain why ionisation energy is different for different elements

- Understand electronegativity, and discuss its relationship with ionisation energy

- Understand and give reasons for differing melting and boiling points of elements

10.1 The periodic table: periods, groups, and periodicity

In Chapter 2 we saw that elements in the periodic table are arranged such that those with similar properties are in columns called *groups*. Elements with electrons in the same outer shell are arranged in horizontal rows or *periods*. Chemical reactivity is underpinned by elements either gaining or losing valence electrons to attain a full outer shell. Hydrogen and helium in the first period attain a full outer shell with just two electrons in the 1s orbital. Elements in rows 2 and 3 attain a full outer shell with eight electrons. Heavier elements in lower periods require larger numbers of valence electrons to complete their outer shells, as their outer shells possess more atomic orbitals.

The periodic table is divided into several areas that are shaded different colours in Figure 10.1. These main areas are called the s-block, p-block, d-block, and f-block.

Elements in a single column are in the same **group** of the periodic table; elements in a single row are in the same **period**.

Reminder: Electrons are arranged around the nucleus of the atom in separate shells. Each principal quantum shell is labelled by the number n and has a specific set of atomic orbitals or sub-shells. Each sub-shell can hold a maximum of two electrons.

Foundations of Chemistry: An Introductory Course for Science Students, First Edition.
Philippa B. Cranwell and Elizabeth M. Page.
© 2021 John Wiley & Sons Ltd. Published 2021 by John Wiley & Sons Ltd.
Companion website: www.wiley.com/go/Cranwell/Foundations

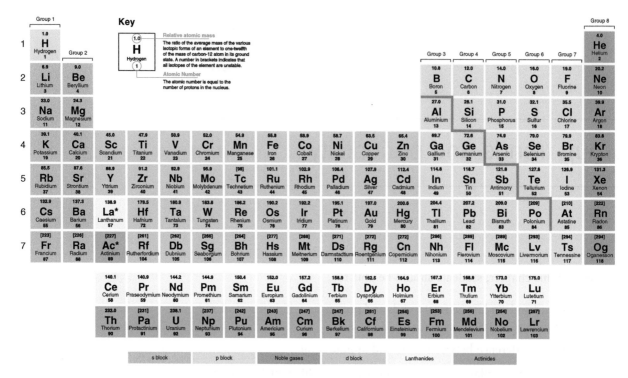

Figure 10.1 Periodic table showing periods and groups with different areas shaded different colours.

- The s block is formed by elements in Groups 1 and 2 on the left of the table where outer electrons are filling the s orbital.

- The p block is made up of elements in Groups 3 (13) – 8 (18) where outer electrons are filling the p orbitals.

- The d block is situated between the s and p blocks where electrons are filling the d orbitals. This block contains the transition elements.

- The f block constitutes two rows of 14 elements where electrons are filling the f orbitals. These two rows are known as the *lanthanides* and *actinides*.

There are seven 4f orbitals and seven 5f orbitals that can each hold a maximum of two electrons, thus accounting for the 14 elements in each row of the f block.

This arrangement of the periodic table leads to similar trends in properties of elements that are in the same groups and periods. These trends are referred to as **periodicity**. In this chapter we will first consider group trends and then trends on moving across a period.

10.2 Trends in properties of elements in the same vertical group of the periodic table

In this section we will look at how properties of the elements vary on going down a vertical group in the periodic table. First we will consider the electron configuration of the elements, which is the factor that controls most of these properties.

10.2.1 Electron configuration

Elements that are in the same group have the same number of electrons in their outer shells. For example, elements that are in Group 2 (e.g. magnesium and calcium) have two electrons in their outer shell. The electronic configurations of Mg and Ca are:

$$Mg\ 1s^2\ 2s^2\ 2p^6\ 3s^2$$

$$Ca\ 1s^2\ 2s^2\ 2p^6\ 3s^2\ 3p^6\ 4s^2$$

Elements in Group 7 (Group 17) have seven electrons in their outer shell:

$$F\ 1s^2\ 2s^2\ 2p^5$$

$$Cl\ 1s^2\ 2s^2\ 2p^6\ 3s^2\ 3p^5$$

The chemical reactivity of an element is determined by its outer electron configuration so that elements in the same group have similar chemical properties.

When elements react they tend to either lose or gain electrons to attain a full outer shell of electrons. Elements in the same group with the same number of outer electrons tend to react in the same way.

For a refresher on electronic configuration, see Chapter 1.

10.2.2 Effective nuclear charge, Z_{eff}

We are familiar with the concept of nuclear charge from Chapter 1. The number of positively charged protons in the nucleus is given by Z, the atomic number, and is also known as the *nuclear charge*. This creates a force of attraction between the protons in the nucleus and electrons in the outer shell. The greater the number of protons in the nucleus, the higher the nuclear charge. Between the electrons in the outer shell and the protons in the nucleus are inner shells of negatively charged electrons that screen the nuclear charge from the outer electrons. The effect of this screening is to reduce the positive pull of the protons on the outer electrons thus reducing the effect of nuclear charge on these outermost electrons. The screening of the valence electrons from the nuclear charge is called **shielding**, and the resultant pull that the shielded outer electrons experience is called the **effective nuclear charge**. The effective nuclear charge has a special symbol, Z_{eff}, and is less than the actual nuclear charge.

This can be seen using the examples of magnesium and calcium, whose electron arrangements are shown in Figure 10.2. Both magnesium and calcium have two electrons in their outermost (valence) shell and so are both in Group 2. Calcium is in Period 4, whereas magnesium is in Period 3, therefore calcium is lower down the periodic table and has an additional filled shell of electrons. This extra shell of electrons screens the positive charge of the nucleus from the outermost electrons, so each outer electron of Ca experiences less of the nuclear charge than those of Mg. When the valence electrons experience a weaker nuclear pull, they are held less strongly to the atom, and this makes them easier to remove.

The effective nuclear charge impacts upon both the atomic radius of the atom and the amount of energy required for the atom to lose its valence electrons.

The effective nuclear charge, Z_{eff}, is the force exerted by the nucleus on an electron when the effect of shielding has been taken into consideration.

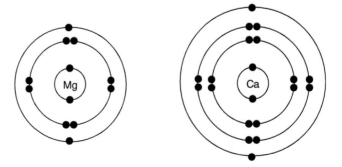

Figure 10.2 Electron filling of shells in magnesium and calcium.

10.2.3   Atomic radius

The atomic radius of an atom is a measure of the size of the atom. However, it's not possible to measure exactly where one atom ends and the next atom starts. To account for this, the **atomic radius** is defined as one half the distance between the nuclei of neighbouring atoms in the pure element. If the element is a metal, then this distance is known as the *metallic radius* (Figure 10.3).

On descending a group in the periodic table, the atomic radius of the elements in the group increases, as shown in Figure 10.4 for the Group 2 elements.

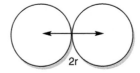

Figure 10.3 The atomic radius (*r*) of an atom is defined as half the distance between the nuclei of neighbouring atoms in the pure element.

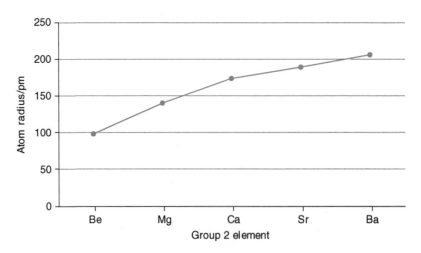

Figure 10.4 The atomic radii of Group 2 elements.

There are two reasons for this. The first reason is that the increasing number of shells of electrons occupy more space, so the atoms become bigger. The second reason is due to the decrease in effective nuclear charge on going down the group due to shielding, which results in a reduced pull on the outer electrons and therefore causes the radius to be larger.

On descending a group, the atomic radius increases.

10.2.4   Ionisation energies

The **ionisation energy**, *IE*, of an element is a measure of the amount of energy required to remove one mole of electrons from one mole of gaseous atoms of the element. The first ionisation energy, IE_1, is the energy required to remove the first electron from its outer (valence) shell and is represented by the following equation, where M is any element:

$$M(g) \rightarrow M(g)^+ + e^- \qquad IE_1$$

The first ionisation energies of the elements decrease on going down a group. The size of the first ionisation energy of an element depends upon the atomic radius of the element **and** the effective nuclear charge:

- The larger the atomic radius, the greater the distance between the nucleus and the outer electrons, and so less energy is required to remove an electron and the ionisation energy therefore is smaller.

- The weaker effective nuclear charge exerts a smaller pull on the valence electrons, and so the ionisation energy is lower.

On going down a group in the periodic table the effective nuclear charge gets smaller and less energy is required to remove an outer electron. Figure 10.5 shows a plot of the first ionisation energies of the elements in Group 2 and illustrates how the first ionisation energy decreases on going down the group.

The first ionisation energy decreases on going down a group in the periodic table.

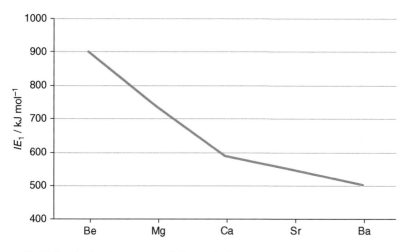

Figure 10.5 First ionisation energy of Group 2 elements.

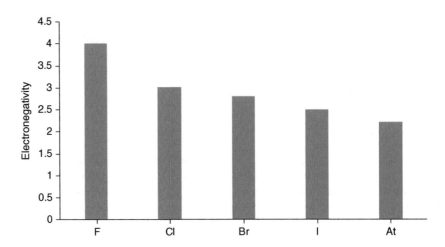

Figure 10.6 Electronegativity values of Group 7 (17) elements (Pauling scale).

10.2.5   Electronegativity

The electronegativity of an element is the power of an atom to pull a pair of bonded electrons towards itself. Electronegativity arises through the positive charge of the nucleus attracting electrons from surrounding atoms. The more electronegative the element, the stronger the pull on bonded electrons. On going down a group in the periodic table the atoms get larger and there is a bigger distance between the nucleus and any bonded electrons. The electronegativity of elements therefore decreases on going down a group. In any group the most electronegative element is the top member of the group. Figure 10.6 shows how the electronegativity of the Group 7 (17) elements decreases down the group.

10.2.6   Electron affinity (electron gain enthalpy)

You may come across the term *electron gain enthalpy* in some textbooks or chemistry courses. This name also describes the process defined by electron affinity. However, electron gain enthalpy and electron affinity mean exactly the same thing. We will use the term *electron affinity* in this book.

The electron affinity of an element is the standard enthalpy change that occurs when one mole of gaseous atoms of the element accepts one mole of electrons to form one mole of gaseous ions with a -1 charge. This process is the opposite of ionisation where the atom loses an electron to form a positive ion. Electron affinity differs from electronegativity in that it represents the process of gaining an electron and forming a negative ion rather than simply attracting electrons from within a bond. Electron affinity values are really only useful for elements in Groups 6 (16) and 7 (17) that readily form negatively charged ions. The symbol for the first electron affinity for an element is $\Delta_{EA1}H^{\ominus}$. Values of first electron affinities are generally exothermic (associated with the release of energy).

The first electron affinity for chlorine is represented by:

$$Cl(g) + e^{-} \rightarrow Cl^{-}(g) \qquad \Delta_{EA1}H^{\ominus} = -364 \, kJ \, mol^{-1}$$

The $^{\ominus}$ (plimsoll) symbol means that the experiment is conducted under standard conditions: i.e. a pressure of 10^{5} Pa (100 kPa, 1 bar) with reactants and products in their standard states. Values are quoted at a specific temperature, usually 298 K.

Electron affinity is a measure of the attraction between an incoming electron and the nucleus and depends to a large extent on atomic radius and shielding.

Box 10.1 The energy change for each successive electron gained has an associated electron affinity value. For example, the second electron affinity, $\Delta_{EA2}H^{\ominus}$, is the enthalpy change when one mole of electrons is added to one mole of gaseous 1– ions to form one mole of gaseous 2– ions under standard conditions. Second electron affinities are always endothermic. The electron affinity values for oxygen are:

$$O(g) + e^- \rightarrow O^-(g) \qquad \Delta_{EA1}H^{\ominus} = -142 \, kJ \, mol^{-1}$$

$$O^-(g) + e^- \rightarrow O^{2-}(g) \qquad \Delta_{EA2}H^{\ominus} = +844 \, kJ \, mol^{-1}$$

Thus the overall electron affinity for the formation of an O^{2-} (g) ion is:

$$-142 \, kJ \, mol^{-1} + \left(+844 \, kJ \, mol^{-1}\right) = +702 \, kJ \, mol^{-1}$$

On descending a group, in general, electron affinity values decrease (i.e. become less negative and less energy is released) as the atomic radius increases and the shells of electrons screen the nuclear charge from the incoming electron.

Table 10.1 shows electron affinity values for the Group 7 (17) elements.

Table 10.1 Electron affinity values for Group 7 (17) elements.

Element	$\Delta_{EA1}H^{\ominus}/kJ \, mol^{-1}$
F	–348
Cl	–364
Br	–342
I	–314

Box 10.2 Why is the electron affinity of fluorine out of step? From Table 10.1, it is clear that F does not obey the group trend and the value of $\Delta_{EA1}H^{\ominus}$ for F is actually lower than that of Cl, the next element down the group. The reason for the less negative value for F is that the atomic radius of F is so small that the incoming electron is forced into a shell that is already crowded with electrons: this results in a certain amount of electron–electron repulsion, which means that the process is less energetically favourable. Although the values of electron affinity appear to be getting larger (i.e. less negative) on going down the group, the actual amount of energy being released is getting smaller.

Worked Example 10.1 Sketch graphs of (a) atomic radius and (b) ionisation energy for the Group 1 elements using the data in the following table. Explain the trends you observe in both properties on going down the group.

Element	Atomic radius/pm[a]	First ionisation energy/kJ mol^{-1}
Li	145	520
Na	180	496
K	220	419
Rb	235	408
Cs	260	376

[a]A picometer (pm) is equivalent to 1×10^{-12} m.

Solution

a. Atomic radius

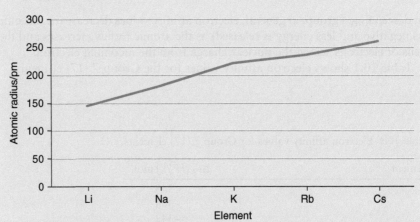

On going down the group the atomic radius of the Group 1 elements increases. This is because the number of filled shells increases as the total number of electrons increases. In addition, the shielding of the nuclear charge by the inner (core) electrons prevents the valence electrons feeling the full nuclear charge and so causes the atomic radius to increase.

b. Ionisation energy

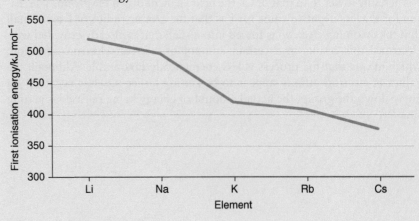

The first ionisation energy refers to the energy required to remove the first electron from a gaseous atom of the element. On going down a group in the periodic table the outer electrons become further from the nucleus as the atomic radius increases (as we saw in the previous part of the question). As the atomic radius increases, the pull on the outer electrons by the nuclear charge becomes weaker as the outer electrons are shielded by the increasing numbers of electron shells. This is a measure of the effective nuclear charge, Z_{eff}. It therefore requires less energy to remove the first electron, and the first ionisation energy therefore decreases on going down the group.

Worked Example 10.2

a. Define the term *electronegativity*.

b. Explain the trend in electronegativity seen on going down Group 6 (16) of the periodic table.

Element	O	S	Se	Te	Po
Electronegativity	3.5	2.5	2.4	2.1	2.0

Solution

a. Electronegativity is the ability of an atom in a bond to attract a pair of bonding electrons towards itself.

b. On going down a group, the atomic radius increases, so the protons in the nucleus exert a smaller pull on electrons within a covalent bond because the distance increases. The electronegativity therefore decreases down a group.

10.3 Trends in properties of elements in the same horizontal period

In this section we will see how properties of the elements differ on going across the row and how they depend upon the electron configuration.

10.3.1 Electron configuration

On going from left to right across the periodic table each atom has one more electron in its outer shell and one more proton in the nucleus. Elements in the same period have the same number of inner shells of electrons but a different number of valence electrons. Because chemical reactivity depends upon the number and arrangement of electrons in the valence shell, elements in the same period will have quite different physical and chemical properties from each other.

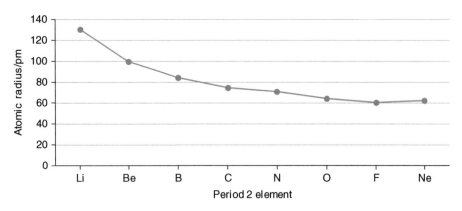

Figure 10.7 Atomic radii of the elements in Period 2.

10.3.2   Atomic radius

The atomic radii of the elements decrease on going from left to right across the periodic table up to Group 7 (17). This is shown in Figure 10.7 for Period 2. On going from Li to F, protons are added to the nucleus and electrons to the outer shell one by one. The outer electrons are poor at shielding the nucleus, so the effective nuclear charge increases. The increased effective nuclear charge causes the radius of the atom to contract on going across the period from left to right.

> On going from left to right across a period, the atomic radius decreases.

Consider three elements in the second period: lithium, carbon, and fluorine. The electron configurations and the number of protons in the nucleus of each element are given in Table 10.2. On going across the row, the radius of the atoms decreases as the number of electrons increases and the effective nuclear charge increases. Unlike on going down a group, there is *no* additional shielding by inner shells of electrons, so the valence electrons experience the full nuclear charge.

Figure 10.8 shows the trend in atomic radii on going across and down the periodic table. It can be seen that atomic radii increase down a group and decrease across a period, apart from Group 8 (18), where the values increase (see the following box).

> **Box 10.3** Atomic radii values are difficult to determine accurately for Group 8 (18) elements (the inert gases) as they form few stable covalent compounds. However, for all periods, the van der Waals radius of a Group 8 (18) element is greater than that of the preceding element and so doesn't fit in the series. This is because at Group 8 (18), we have a full outer shell of electrons, and electron–electron repulsions increase the radius.

Table 10.2 The electronic configurations and number of protons for selected elements in period 2.

Element	Electronic configuration	Number of protons in the nucleus (Z)	Atomic radius
Lithium	$1s^2\ 2s^1$	3	123 pm
Carbon	$1s^2\ 2s^2\ 2p^2$	6	77 pm
Fluorine	$1s^2\ 2s^2\ 2p^5$	9	72 pm

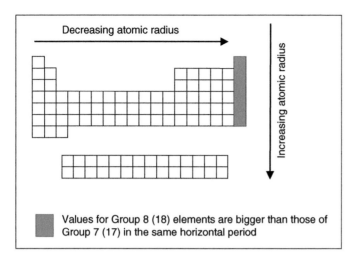

Figure 10.8 Trends in atomic radius across and down the periodic table.

10.3.3 Ionisation energy

The overall trend in first ionisation energy on going across a period is that it increases due to the decrease in atomic radius and increased effective nuclear charge, Z_{eff}. It takes more energy to remove an electron from the outer shell when it is closer to the nucleus. Elements in Group 1 have the largest atomic radius and they also have the smallest first ionisation energy. Figure 10.9 shows the variation in the ionisation energy of the elements in the second and third periods. However, the increase in ionisation energy across a row is not linear, and there are a few peaks and dips, which can be explained by considering the specific orbitals from which the electrons are being lost.

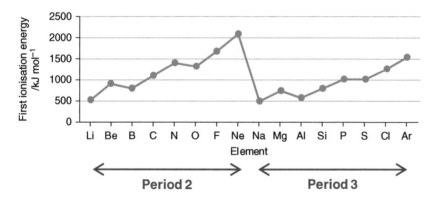

Figure 10.9 First ionisation energy of the elements in Periods 2 and 3.

Group 2 to Group 3 (13)

When moving from Be to B and Mg to Al, there is a small drop in first ionisation energy. The actual values and the equations that represent the changes for the Period 2 elements are as follows:

$$Be(g) \rightarrow Be^+(g) + e^- \quad IE_1 = +900 \, kJ \, mol^{-1}$$
$$1s^2 2s^2 \rightarrow 1s^2 2s^1 + e^-$$
$$B(g) \rightarrow B^+(g) + e^- \quad IE_1 = +799 \, kJ \, mol^{-1}$$
$$1s^2 2s^2 2p^1 \rightarrow 1s^2 2s^2 + e^-$$

The electronic configurations of the atoms involved and the ions formed are also given. The values of first ionisation energies indicate that it is easier to remove the first electron from boron than beryllium. There are two reasons for the decrease in first ionisation energy from Be to B:

i.   The outer electron in B is in a higher energy level than the 2s electron in Be as it is in a 2p atomic orbital. It is therefore further from the nucleus on average than the electrons in the 2s orbital. This results in a reduced attraction between the 2p electron and the nucleus.

ii. The electron in B is shielded to a certain extent by the 2s electrons, so the effective nuclear charge is therefore lower, which also results in a lower ionisation energy.

Similar trends are seen between Mg and Al in Period 3.

Group 5 (15) to Group 6 (16)

A similar dip in values for the first ionisation energies occurs between N and O, and P and S, in Groups 5 (15) and 6 (16), respectively. The equivalent equations and electronic configurations for the Period 2 elements are as follows:

$$N(g) \rightarrow N^+(g) + e^- \quad IE_1 = +1400 \, kJ \, mol^{-1}$$
$$1s^2 2s^2 2p^3 \rightarrow 1s^2 2s^2 2p^2 + e^-$$
$$O(g) \rightarrow O^+(g) + e^- \quad IE_1 = +1310 \, kJ \, mol^{-1}$$
$$1s^2 2s^2 2p^4 \rightarrow 1s^2 2s^2 2p^3 + e^-$$

In this case, both electrons are being removed from the same 2p sub-shell. However, the electron being removed from N is in a singly-occupied orbital, whereas the electron being removed from O is sharing the 2p orbital with another electron. It takes less energy to remove an electron from an orbital where the electrons are paired because there is a slight repulsion between the negatively charged electrons in the same orbital. In addition, the electron being removed from N is in a half-filled p sub-shell, which lends additional stability. This results in a slightly lower ionisation energy for O, as is observed.

Similar trends are seen between P and S in Period 3.

For a refresher on orbitals and how they hold electrons, see Chapter 1.

Group 8 (18) to Group 1

Perhaps the most significant dip in first ionisation energies is seen between neon and sodium. The relevant equations and electronic configurations are as follows:

$$Ne(g) \rightarrow Ne^+(g) + e^- \quad IE_1 = 2080 \text{ kJ mol}^{-1}$$

$$1s^2 2s^2 2p^6 \rightarrow 1s^2 2s^2 2p^5 + e^-$$

$$Na(g) \rightarrow Na^+(g) + e^- \quad IE_1 = 494 \text{ kJ mol}^{-1}$$

$$1s^2 2s^2 2p^6 3s^1 \rightarrow 1s^2 2s^2 2p^6 + e^-$$

At Ne, the electron is being removed from a filled 2p sub-shell, and this electron configuration is especially stable, as we have seen. It therefore requires a large amount of energy to disrupt this evenly spread cloud of electron density. The next electron, at sodium, occupies a 3s orbital. This orbital is at a higher energy, further from the nucleus and evenly shielded by the core electrons in the first and second shells. It is therefore relatively easy to remove this single electron, and the ionisation energy for Na is significantly lower than that of Ne.

Similar trends are seen between Ar and K in Periods 3 and 4.

10.3.4   Electronegativity

Electronegativity is the tendency of an atom in a bond to attract a pair of electrons. As atomic radius decreases across a period, the attractive pull of the nucleus for electrons increases, so electronegativity increases on going from left to right across the table up to and including Group 7 (17). The increase in electronegativities across Period 2 and Period 3 can be seen in Figure 10.10.

In the previous section we saw that electronegativities decrease down a group. Figure 10.11 shows the variation in electronegativity values both across and down the periodic table. Fluorine is the most electronegative element in the periodic table with an electronegativity value of 4.0 on the Pauling scale. It is the first member of Group 7 (17) and the element with the highest electronegativity in Period 2. On Pauling's scale of electronegativity, which is the one used here, no values are defined for the noble gases, so we have not included them in this trend.

> Electronegativity values decrease down a group and increase from left to right across a period up to and including Group 7 (17).

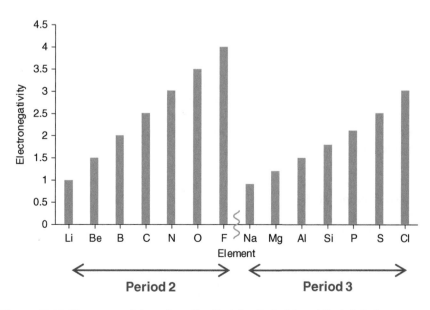

Figure 10.10 Electronegativity values (Pauling) for Period 2 and Period 3 elements excluding Group 8 (18).

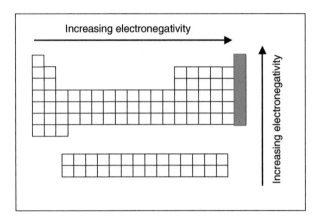

Figure 10.11 Trends in electronegativity across the periodic table. Note that the noble gases are not included in the trend.

10.3.5   Electron affinity, $\Delta_{EA}H^{\ominus}$

On going across a period, the number of protons in the nucleus and the number of electrons in the valence shell both increase. The attractive force of the nucleus for an incoming electron therefore increases across a row because there are the same number of core electrons but a greater nuclear charge. Therefore the electron affinity would also be expected to increase across a period. To a large extent this is true, although the increase isn't smooth. The symbol for the first electron affinity is $\Delta_{EA1}H^{\ominus}$. The $\Delta_{EA1}H^{\ominus}$ values for lithium (Group 1) and fluorine (Group 7 (17)) are given here along with the associated equations:

$$\text{Li(g)} + \text{e}^- \rightarrow \text{Li}^-\text{(g)} \quad \Delta_{EA1}H^{\ominus} = -52\,\text{kJ}\,\text{mol}^{-1}$$

$$\text{F(g)} + \text{e}^- \rightarrow \text{F}^-\text{(g)} \quad \Delta_{EA1}H^{\ominus} = -348\,\text{kJ}\,\text{mol}^{-1}$$

The addition of an electron to fluorine (a Group 7 (17) element) releases far more energy than the same process for lithium (a Group 1 element) because fluorine has a greater number of protons in the nucleus than lithium and a smaller atomic radius. This reflects the overall pattern of electron affinities on moving across a period. Elements on the left-hand side (Groups 1–3 (13)) are reluctant to accept electrons and form negatively charged ions and prefer to lose electrons to form cations. The non-metallic elements (Groups 5 (15) to 7 (17)) on the right of the periodic table accept electrons far more readily and tend to form negatively charged anions. However, as mentioned earlier, the variation across the table isn't smooth.

Elements on the left of the periodic table prefer to form positively charged cations by losing electrons. Elements on the right of the periodic table (except noble gases) prefer to form negatively charged anions and gain electrons.

Worked Example 10.3

a. Explain what is meant by the first electron affinity of an element, and give a general equation to represent the process.

b. When an electron is added to a non-metallic element, is energy released or absorbed? Explain your answer.

c. The first electron affinities for some elements in Period 2 are given in the table. Explain the variation in the values in terms of the filling of atomic orbitals in the atoms by electrons.

Element	Li	Be	B	C	N	O	F
First electron affinity/kJ mol^{-1}	-52		-29	-120	-3	-142	-348

Solution

a. The first electron affinity is the energy change when one mole of gaseous atoms of an element accepts one mole of electrons to form one mole of singly negatively charged ions: $X(g) + e^- \rightarrow X^-(g)$.

b. When an electron is added to a non-metallic element, energy is released because these elements prefer to form negative ions due to the greater attractive force of the nucleus (Z_{eff}), which increases towards the right of the periodic table.

c. First electron affinity values are negative for Li, B, C, O, and F and close to zero for N. In general, first electron affinities are exothermic because of the attractive force of the nucleus for incoming electrons. In all these cases an electron is being added to the same outer shell, even though it may be added to a different sub-shell. However, in the case of N, the additional electron is being added to a half-filled 2p orbital and so must pair up with another negatively charged electron and overcome the pairing energy. This process requires energy. In the case of O and F, the effective nuclear charge is increasing, so the electron affinity becomes more exothermic as the process becomes more favourable.

Worked Example 10.4 State and explain:

a. Which element is the most electronegative in the periodic table.

b. Which element you would expect to have the lowest first ionisation energy in the periodic table.

Solution

a. F is the most electronegative element in the periodic table as it is at the top of Group 7 (17) and is the smallest element in Period 2. Electronegativity increases across a period as the atoms become smaller and the nuclear charge increases up to Group 7 (17). Electronegativities increase up a group because elements at the top of any group have the smallest atomic radii.

b. We would expect an element in Group 1 to have the lowest first ionisation energy because Group 1 elements have the largest atomic radii in their period and ionisation energy decreases with increasing atomic radius. We would also expect the element with the lowest first ionisation energy to be at the bottom of the group, again because these elements have the largest atomic radii. Therefore Fr (francium) at the bottom of Group 1 would be expected to have the smallest ionisation energy.

10.3.6   Ionic radius

The **ionic radius** of an element is determined from the distance between neighbouring ions in an ionic structure. When an atom loses or gains one or more electrons to form an ion, its radius changes in size.

Ionic radius: cations

A positively charged ion is called a **cation** and negatively charged ion is called an **anion**.

When elements from Groups 1, 2, or 3 (13) lose all their outer shell electrons they form cations with charges of +1, +2, and +3, respectively. Because these ions have effectively lost all their outer shell electrons and have more protons than electrons, the radius of the cation is smaller than that of the uncharged atom. This is shown in Figure 10.12 for sodium. The radius of Na^+ is smaller than that of a sodium atom.

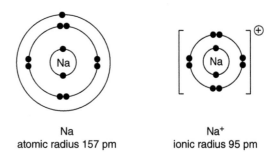

Na
atomic radius 157 pm

Na⁺
ionic radius 95 pm

Figure 10.12 Comparative sizes of Na and Na^+ (not to scale).

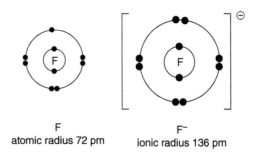

F
atomic radius 72 pm

F⁻
ionic radius 136 pm

Figure 10.13 Comparative sizes of F and F^- (not to scale).

Ionic radius: anions

When an atom gains one or more electrons it forms an anion, and the radius of the anion is always larger than the radius of the atom from which it is formed. This is because the anion has additional electrons in its outer shell and therefore increased repulsion between the electrons. The nuclear charge remains constant, so the attractive pull of the nucleus on the outer electrons is lower. This is shown in Figure 10.13 for fluorine. The atomic radius of the fluoride ion, F^-, is larger than that of a fluorine atom.

Table 10.3 Names and charges of monatomic anions.

Element name	Symbol	Anion name	Symbol
Hydrogen	H	Hydride	H^-
Fluorine	F	Fluoride	F^-
Chlorine	Cl	Chloride	Cl^-
Bromine	Br	Bromide	Br^-
Iodine	I	Iodide	I^-
Oxygen	O	Oxide	O^{2-}
Sulfur	S	Sulfide	S^{2-}
Nitrogen	N	Nitride	N^{3-}

Anions generally take a slightly different name to the parent atom from which they are formed. For example, the ion formed when a fluorine atom accepts an electron is called a **fluoride** ion. Table 10.3 lists the names of some common monatomic anions. It is important to distinguish clearly between atoms and ions.

Ionic radius: general trends across the table

The ions formed when the first four elements of Period 3 lose all their valence electrons are Na^+, Ca^{2+}, Al^{3+}, and Si^{4+}, respectively. The radii of these ions decrease as we go across the period as the nuclear charge increases and exerts a greater pull on the valence electrons.

Because the radii of anions are generally larger than the radii of cations with a similar number of protons, there is a stepwise increase in ionic radius between the Group 4 (14) and 5 (15) elements as shown in Figure 10.14. Up to Group 3 (13), the elements tend to lose electrons and form cations, whereas in Groups 5 (15) to 7 (17), the elements prefer to gain electrons and form anions. Group 4 (14) elements of Periods 3 and 4 rarely form ions because of the large amounts of energy needed to either lose or gain 4 electrons. When moving from P^{3-} to Cl^-, the ionic radius again decreases, because the number of protons in the nucleus and the effective nuclear charge are both increasing; therefore, the electrostatic attraction for outer electrons increases.

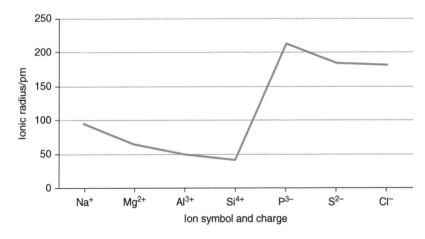

Figure 10.14 Variation in the radius of stable ions of Period 3 elements.

Table 10.4 The bonding, structures, and melting and boiling points of the Period 2 elements.

Element	Lithium	Beryllium	Boron	Carbon[a]	Nitrogen	Oxygen	Fluorine	Neon
Bonding	Metallic	Metallic	Metalloid	Covalent	Covalent	Covalent	Covalent	—
Structure	Giant metallic	Giant metallic	Giant covalent	Giant covalent	Simple covalent	Simple covalent	Simple covalent	Single atoms
Melting point/K	453	1550	2600	4000	63.0	54.4	53.6	24.6
Boiling point/K	1600	2750	4200	5100	77.4	90.2	85.0	27.1

[a]graphite

Kelvin (K) is the usual unit for temperature in the physical sciences. 0 K is equivalent to –273.15 °C (often rounded to –273 °C).

For a refresher on bonding, see Chapter 2.

A *metalloid* is an element that exhibits both metallic and non-metallic properties. Elements that sit next to the metal/non-metal dividing line on the periodic table often have properties that are exhibited by both metals and non-metals.

For a refresher on intermolecular forces, see Section 2.4.

10.3.7   Melting point and boiling point

Physical properties of elements such as melting and boiling points and electrical conductivity vary across a period. The variations in melting and boiling points largely depend upon the structure and type of bonding in the element. For example, if we consider Period 2 from lithium to neon, the type of bonding and the structure of the elements change significantly across the row (Table 10.4).

Lithium and beryllium, at the left of Period 2, are both metallic elements with metallic bonding. Lithium can contribute one electron and beryllium two to the sea of delocalised electrons; therefore, in beryllium, there is greater electrostatic interaction between the Be^{2+} centre and the two delocalised electrons than in lithium. The greater the number of delocalised electrons from the valence shells of the elements, the stronger the interatomic forces and the higher the melting point. This is illustrated by the melting points of lithium (453 K) and beryllium (1560 K).

Boron sits on the borderline between metals and non-metals, so it exhibits some metallic and some non-metallic properties; it is called a **metalloid**. Pure boron and pure carbon both have a giant covalent structure where the boron and carbon atoms are held together by very strong covalent bonds, which extend throughout the three-dimensional structure. These strong covalent bonds mean that the boron and carbon atoms are very hard to pull apart and therefore their melting points and boiling points are very high.

The three elements on the right-hand side of Period 2 (nitrogen, oxygen, and fluorine) exist as simple covalent molecules. The instantaneous dipole–induced dipole forces between these molecules are weak, which means that the melting and boiling points for these elements are low.

The very weak instantaneous dipole–induced dipole forces between atoms of the noble gases mean that the melting and boiling points of these elements are extremely low. The melting point of helium (–272.2 °C) is close to absolute zero.

10.3.8   Trends in chemical properties across a period

All chemical reactions involve the movement of electrons between the reacting species. Therefore the chemical reactivity of different elements depends upon their electronic structures.

On going down a group in the periodic table the outer electron arrangements are the same for all elements in that group. Therefore, elements in the same group have similar chemical properties.

On going across a row in the periodic table, electrons are being added to the valence shell of the atoms, which therefore have different outer electron configurations. This means that both the physical and chemical properties of the elements change and depend largely upon the group number and not the row number. In Chapter 11, we will look at the chemical properties of some specific groups in the periodic table.

Quick-check summary

- Elements in the same vertical group of the periodic table have the same outer electron configuration. The number of filled shells of electrons in an atom increases on going down a group.

- Elements in the same vertical group of the periodic table have similar chemical properties as they have the same outer electron configuration.

- The atoms of neighbouring elements in a period differ by one electron and one proton, and the total number of electrons increases across a period.

- Elements in the same horizontal row of the periodic table show variations in properties due to the difference in outer electron configuration.

- Physical properties that show periodic variations include atomic and ionic radii, ionisation energy, electronegativity, electron affinity, and melting and boiling points.

- The main factors affecting these properties are

  ○ The distance of the outer electrons from the nucleus (atomic radius)

  ○ The effective nuclear charge exerted by the nucleus

  ○ The shielding effect of the inner electrons

- The effective nuclear charge decreases down a group and increases across a row.

- Atomic radii of the elements increase down a group and decrease across a row.

- Ionisation energies of the elements decrease down a group and increase across a row.

- Electronegativity values decrease down a group and increase across a row up to Group 7 (17).

- Fluorine is the most electronegative element in the periodic table.

- Electron affinity values generally decrease down a group and increase across a row.

- Positive ions are called cations and are smaller than their parent atoms. Negative ions are called anions and are larger than their parent atoms.

- On going across a period, the structures of the elements change from metallic to giant molecular through to simple molecular. Group 8 (18) elements exist as single atoms. Melting points and boiling points of the elements depend upon the structure of the elements.

End-of-chapter questions

1 a. Find the atomic number, Z, and give the electronic configurations of the following elements in the form $1s^2 2s^2$..... :

    i. Potassium, K

    ii. Sulfur, S

    iii. Aluminium, Al

    iv. Bromine, Br

  b. State whether you expect each element in the list to form a cation or anion on ionisation, and explain why. Give the symbol and charge of the ion formed.

2 Of the following pairs of atoms and ions, state which of the pair will experience (i) the greater nuclear pull and why. State also (ii) which has the larger atomic or ionic radius:

  a. Na and Cl

  b. Na and K

  c. Fe^{2+} and Fe^{3+}

3 State which element of the following pairs has the greater first ionisation enthalpy and why:

  a. K and Br

  b. Mg and Sr

  c. Ne and Na

4 State and explain how the electronegativity of elements varies:

  a. Across a period

  b. Down a group in the periodic table

5 Give the electron configurations of the following species: H, H^+, and H^-. State and explain which of the three species has:

  a. The smallest radius

  b. The largest radius

6 The following table shows the ionisation energies of some elements in the s and p blocks.

Element	First ionisation energy/ kJ mol^{-1}	Second ionisation energy/kJ mol^{-1}	Third ionisation energy/ kJ mol^{-1}	Fourth ionisation energy/kJ mol^{-1}
Na	494	4560	6940	9540
Mg	736	1450	7740	10 500
N	1400	2860	4590	7480
O	1310	3390	5320	7450
Ne	2080	3950	6150	9290

Explain the following observations:

a. There is a large increase between the first and second ionisation energies for Na.

b. There is a large increase between the second and third ionisation energies for Mg.

c. The first ionisation energy of O is smaller than that of N.

d. There is a large increase between the third and fourth ionisation energies for N.

e. The first ionisation energy of Ne is extremely high.

7 The melting points for the elements of Period 3 are given in the following table.

Element	Na	Mg	Al	Si	P	S	Cl	Ar
M.pt/°C	97.8	650	660	1410	44	113	−101	−189

Briefly explain the reason for the trend in melting points observed on going across the period.

8 Explain why the ionic radii of Na$^+$, Mg^{2+}, and Al^{3+} are all smaller than their respective atomic radii.

9 Carbon and silicon are both in Group 4 (14):

a. Give the electron configuration of a carbon atom.

b. Give the electron configuration of a silicon atom.

c. The first ionisation energy of carbon is +1086 kJ mol^{-1} and that of silicon is +789 kJ mol^{-1}. Suggest a reason why these values differ.

d. Explain why both C and Si tend to form covalent rather than ionic compounds.

10 The covalent radius of a nitrogen atom is 150 pm. The ionic radius of a N^{3-} ion is 171 pm, and that of a N^{5+} ion is 11 pm. Explain the difference in ionic radii of N^{3-} and N^{5+} compared to that of the nitrogen atom.

11

The periodic table – chemistry of Groups 1, 2, 7 (17), and transition elements

At the end of this chapter, students should be able to:

- Identify elements in different sections of the periodic table

- Explain trends in physical properties and chemical reactivity for elements of Groups 1, 2, and 7 (17)

- Understand the factors affecting solubility and thermal stability of compounds of the Group 1 and 2 elements, and explain the trends observed on going down the groups

- Explain redox reactions of the halogen elements in terms of their standard reduction potentials, and rationalise the trend in oxidising power of the elements on going down the group

- Determine the oxidation states of halogen elements in oxyacids and their anions and within interhalogen compounds, and write equations for disproportionation reactions of the halogens

- Describe chemical tests that permit the identification of different halide ions

- Define a transition element, and give electron configurations of first-row transition metals in their elemental state and in simple complexes

- Describe some characteristic properties of transition elements and the variation of properties across the first row

- Understand and define the terms coordination complex, coordination number, ligand, geometry, and complex ion

- Explain the origin of colour in transition metal complexes

Foundations of Chemistry: An Introductory Course for Science Students, First Edition.
Philippa B. Cranwell and Elizabeth M. Page.
© 2021 John Wiley & Sons Ltd. Published 2021 by John Wiley & Sons Ltd.
Companion website: www.wiley.com/go/Cranwell/Foundations

- Become familiar with the geometries adopted by transition metal ions in complexes with coordination numbers 2, 4, and 6

- Determine the oxidation state of a transition metal ion in a complex, and write balanced half-equations for typical redox reactions

- Be aware that some transition metal complexes can exhibit stereoisomerism leading to different enantiomers

- Understand how ligand displacement reactions occur in transition metal complexes

11.1   Introduction

In Chapter 10 we saw that the periodic table is broadly divided into five different areas, usually depicted by different coloured shading on a copy of the periodic table, as shown in Figure 11.1.

The names of the different areas are:

s-block

p-block

d-block (containing the transition elements)

f-block (comprising the lanthanides and actinides).

Elements in the s, p, and d blocks have valence electrons filling s, p, and d orbitals, respectively. The lanthanides and actinides have valence electrons filling their outer f orbitals.

In this book, we use the numbers 1–8 to depict the vertical groups in the periodic table in the s- and p-blocks. This is a helpful numbering system because the group numbers 1–8 give you information about the number of outer electrons that elements in each group possess. However, you may come across periodic tables in which the vertical groups take numbers from 1 to 18. This numbering system is more useful when studying the transition elements.

In this chapter, we shall look at the physical and chemical properties of elements in three vertical groups of the periodic table: Groups 1, 2, and 7 (17), and elements in the first row of the d block.

The lanthanide and actinide elements are usually shown at the bottom of the periodic table. These are elements with high atomic and mass numbers, and many are radioactive. We will not be studying their chemistry in this book, but they are becoming increasingly important in technological applications. The elements from lanthanum (La, $Z = 57$) to lutetium (Lu, $Z = 71$), along with scandium (Sc, $Z = 21$) and yttrium (Y, $Z = 39$), are known as the *rare earth elements*, and many of them can be found in your mobile phone (see Box 11.1).

Elements in each area of the periodic table have broadly similar properties to each other, as we will see in this chapter.

The *rare earth* elements are the 14 lanthanides from lanthanum ($Z = 57$) to ytterbium ($Z = 70$) along with scandium ($Z = 21$), yttrium ($Z = 39$), and lutetium ($Z = 71$). Despite their name, they are actually distributed quite widely in the earth's crust in relatively high concentrations in places. They are economically highly important elements.

11.2   Group 1 – the alkali metals

Elements in Group 1 are found on the far left of the periodic table and consist of the metals lithium, sodium, potassium, rubidium, caesium, and francium. Although hydrogen is often placed at the top of Group 1 in the periodic table, it has little similarity to the elements in Group 1 apart from its tendency to form +1 ions. The heaviest element, francium, at the bottom of the group, is radioactive. As a result, very little is known about the chemical and physical properties of francium. It is a liquid at temperatures between 21 and 650 °C,

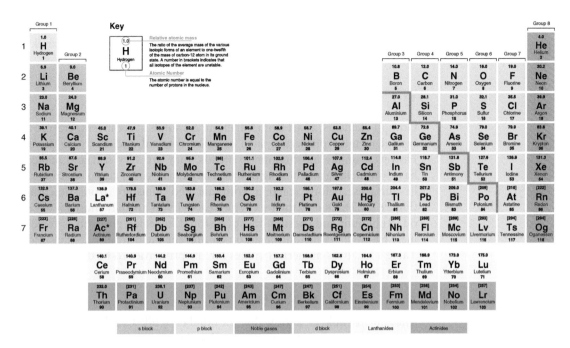

Figure 11.1 The periodic table with the different areas shaded.

Box 11.1

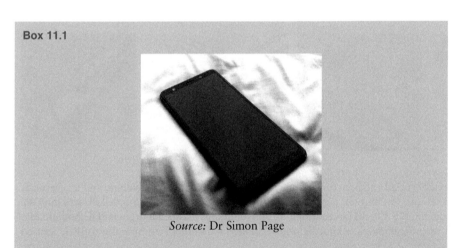

Source: Dr Simon Page

Most smart phones contain around 70 different elements. That's 84% of all the stable, non-radioactive elements. A typical smart phone may contain 62 different metals in tiny amounts, including the precious metals silver, gold, palladium, and platinum. Many of the metals used are rare earth elements. All of these 17 elements, apart from radioactive promethium, have found their way into at least one type of mobile phone. Most mobile phones contain at least 8 different rare earth elements. Elements such as terbium, dysprosium, and neodymium are used to make your phone vibrate. Terbium and dysprosium are also used to produce the bright colours in a phone's display.

Unfortunately, our close relationship with our smart phones is depleting the world's resources of these 'rare' earth elements. Most rare earth metals are found in China, giving this country even greater financial power as other sources of the elements become depleted. A lot of effort is being invested into finding economic ways of extracting these valuable metals from unused mobile phones, as there are no replacements for some of the elements. This is a good reason for digging out your old devices and recycling them in an appropriate manner.

which is unusual for a metallic element. The Group 1 elements react readily with water and form solutions that are alkaline in pH, and hence they are known as the **alkali metals**.

11.2.1   Physical properties of Group 1 elements

Group 1 elements, apart from francium, are solids at room temperature and have typical metallic properties such as good thermal and electrical conductivity. However, unlike most common metals, they are very soft and malleable and have low melting and boiling points. They can be easily cut to reveal shiny metallic surfaces, before reacting with air and moisture from the atmosphere (Figure 11.2). Due to their extreme reactivity in air they must be stored under oil. Rubidium and caesium must be stored in sealed glass tubes filled with an inert gas such as helium or argon to prevent them from bursting into flames. Table 11.1 gives numerical values for the properties to be discussed.

Atomic radii and ionisation energies

The data in Table 11.2 show that the number of filled shells of electrons increases on going down Group 1, which results in an increase in atomic radius as seen in Figure 11.3. The increase in atomic radius and decrease in effective nuclear

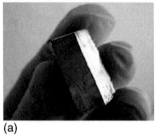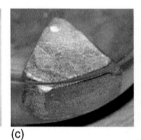

(a)   (b)   (c)   (d)   (e)

Figure 11.2 The elements (a) lithium (freshly cut), (b) sodium (under mineral oil), (c) potassium, (d) rubidium, and (e) caesium, showing different storage methods. *Sources:* (a), (d), and (e), Dennis S.K, https://commons.wikimedia.org/wiki/User:Dnn87#/media/File:Lithiumcut.JPG, licensed under CC BY 3.0. (b) Greenhorn1, https://commons.wikimedia.org/wiki/File:Sodium.jpg, licensed under PD. (c) Schmid & Rauch, https://commons.wikimedia.org/wiki/File:KaliumunterTetrahydrofuran.JPG, licensed under PD.

Table 11.1 Physical properties of Group 1 metals.

Group 1 element	Z	Atomic (metallic) radius/pm[a]	Ionic radius/pm	First ionisation energy/kJ mol^{-1}	Melting point/°C	Boiling point/°C	Electronegativity[b]
Lithium	3	152	60	519	180	1327	1.0
Sodium	11	186	95	494	98	890	0.9
Potassium	19	231	133	418	64	774	0.8
Rubidium	37	244	148	402	39	688	0.8
Caesium	55	262	169	376	29	690	0.7

[a] 1 pm = 10^{-12} m.
[b] Electronegativity values on the Pauling scale (0–4.0).

charge (Z_{eff}) on going down the group result in a decrease in first ionisation energy of the elements, as shown in Figure 11.4.

For a reminder about atomic and ionic radii, check back to Chapter 10.

The first ionisation energy of a Group 1 element involves removing the single valence electron from the half-filled s orbital to leave a positively charged ion with a complete outer shell of electrons:

e.g. $\qquad$ $Na(g) \rightarrow Na^{+}(g) + e^{-}$ $\quad$ $IE_1 = 494 \, kJ \, mol^{-1}$

$$1s^2 2s^2 2p^6 3s^1 \rightarrow 1s^2 2s^2 2p^6 + e^-$$

Ionisation energies of Group 1 elements are therefore very low due to the relative ease of removing this electron. As a result, the Group 1 metals are highly

Table 11.2 Electron configurations of the elements of Group 1.

Lithium (Li)	$1s^2 2s^1$
Sodium (Na)	$1s^2 2s^2 2p^6 3s^1$
Potassium (K)	$1s^2 2s^2 2p^6 3s^2 3p^6 4s^1$
Rubidium (Rb)	$1s^2 2s^2 2p^6 3s^2 3p^6 3d^{10} 4s^2 4p^6 5s^1$
Caesium (Cs)	$1s^2 2s^2 2p^6 3s^2 3p^6 3d^{10} 4s^2 4p^6 4d^{10} 5s^2 5p^6 6s^1$

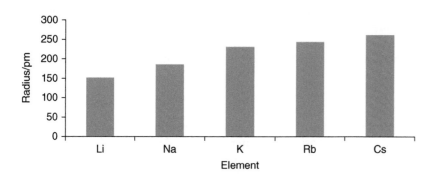

Figure 11.3 Atomic radii for the Group 1 metals, showing increase down the group.

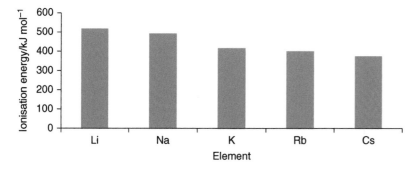

Figure 11.4 First ionisation energies of the Group 1 metals showing decrease down the group.

reactive and form positive ions and stable ionic salts with highly exothermic enthalpies of formation.

The second ionisation energy of the Group 1 elements is almost 10 times higher than the first as it involves removing an electron from a filled shell and relates to the following process:

$$Na^+(g) \rightarrow Na^{2+}(g) + e^- \quad IE_2 = 4560\,kJ\,mol^{-1}$$
$$1s^2 2s^2 2p^6 \rightarrow 1s^2 2s^2 2p^5$$

The electron being removed here is closer to the nucleus and at higher energy and experiences a much greater nuclear charge than the 3s electron. Because of this, Group 1 elements do not easily form M^{2+} ions.

Melting points and boiling points

The melting points of Group 1 elements are very low for metallic materials. In fact caesium melts just above room temperature (29 °C) (Figure 11.5). The low melting points are due to the weak metallic bonding in the structures. Because Group 1 elements contribute only one electron per atom to the electron sea, the bonding is fairly weak. As the metal atoms get larger on descending the group, the distance between the atomic nuclei and the delocalised electrons increases; thus the attractive forces decrease, making the melting points lower on going down the group. A similar argument can be used to explain the boiling points of the metals, which generally also decrease on going down the group. These weak metallic bonds explain why alkali metals are so soft and easily cut, unlike most other metals.

For a reminder of metallic bonding, see Chapter 2.

11.2.2 Chemical properties of Group 1 elements

The chemical properties of Group 1 elements are dominated by their tendency to form M^+ cations by losing the outer electron. As the first ionisation energy decreases down the group the tendency to form positive ions increases down the group, so the reactivity of the elements increases from Li to Cs.

Reaction with water

The reactions of Li, Na, and K with water should only be carried out by experienced chemists wearing appropriate personal protective equipment (PPE). The reaction should be well-screened by conducting it in a fume hood or behind a safety screen, and a full risk assessment should be written and approved. The reactions of Rb and Cs can be enjoyed on YouTube.

This is a classic reaction that can be demonstrated *with care* in the lab for lithium, sodium, and potassium but should *not* be attempted with either rubidium or caesium.

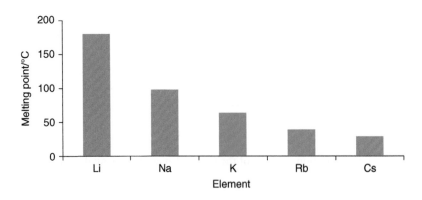

Figure 11.5 Melting points of the Group 1 metals.

All the elements react with water to form the soluble metal hydroxide, and hydrogen gas. The general equation for the reaction is:

$$2M(s) + 2H_2O(l) \rightarrow 2MOH(aq) + H_2(g)$$

Lithium tends to react slowly and steadily with water. Sodium floats on water: the reaction melts the metal, and the hydrogen gas released causes the molten metal to skate around the surface of the water. When potassium is placed in a bowl of water, it reacts quickly, releasing heat and igniting the hydrogen gas. The resulting solutions are basic due to the presence of the metal hydroxides.

Reaction with oxygen

All Group 1 metals react with oxygen to give the metal oxide, M_2O. Because the metals react so readily with both air and water, they are stored under oil in a sealed container. The general equation for the reaction with oxygen is:

$$4M(s) + O_2(g) \rightarrow 2M_2O(s)$$

Lithium metal also reacts with nitrogen in the atmosphere to give lithium nitride as a purple solid:

$$6Li(s) + N_2(g) \rightarrow 2Li_3N(s)$$

Worked Example 11.1

  a. What is the trend in ionisation energies down Group 1? How does this affect the reactivities of the elements with water and air?
  b. Give two reasons to explain why the second ionisation energy of sodium is much higher than the first ionisation energy.
  c. How would you expect the melting points of Group 1 elements to vary on going down the group? Explain your answer.

  d. Lithium metal reacts with both oxygen and nitrogen when heated in air. Predict a formula for lithium nitride, and give balanced chemical equations for the reactions taking place.

Solution
  a. The ionisation energies of the Group 1 elements decrease down the group as the number of filled outer shells of electrons increases. The filled shells of electrons shield the increasing nuclear charge from the valence electrons, making them easier to remove, so the ionisation energies decrease. This means that the elements react more rapidly with air and water on going down the group.
  b. The main reason the second ionisation energy of sodium is much higher than the first is that it is relatively easy to remove the first valence electron because it is in an s orbital that is well-shielded from the nucleus by the filled shells of electrons. Removing the second electron means breaking into the next shell, which is at higher energy, is closer to the nucleus, and has filled s and p orbitals. The second reason is that the second electron must be removed from a positively charged ion, which has a stronger attractive force for outer electrons than an uncharged atom.
  c. The melting points of the Group 1 elements decrease on going down the group. Although the number of electrons in the sea of charge holding

the positive ions together is the same, the metal ions are further from each other and therefore easier to separate by heat.

d. Lithium reacts with air to give lithium oxide, Li_2O, and lithium nitride, Li_3N.

$$4Li + O_2 \rightarrow 2Li_2O \qquad 6Li + N_2 \rightarrow 2Li_3N$$

11.3   Group 2 – the alkaline earth metals

The elements in Group 2 have two outer electrons and are known as the *alkaline earth metals*. The term *alkaline* comes from the pH of the compounds formed from the elements when allowed to react with air or water. The term *earth* is a historical name given to substances found as ores (for example, metal salts) that are insoluble in water and stable on heating. Many compounds of Group 2 elements fit this description. The periodic table shows the members of Group 2 to consist of beryllium, magnesium, calcium, strontium, barium, and radium. Calcium and magnesium are the sixth and eighth most abundant elements, respectively, in the earth's crust.

Box 11.2 As with Group 1, the heaviest member of this group, radium, is radioactive. Radium was discovered by Pierre and Marie Curie in 1898. They separated 1 mg of radium from 10,000 kg of pitchblende – an ore of uranium. Its name comes from the faint blue glow produced by the element as it decays. The property was used for many years in luminous paint for clock and watch hands and dials and also in treating some forms of cancer. However, its extreme radioactive nature means that it is rarely used today.

Pierre Curie (1859–1906) and Marie Sklodowska Curie (1867–1934), c. 1903. Source: Magnus Manske, https://commons.wikimedia.org/wiki/File:Pierre_Curie_ (1859-1906)_and_Marie_Sklodowska_Curie_(1867-1934),_c._1903_ (4405627519).jpg, licensed under CC0 1.0 Universal (CC0 1.0).

11.3.1 Physical properties of Group 2 elements

The elements in Group 2 are all metallic in nature with typical properties of metals, although the first member of the group, beryllium, shows some differences. The metals are all silvery grey when pure, although magnesium and beryllium react slowly with air to become coated with a thin layer of the white metal oxide, MO, that prevents further reaction. Table 11.3 gives some physical properties of the elements of Group 2.

Atomic radii and ionisation energy

The metallic radii of the elements increase on going down the group. The radii of the Group 2 elements are all smaller than those of the equivalent Group 1 element in the same row of the periodic table, as can be seen in Figure 11.6, due to the increase in **effective nuclear charge, Z_{eff},** on going across the row. As the atomic radii increase down the group the ionisation energies of the elements decrease.

Table 11.3 Physical properties of the Group 2 elements.

Group 2 element	Z	Atomic (metallic) radius/pm[a]	Ionic radius (M^{2+})/pm	First ionisation energy/ kJ mol^{-1}	Second ionisation energy/kJ mol^{-1}	Melting point/°C	Boiling point/ °C	Electronegativity[b]
Be	4	112	31	900	1760	1277	2477	1.5
Mg	12	160	65	736	1450	650	1107	1.2
Ca	20	197	99	590	1150	850	1487	1.0
Sr	38	215	113	548	1060	768	1377	1.0
Ba	56	217	135	502	966	714	1637	0.9
Ra	88	220	140	510	979	697	1137	0.9

[a] 1 pm = 10^{-12} m.
[b] Electronegativity values on the Pauling scale (0–4.0).

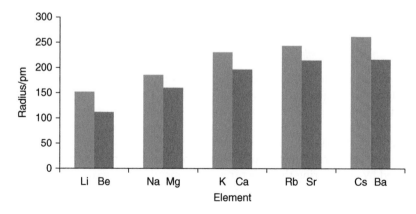

Figure 11.6 Comparison of the atomic radii of Group 1 and Group 2 metals.

Table 11.4 Electron configurations of Group 2 elements.

Beryllium (Be)	$1s^2 2s^2$
Magnesium (Mg)	$1s^2 2s^2 2p^6 3s^2$
Calcium (Ca)	$1s^2 2s^2 2p^6 3s^2 3p^6 4s^2$
Strontium (Sr)	$1s^2 2s^2 2p^6 3s^2 3p^6 3d^{10} 4s^2 4p^6 5s^2$
Barium (Ba)	$1s^2 2s^2 2p^6 3s^2 3p^6 3d^{10} 4s^2 4p^6 4d^{10} 5s^2 5p^6 6s^2$

The electron configurations of the elements are given in Table 11.4, from which it can be seen that Group 2 atoms have two outer s electrons. Reactions of the metals are dominated by loss of these two electrons to form 2+ ions with a filled inner shell of electrons:

$$Mg(g) \rightarrow Mg^{2+}(g) + 2e^-$$

$$1s^2 2s^2 2p^6 3s^2 \rightarrow 1s^2 2s^2 2p^6 + 2e^-$$

This process occurs in a step-wise manner. The energy changes involved are the first and second ionisation energies, and the steps are as follows for calcium:

$$Ca(g) \rightarrow Ca^+(g) + e^- \quad IE_1 = 590 \, kJ \, mol^{-1}$$

$$Ca^+(g) \rightarrow Ca^{2+}(g) + e^- \quad IE_2 = 1150 \, kJ \, mol^{-1}$$

The total ionisation energy for the formation of Ca^{2+} ions in the gas phase from gaseous calcium atoms is $(590 + 1150) \, kJ \, mol^{-1} = 1740 \, kJ \, mol^{-1}$, and the overall equation is:

$$Ca(g) \rightarrow Ca^{2+}(g) + 2e^-$$

This value is significantly higher than the equivalent value for the formation of K^+ ions ($418 \, kJ \, mol^{-1}$), suggesting that the formation of Ca^{2+} ions is less easily achieved. This directly affects the chemical reactivity of the elements. Group 2 metals are far less reactive than their Group 1 neighbours. The sum of the first and second ionisation energies of the Group 2 elements are shown graphically in Figure 11.7, where it can be seen that the values decrease on descending the group as the atomic radii get larger.

Melting point and boiling point

In general, the melting points and boiling points of the Group 2 metals decrease on going down the group, as shown in Figure 11.8. The increase in atomic radius on going down the group due to the weaker interatomic forces means that the metal atoms can be more easily separated by heating. A clear trend down the group is seen in Figure 11.8, where only the value for Mg is out of step.

The deviation of the melting point of magnesium from the rest of the group is explained by considering the structures of the metals. The two smaller elements Be and Mg have the same arrangement of atoms in the solid state, whereas the larger elements have a different structure. When we compare melting points and densities of metals we should compare those with the same arrangement of atoms (solid state structures) and the same number of nearest neighbours in the lattice.

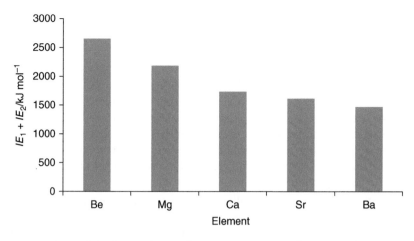

Figure 11.7 Sum of the first and second ionisation energies of Group 2 elements.

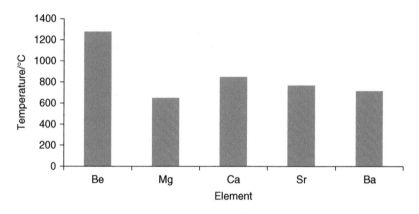

Figure 11.8 Melting points of the Group 2 elements.

Worked Example 11.2   Sodium and magnesium are adjacent metallic elements in Period 3. Some data are given here for the elements.

	Atomic radius/pm	Melting point/°C	Density/ g cm^{-3}	Electrical conductivity/ S^a m^{-1}
Sodium	186	97.8	0.97	2.18×10^7
Magnesium	160	650	1.74	2.24×10^7

^{a} S is the SI unit *siemens* of electrical conductivity.

a. Explain why magnesium has a smaller atomic radius than sodium.

b. Suggest why sodium has a lower melting point and lower density than magnesium.

c. Explain why both sodium and magnesium are good conductors of electricity, and suggest why sodium has a lower electrical conductivity than magnesium.

Solution

a. A Mg atom has one more proton in the nucleus than a Na atom. The valence electrons in Mg therefore experience a greater nuclear charge than those of Na, so the atomic radius is smaller.

b. Sodium atoms have one valence electron, whereas magnesium atoms have two. The greater the number of valence electrons the metal atoms lose to the sea of electrons, the stronger the metal–metal bonding. Magnesium atoms are therefore held together more strongly than sodium atoms, so the density is higher. As Mg atoms are held more strongly, the melting point is also higher than that of Na.

c. Both Na and Mg are metallic elements in which the valence electrons are found delocalised between the metal ions in the solid state. When a potential difference is applied, the electrons move through the metal to the positive electrode and so conduct electricity. As Mg has two valence electrons to contribute to the sea of electrons, it has a higher electrical conductivity than Na.

Worked Example 11.3   Use data from Tables 11.1 and 11.3 and information from Chapter 10 to answer the following questions about trends in properties of Group 2 elements:

a. Give the electron configurations of K^+ and Ca^{2+} ions.

b. Explain the difference in ionic radius between K^+ and Ca^{2+}.

c. Explain the difference in electronegativity values between potassium and calcium.

Solution

a. Both K^+ and Ca^{2+} have the electron configuration $1s^2 2s^2 2p^6 3s^2 3p^6$.

b. The ionic radius of Ca^{2+} is smaller (99 pm) than that of K^+ (133 pm). Both ions have the same electron configuration, but Ca^{2+} has lost two electrons, giving it a +2 charge. The nuclear pull per electron in Ca^{2+} is therefore greater than in K^+ as the number of protons in Ca is one more than in K, resulting in a smaller ionic radius.

c. On going across the period, the number of protons in the nucleus increases and the atomic radius decreases. The effective nuclear charge increases from K to Ca, and therefore the atoms become more electronegative.

11.3.2   Chemical properties of Group 2 elements

As with Group 1 elements, the ionisation energies of Group 2 elements decrease on going down the group, meaning that the metals become more reactive. The ionisation energies for the formation of the Group 2, M^{2+} ions (M = Be to Ba) are higher than those for the formation of the equivalent Group 1, M^+ ions (M = Li to Cs), so the Group 2 metals are *less reactive* than Group 1 metals.

Reaction with water

Reaction of Group 2 metals with water gives the metal hydroxide and hydrogen, according to this general equation:

$$M(s) + 2H_2O(l) \rightarrow M(OH)_2(aq) + H_2(g) \; (M = Mg, Ca, Sr, Ba)$$

On going down the group the elements become more reactive with water as the ionisation energy decreases and the hydroxides formed increase in solubility (see Section 11.3.3). This has the effect of making the solutions more alkaline with a higher pH.

Beryllium does not react with water, and magnesium reacts only very slowly with cold water to give a weakly alkaline solution of magnesium hydroxide. Magnesium reacts vigorously with steam to give magnesium oxide and hydrogen gas:

$$Mg(s) + H_2O(g) \rightarrow MgO(s) + H_2(g)$$

Calcium reacts more vigorously with water than magnesium but the product, calcium hydroxide, is only slightly soluble in water and forms as a white precipitate.

Reaction with oxygen

The Group 2 elements react with oxygen to form the metal oxide, MO. The general equation is:

$$2M(s) + O_2(g) \rightarrow 2MO(s)$$

Beryllium is very unreactive unless in powder form. When stored in air both beryllium and magnesium become coated with a thin film of oxide, which reduces their reactivity. When the coating is removed by rubbing with sandpaper the shiny magnesium metal revealed reacts rapidly with oxygen when ignited and burns with a characteristic bright white light.

Calcium metal also burns vigorously in air with an intense white flame.

11.3.3   Some s block compounds and their properties

Factors determining solubility

For a salt to be soluble the overall Gibbs energy change must be negative. The Gibbs energy is related to the enthalpy change by $\Delta G = \Delta H - T\Delta S$ and so is favoured by an exothermic value for the enthalpy of solution for the compound. The Gibbs energy also depends on the entropy change (ΔS) for the reaction and the absolute temperature (T). Therefore, a salt with an endothermic value for the enthalpy of solution can dissolve if the temperature is high enough and there is a sufficiently large increase in entropy. This is why the solubility of salts generally increases with increase in temperature.

Enthalpy changes involved when a metal salt dissolves

When an ionic compound dissolves in water, two distinct enthalpy changes take place, and together they equal the enthalpy of solution, $\Delta_{sol}H$. The overall process can be represented by the following general equation for an ionic salt, MX, where M is a Group 1 metal:

$$MX(s) \rightarrow M^+(aq) + X^-(aq) \quad \Delta_{sol}H$$

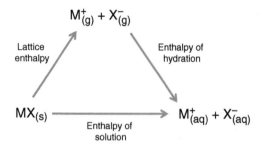

Figure 11.9 Enthalpy changes occurring when an ionic solid dissolves in water.

For a reminder on lattice enthalpy see Chapter 6, section 6.1.8. For a reminder on Hess's Law see section 6.1.6.

The two enthalpy changes are the lattice enthalpy ($\Delta_{lat}H$) and the hydration enthalpy ($\Delta_{hyd}H$). These can be combined in an enthalpy triangle using Hess's Law, as shown in Figure 11.9.

The first step in the process is to break the lattice into gaseous ions. The enthalpy change for this step is the **lattice enthalpy of dissociation**, $\Delta_{lat}H$, and is an **endothermic** process as energy is required to break the electrostatic forces between the oppositely charged ions:

$$MX(s) \rightarrow M^+(g) + X^-(g) \quad \Delta_{lat}H$$

In general, a high lattice enthalpy exists between small, highly charged ions that can pack closely together.

In the second step the free ions are surrounded by water to give hydrated ions:

$$M^+(g) + X^-(g) \rightarrow M^+(aq) + X^-(aq) \quad \Delta_{hyd}H$$

The enthalpy change for this step is the enthalpy of hydration of the salt, $\Delta_{hyd}H$. The M^+ cations are surrounded by the slightly negative oxygen atoms of the water molecules and the A^- anions ions are surrounded by the slightly positive hydrogen atoms of the water molecules, as in Figure 11.10 for NaCl. Because this step involves forming new intermolecular forces between the ions and water molecules, energy is released, and the process is overall **exothermic**. The smaller and more highly charged the ions, the more energy is released upon hydration.

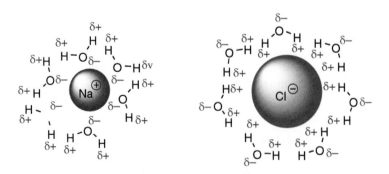

Figure 11.10 Hydrated sodium and chloride ions, as in an aqueous solution of sodium chloride.

The overall enthalpy change for dissolving the salt in water is obtained by adding the lattice ethalpy for MX to the enthalpies of hydration of the ions. An exothermic, or negative, value for the enthalpy of solution is favoured by a small lattice enthalpy and a high enthalpy of hydration:

$$\Delta_{lat}H + \Delta_{hyd}H = \Delta_{sol}H$$

A general rule to remember is that all Group 1 salts are soluble and Group 2 salts tend to be less soluble.

In the following sections we will see how the solubilities of different ionic solids affect their chemical behaviour.

Group 1 and 2 oxides

Group 1 oxides have the general formula M_2O. They react readily with water to give the metal hydroxide:

$$M_2O(s) + H_2O(l) \rightarrow 2MOH(aq)$$

Group 1 metal hydroxides are all soluble in water and produce alkaline solutions.

The oxides of Group 2 metals have the general formula MO. They are much less soluble in water because of their high lattice enthalpies. This is due to the smaller size of the M^{2+} ion and the double positive charge on the metal. The general equation for the reaction is given by:

$$MO(s) + H_2O(l) \rightarrow M(OH)_2(aq)$$

Group 1 and 2 hydroxides

The hydroxides of Group 1 metals have the general formula MOH (M = Li to Cs). They are all soluble in water and dissolve to give basic (alkaline) solutions with high pH values. They are classed as **strong bases**.

The Group 2 metal hydroxides are not very soluble in water. Their solubility increases on going down the group with the lowest member, barium hydroxide, $Ba(OH)_2$, being fairly soluble. The solubilities of the Group 2 metal hydroxides are shown in Table 11.5. The increase in solubility on going down the group can be explained by a decrease in lattice enthalpy as the M^{2+} ion gets larger and the distance between cations and anions increases. Although the hydration enthalpy of the metal ion decreases as the ion gets larger, this energy change does not decrease as much as the lattice enthalpy.

On going down Group 2, the hydroxides become more basic as their solubilities increase, and hence the pH of the solutions formed also increases.

A strong base is totally dissociated in water to give high concentrations of hydroxide ions. See Chapter 7 for a reminder about strong acids and bases.

Most Group 2 metal hydroxides are weak bases as they are only partially dissociated in water because of their lower solubility.

Table 11.5 The solubilities of Group 2 metal hydroxides increase down the group.

Group 2 hydroxide	Solubility/g per 100 g water
$Be(OH)_2$	Sparingly soluble
$Mg(OH)_2$	0.000 9
$Ca(OH)_2$	0.156
$Sr(OH)_2$	0.8
$Ba(OH)_2$	3.9

Most are classed as **weak bases** as they dissociate only partially in water according to the following equilibrium:

$$M(OH)_2(s) \rightleftharpoons M^{2+}(aq) + 2OH^-(aq) \; (M = Mg, Ca, Sr)$$

Barium hydroxide is fully dissociated in water and therefore is a strong base. Because of its low solubility, magnesium hydroxide is only weakly basic and is used in indigestion tablets in formulations such as milk of magnesia to counteract stomach acid.

Solubility of Group 1 and Group 2 carbonates and sulfates

Group 1 metal carbonates (M_2CO_3) and sulfates (M_2SO_4) are all soluble in water apart from lithium carbonate, which is sparingly soluble. Group 2 metal carbonates have the formula MCO_3 and are all insoluble in water due to their high lattice enthalpies resulting from the relatively high charges on the ions. Group 2 metal sulfates decrease in solubility on going down the group. Magnesium sulfate is soluble, calcium sulfate is sparingly soluble, and barium sulfate is insoluble. This is opposite to the behaviour shown by the hydroxides. The lattice enthalpies of all three salts are reasonably similar because the sulfate ion is very large, so changing the size of the metal cation has little effect upon the lattice enthalpies. However, the hydration enthalpy of the M^{2+} ions decreases as the metal ion gets larger, so less energy is available to break the lattice on going down the group. The salts therefore decrease in solubility.

Thermal stability of Group 1 and Group 2 carbonates

All the Group 1 metal carbonates, apart from lithium carbonate, are thermally stable. Lithium carbonate is more covalent in nature and decomposes to give carbon dioxide and lithium oxide:

$$Li_2CO_3(s) \rightarrow Li_2O(s) + CO_2(g)$$

The other Group 1 metal carbonates are often found in the hydrated form. This means their bulk crystals contain water molecules, known as **water of crystallisation**. For example, hydrated sodium carbonate has the formula $Na_2CO_3.10H_2O$ (sodium decahydrate). On heating, the hydrated metal carbonates lose their water of crystallisation to give anhydrous salts:

$$Na_2CO_3.10H_2O(s) \rightarrow Na_2CO_3(s) + 10H_2O(g)$$

The Group 2 metal carbonates are all thermally unstable and decompose to give the metal oxide and carbon dioxide, as shown here for calcium carbonate:

$$CaCO_3(s) \rightarrow CaO(s) + CO_2(g)$$

The thermal stability of Group 2 metal carbonates increases down the group. Magnesium carbonate starts to decompose at around 350 °C, whereas barium carbonate decomposes at around 1350 °C. The order of thermal stability is given by

$$MgCO_3 < CaCO_3 < SrCO_3 < BaCO_3$$

The energy required to decompose a metal salt depends, to a large extent, upon the lattice enthalpy of the solid. The stronger the electrostatic forces between the ions, the more energy is required to break the lattice. A simple consideration of the lattice enthalpies of the metal carbonates might suggest that $BaCO_3$ would have the smallest lattice enthalpy because it contains the largest ions, and this is true. However, this would be expected to make $BaCO_3$ the most thermally *unstable* of the Group 2 carbonates, which is not the case. This is because we also have to consider the lattice enthalpy of the metal oxide product formed, in this case BaO. The lattice enthalpy of the metal oxide decreases down the group, but at a faster rate than that of the metal carbonate. Thus, there is a decrease in the 'pay-back' energy obtained from the formation of the metal oxide, which means that the reaction becomes less favourable and requires greater energy input. The smaller the metal cation, the stronger the metal oxide lattice.

Another factor also comes into play here: the **polarization** of the carbonate ion by the metal cation. Pure ionic bonding assumes ions in solids are spherical in shape with an even distribution of electron density (Figure 11.11a). In practice, an ion with a high charge density can polarize a neighbouring ion of opposite charge, as shown in Figure 11.11b. The electron density is no longer evenly distributed, and the anion no longer spherical. There is a degree of covalency in the M—X bond.

The electrons in the carbonate ion are delocalised, making each C—O bond equivalent (Figure 11.12a).

When the carbonate ion is close to a metal cation, the cation can **polarize** the carbonate ion, pulling electron density towards the positively charged ion and weakening the C—O bond (shown in red) (Figure 11.12b). Heating the metal carbonate provides enough energy to break the C—O bond and form the metal oxide (Figure 11.12c). The smaller the metal cation, the more strongly polarizing the effect. Thus, metal carbonates at the top of Group 2 decompose more readily on heating than those larger ones at the bottom of the group.

Delocalisation of electrons will be explained in more detail in Chapter 15.

For a reminder on polarization and polarizability, see Chapter 2. Small, highly charged cations have the greatest polarizing power. Mg^{2+} is more polarizing than Na^+ because it is smaller and has a higher charge. The larger the anion and the greater the charge, the easier it will be to polarise. Small, singly charged anions are not easy to polarize, so F^- is less polarizable than Cl^-; I^- is quite readily polarized, leading to a high degree of covalent character in iodide compounds. Larger anions with delocalised electrons, such as CO_3^{2-}, are readily polarizable.

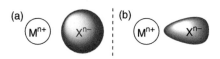

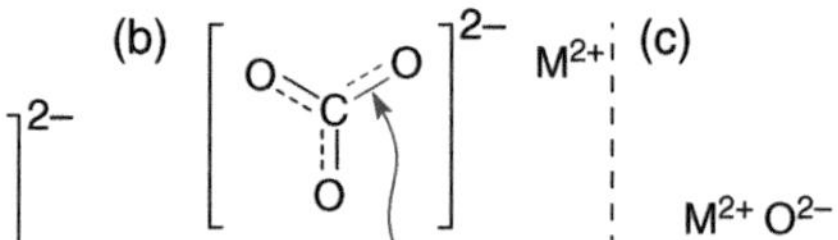

Figure 11.11 (a) Pure ionic bonding between a metal cation and anion; (b) anion, X^{n-}, polarised by cation, M^{n+}, induces covalent character into an M—X bond.

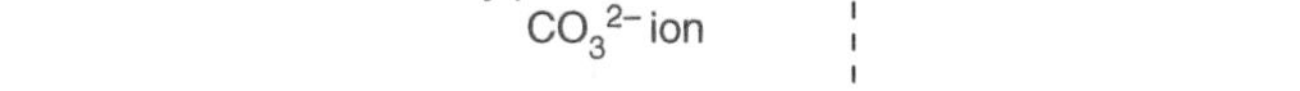

Figure 11.12 (a) Delocalisation in the carbonate ion makes all C—O bonds equal in length. (b) The presence of a metal cation polarises the carbonate ion, pulling electron density towards the cation and weakening the C—O bond (red) of the oxygen adjacent to the cation. (c) On heating the metal carbonate, the C—O bond (red) is broken, the metal oxide formed, and CO_2 released.

Thermal stability of metal nitrates

Group 1 metal nitrates, apart from lithium, all decompose to give the nitrite (MNO_2) and release oxygen according to the following equation:

$$MNO_3(s) \rightarrow MNO_2(s) + \tfrac{1}{2}O_2(g)$$

If they are heated more strongly, the metal oxide is formed and nitrogen dioxide produced. Lithium nitrate decomposes on heating to form the oxide directly, along with nitrogen dioxide and oxygen:

$$2LiNO_3(s) \rightarrow Li_2O(s) + 2NO_2(g) + \tfrac{1}{2}O_2(g)$$

Group 2 metal nitrates also decompose directly to the oxide, nitrogen dioxide, and oxygen. The thermal stability of both Group 1 and 2 nitrates increases down the group, so higher temperatures are needed for the heavier metal nitrates:

$$2Ca(NO_3)_2(s) \rightarrow 2CaO(s) + 4NO_2(g) + O_2(g)$$

The difference in behaviour of the Group 1 and 2 metal nitrates is due to the increased stability of the metal oxides formed in each case. Group 2 metal oxides have very high negative standard enthalpies of formation and high lattice enthalpies, and thus there is a large energy gain when these oxides are formed.

> Group 1 metal nitrates decompose to the metal nitrite and oxygen, apart from lithium nitrate. Group 2 metal nitrates decompose to the metal oxide, nitrogen dioxide, and oxygen. The stability of metal nitrates increases down the groups:
>
> $$NaNO_3 < KNO_3 < RbNO_3 < CsNO_3$$
>
> and
>
> $$Mg(NO_3)_2 < Ca(NO_3)_2 < Sr(NO_3)_2 < Ba(NO_3)_2$$

Worked Example 11.4

a. Write equations for the reactions of sodium and calcium with oxygen. What type of bonding is present in the oxides formed?

b. Write equations for the reactions of the oxides formed with water. Which oxide would react more vigorously with water? Explain your answer. What would be the approximate pH of the resulting solutions?

c. Give two possible reasons why the reactivity of the Group 2 oxides with water increases down the group.

Solution

a. $2Na(s) + \tfrac{1}{2}O_2(g) \rightarrow Na_2O(s)$, $2Ca(s) + O_2(g) \rightarrow 2CaO(s)$
   Both oxides would be ionic.

b. $Na_2O(s) + H_2O(l) \rightarrow 2NaOH(aq)$, $CaO(s) + H_2O(l) \rightarrow Ca(OH)_2(aq)$
   Na_2O would react more vigorously with water as the sodium hydroxide formed is a strong base and completely dissociated in water. $Ca(OH)_2$ is a weak base and only partially dissociated in water.
   The pH of both solutions would be high: around 12–14.

c. The reactivity of the Group 2 oxides with water increases down the group because the lattice enthalpy of the metal oxides decreases as the metal ion gets larger. In addition, the solubility of the metal hydroxide formed increases on going down the group due to the decrease in lattice enthalpy of the $M(OH)_2$ salts.

Worked Example 11.5 Identify the solids remaining when each of the following salts is heated, and write balanced chemical equations for the reactions. Explain the difference in behaviour on heating between the Group 1 and Group 2 metal nitrates:

Potassium nitrate, KNO_3

Calcium nitrate, $Ca(NO_3)_2$

Solution

$$KNO_3(s) \rightarrow KNO_2(s) + \tfrac{1}{2}O_2(g)$$

$$Ca(NO_3)_2(s) \rightarrow CaO(s) + 2NO_2(g) + \tfrac{1}{2}O_2(g)$$

The Group 2 metal nitrates decompose to the metal oxide, nitrogen dioxide, and oxygen, whereas Group 1 metal nitrates form the metal nitrite and oxygen only. The difference in behaviour is due to the more exothermic enthalpy of formation of the Group 2 metal oxide formed. The small size of the Ca^{2+} ion and the doubly charged Ca^{2+} and O^{2-} ions result in a stable lattice, $\Delta_f H$ (CaO = –635 kJ mol^{-1}). The equivalent oxide of potassium, K_2O, has a smaller enthalpy of formation $\Delta_f H$ (K$_2$O = –362 kJ mol^{-1}) because of its lower lattice enthalpy; therefore, its formation is not as thermodynamically favourable.

Worked Example 11.6

a. Explain the factors that influence solubility and suggest an explanation for the fact that the solubilities of the Group 2 metal hydroxides increase as you go down the group but the solubilities of the Group 2 metal sulfates decrease on going down the group.

b. Explain why the thermal stability of Group 2 metal carbonates increases down the group despite the lattice enthalpy of the carbonates decreasing down the group.

Solution

a. The solubility of a salt in water can be explained in terms of its lattice enthalpy of dissociation, which is endothermic, and the hydration enthalpy of its ions, which is exothermic. The Group 2 metal hydroxides increase in solubility on going down the group. This can be explained because the lattice enthalpies of the metal hydroxides decrease on going down the group, so less energy is required to break the lattice because the metal ions become larger and the electrostatic forces become smaller as the distance between the ions increases.

For the sulfates, the solubilities decrease on going down the group, which is opposite to the behaviour of the hydroxides. Because the sulfate ion is so large compared to the hydroxide ion, the lattice enthalpies of the Group 2 metal sulfates are fairly similar. Here it is the hydration enthalpies of the metal cations that influence the solubility of the sulfates, with a smaller hydration enthalpy for the larger cations, so we have less energy to 'pay back' and use to break the lattice; the result is lower solubility as we go down the group.

b. The lattice enthalpies of the Group 2 metal carbonates decrease down the group due to the increasing size of the cation. This would be expected to make the salts less thermally stable. However, the opposite trend is observed. The smaller metal carbonates decompose more readily. This can be explained by two effects. The smaller cations are more strongly polarizing and polarize the charge in the carbonate ions, thus weakening the C—O bond and enabling the lattice to break down more easily. In addition, the lattice enthalpies of formation of the metal oxide (MO) product are more exothermic when the cation is smaller and so are higher than those lower down the group. Therefore, the enthalpy change for decomposition of the metal carbonates higher in the group is less endothermic and decomposition temperatures are lower.

11.4   Group 7 (17) – the halogens

The halogens are the elements in Group 7 (17) towards the right-hand side of the periodic table. They are p-block elements with seven electrons in their outer shells. They are among the most reactive of the p-block elements, and their compounds are important in many industrial applications and processes.

11.4.1   Physical properties of Group 7 (17) elements

All the halogens exist as diatomic molecules with general formula X_2 (where X = F, Cl, Br, I). Each halogen atom has seven electrons in its outer shell, with the outer electron configuration ns^2np^5 (Table 11.6). In its elemental form the halogen atom completes its octet by sharing its one unpaired electron with the unpaired electron of another halogen atom to form a single covalent bond, as shown in Figure 11.13 for Cl_2.

On going down Group 7 (17), the total number of electrons possessed by each halogen atom increases, and this results in an increase in the instantaneous dipole–induced dipole forces between the halogen molecules. The increase in

Table 11.6 Electron configurations of the halogen elements.

Fluorine (F)	Z = 9	$1s^2 2s^2 2p^5$
Chlorine (Cl)	Z = 17	$1s^2 2s^2 2p^6 3s^2 3p^5$
Bromine (Br)	Z = 35	$1s^2 2s^2 2p^6 3s^2 3p^6 3d^{10} 4s^2 4p^5$
Iodine (I)	Z = 53	$1s^2 2s^2 2p^6 3s^2 3p^6 3d^{10} 4s^2 4p^6 4d^{10} 5s^2 5p^5$

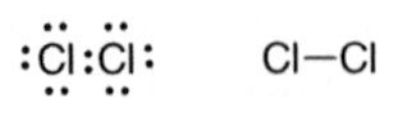

$$\overset{..}{:}\!\overset{..}{Cl}\!:\!\overset{..}{Cl}\!\overset{..}{:} \qquad Cl-Cl$$

Figure 11.13 Sharing of electrons in Cl_2 and covalent bond formation.

intermolecular forces on going down the group results in an increase in melting and boiling points of the elements.

Fluorine, F_2, is the first member of the group. It is a gas and an extremely powerful oxidising agent. In fact, it is so reactive that it cannot be stored in glass containers but must be kept in stainless steel vessels. Fluorine gas is colourless. Chlorine is the second element in the group and exists as a pale greenish-yellow gas. Chlorine is poisonous and notorious for its use in the First World War. Bromine, Br_2, is a corrosive brown liquid; and iodine, I_2, is a shiny, crystalline black solid that sublimes to give a purple vapour. The lowest member of the group is astatine, which is radioactive. There are 39 known isotopes of astatine, all with very short half-lives, so the element has no practical applications.

Fluorine reacts with water, so its solubility cannot be quoted. Chlorine, bromine, and iodine have very limited solubility in water. Being non-polar molecules, they form only weak intermolecular forces with polar water molecules. The halogens are far more soluble in non-polar solvents such as hexane and cyclohexane. In non-polar solvents, weak instantaneous dipole-induced dipole forces are established between the halogen molecules and the solvent. In aqueous solutions the halogens have slightly different colours than in non-polar solvents, as can be seen in Table 11.7.

The electronegativity of the halogen elements is high due to their small atomic radii and high effective nuclear charge. In fact, fluorine, F, is the most electronegative element in the periodic table, with a value of 4.0 on the Pauling scale. As the atomic radius of the halogens increases on descending the group, the electronegativity of the halogens decreases (Figure 11.4).

Table 11.7 Colours of elemental halogens in polar and non-polar solvents.

Halogen	Physical state at room temperature	M.pt/ °C	B.pt/ °C	Colour of the element	Colour in hexane	Colour in aqueous solution
Fluorine, F_2	Gas	−220	−188	Pale yellow		
Chlorine, Cl_2	Gas	−101	−34.7	Greenish-yellow	Greenish-yellow	Very pale green
Bromine, Br_2	Liquid	−7.2	58.8	Orange-brown	Reddish-orange	Dark brown
Iodine, I_2	Solid	114	184	Black	Purple-pink	Dark red-brown[a]

[a] Iodine is dissolved in aqueous KI as it is very insoluble in water. Colour is due to I_3^- ions.

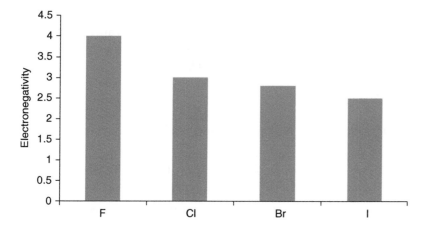

Figure 11.14 Electronegativities of the Group 7(17) elements.

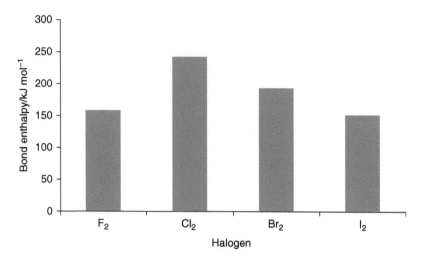

Figure 11.15 Bond enthalpies of the Group 7(17) elements.

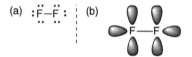

Figure 11.16 (a) Lone pairs in a F_2 molecule; (b) repulsion between lone pairs in a F_2 molecule leads to a weaker F—F bond.

Halogens tend to react by gaining an electron to form compounds containing the halide ion, X^-. This is reflected in their electron affinity values, which are high. Removal of electrons on ionisation to form positive ions is not an important process for halogen atoms and so their ionisation energies are very high.

The bond dissociation enthalpies of the halogens are plotted in Figure 11.15. They are all endothermic as breaking bonds requires energy to be supplied. The trend from chlorine to iodine is for the bond enthalpy to decrease, which is what we would expect from a consideration of the increasing size of the atoms and length of the bond as we go down the group. However, fluorine doesn't fit this trend, and the value of the bond enthalpy for fluorine is lower than those of chlorine and bromine.

The F—F bond length is quite short, and fluorine atoms are quite small. This results in considerable repulsion between the lone pairs on the two bonded fluorine atoms and hence a weaker F—F bond (Figure 11.16).

> **Worked Example 11.7**
>
> a. Describe the type of bonding that exists in the molecule Cl_2.
>
> b. Chlorine is a gas at room temperature, but the next member down Group 7 (17) is a liquid. Give reasons for the difference in physical states of the molecules.
>
> c. Explain why the electronegativities of the halogen elements decrease down the group.

d. Explain why the enthalpies of atomisation, $\Delta_{at}H$, of the halogens decrease as we go from Cl to I.

Solution

a. Each halogen atom has seven valence shell electrons and requires one more electron to attain a filled shell. In the molecule Cl_2, each Cl atom shares its unpaired electron with another Cl atom to form a covalent bond.

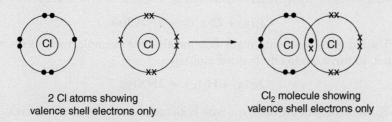

2 Cl atoms showing valence shell electrons only

Cl_2 molecule showing valence shell electrons only

b. A chlorine molecule has a total of 34 electrons, whereas a bromine molecule has 70. Bromine molecules therefore have greater intermolecular forces. This causes the melting point and boiling point of bromine to be higher than those of chlorine and results in chlorine being a gas at room temperature, whereas bromine is a liquid.

c. The electronegativities of the halogen atoms decrease down the group because the atomic radius increases and the effective nuclear charge decreases. This means the ability of the atom to attract electrons, or the electronegativity, must decrease on going down the group.

d. The enthalpy of atomisation of an element is the energy required to create one mole of gaseous atoms of the element. The halogens exist as diatomic molecules at room temperature. The enthalpy of atomisation is directly related to the bond dissociation enthalpy of the X—X bond, as can be seen from the equation:

$$X—X(g) \rightarrow 2X(g) \ E(X—X)$$

The enthalpy change is twice the enthalpy of atomisation. The bond enthalpies of the halogens decrease from Cl to I as the atoms get larger; therefore, the enthalpies of atomisation also decrease.

11.4.2 Reactions of Group 7 (17) elements

Reactions as oxidising agents

The elemental halogens generally react as oxidising agents, being reduced themselves to the respective halide ion. Halogen atoms require just one more electron to attain the noble gas configuration. On gaining one electron the halogen atoms are reduced, and a **halide** ion is formed. The oxidation number of the halogen atom changes from 0 to –1.

Remember: Oxidation is loss of electrons. Reduction is gain of electrons.

The symbol X is generally used to represent any halogen atom.

Halogen atoms have seven electrons in their outer shells and so require just one more electron to attain the noble gas electron configuration:

$$X(g) + e^- \rightarrow X^-(g) \ (X = F, Cl, Br, I)$$

The halogens can oxidise both metals and non-metals, and also other halide ions.

The reaction of halogens with metals gives ionic salts containing the halide ion, where the metal atom has been oxidised and the halogen reduced. For example, sodium metal reacts vigorously in chlorine to give sodium chloride:

$$Cl_2(g) + 2Na(s) \rightarrow 2NaCl(s)$$

The halogens react with many non-metals – for example, hydrogen – when heated, to form covalently bonded molecules:

$$X_2(g) + H_2(g) \rightarrow 2HX(g)$$

The reactivity of the halogens with both metals and non-metals decreases on going down the group. This is due to the decrease in both the electronegativity and oxidising power of the halogens. The electronegativity values for the halogens are given in Figure 11.14. The oxidising ability of the halogens is measured by the standard reduction potential, $E^\ominus$, which was described in Chapter 9, and is a measure of the tendency of the species to be reduced. Standard reduction potentials, measured in volts, are given for the halogens in Table 11.8. The larger and more positive the value, the stronger the oxidising agent. Fluorine, at the top of the group, has a standard reduction potential of +2.87 V, meaning it is a much stronger oxidising agent than iodine, which has a standard reduction potential of +0.54 V.

These factors mean that fluorine is the strongest oxidising agent in Group 7 (17) and iodine is the weakest.

Displacement reactions of halogens

The trend in oxidising power of the halogens can be observed by studying the reactions of aqueous solutions of the halogens with different halide ions. This type of reaction, where one halogen molecule displaces another halogen molecule from a solution of its halide ions is called a *displacement reaction*. Displacement reactions are examples of redox reactions and can be broken down into two half-equations. These are given here using X and Y as general symbols for the halogen atoms and omitting physical states:

$$X_2 + 2e^- \rightarrow 2X^- \text{ reduction half-equation}$$

$$2Y^- \rightarrow Y_2 + 2e^- \text{ oxidation half-equation}$$

$$\text{Overall equation: } X_2 + 2Y^- \rightarrow 2X^- + Y_2$$

Table 11.8 Standard reduction potentials of the halogens.

Electrode reaction	Standard reduction potential, $E^\ominus$/V
$\frac{1}{2}F_2(g) + e^- \rightarrow F^-(g)$	+2.87
$\frac{1}{2}Cl_2(g) + e^- \rightarrow Cl^-(g)$	+1.36
$\frac{1}{2}Br_2(l) + e^- \rightarrow Br^-(g)$	+1.07
$\frac{1}{2}I_2(s) + e^- \rightarrow I^-(g)$	+0.54

Table 11.9 Displacement reactions of halogens.

Halogen	Halide ion	Overall equation	Change in colour of hexane
Cl_2	Br^-	$Cl_2 + 2Br^- \rightarrow 2Cl^- + Br_2$	Colourless to orange
Cl_2	I^-	$Cl_2 + 2I^- \rightarrow 2Cl^- + I_2$	Colourless to purple
Br_2	I^-	$Br_2 + 2I^- \rightarrow 2Br^- + I_2$	Orange to purple

Table 11.9 lists the combinations of halogen and halide ion that lead to reaction and the associated equations. Because the halogens have characteristic colours in covalent solvents such as hexane, reactivity can be assessed from the change in colour during reaction. The reactions of fluorine have been excluded from this table as fluorine and its compounds are particularly hazardous to work with.

Table 11.9 shows that a halogen higher up Group 7 (17) will displace a halogen lower in the group from a solution containing the halide ion.

Standard reduction potentials can be used to determine whether the redox reaction will proceed in a specific direction. For example, the displacement reaction of iodine by chlorine is made up of two half-equations:

$$Cl_2 + 2e^- \rightarrow 2Cl^- \quad E^\ominus = +1.36\,V \quad \text{Reduction}$$

$$2I^- \rightarrow I_2 + 2e^- \quad E^\ominus = -0.54\,V. \quad \text{Oxidation}$$

The overall electrode potential for the reaction $Cl_2 + 2I^- \rightarrow 2Cl^- + I_2$ is:

$$E^\ominus = +1.36\,V - 0.54\,V = +0.82\,V$$

This positive value suggests that the reaction should proceed in the direction as written because Cl_2 is a stronger oxidising agent than I_2, as shown by their standard reduction potentials.

> The order of oxidising ability of the halogens is $Cl_2 > Br_2 > I_2$.
>
> Chlorine is more readily reduced to the chloride ion, Cl^-, so it oxidises halide ions lower in the group to the halogen. Iodine would prefer to exist as I_2, so iodide ions, I^-, are readily oxidised.

Oxidation states of the halogens

Although the most common oxidation states of the halogens are 0 (in the elemental state) and –1 (in halide ions), the elements can also exist in a variety of higher oxidation states by losing one or more valence electron. Table 11.10 gives the oxidation state of chlorine in some different oxyanions. When chlorine is combined with a strongly oxidising element, such as oxygen, it can have a positive oxidation state. Note that in each of these oxyanions, the oxygen atom has an oxidation state of –2. The chlorine atom can lose up to seven valence electrons to form oxyanions in which the halogen atom has a maximum oxidation state of +7. In the perchlorate ion, ClO_4^-, the chlorine atom has lost all seven of its valence electrons.

Table 11.10 Oxyanions of chlorine.

Species	Common name	Formal name	Oxidation state of chlorine	Oxidation state of oxygen
ClO^-	Hypochlorite	Chlorate(I)	+1	–2
ClO_2^-	Chlorite	Chlorate(III)	+3	–2
ClO_3^-	Chlorate	Chlorate(V)	+5	–2
ClO_4^-	Perchlorate	Chlorate(VII)	+7	–2

Box 11.3 An oxyanion is an anion containing one or more oxygen atom bonded to another element, in this case a halogen atom. The oxidation state of oxygen, O, is -2; but the oxidation state of the other element can vary, thus leading to different formulations of oxyanions with the halogen in different oxidation states. The charge on the oxyanion is determined by the number of halogen and oxygen atoms in the anion and the oxidation state of the halogen element.

The Group 7 (17) elements can also form covalently bonded molecules with each other, called **interhalogen compounds**. The compound iodine monochloride, ICl, is a typical example of an interhalogen compound. It is composed of red crystals that melt at $27\,°C$ to form a choking brown gas. Due to the difference in electronegativities between the iodine and the chlorine atoms, the molecule is polar covalent, with stronger intermolecular forces than in the parent halogen molecules due to permanent dipole–permanent dipole interactions:

$$\overset{\delta+}{I} - \overset{\delta-}{Cl}$$

Disproportionation reactions

A *disproportionation reaction* is a reaction that occurs with the simultaneous oxidation and reduction of an element. Chlorine undergoes a disproportionation reaction with aqueous sodium hydroxide. In cold dilute sodium hydroxide, sodium chloride and sodium hypochlorite are formed as products:

$$Cl_2(aq) + 2NaOH(aq) \rightarrow NaCl(aq) + NaOCl(aq) + H_2O(l)$$

The half-equations for the redox reactions are:

Reduction		$\frac{1}{2}Cl_2 + e^- \rightarrow Cl^-$
Oxidation states	0	-1
Oxidation		$\frac{1}{2}Cl_2 \rightarrow Cl^+ + e^-$
Oxidation states	0	$+1$

In hot concentrated sodium hydroxide, chlorine is further oxidised to the $+5$ oxidation state, with the formation of sodium chlorate:

$$3Cl_2(aq) + 6NaOH(aq) \rightarrow 5NaCl(aq) + NaClO_3(aq) + 3H_2O(l)$$

Identification of halide ions

The elemental halogens are recognisable by their characteristic colours and different physical states at room temperature. Aqueous solutions of chloride,

Box 11.4 Writing the half-equations for the oxidation and reduction reactions helps in balancing the overall equation.

Reduction	$\frac{1}{2}Cl_2 + e^- \rightarrow Cl^-$
Oxidation states	0 −1
Oxidation	$\frac{1}{2}Cl_2 \rightarrow Cl^{5+} + 5e^-$
Oxidation states	0 +5

Once the half-equations have been written for the reactions the electrons can be cancelled and the overall equation obtained. In this case, the number of electrons transferred must be five for every Cl atom that reacts. So, we multiply the reduction half-equation by 5:

$$\frac{5}{2}Cl_2 + 5e^- \rightarrow 5Cl^-$$

Add this half-equation to the oxidation half-equation:

$$\frac{6}{2}Cl_2 + 5e^- \rightarrow 5Cl^- + Cl^{5+} + 5e^-$$

Cancelling the electrons and adding the sodium and hydroxide ions gives:

$$3Cl_2 + 6NaOH \rightarrow 5NaCl + NaClO_3 \quad \text{UNBALANCED EQUATION}$$

To balance the equation, we need to add three moles of water to the right-hand side:

$$3Cl_2 + 6NaOH \rightarrow 5NaCl + NaClO_3 + 3H_2O \quad \text{BALANCED EQUATION}$$

---

bromide, and iodide ions are generally colourless, but it is possible to identify them by a simple chemical test. The metal halide (say, potassium bromide) is dissolved in water and dilute nitric acid added. To this solution is added a few drops of aqueous silver nitrate. A precipitate of silver halide is formed, and the colour of the precipitate depends upon the nature of the halide ion. The general equation for the formation of the precipitate from a Group 1 metal halide is:

$$MX(aq) + AgNO_3(aq) \rightarrow AgX(s) + MNO_3(aq) \quad (M = \text{Group 1 metal}, X = Cl, Br, I)$$

The overall ionic equation is: $X^-(aq) + Ag^+(aq) \rightarrow AgX(s)$.

Silver chloride (AgCl) is formed as a white precipitate, silver bromide (AgBr) is cream, and silver iodide (AgI) is yellow. However, the colour of the precipitate cannot conclusively determine the nature of the halide ion as the colours are fairly similar to each other. A further test that can be carried out to fully identify the halide ion present is to add concentrated aqueous ammonia to the precipitate after decanting (pouring) off some of the liquid from above it. The different silver halide salts have differing solubilities in concentrated ammonia, as shown in Table 11.11. The concentrated ammonia solution reacts with the silver chloride and silver bromide precipitates to give a soluble silver ion, $[Ag(NH_3)_2]^+$:

$$AgX(s) + 2NH_3(aq) \rightleftharpoons \left[Ag(NH_3)_2\right]^+ (aq) + X^- (aq) \quad (X = Cl, Br)$$

Table 11.11 Results from tests to identify halide ions.

Halide ion	Precipitate formed with $AgNO_3$	Colour of silver halide precipitate	Action of concentrated ammonia solution
Cl^-	AgCl	White	Dissolves
Br^-	AgBr	Cream	Slightly soluble
I^-	AgI	Pale yellow	Insoluble

Worked Example 11.8   Chlorine gas is bubbled into an aqueous solution of sodium bromide and the solution shaken:

　　a. Explain the reaction that takes place, and describe what you would observe.

　　b. A small amount of hexane is added to the test tube and the contents shaken. Describe what you would see happen.

　　c. Use the standard reduction potentials in Table 11.8 to explain your observations, and write appropriate redox half-equations for the reaction.

　　d. Explain what is happening in (b) when hexane is added.

　　e. What would you expect to see if bromine was added to a solution of sodium chloride?

Solution

　　a. Chlorine, Cl_2, oxidises the bromide ion, Br^-, to Br_2 and is reduced itself to Cl^-. The solution turns from colourless to yellow-orange as bromine is formed.

　　b. Two layers are formed. The top layer is a deep orange colour.

　　c. The standard reduction potentials for Cl and Br show that the value for Cl is greater (+1.36 V) than that of Br (+1.06 V). This means that Cl is more readily reduced to Cl^- than Br is reduced to Br^-. Chlorine is therefore a stronger oxidising agent than bromine and oxidises Br^- ions to Br_2 molecules:

$$Cl_2(g) + 2e^- \rightarrow 2Cl^-(aq)$$

$$Br^-(aq) \rightarrow \tfrac{1}{2}Br_2(aq) + e^-$$

Overall equation: $Cl_2(g) + 2Br^-(aq) \rightarrow 2Cl^-(aq) + Br_2(aq)$

　　d. On addition of hexane, the non-polar bromine molecules dissolve preferentially in the non-polar organic solvent, which is lighter than water, so the top layer becomes coloured with brown-orange bromine molecules and the bottom layer is very pale yellow.

　　e. No reaction would be expected between Br_2 and Cl^- ions as Br_2 is a weaker oxidising agent than Cl_2 and cannot oxidise Cl^-.

Worked Example 11.9 The dissociation constants and pK_a values of some inorganic oxyacids of chlorine are given here:

Formula	K_a	pK_a
$HClO_3$	10	−1
$HClO_2$	1×10^{-2}	2
$HClO$	3×10^{-8}	7.5

a. Write an equation for the dissociation of $HClO_3$ in aqueous solution and an expression for K_a for this compound.

b. Give the relationship between pK_a and K_a.

c. Give the oxidation state of Cl in each of the oxyacids in the table.

d. State which oxyacid is the strongest acid, and explain why this is the case.

Solution

a. $HClO_3(aq) + H_2O(l) \rightarrow H_3O^+(aq) + ClO_3^-(aq)$ $K_a = \frac{[H_3O^+][ClO_3^-]}{[HClO_3]}$

b. $pK_a = -\log_{10}K_a$

c. $HClO_3$ has Cl = +5
   $HClO_2$ has Cl = +3
   $HClO$ has Cl = +1

d. $HClO_3$ is the strongest acid because it has the lowest pK_a value. The acids dissociate by loss of a proton (H^+): $HClO_3 \rightleftharpoons H^+ + ClO_3^-$.
The structure of the ClO_3^- ion is similar to the CO_3^{2-} ion in that the negative charge is delocalised over the electronegative oxygen atoms, as shown in the following diagram. This stabilises the ClO_3^- ion because the negative charge can be spread over three atoms in ClO_3^-, which allows the equilibrium position for the acid dissociation to lie over to the right.

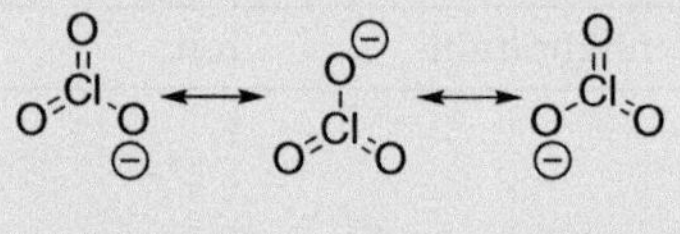

In the ClO_2^- and ClO^- ions, there are two and one O^{2-} ions, respectively. Therefore, the negative charge cannot be delocalised to the same extent.

11.5 The transition elements

The large central section of the periodic table is occupied by the d-block elements. These are the elements in Groups 3–12, inclusive, of the table. In this book, we will only be concerned with the elements in the first row of the d block, i.e. the row beginning with scandium ($Z = 21$) and ending with zinc ($Z = 30$). As we move from scandium to zinc, each element has one more proton in the nucleus and one more electron in its 3d orbitals. As there are five d orbitals in total and each orbital can accommodate a maximum of two electrons, this accounts for the 10 electrons and the 10 elements across the row.

The Groups 1–18 numbering system is more useful when discussing transition elements.

d_{xz} d_{yz} d_{xy} d_{z^2} $d_{x^2-y^2}$

The term **transition element** refers to any element 'whose atom has an incomplete d sub-shell, or which can give rise to cations with an incomplete d sub-shell'.[1] Using this definition, the elements in Group 12 (headed by Zn) are therefore *not* transition elements, as they have a filled set of d orbitals in their atoms and only form cations with a +2 charge in which the ions retain a filled set of d orbitals. However, they are in the d block.

The transition elements and their compounds are used widely in industry and medicine, and several are found in the body as trace elements. They have an enormous number of applications and are extremely valuable materials. The transition elements are all metals with typical metallic properties, but the easy availability of their 3d and 4s electrons means they react readily to form a huge number of complexes, each with particular properties. In this book we have space for only a brief overview of these versatile and exciting elements.

In many textbooks and courses, you will read that scandium is not a transition element as it forms only stable ions with a 3+ charge in which it has no d electrons. However, atoms of the element have one d electron, which means that it can be defined as a transition element under the International Union of Pure and Applied Chemistry (IUPAC) rules. These are the rules chemists use.

11.5.1   Physical properties of transition elements

Electron configurations

As mentioned, the transition elements are those elements in Groups 3–11. In this chapter, we will focus on the first-row elements Sc to Cu.

The ground-state electron configurations of the first 12 elements in Period 4 and those of their ions are given in Table 11.12.

The *ground-state electron configuration* of an element is the electron configuration when the electrons occupy the lowest energy levels possible..

Table 11.12 The electron configurations of Period 4 elements and stable ions. The symbol [Ar] is used to represent the electron configuration of Ar, which is $1s^2 2s^2 2p^6 3s^2 3p^6$.

Element	Electron configuration	Ion	Electron configuration
K	[Ar] $4s^1$	K^+	[Ar]
Ca	[Ar] $4s^2$	Ca^{2+}	[Ar]
Sc	[Ar] $3d^1$ $4s^2$	Sc^{3+}	[Ar]
Ti	[Ar] $3d^2$ $4s^2$	Ti^{2+}	[Ar] $3d^2$
V	[Ar] $3d^3$ $4s^2$	V^{2+}	[Ar] $3d^3$
Cr	[Ar] $3d^5$ $4s^1$	Cr^{2+}	[Ar] $3d^4$
Mn	[Ar] $3d^5$ $4s^2$	Mn^{2+}	[Ar] $3d^5$
Fe	[Ar] $3d^6$ $4s^2$	Fe^{2+}	[Ar] $3d^6$
Co	[Ar] $3d^7$ $4s^2$	Co^{2+}	[Ar] $3d^7$
Ni	[Ar] $3d^8$ $4s^2$	Ni^{2+}	[Ar] $3d^8$
Cu	[Ar] $3d^{10}$ $4s^1$	Cu^{2+}	[Ar] $3d^9$
Zn	[Ar] $3d^{10}$ $4s^2$	Zn^{2+}	[Ar] $3d^{10}$

[1] IUPAC. *Compendium of Chemical Terminology*, 2nd ed. (the "Gold Book"). Compiled by A. D. McNaught and A. Wilkinson. Blackwell Scientific Publications, Oxford (1997). Online version (2019) created by S.J. Chalk. ISBN 0-9678550-9-8. https://doi.org/10.1351/goldbook.

At scandium ($Z = 21$) the 4s shell is full, and the next 10 electrons fill the 3d orbitals. However, there is an anomaly at Cr ($Z = 24$), which has a half-filled set of d orbitals, and Cu ($Z = 29$), which has a filled set of d orbitals and just one 4s electron. There are several possible explanations why Cr and Cu don't obey the Aufbau principle, due largely to the closeness in energy of the 3d and 4s orbitals and the extra energy required to pair electrons in the 4s orbital.

Oxidation states

When a transition element reacts it generally loses electrons and becomes oxidised, forming a positive ion. Because the energies of electrons in the 3d and 4s orbitals are very close, they can lose varying numbers of electrons and so form complexes with a range of oxidation states. The first electrons to be lost are the 4s electrons. Table 11.12 gives the electron configurations of some Period 4 ions in the +2 oxidation state, from which it can be seen that all the atoms from Ti to Zn lose the outer 4s electrons first. The Zn^{2+} ion has a filled set of d orbitals with 10 electrons, and therefore zinc doesn't qualify as a transition element.

The maximum oxidation states and the most common oxidation states of the transition elements are given in Table 11.13. You can see that the early transition elements (from Sc to Mn) can lose all their 4s and 3d electrons to form stable ions in high oxidation states. They tend to form complexes with strongly electronegative elements such as oxygen, e.g. $Na_2Cr_2O_7$ and $KMnO_4$. Elements at the right-hand side of the row tend to form complexes in lower oxidation states, +2 being the most common oxidation state.

The ability of transition elements to exist in a variety of oxidation states allows them to behave as catalysts, where electron transfer between molecules and other species is an important process.

The different electron configurations of ions in different oxidation states and the similar energies of the d orbitals causes complexes of transition elements to exhibit a range of different colours.

Atomic radius

On going from Sc to Zn the atomic radius of the elements decreases only slightly, as seen in Figure 11.17, due to a slight increase in effective nuclear charge across the row. There is a slight increase in atomic radius at Cu and Zn due to interelectron repulsions in the d orbitals.

Table 11.13 Maximum stable oxidation states and most common oxidation states of first-row transition elements.

Element	Maximum stable oxidation state	Electron configuration of ion	Common oxidation states
Sc	Sc^{3+}	[Ar]	+3
Ti	Ti^{4+}	[Ar]	+4, +3
V	V^{5+}	[Ar]	+5, +4, +3, +2
Cr	Cr^{6+}	[Ar]	+6, +3
Mn	Mn^{7+}	[Ar]	+7, +6, +4, +2
Fe	Fe^{3+}	[Ar] $3d^5$	+3, +2
Co	Co^{3+}	[Ar] $3d^6$	+3, +2
Ni	Ni^{2+}	[Ar] $3d^8$	+2
Cu	Cu^{2+}	[Ar] $3d^9$	+2, +1

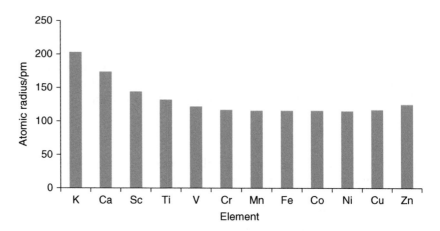

Figure 11.17 Atomic radii of the elements in Period 4.

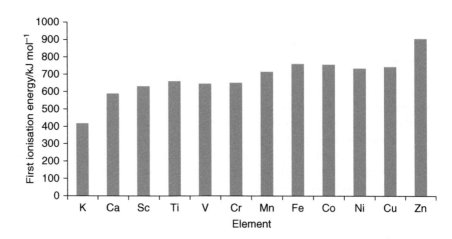

Figure 11.18 First ionisation energy of the elements in Period 4.

Ionisation energies

As the effective nuclear charge increases slightly across the row, the first ionisation energy of the transition elements increases slightly. However, there is very little variation in values until the end of the row, where there is a steep increase at zinc, as can be seen in Figure 11.18. Successive ionisation energies of the transition elements follow the same pattern with little variation on going from left to right. The small difference in ionisation energies accounts for the variable oxidation states shown by the transition elements in their compounds.

Metallic properties

The transition elements are all typical metals with high melting and boiling points. They find uses in many applications requiring very hard, stable, temperature-resistant materials.

As metals the elements have high densities and are very hard and rigid, so they are useful as construction materials. The elements have typical metallic properties of high thermal and electrical conductivity. The six elements rhodium and iridium, palladium and platinum, and silver and gold are known as the **noble metals** as they are highly resistant to corrosion and attack by other chemicals.

Summary

Characteristic properties of transition elements include:

- Ability to exist in variable oxidation states
- Formation of complexes and complex ions with distinctive colours
- Catalytic behaviour
- Typical metallic properties
- Magnetic properties

Worked Example 11.10

a. Manganese can form complexes in the following oxidation states: +7, +6, +4, and +2. Give the electron configurations of manganese in the metal and in its ions that have the oxidation states listed.

b. Figure 11.18 gives a plot of first ionisation energies for the transition elements Sc to Cu. What is the general trend on going across the row? Explain this trend in terms of the electron configurations of the elements.

c. Table 11.13 gives the maximum oxidation states of the first-row transition elements. What do you notice about the maximum common oxidation states of elements after Mn? Can you suggest reasons for this trend?

Solution

a. Mn : $[Ar]3d^5 4s^2$
Mn^{2+}: $[Ar]3d^5$
Mn^{4+}: $[Ar]3d^3$
Mn^{6+}: $[Ar]3d^1$
Mn^{7+}: $[Ar]$

b. The first ionisation energies of the elements increase slightly on going across the row. The number of protons in the nucleus increases on going across the row, but the d electrons being added to the valence shell are not efficient at shielding the electrons from the nuclear charge, so the effective nuclear charge, Z_{eff}, increases. This results in an increase in first ionisation energy of the elements.

c. For the elements after Mn, the maximum common oxidation states are low relative to those of the earlier elements. The most common oxidation states are +2 and +3 (although higher oxidation states are possible for some elements they are difficult to attain). This is because the ionisation energies of the elements increase across the row. The 4s electrons are reasonably easy to remove as they are more distant from the nucleus; but the energy required to remove electrons from the 3d shell is higher, so lower oxidation states are more common.

11.5.2   Complexes of transition elements

Bonding in complexes of transition elements

Because of their ability to exist in variable oxidation states and their relatively large metallic radii, transition elements are able to form *complexes*, where the metal ion is surrounded by a number of other atoms or molecules. These complexes have the general name **coordination complexes**. The bonding between the metal ion and the surrounding groups is dative covalent bonding. The groups surrounding the central metal ion are known as **ligands**, and these can be ions, for example Cl^-, or neutral molecules, for example H_2O. The ligand donates a pair of electrons to the central metal ion and a coordinate covalent bond is formed, where the electrons are shared between the bonded atoms (the metal atom and the **donor atoms** of the ligand). This is often referred to as **coordinate bonding** in transition metal complexes. The complex formed may be a neutral compound, e.g. $TiCl_4$ (Figure 11.19a), or a **complex ion**, e.g. $[Fe(H_2O)_6]^{2+}$ (Figure 11.19c). The direction of coordinate bonding from one lone pair on each ligand to the central atom is depicted by the arrows (Figure 11.19b,d). In complexes such as this, the water ligand and Cl^- ligand act as Lewis bases donating electrons, and the metal acts as a Lewis acid.

For a reminder of Lewis acids and bases, refer back to Chapter 7.

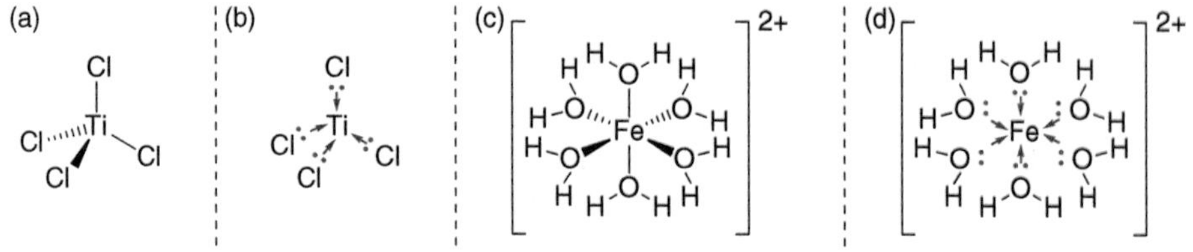

Figure 11.19 Coordination complexes: (a) $TiCl_4$; (b) Cl^- donating a lone pair to Ti (centre); (c) $[Fe(OH_2)_6]^{2+}$; (d) oxygen in water donating lone pair to Fe (centre). Direction of arrows represents donation of lone pairs.

Box 11.5 A **complex ion** consists of a central metal ion surrounded by ligands that may be charged or uncharged. The formula for a complex metal ion is written in square brackets, and the charge on the ion given as a superscript outside the right-hand bracket. For example, $[Co(NH_3)_6]^{2+}$ (a) contains a Co^{2+} ion surrounded by six ammonia molecules. In this case, the overall charge on the complex ion is the same as the charge on the cobalt ion and is +2. In $[CoCl_4]^{2-}$, (b) the Co^{2+} ion is surrounded by four chloride ions, and the charge on the complex ion is –2.

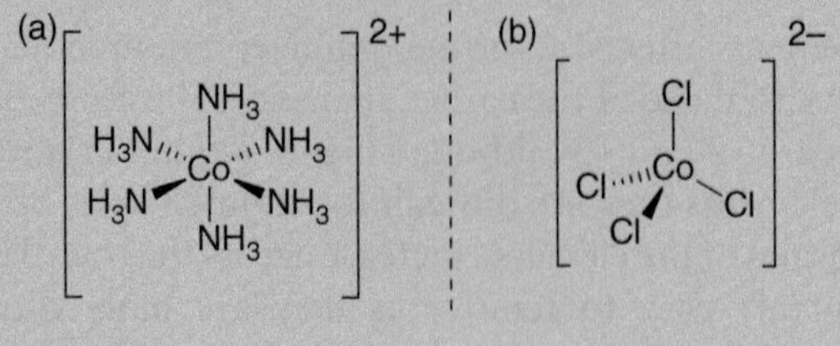

Shapes of transition metal complexes

The number of atoms donating electron pairs to the central metal ion in a transition metal complex is called the **coordination number**. Many transition metal complexes contain a central metal ion surrounded by six ligands and have a coordination number of 6. The arrangement of the six ligands around the metal is said to be *octahedral* as the three-dimensional shape that is formed has eight sides (Figure 11.20).

Although the formulae of some metal–ligand complexes such as iron(II) chloride are written simply as $FeCl_2$, in the solid state the metal ion is actually surrounded by six chloride ions, again giving an octahedral arrangement around the metal.

Another common coordination number for transition metal complexes is 4. In this case the metal is surrounded by four ligands. There are two possible arrangements for four coordinate complexes, which can be **tetrahedral** or **square planar**, as shown in Figures 11.21 and 11.22.

(a)   (b)   (c)

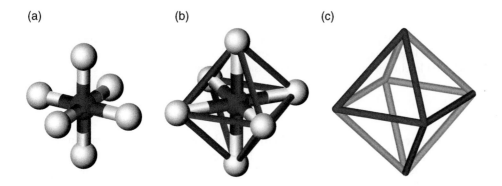

Figure 11.20 (a) A complex containing a central metal atom with six ligands; (b) the complex drawn inside an octahedral shape with eight triangular sides; (c) the 3-D shape of the octahedral complex. Source: (c), based on https://commons.wikimedia.org/wiki/File:Octahedron-3-3D-balls.png.

(a)   (b)   (c)

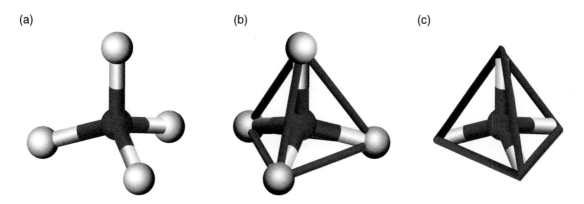

Figure 11.21 (a) A complex containing a central metal atom with four ligands in a tetrahedral arrangement; (b) the complex drawn inside a tetrahedral shape with four triangular sides; (c) the 3-D shape of the tetrahedral complex. Source: (c), based on https://commons.wikimedia.org/wiki/File:Tetrahedron-2-3D-balls.png.

Figure 11.22 A complex containing a central metal atom with four ligands in a square planar arrangement. Source: Based on https://en.wikipedia.org/wiki/Square_planar_ molecular_geometry.

(a)   (b)

$$[\text{H}_3\text{N–Ag–NH}_3]^+$$

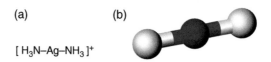

Figure 11.23 (a) $[\text{Ag(NH}_3)_2]^+$ a linear coordination complex with coordination number 2; (b) the linear shape of the molecule. Source: (b), based on https://revisionscience.com/ a2-level-level-revision/chemistry-level-revision/bonding-and-structure/shapes-molecules.

Another, less common, coordination number is 2, where the metal ion is connected to only two ligands and gives a **linear** complex (Figure 11.23). An example of this type of complex is $[\text{Ag(NH}_3)_2]^+$, which is the soluble silver complex obtained when AgCl or AgBr dissolves in concentrated aqueous ammonia. The halide ligands are replaced by ammonia molecules to give this linear complex ion.

Coordination complexes can have a variety of coordination numbers from 1 to 12, but the most common are 2, 4, and 6.

Oxidation states and charge on the central metal ion

The central ion has an associated charge dependent upon its oxidation state. This is calculated from the overall charge on the complex and the charges on the ligands. If the ligands are neutral, such as H_2O, then the charge on the complex is the same as the charge on the central ion. For example, in $[\text{Fe(H}_2\text{O})_6]^{2+}$, the central iron ion has a charge of +2 and therefore an oxidation state of +2.

Tetrachlorotitanium, $TiCl_4$, is uncharged, but each Cl ligand has a charge of –1. As there are four Cl^- ligands, the charge on the titanium and its oxidation state must be +4.

Naming transition metal complexes

There is a set of rules for correctly naming coordination complexes of transition metals. The ligands are always named first and their number given using the system *mono*, *di*, *tri*, *tetra*, etc. The name of the metal follows, with the oxidation state of the metal given in parentheses after the name. See the following examples:

VCl$_4$ is named tetrachlorovanadium(IV).

[Ag(NH$_3$)$_2$]$^+$ is named diamminesilver(I).

CrCl$_6$ is named hexachlorochromium(VI).

[Cu(H$_2$O)$_6$]$^{2+}$ is named hexaaquacopper(II).

If the complex ion is negatively charged, the Latin name of the metal is used. For example:

$$[Fe(CN)_6]^{4-} \quad \text{hexacyanoferrate(II)}$$

The colour of transition metal complexes

Aqueous solutions of transition metal complexes are often coloured. Figure 11.24 illustrates the colours of some first-row transition metal complexes in aqueous solution. How many colours do you recognise? The origin of colour in transition metal complexes will be explained in more detail in Section 11.5.4.

11.5.3 Redox reactions

Many transition metal complexes undergo redox reactions readily because of the ability of the metal ions to exist in different oxidation states and therefore be easily oxidised or reduced. This property makes transition metal ions good oxidising and/or reducing agents.

When a transition metal ion is *oxidised* it *loses* electrons, and its oxidation state *increases*.

When a transition metal ion is *reduced* it *gains* electrons, and its oxidation state *decreases*.

Remember: Oxidation Is Loss. Reduction Is Gain. **OIL RIG.**

Figure 11.24 Colours of some first-row transition metal complexes in aqueous solution
Photo credit: Elizabeth Page.

For example, the Fe^{2+} ion is readily oxidised by $KMnO_4$ in acidic solution to Fe^{3+}:

$$Fe^{2+}(aq) \rightarrow Fe^{3+}(aq) + e^- \text{ Oxidation}$$

The MnO_4^- ion is reduced to Mn^{2+} by gaining five electrons. The oxygen from MnO_4^- combines with the hydrogen ions to produce water molecules. The MnO_4^- ion acts as an oxidising agent. Manganese has an oxidation state of $+7$ in MnO_4^- and is reduced to Mn^{2+} with a gain of five electrons:

$$MnO_4^-(aq) + 8H^+(aq) + 5e^- \rightarrow Mn^{2+}(aq) + 4H_2O(l) \text{ Reduction}$$

By multiplying the oxidation reaction by 5 and adding the two equations we can obtain the overall equation:

$$MnO_4^-(aq) + 8H^+(aq) + 5Fe^{2+}(aq) \rightarrow Mn^{2+}(aq) + 5Fe^{3+}(aq) + 4H_2O(l)$$

Refer back to Chapter 5 for a reminder on balancing redox equations.

Values of standard reduction potentials allow us to predict whether such redox reactions should take place, as seen in Chapter 9.

Box 11.6 This reaction forms the basis of a useful titration to determine the amount of iron in iron tablets, for example. The titration is self-indicating as the manganate(VII) ion is purple and the Fe^{2+} aqueous ions colourless at low concentrations. When adding purple manganate solution from a burette to an unknown concentration of acidified iron(II) ions in a conical flask the manganate ions are decolourised to Mn^{2+} on reaction with the Fe^{2+}. When all the Fe^{2+} ions have reacted the solution in the flask will turn pink, indicating the presence of unreacted manganate ions. This is the end point of the reaction and is easily observed by the distinctive colour change.

Worked Example 11.11 An acidified solution of potassium dichromate, $K_2Cr_2O_7$, reacts with iron(II) ions in solution in a redox reaction. The iron(II) ions are oxidised to iron(III) and the dichromate ions reduced to chromium(III):

  a. Write half-equations to represent the oxidation and reduction reactions.

  b. Balance the electrons and add the half-equations to obtain the overall equation for the redox reaction.

  c. State which species is the oxidising agent and which is the reducing agent.

  d. Given that the standard electrode potential for the reduction of dichromate to chromium(III) in acidic solution is $+1.33$ V and the equivalent standard electrode potential for the reduction of iron(III) to iron(II) is $+0.77$ V, determine whether the reaction is likely to proceed in the forward direction.

Solution

a. First write the two half-equations for the reactions. In the case of the dichromate reduction we must determine the oxidation state of the chromium in $Cr_2O_7^{2-}$ in order to calculate the electron change.

Let the charge on the chromium ions = n. The charge on each O atom is –2:

For $\qquad Cr_2O_7^{2-} : 2 \times n + (7 \times -2) = -2$

$$2n = -2 + 14 = +12$$

$$n = +6$$

The $Cr_2O_7^{2-}$ ion contains Cr(VI), which is reduced to chromium(III) by gaining three electrons. As there are two chromium ions in each $Cr_2O_7^{2-}$ ion, a total of six electrons must be required. The reduction reaction takes place in acidic solution, so we must also add hydrogen (H^+) ions to the left-hand side of the equation and water molecules to the right-hand side:

$$Cr_2O_7^{2-}(aq) + H^+(aq) + 6e^- \rightarrow 2Cr^{3+}(aq) + H_2O(l)$$

UNBALANCED HALF-EQUATION

The half-equation must be balanced by first adding seven water molecules to the right-hand side to account for the seven oxygen atoms in $Cr_2O_7^{2-}$. Seven water molecules require 14 hydrogen ions on the left-hand side of the equation:

$$Cr_2O_7^{2-}(aq) + 14H^+(aq) + 6e^- \rightarrow 2Cr^{3+}(aq) + 7H_2O(l) \quad \text{Reduction}$$

The half-equation for the oxidation of the Fe^{2+} ions is simpler as it involves a loss of one electron from each Fe^{2+} ion:

$$Fe^{2+}(aq) \rightarrow Fe^{3+}(aq) + e^- \text{ Oxidation}$$

b. To balance the half-equations we must equalise the number of electrons lost and gained. Each dichromate ion requires six electrons to reduce the chromium ions, so we must multiply the half-equation for the oxidation of iron(II) by 6 to obtain a total of six electrons:

$$6Fe^{2+}(aq) \rightarrow 6Fe^{3+}(aq) + 6e^-$$

This half-equation can now be added to the half-equation for the reduction reaction to obtain the overall equation:

$$6Fe^{2+}(aq) + Cr_2O_7^{2-}(aq) + 14H^+(aq) + 6e^-$$
$$\rightarrow 6Fe^{3+}(aq) + 6e^- + 2Cr^{3+}(aq) + 7H_2O(l)$$

After cancelling the electrons on each side we obtain

$$6Fe^{2+}(aq) + Cr_2O_7^{2-}(aq) + 14H^+(aq) \rightarrow 6Fe^{3+}(aq) + 2Cr^{3+}(aq) + 7H_2O(l)$$

c. The oxidising agent is the species being reduced, so in this case the dichromate ion is the oxidising agent.

The reducing agent is the species being oxidised, so in this case the iron(II) ion is the reducing agent.

You should always check that an equation is balanced by ensuring that the number and type of each atom on each side of the equation is the same and that the overall charge on each side of the equation is the same. In this case the overall charge on each side is +24.

It is important that you specify that the iron(II), or Fe^{2+}, ions are the reducing agent. Stating the iron is the reducing agent is not specific enough, as this could imply either Fe^{2+} or Fe^{3+}.

d. We have been given the standard reduction potentials for each half-equation as

$$Cr_2O_7^{2-}(aq) + 14H^+(aq) + 6e^- \rightarrow 2Cr^{3+}(aq) + 7H_2O(l) \; E^\ominus = +1.33 \text{ V}$$

$$Fe^{3+}(aq) + e^- \rightarrow Fe^{2+}(aq) \quad E^\ominus = +0.77 \text{ V}$$

As the overall redox reaction involves the oxidation of iron(II) ions, the half-equation and related $E^\ominus$ value must be reversed:

$$Fe^{2+}(aq) \rightarrow Fe^{3+}(aq) + e^- \; E^\ominus = -0.77 \text{ V}$$

To equalise the numbers of electrons this equation is multiplied by 6, but the $E^\ominus$ value for the cell is constant and does not depend upon the amount in moles reacting:

$$6Fe^{2+}(aq) \rightarrow 6Fe^{3+}(aq) + 6e^- \; E^\ominus = -0.77 \text{ V}$$

To obtain the $E^\ominus$ value for the redox reaction the individual $E^\ominus$ values are added:

$$+1.33\text{V} + (-0.77\text{V}) = +0.56\text{V}$$

The overall electrode potential is positive, which implies that the reaction is favourable in the direction given:

$$6Fe^{2+}(aq) + Cr_2O_7^{2-}(aq) + 14H^+(aq) \rightarrow 6Fe^{3+}(aq) + 2Cr^{3+}(aq) + 7H_2O(l)$$

This means that potassium dichromate in acidic solution can be used to oxidise iron(II) ions to iron(III).

> Note that standard electrode potentials are always quoted as reduction reactions. In the reaction with dichromate ions, the iron(II) ions are being oxidised, so the sign of the $E^\ominus$ is reversed.

11.5.4   Origin of colour in transition metal complexes

Figure 11.24 shows the variety of colours of some first-row transition metal complexes. Colour arises when a complex absorbs some of the wavelengths of visible light and transmits the remainder. The colour we observe is due to the wavelengths that are transmitted. Figure 11.25 depicts a tube containing a blue solution. The light incident on the solution contains all wavelengths in the visible region. On passing through the solution wavelengths from the red (low-energy) end of the spectrum are absorbed. The transmitted light is missing these red wavelengths and therefore appears blue.

A property of transition metal complexes is that they have electrons in d orbitals. In the ground state, the energies of the d orbitals are all the same. They are said to be **degenerate**. When a transition metal ion is surrounded by a set of six ligands in an octahedral arrangement, the electron field from the ligands causes the energies of the d orbitals to be split into two sets. This is because the d orbitals point in different directions and don't all experience the electric field of the ligands equally. This is shown in Figure 11.26 for $[Ti(H_2O)_6]^{3+}$. Figure 11.26a shows the five d orbitals in the free Ti^{3+} ion that are all equal in energy with a single electron occupying one of the orbitals. When the Ti^{3+} ion is surrounded by six water molecules in $[Ti(H_2O)_6]^{3+}$ the d orbitals are no longer degenerate and split into two groups. We now have three d orbitals that are lower in energy and two that

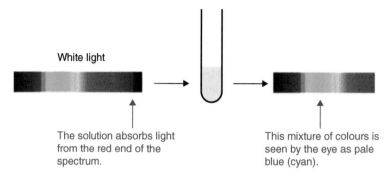

Figure 11.25 Absorption of red wavelengths from visible light by a solution. Source: Jim Clark, 'The colours of complex metal ions', https://www.chemguide.co.uk/inorganic/complexions/colour.html.

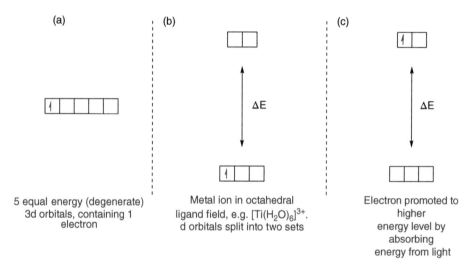

Figure 11.26 (a) Five d orbitals have equal energy in a free ion. (b) When surrounded by six water molecules, the d orbitals are no longer degenerate. (c) The single electron absorbs energy from visible light and is promoted to a higher-energy d orbital. ΔE is the energy difference between the two sets of d orbitals.

are higher. The electron in the lower-energy orbital can gain enough energy from incident visible light to be *promoted* to a higher-energy orbital. The light transmitted through the solution is no longer 'white' but is missing the absorbed wavelengths. Thus the complex will appear to be coloured, depending upon the energies of the wavelengths absorbed. In the case of $[Ti(H_2O)_6]^{3+}$, green-yellow light of around 500 nm is absorbed, and the resulting solution is a pale violet colour.

Promotion of an electron from a lower to a higher-energy orbital can only occur when there are fewer than 10 electrons occupying the d orbitals. Thus, complexes of Zn^{2+} that have a filled set of five d orbitals are colourless.

The colour of the complex depends upon a number of factors including:

- The number and nature of the ligands

- The spatial arrangement of the ligands

- The oxidation state of the metal and hence number of d electrons.

11.5.5   Isomerism in coordination complexes

Isomers are complexes with the same chemical formula but a different arrangement of atoms. You will meet isomerism in organic chemistry in Chapter 13, where it is a very important feature. Here we will look at some examples of **stereoisomerism**, where the atoms are bonded together in the same way but occupy different positions in space. Stereoisomers can be **geometric isomers** or **optical isomers**.

Geometric isomers

An important type of geometric isomerism is *cis-/ trans-* isomerism, which can occur in some square-planar and octahedral complexes. The best known example of *cis-/trans-* isomerism is seen in the anti-cancer agent *cis*-platin (*cis*-$[PtCl_2(NH_3)_2]$). Two forms of the complex are known. The one in which the Cl ligands and NH_3 ligands are on the same sides of the square adjacent to each other is the *cis*- form. The second in which the Cl ligands and NH_3 ligands are opposite each other is the *trans*- form. Only the *cis*- form has anti-cancer activity.

cis-$[PtCl_2(NH_3)]$ *trans*-$[PtCl_2(NH_3)_2]$

Cis- and *trans*- isomers are also possible for octahedral complexes with general formula MA_2B_4 (where A and B are different ligands with only one donor atom), as shown for $[CoCl_2(NH_3)_4]^+$.

cis-$[CoCl_2(NH_3)_4]^+$ *trans*-$[CoCl_2(NH_3)_4]^+$

Optical isomers

Optical isomers are compounds having the same molecular formula with atoms arranged differently in space such that they rotate the plane of polarised light in opposite directions. A description of optical isomerism in organic chemistry is given in Chapter 12. Optical isomers are non-superimposable mirror images and the two forms of the compound are called **enantiomers**. An example of a compound that shows optical isomerism is $[CoCl_2(en)_2]^+$. The ligand ethylenediamine, $NH_2CH_2CH_2NH_2$, abbreviated to en, is a **bidentate** ligand, which means it can coordinate to a metal ion through two different atoms. Here it is coordinating through the two N atoms on the molecule. Despite having very similar structures, the two isomers are geometrically different and cannot be superimposed because of the three-dimensional geometry.

> Ligands that bond through one donor atom are known as *monodentate* ligands, e.g. H_2O and NH_3. Ligands that bond through two donor atoms are *bidentate* ligands, e.g. $NH_2CH_2CH_2NH_2$. Tridentate (three donor atoms), quadridentate (four donor atoms), and hexadentate (six donor atoms) ligands also exist.

Mirror plane

11.5.6 Ligand substitution in transition metal complexes

The ligands surrounding a central transition metal ion can be replaced by other ligands in ligand substitution reactions.

The hydrated Ni^{2+} ion is pale green in colour and has the formula $[Ni(H_2O)_6]^{2+}$. When a solution of aqueous ammonia is added to the green solution the water ligands are displaced by ammonia molecules according to this equation:

$$[Ni(H_2O)_6]^{2+}(aq) + 6NH_3(aq) \rightleftharpoons [Ni(NH_3)_6]^{2+}(aq) + 6H_2O(l)$$

Pale green $\qquad\qquad\qquad$ Violet

The new solution is violet in colour. Although the equation has been written as a single step, ligand displacement reactions actually occur in a step-wise fashion, where one ligand is replaced at a time. Each of the steps is an equilibrium process and can be reversed by changing the concentration of the ligands. The first step in the process is:

$$[Ni(H_2O)_6]^{2+}(aq) + NH_3(aq) \rightleftharpoons [Ni(H_2O)_5(NH_3)]^{2+}(aq) + H_2O(l)$$

The remaining H_2O ligands are replaced until the final step occurs:

$$[Ni(H_2O)(NH_3)_5]^{2+}(aq) + NH_3(aq) \rightleftharpoons [Ni(NH_3)_6]^{2+}(aq) + H_2O(l)$$

The overall substitution reaction has an equilibrium constant called the *stability constant*, K_{stab}, which is a measure of how stable the complex is. The stability constant for the substitution of water ligands by ammonia in Ni^{2+} is expressed as:

$$K_{stab} = \frac{\left[[Ni(NH_3)_6]^{2+}\right]}{\left[[Ni(H_2O)_6]^{2+}\right][NH_3]^6}$$

The stability constant for the formation of $[Ni(NH_3)_6]^{2+}$ is the equilibrium constant for the reaction in which it is formed. This is represented by the first equation in 11.5.6. Note that the concentration of water is not included as it is a pure liquid and so constant.

The value of K_{stab} is around 4×10^8, showing the reaction lies well over to the right. As values for stability constants are typically very high, they are often expressed as $\log_{10}$ values in order to make the numbers easier to manage. So in this case $\log_{10}K_{stab} = \log_{10}(4 \times 10^8) = 8.6$.

Some substitution reactions occur with a change in geometry of the complex ion. For example, if concentrated HCl is added to the Ni^{2+} solution the chloride ions displace the water molecules to produce a four-coordinate complex, $[NiCl_4]^{2-}$, which is tetrahedral in shape.

Worked Example 11.12

a. Complete the following electron configurations:

Cr: [Ar]....

Cr^{3+}: [Ar]....

b. Consider the following reaction scheme:

$$[Cr(H_2O)_6]^{3+} \xrightarrow[\text{Step 1}]{NH_3(aq)} [Cr(H_2O)_3(OH)_3] + 3H_2O \xrightarrow[\text{Step 2}]{\text{Excess conc.} NH_3(aq)} \text{Solution X}$$

  i. Name the two types of bond present in the $[Cr(H_2O)_6]^{3+}$ ion.

  ii. Describe what is happening in Step 1, and write a balanced chemical equation for the step.

  iii. Name the type of reaction occurring in Step 2, and write an equation for the reaction. What is the name of the chromium-containing complex in solution X?

Solution

a. Cr: [Ar] $3d^5 4s^1$

Cr^{3+}: [Ar] $3d^3$

b. i. Covalent bonding exists in the H_2O molecule between the O and H atoms.
    Coordinate bonding exists between the O atoms in the water molecules and the Cr^{3+} ion.

   ii. In Step 1 the water molecules are partially replaced by OH^- ions of the aqueous ammonia. The hydroxide ions from the aqueous ammonia solution (NH_4OH) deprotonate the ligand water molecules according to:

$$HOH + OH^- \rightleftharpoons OH^- + HOH$$

The red hydrogen ion is pulled away from the ligand water molecule by the OH^- ion (blue) from the ammonia to form a new water molecule.
Three of the water molecules are deprotonated by OH^- ions, and the complex becomes neutral and forms an insoluble precipitate:

$$[Cr(H_2O)_6]^{3+} + 3OH^- \rightleftharpoons [Cr(H_2O)_3(OH)_3] + 3HOH$$

iii. When concentrated ammonia is added the high concentration of NH_3 molecules shifts the equilibrium to the right and causes the water and hydroxide ligands to be replaced by NH_3 molecules, forming the hexaamminechromium(III) ion:

$$\left[Cr(H_2O)_3(OH)_3\right] + 6NH_3 \rightleftharpoons \left[Cr(NH_3)_6\right]^{3+} + 3OH^- + 3H_2O$$

Worked Example 11.13   The complex $[CoCl_2(en)_2]^+$ can exist as *cis-* and *trans-* isomers as well as optical isomers. Sketch the *cis-* and *trans-* forms, and state which form can show optical activity.

Solution

cis-[CoCl$_2$(en)$_2$]$^+$ *trans*-[CoCl$_2$(en)$_2$]$^+$

Only the *cis-* form can show optical activity, as no other arrangements are possible for the *trans-* form.

Quick-check summary

- The Group 1 and Group 2 metals constitute the s block of the periodic table.

- The Group 1 elements are soft metals with relatively low melting and boiling points.

- The Group 2 elements are typical metals with high melting and boiling points. All elements in the s block are good conductors of electricity.

- Going down both Group 1 and Group 2, the atomic radii of the elements increase and the first ionisation energies and electronegativities decrease. This is due to the increasing numbers of filled shells of electrons for each element as the groups are descended.

- The melting and boiling points of the Group 1 and 2 elements decrease on going down a group and from left to right across a row, as metal-metal bonding strength decreases.

- Group 1 elements react rapidly with air and moisture, and reactivity increases on going down the group due to the lower ionisation energies and reduced metal–metal bonding.

- Group 2 elements are less reactive than Group 1 elements, and reactivity increases on going down the group.

- Reaction of s block metals with air gives metal oxides that dissolve in water to give hydroxides that are basic.

- When reaction of the metals with water occurs, hydrogen gas is produced along with the metal hydroxide. The solubility of the Group 2 hydroxides increases on going down the group.

- Salts of Group 1 metals are all soluble in water.

- Group 2 metal salts are less soluble than Group 1, and solubility generally decreases on going down the group, apart from the hydroxides.

- Group 1 and Group 2 metal carbonates and nitrates decompose on heating. The thermal stability of the salts increases on going down each group.

- The halogen elements form Group 7 (17) of the periodic table and are positioned in the p block.

- The elements in Group 7 (17) are all covalent, diatomic molecules with single bonds between the atoms.

- The melting points and boiling points of the halogens increase on going down the group as instantaneous dipole-induced dipole forces increase.

- The covalent radii of the elements increase on going down the group, and electronegativity decreases.

- All the halogens act as oxidising agents, but oxidising ability decreases on going down the group, as does reactivity.

- The halogen elements can exist in a variety of different oxidation states in oxyacids and interhalogen compounds. Fluorine can only exist in oxidation state −1, apart from in elemental fluorine, where it has oxidation state of 0.

- The halogens undergo displacement reactions, where a halogen element higher up the group will displace a halogen lower in the group from a solution of its halide ions.

- The transition elements are found in the d block of the periodic table.

- The first-row transition elements are the elements from Sc to Cu. Either the elements or some of their cations have incomplete d subshells.

- The transition elements are typical metals with very high melting and boiling points and are good conductors of electricity.

- Some of the transition metals have magnetic properties and can act as catalysts.

- The atomic radii of the transition elements decrease slightly across the row as the effective nuclear charge increases.

- The first ionisation energies increase slightly across the row.

- Transition elements form complexes with ligands in which the ligand acts as an electron donor and shares a pair of electrons with the central metal ion, resulting in coordinate bonding. These complexes can be neutral or charged, and the coordination number of the central metal ion can vary, giving complexes with a range of different geometries.

- The metal can have a variety of oxidation states in such complexes. The earlier transition metals tend to exhibit high oxidation states, whereas the later ones tend to form compounds with the metals in the +2 or +3 oxidation state.

- Transition metals form highly coloured complexes due to the splitting of energies of the d orbitals giving energy separations in the visible region.

- Redox reactions are important for transition elements because of the easy accessibility of a range of oxidation states.

- Transition metal complexes can exhibit stereoisomerism.

- Transition metal complexes undergo ligand displacement reactions in a step-wise fashion.

End-of-chapter questions

1 a. Define the term *ionisation energy*.

   b. Write an equation, including state symbols, that represents the process related to the first ionisation energy of lithium.

   c. The first ionisation energies of the Group 1 elements are listed here.

Element	Ionisation energy/kJ mol^{-1}
Li	+519
Na	+494
K	+418
Rb	+402
Cs	+376

   Explain the trend in ionisation energies with increasing atomic number down the group.

   d. Lithium reacts with iodine to form the compound lithium iodide, LiI. Predict and explain the type of bonding in LiI.

2 a. Write the full electron configuration of calcium in the form $1s^2$....

   b. Write the two equations that represent the step-wise formation of a Ca^{2+} ion from a gaseous calcium, Ca, atom. Explain the energy terms involved.

   c. Write an equation to show the reaction of calcium metal with water.

   d. Predict with reasoning whether potassium, K, the element next to calcium, Ca, in the periodic table, will react more or less vigorously with water than calcium, Ca.

   e. Predict with reasoning whether strontium, Sr, the element below calcium, Ca, in the periodic table will react more or less readily with water than calcium, Ca.

3   The elements fluorine, chlorine, bromine, and iodine are in Group 7 (17) of the periodic table.

   a. Give the electron configuration of chlorine.

   b. Explain why the halogens tend to form −1 ions.

   c. State and explain the trend in electronegativities of the halogens on going down the group.

   d. Give a chemical test that would help to distinguish between sodium chloride and sodium iodide.

   e. Chlorine gas is passed into a colourless aqueous solution of sodium bromide in a test tube. State what you would expect to observe, and give an equation for the reaction taking place.

   f. If a small quantity of cyclohexane, a non-polar solvent, is added to the test tube in (e) and the mixture shaken, what would you observe? Explain your observations. (Density of water = $1 \, g \, cm^{-3}$; density of cyclohexane = 0.8 $g \, cm^{-3}$.)

   g. A few drops of brown bromine liquid are carefully added to a solution of sodium chloride in a test tube, followed by a small quantity of cyclohexane. The contents of the tube are shaken. The appearance of the contents of the tube is very similar to that in (f). Predict whether a chemical reaction has taken place, and explain your prediction.

4   The table shows the successive ionisation data for the atoms of three different elements A, B, and C. The elements are calcium, scandium and vanadium but not in that order.

	$IE_1/$ $kJ \, mol^{-1}$	$IE_2/$ $kJ \, mol^{-1}$	$IE_3/$ $kJ \, mol^{-1}$	$IE_4/$ $kJ \, mol^{-1}$	$IE_5/$ $kJ \, mol^{-1}$	$IE_6/$ $kJ \, mol^{-1}$
A	648	1370	2870	4600	6280	12,400
B	590	1150	4940	6480	8120	10,700
C	632	1240	2390	7110	8870	10,700

   a. Identify the elements calcium and vanadium, and explain your answer.

   b. Explain the ionisation energy data for the remaining element, scandium. What oxidation state do you expect for its complexes?

   c. Explain why solutions of calcium ions are colourless, whereas solutions of vanadium ions are coloured. Would you expect complexes of scandium to be coloured?

   d. When zinc metal is added to a yellow solution of ammonium vanadate, NH_4VO_3, in acidic solution, a series of colour changes takes place. The first species to be formed is the blue VO^{2+} ion. The solution next turns green, and eventually the purple $[V(H_2O)_6]^{2+}$ ion is formed. Determine the oxidation state of vanadium in these three complexes, and write half-equations to represent their formation, starting from NH_4VO_3.

5 The first electron affinities of the halogens are given here.

F	-348 kJ mol^{-1}
Cl	-364 kJ mol^{-1}
Br	-342 kJ mol^{-1}
I	-314 kJ mol^{-1}

   a. Define the term electron affinity, and write a general equation to represent the first electron affinity of the halogens using X to represent a halogen atom.

   b. Explain the variation in electron affinity values on going down Group 7 (17). Suggest why the value for F does not fit this trend.

6 a. Complete the electronic structures of the following atoms and ions:

Cu	Cu^{2+} [Ar]......
Zn	Zn^{2+} [Ar]......

   b. Use the electronic structures to explain why aqueous solutions of copper(II) salts are coloured, whereas those of zinc(II) are usually colourless.

   c. Copper(II) ions can be reduced by iodine to copper(I) iodide. Write half-equations for the oxidation and reduction equations occurring. Use the half-equations to obtain the overall equation for the reaction.

7 a. State three distinctive properties of transition metals.

   b. Give electron configurations for Fe, Fe^{2+}, and Fe^{3+}.

   c. Give the names of the following iron complexes: $[Fe(H_2O)_6]^{2+}$ and $[Fe(CN)_6]^{4-}$.

   d. Write an equation for the reaction between aqueous $[Fe(H_2O)_6]^{2+}$ and aqueous potassium cyanide, KCN, in which $[Fe(CN)_6]^{4-}$ is formed.

   e. Give an expression for the stability constant for the reaction.

8 a. The general formula for Group 2 metal chlorides is MCl$_2$ (M = Be, Mg, Ca, Sr, Ba). Which of these chlorides would you expect to have the most covalent character? Explain your answer.

   b. The solubility of the Group 2 metal hydroxides increases down the group. Suggest a reason for this behaviour.

   c. The decomposition temperatures of the Group 2 metal carbonates increase down the group. Write a balanced equation to represent the decomposition of calcium carbonate, and suggest why the metal carbonates become more stable on going down the group.

9   Three test tubes were labelled A, B, and C and quantities of sodium halide, NaX, sodium halate, $NaXO_3$, and hydrochloric acid added according to the following table. The solutions were mixed and then $1 \, cm^3$ cyclohexane added to each tube. The observations made are shown in the table.

	Colour of upper layer	Colour of lower layer
$5cm^3$ NaCl, $5cm^3$ $NaClO_3$, and $1 \, cm^3$ 0.1 M HCl	Colourless/very pale green	Colourless
$5cm^3$ NaBr, $5cm^3$ $NaBrO_3$, and $1 \, cm^3$ 0.1 M HCl	Orange	Pale orange/brown
$5cm^3$ NaI, $5cm^3$ $NaIO_3$, and $1 \, cm^3$ 0.1 M HCl	Pink	Pale brown

a. Explain what the colour is due to in each test tube and why the colour observed in the upper layer is more intense in each case.

b. Determine the oxidation states of the halogen atoms in X^-, XO_3^-, and X_2.

c. Write half-equations for the reactions, using X to represent a halogen atom. Note that X_2 molecules are the only halogen-containing product formed.

d. What type of reaction is occurring between the halide and halate ions? Determine the change in oxidation states of the halogen atoms.

e. Combine the half-equations to obtain a general equation for the reactions.

10 a. Iodine monofluoride (IF) is an example of an interhalogen compound. It disproportionates into IF_5 and I_2. Give the oxidation states of I and F in IF, IF_5, and I in I_2.

b. Explain what is meant by disproportionation, and give half-equations for the oxidation and reduction processes occurring here.

c. Write an overall equation for the disproportionation reaction.

d. Given that the enthalpies of formation ($\Delta_f H^\ominus$) of IF, IF_5, and I_2 are –95, –840, and $0 \, kJ \, mol^{-1}$, respectively, calculate $\Delta_r H^\ominus$ for the disproportionation reaction you gave in (c).

11 The standard reduction potentials for MnO_4^-/Mn^{2+} and Fe^{3+}/Fe^{2+} are as follows:

$$MnO_4^- \, (aq) + 8H^+ \, (aq) + 5e^- \rightarrow Mn^{2+} \, (aq) + 4H_2O(l) \; E^\ominus = +1.52 \, V$$

$$Fe^{3+} \, (aq) + e^- \rightarrow Fe^{2+} \, (aq) \quad E^\ominus = +0.77 \, V$$

Use the values to determine whether MnO_4^- will oxidise Fe^{2+} ions in acidic solution. Write a balanced equation for the redox reaction that would take place.

12 a. The enthalpy of hydration of the potassium ion is $-322\,\text{kJ}\,\text{mol}^{-1}$, and the enthalpy of hydration of the sodium ion is $-406\,\text{kJ}\,\text{mol}^{-1}$. By describing the nature of the forces involved, explain why the enthalpy of hydration of sodium is more negative than that of potassium.

   b. The enthalpy of hydration of the bromide ion is $-335\,\text{kJ}\,\text{mol}^{-1}$. The lattice enthalpy of dissociation for potassium bromide is $+670\,\text{kJ}\,\text{mol}^{-1}$. Calculate the enthalpy of solution of potassium bromide.

   c. Potassium bromide is found to be reasonably soluble in water at room temperature. Does this agree with your answer from (b)? If not, suggest why this may be.

   d. When calculated in a similar manner, the enthalpy of solution of sodium bromide is found to be $-8\,\text{kJ}\,\text{mol}^{-1}$. Give reasons for the difference between this value and that for KBr.

12

Core concepts and ideas within organic chemistry

At the end of this chapter, students should be able to:

- Explain the difference between empirical and molecular formulae
- Name simple organic compounds
- Be able to recognise key functional groups
- Understand the different types of isomerism that exist
- Understand and use curly arrows
- Explain the difference between radicals, nucleophiles, and electrophiles

12.1 Types of molecular formulae

There are six types of formulae that you will commonly encounter. These are:

- *Molecular formula* – The number and type of atoms of each element in a molecule

- *Empirical formula* – The simplest whole-number ratio of atoms of each element present in a compound

- *General formula* – The simplest algebraic formula of a member of a homologous series

- *Structural formula* – The minimal detail that shows the arrangement of atoms in a molecule

The terms *molecule* and *compound* are often used interchangeably.

Foundations of Chemistry: An Introductory Course for Science Students, First Edition.
Philippa B. Cranwell and Elizabeth M. Page.
© 2021 John Wiley & Sons Ltd. Published 2021 by John Wiley & Sons Ltd.
Companion website: www.wiley.com/go/Cranwell/Foundations

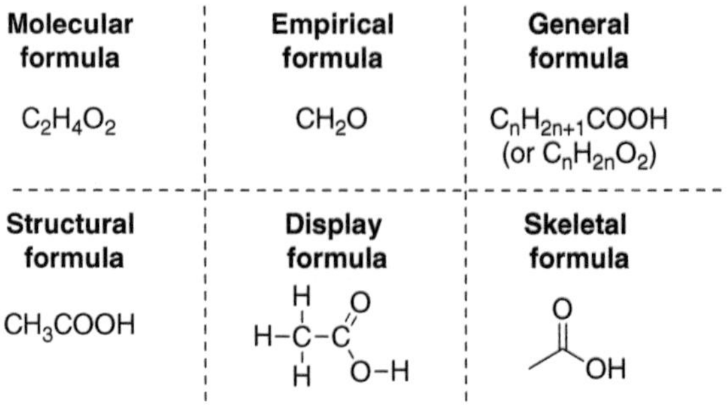

Molecular formula	Empirical formula	General formula
$C_2H_4O_2$	CH_2O	$C_nH_{2n+1}COOH$ (or $C_nH_{2n}O_2$)

Structural formula	Display formula	Skeletal formula
CH_3COOH		

Figure 12.1 The different ways that ethanoic acid is represented by each type of formula.

- *Display formula* – The relative positioning of atoms and the bonds between them

- *Skeletal formula* – The simplified organic formula, shown by removing hydrogen atoms from alkyl chains, leaving just a carbon skeleton and associated functional groups

This may seem like rather a long list, but all of these ways of representing a compound show slightly different things and are useful for different applications.

Figure 12.1 shows how each of the different formulae listed would be used for the compound ethanoic acid. In this case, the **molecular formula** and **empirical formula** are not particularly useful because they do not show the connectivity of the atoms. The **general formula** does, and it can quickly be seen that there is a carboxylic acid present. The **structural, display**, and **skeletal** formulae provide a great deal more detail about the actual structure of ethanoic acid and are often the most useful. Skeletal formulae are commonly used throughout this book.

Carboxylic acids are a class of molecule that will be explored further in Chapter 14

12.1.1   Empirical and molecular formulae

The *empirical formula* represents the simplest whole-number ratio of atoms of each element in a compound. The *molecular formula* represents the actual number of atoms of each element in a compound. For example, benzene, the simplest aromatic molecule, has molecular formula C_6H_6; there are six carbon atoms present, and each is bonded to one hydrogen atom. The lowest ratio of atoms of each element in benzene is $1:1$, so the empirical formula is CH.

For a refresher, see Chapter 3.

The empirical formula represents the simplest whole-number ratio of atoms of each element in a compound. The molecular formula represents the actual number of atoms of each element in a compound.

12.1.2   Skeletal formula

When studying organic chemistry, one of the biggest hurdles that students have to overcome when transitioning beyond their previous studies is the way that chemists draw out structures of molecules. In prior study, this was likely done by using display formulae where all of the atoms that were bonded were shown. Unfortunately, when discussing more advanced chemistry, the structures can become quite complex, so showing all of the atoms is cumbersome and actually leads to a very confusing diagram. In more complex situations, chemists often use a *skeletal formula*, where the carbon chain is shown as a zig-zag line.

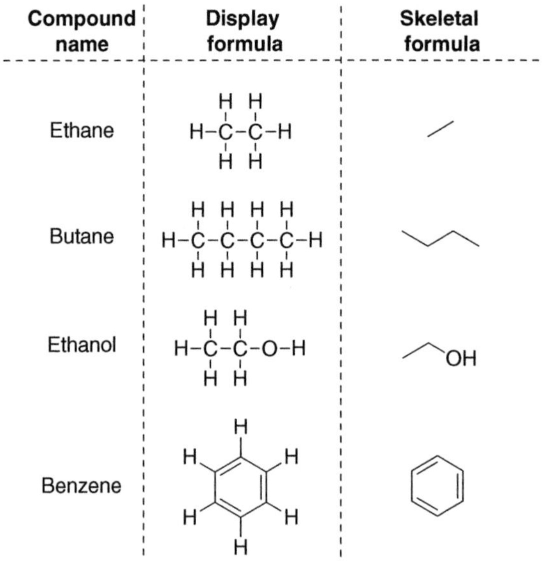

Compound name	Display formula	Skeletal formula
Ethane	H–C–C–H (with H atoms above and below each C)	(skeletal zig-zag)
Butane	H–C–C–C–C–H (with H atoms above and below each C)	(skeletal zig-zag)
Ethanol	H–C–C–O–H (with H atoms above and below each C)	(skeletal with OH)
Benzene	(benzene ring with H atoms)	(skeletal hexagon with circle)

Figure 12.2 Display and skeletal formulae for some simple organic compounds.

Figure 12.2 shows the comparative display and skeletal formulae of some organic compounds.

Skeletal formulae are often viewed with horror by students, but once you have become used to them, they are very easy to use. Here are some guidelines to drawing them:

1. Hydrogen atoms that are bonded to a carbon atom are not shown. There are too many, and they clutter the diagram.

2. Chains of carbon atoms are simply drawn as zig-zags in which the point of the zig-zag represents a carbon atom that is also bonded to hydrogen atoms, which are not shown.

3. Double and triple bonds are represented by two or three lines between connecting points, respectively.

4. Heteroatoms (e.g. N, O, P, S) and the halogens (F, Cl, Br, I) are all written using their atomic symbol. If they are bonded to a hydrogen atom, it is also shown.

The term *heteroatom* is used for elements other than C or H in an organic molecule.

5. Lone pairs of electrons are not shown on an element unless required.

Worked Example 12.1 The display formulae and names for three compounds are given here. Redraw these structures in their skeletal form:

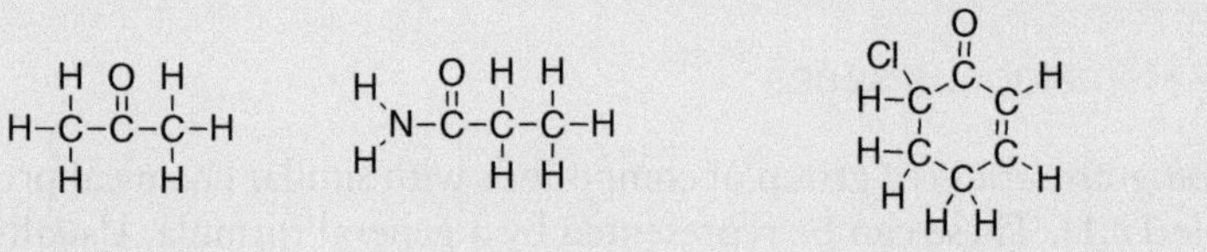

propan-2-one   propionamide   6-chlorocyclohex-2-en-1-one

Solution

When drawing a skeletal formula, it is often best to start by drawing a zig-zag line that contains the same number of points as there are elements in the longest chain, and then add the requisite functional groups. For example, the first structure, propan-2-one (also known as acetone), is reasonably simple as it only contains three carbon atoms in a line. The easiest way of solving this is to draw a zig-zag line containing three points (step 1) and to add a C=O to the central atom (step 2):

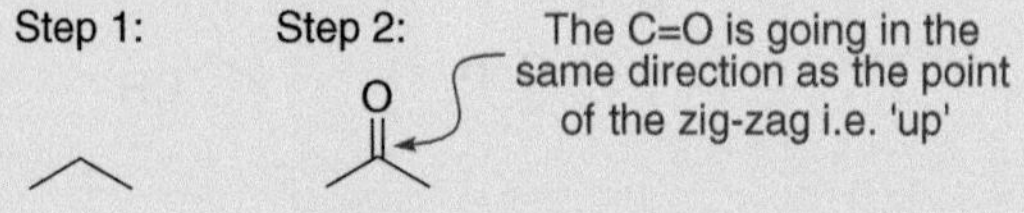

Step 1:   Step 2:   The C=O is going in the
                              same direction as the point
                              of the zig-zag i.e. 'up'

In the second example, propionamide, there are three carbon atoms and one nitrogen atom in a chain, so a zig-zag line with four points is needed (step 1). The first point is a nitrogen, but this also requires the two hydrogen atoms that are bonded to it to be shown, and hence these are added as a continuation of the zig-zag chain (step 2). Finally, the double bond to oxygen is added on the carbon atom next to the nitrogen (step 3).

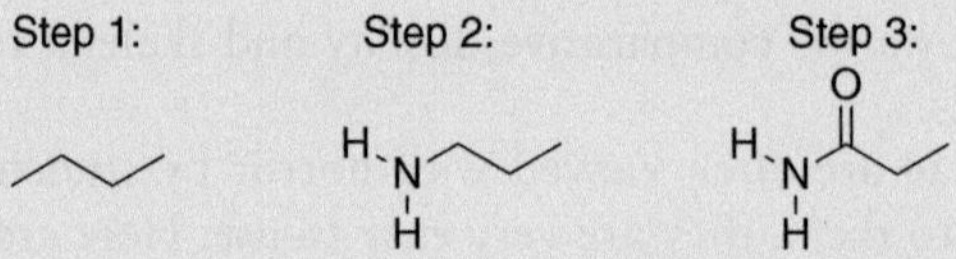

Step 1:   Step 2:   Step 3:

The final example, 6-chlorocyclohex-2-en-1-one, looks a lot more complex, but if it is tackled logically, there should be few difficulties. As before, count how many atoms are in the ring. In this example, there are six, so we draw a hexagon (step 1). For ease of explanation, the carbon atoms have been numbered, but this is not strictly necessary. In step 2, add the double bond between carbon atoms 2 and 3, and the double bond to oxygen at carbon-1. Finally, add the chlorine atom at carbon 6 (step 3).

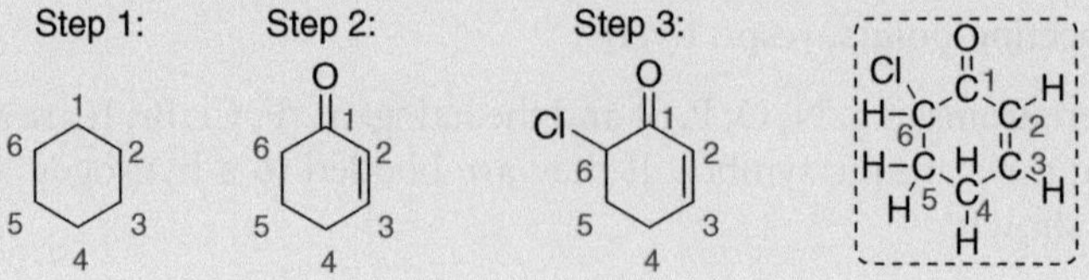

Step 1:   Step 2:   Step 3:

Throughout the book, there will be ample opportunities to see skeletal formulae being used. Although they will be tricky to understand at first, they should start to become second nature over time.

12.1.3   Homologous series

An *homologous series* is a group of compounds with similar chemical properties (see Table 12.1). These can be represented by a general formula. Usually, compounds with similar chemical properties contain the same functional group(s), and the homologous series is named after the dominant functional group.

Table 12.1 Some commonly encountered homologous series and examples. Note: R and R′ denote alkyl groups; X denotes a halogen.

Group	Suffix/prefix	General formula	Example Formula	Example Name
Alkane	-ane	C_nH_{2n+2}	C_2H_6	Ethane
Alkene	-ene	C_nH_{2n}	C_2H_4	Ethene
Halogenoalkane	halo-	RX	C_2H_5Cl	Chloroethane
Alcohol	-ol	ROH	C_2H_5OH	Ethanol
Aldehyde	-al	RCHO	CH_3CHO	Ethanal
Ketone	-one	RC(O)R′	CH_3COCH_3	Propanone
Carboxylic Acid	-oic acid	RCOOH	CH_3COOH	Ethanoic acid
Ester	-oate	RCOOR′	CH_3COOCH_3	Methyl ethanoate
Amine	-amine	RNH_2	$C_2H_5NH_2$	Ethylamine
Amide	-amide	$RCONR′_2$	$C_2H_5CONH_2$	Propylamide

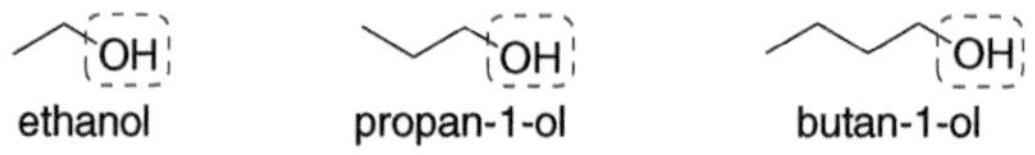

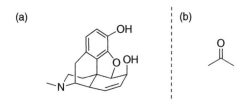

ethanol   propan-1-ol   butan-1-ol

Figure 12.3 Some members of the alcohol homologous series.

For example, ethanol contains an alcohol functional group (Figure 12.3). Propan-1-ol and butan-1-ol are said to be members of the same homologous series because they both contain an alcohol group.

12.2 Nomenclature of simple alkanes

For chemists to be able to determine the structure of a compound from its name, the International Union of Pure and Applied Chemistry (IUPAC) devised some rules. The naming of compounds is often referred to as **nomenclature**. A chemical name, as devised by IUPAC, has three parts: Prefix – Parent – Suffix.

- *Prefix:* Where are substituents and functional groups located?

- *Parent:* How many carbon atoms are in the main chain?

- *Suffix:* What functional groups are present?

The name of an organic compound can rapidly become extremely complicated: for example, the drug molecule morphine (Figure 12.4) has a long IUPAC

(a)

(b)

Figure 12.4 (a) The structure of morphine; (b) the structure of propan-2-one, also called acetone.

name (7,8-didehydro-4,5-epoxy-17-methylmorphinan-3,6-diol); therefore, when naming some more complicated compounds, chemists often use trivial names. In addition, some smaller compounds also have trivial names that are regularly used; for example, propan-2-one (Figure 12.4) is often referred to as acetone. You must always follow four steps to name an organic compound, as outlined in Box 12.1.

Box 12.1

Step 1: **Find the parent hydrocarbon.** This will be the longest continuous carbon chain. If there are no branches in the chain, it will simply be named according to the number of carbon atoms in the chain (methyl, ethyl, etc.). In the case of cyclic species, *cyclo-* is added: e.g. cyclohexane, cyclopentane, etc.

Step 2: **Number the atoms in the main chain.** If branching is present, number the carbon atoms in the chain, ensuring that the first alkyl substituent (branch) has the lowest number possible.

Step 3: **Identify and number the substituents.** Use the numbering system arrived at in step 2.

Step 4: **Write out the name.** The name will be a single word; use hyphens to separate different prefixes and commas to separate numbers. If two or more substituents are present, list these in alphabetical order. If two or more identical substituents are present, then use the prefixes *di-*, *tri-*, and *tetra-* as applicable.

Before we can start naming compounds, we need to know how to name the hydrocarbon framework. The names of the first five simple hydrocarbon chains, the display formulae, and the skeletal formulae are given in Table 12.2. The common names and prefixes of the first 10 hydrocarbon chains are given in Table 12.3.

Further explanation about skeletal formulae is in Section 12.1.2.

The easiest way of remembering the names of the first four alkanes is by using a mnemonic: for example, Monkeys Eat Peeled Bananas.

Table 12.2 The names and structures of the first five hydrocarbon chains.

Number of carbon atoms	Name	Display formula	Skeletal formula
1	Methane	$H-\overset{H}{\underset{H}{C}}-H$	CH_4
2	Ethane	$H-\overset{H}{\underset{H}{C}}-\overset{H}{\underset{H}{C}}-H$	/
3	Propane	$H-\overset{H}{\underset{H}{C}}-\overset{H}{\underset{H}{C}}-\overset{H}{\underset{H}{C}}-H$	⌢
4	Butane	$H-\overset{H}{\underset{H}{C}}-\overset{H}{\underset{H}{C}}-\overset{H}{\underset{H}{C}}-\overset{H}{\underset{H}{C}}-H$	⌇
5	Pentane	$H-\overset{H}{\underset{H}{C}}-\overset{H}{\underset{H}{C}}-\overset{H}{\underset{H}{C}}-\overset{H}{\underset{H}{C}}-\overset{H}{\underset{H}{C}}-H$	⌇⌢

Table 12.3 The name, structural formulae, and prefixes of the first 10 hydrocarbon chains.

Number of carbon atoms	Structural formula	Hydrocarbon	Prefix
1	CH_4	Methane	Meth-
2	CH_3CH_3	Ethane	Eth-
3	$CH_3CH_2CH_3$	Propane	Prop-
4	$CH_3CH_2CH_2CH_3$	Butane	But-
5	$CH_3CH_2CH_2CH_2CH_3$	Pentane	Pent-
6	$CH_3CH_2CH_2CH_2CH_2CH_3$	Hexane	Hex-
7	$CH_3CH_2CH_2CH_2CH_2CH_2CH_3$	Heptane	Hept-
8	$CH_3CH_2CH_2CH_2CH_2CH_2CH_2CH_3$	Octane	Oct-
9	$CH_3CH_2CH_2CH_2CH_2CH_2CH_2CH_2CH_3$	Nonane	Non-
10	$CH_3CH_2CH_2CH_2CH_2CH_2CH_2CH_2CH_2CH_3$	Decane	Dec-

Worked Example 12.2 Name the following alkane according to IUPAC nomenclature.

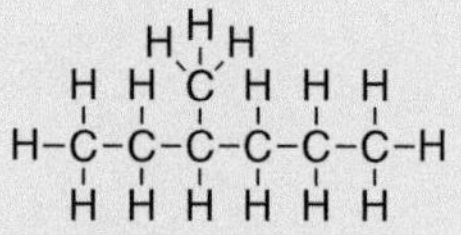

Solution

Step 1: Identify the parent hydrocarbon.

The parent hydrocarbon contains six carbon atoms, so is based upon 'hexane' and therefore has the parent 'hex'.

Step 2: Number the atoms in the chain so that each branch is on the lowest-numbered carbon atom possible.

In this example, the branch is the methyl group, which can be placed on either carbon-3 or carbon-4. Carbon-3 is preferred, as three is a smaller number than four.

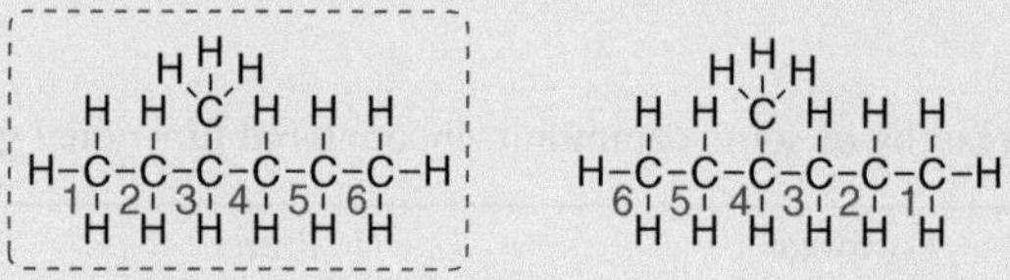

Correct

Step 3: Identify and number the substituents.

In this example, there is one side chain at carbon-3. The side chain at carbon-3 contains only one carbon atom, so it is designated 'methyl'.

Step 4: Write out the name.

- In this case, the longest chain contains six carbon atoms, so the parent name is 'hex'.

- There are only C—H bonds present, so the dominant functional group is an alkane; therefore, the suffix is 'ane'.

- There is one substituent: a methyl group at carbon-3.

- The overall name is therefore

3-methylhexane

When discussing organic chemistry, it is unlikely that you will only encounter molecules that solely contain C and H. The incorporation of other elements, e.g. O, Cl, etc., allows the formation of many different functional groups. Some commonly encountered functional groups are given in Figure 12.5.

Note that alkenes (double bonds) are a little more complicated than this, and their nomenclature is covered in Section 12.3.4. When naming them, the alkene 'starts' at the alkene terminus that is designated with a lower number; so the alkene shown in Figure 12.5 is named but-2-ene, as the alkene 'starts' at the second carbon atom. The same is also true for the alkynes; therefore, in Figure 12.5, the alkyne is but-2-yne.

When naming organic compounds, there is a hierarchy of functional groups. Some commonly encountered functional groups are shown in Table 12.4. Functional groups can be represented by prefixes and suffixes. The higher up the table, the more important a functional group is, which therefore dictates the suffix. It is not necessary to learn this table; just be aware that there is a hierarchy.

> This may seem like a large number of functional groups to learn, but we will cover most of them in detail over the next few chapters.

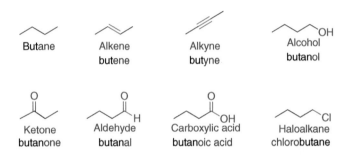

Figure 12.5 The structure and names of some commonly encountered functional groups.

Table 12.4 The hierarchy of some commonly encountered functional groups.

Functional group	Formula	Prefix	Suffix
Ammonium	$-NR_3^+$	-ammonio-	-ammonium
Carboxylic acid	$-COOH$	-carboxy-	-oic acid
Ester	$-COOR$	-oxycarbonyl-	-oate
Acyl halide	$-COOX$	-halidealcanoyl-	-oyl halide
Amide	$-COONR_2$	-carbamoyl-	-amide
Aldehyde	$-CHO$	-formyl-	-al
Ketone	$-C(O)-$	-oxo-	-one
Alcohol	$-OH$	-hydroxy-	-ol
Thiol	$-SH$	-sulfanyl-	-thiol
Amine	$-NR_2$	-amino-	-amine
Alkene	$-CH=CHR$	alkenyl-	-ene
Alkyne	$-C\equiv C-$	alkynyl-	-yne
Alkane	$-CH_2CH_2R$	alkyl-	-ane
Alkyl halide	$-X$ (where X is F, Cl, Br, I)	halo-	-ane
Nitro	$-NO_2$	nitro-	-ane

Worked Example 12.3   Name the following molecule according to IUPAC nomenclature:

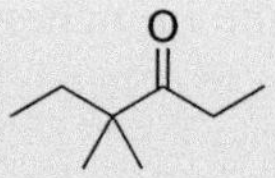

Solution
Step 1: Identify the parent hydrocarbon.

Step 2: Number the atoms in the chain so that each branch, or the dominant functional group, has the lowest number possible.

In this case, the ketone is the highest-priority functional group, as the only other functional group present is an alkane, so it therefore has to have the lowest number when numbering the chain.

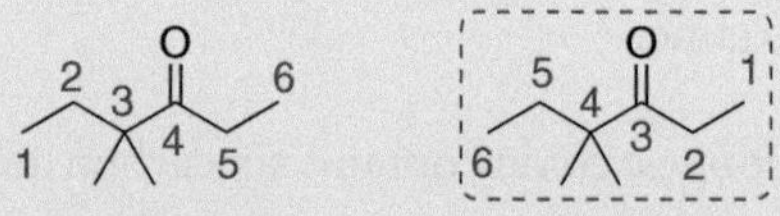

Correct

Step 3: Identify and number the substituents.

In this example, there are two side chains that each contain one carbon atom, so they are both designated 'methyl', located on carbon-4. However, because two methyl groups are present, we need to add the prefix 'di', so it becomes 'dimethyl'.

> If there is more than one of a particular functional group present, the prefix di- (2), tri- (3), or tetra- (4) need to be added, as applicable.

Step 4: Write out the name.

- In this case, the longest chain contains six carbon atoms, so the parent is 'hexan'.

- The dominant functional group is a ketone, which is on carbon-3, so the suffix is 'one' (rhymes with 'tone').

- There are two methyl substituents at carbon-4. These will form the prefix. The ketone is on carbon-3, so this is indicated between the parent and the suffix.

The overall name is therefore:

4,4-dimethylhexan-3-one

Worked Example 12.4   Name the following molecule according to IUPAC nomenclature:

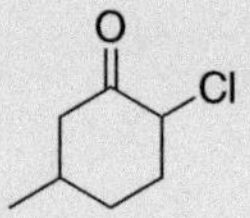

Solution
Step 1: Identify the parent hydrocarbon.

Step 2: Number the atoms in the chain so that each branch is on the carbon atom with the lowest number possible.

The ketone is the highest-priority functional group, so it has to have the lowest number when numbering the chain. Numbering in a clockwise direction leads to the lower total at the branches ([2 + 5] versus [6 + 3]), so it is therefore correct.

Correct

Step 3: Identify and number the substituents.

In this example, there are two substituents: a chlorine atom at carbon-2 and a methyl group at carbon-5.

Step 4: Write out the name.

- In this case, the longest chain contains six carbon atoms, so the parent is 'hexan'; but it is also in a ring, so it is 'cyclohexan'.

- The dominant functional group is a ketone, which, in this case, is on carbon-1 by default, so the suffix is 'one' and the '1' is not needed.

- There are two substituents; these will make the prefix. There is a chlorine atom at carbon-2 and a methyl group at carbon-5.

- When naming the compound, the chlorine atom is discussed first, as alphabetically it is before methyl.

The overall name is therefore

2-chloro-5-methylcyclohexanone

Further nomenclature rules can be applied to aromatic systems. These will be discussed in Chapter 15.

12.2.1 Nomenclature for esters

One class of compound that it is worth discussing further is the ester. You are likely to encounter these during your studies, and naming them requires a slightly different set of rules.

Esters consist of two parts: one side is derived from an alcohol and the other from a carboxylic acid (Figure 12.6). The alcohol part is directly bonded to the oxygen. The carboxylic acid part is bonded to the C=O.

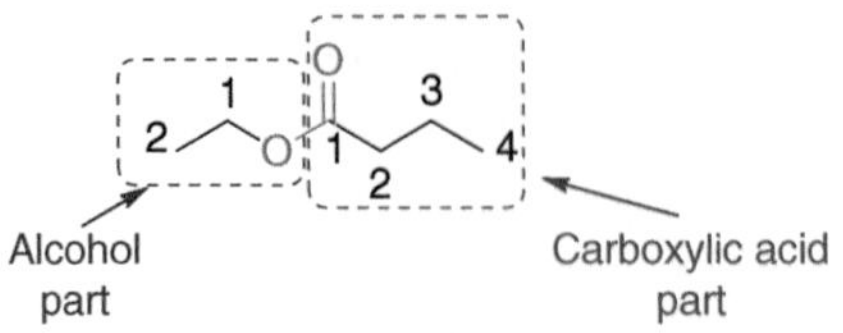

Alcohol
part

Carboxylic acid
part

Figure 12.6 The two component parts of an ester. The ester group itself is highlighted in orange.

When naming esters, there are two parts: the first, from the alcohol, ends with -*yl*; the second, from the carboxylic acid, ends with -*oate*. If a number is required, the numbering starts at the end closest to the ester bond.

Worked Example 12.5   Name the ester shown:

Solution
When naming an ester, the first step is to look for the ester bond and then count the number of carbon atoms back from it. This will give two different carbon chains, which may have different lengths. In this example, there is a portion containing two carbon atoms, so 'eth', and another one containing four carbon atoms, so 'but' (rhymes with 'mute'). To name the ester, the parts that each component came from need to be considered. The 'eth' part came from the alcohol section, so it becomes 'ethyl'. The 'but' part comes from the carboxylic acid section, so it becomes 'butanoate'. The final name is therefore ethyl butanoate.

12.3   Isomers

You may have noticed in the previous sections that in some situations, a compound could have had different connectivity of atoms, but the compound would still have the same molecular formula. This is called **isomerism**. Structural **isomers** have the same molecular formula but a different spatial arrangement of atoms. Structural isomers may exhibit different physical and chemical properties.

> Structural isomers have the same molecular formula but a different spatial arrangement of atoms.

There are five main types of isomerism to discuss at the foundation level:

- Chain isomerism

- Positional isomerism

- Functional group isomerism

- *E* and *Z* isomerism (alkenes only)

- Chirality

Chain isomerism, positional isomerism, and functional group isomerism are all types of structural isomerism. *E* and *Z* isomerism and chirality are both types of stereoisomerism.

12.3.1   Chain isomerism

Chain isomerism occurs when the longest chain length in an alkane varies. For example, butane and 2-methylpropane both have the same molecular formula, C_4H_{10}, but the atoms have different connectivities (Figure 12.7).

butane   2-methylpropane

Figure 12.7 Chain isomers of C_4H_{10}.

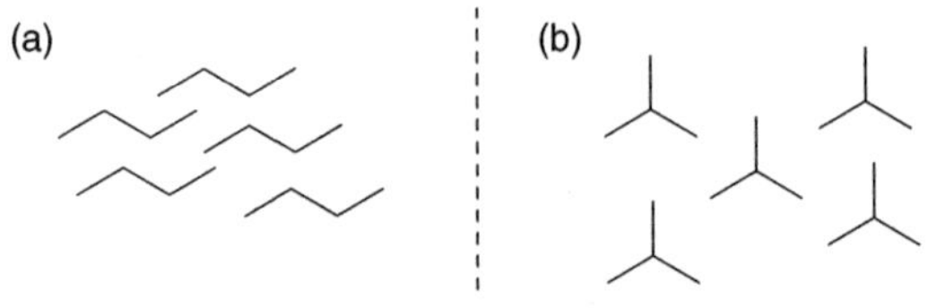

(a)   (b)

Figure 12.8 (a) Butane molecules; (b) 2-methylpropane molecules.

In this case, the different arrangement of atoms has a large impact upon the physical properties of butane and 2-methylpropane, although their chemical reactivity is very similar. Butane has a melting point of –138 °C and boiling point of –0.5 °C, whereas 2-methylpropane has a melting point of –160 °C and boiling point of –12 °C. The reason for the difference is that in butane, there are stronger intermolecular forces between molecules because they can align much more closely (Figure 12.8).

For a refresher on intermolecular forces, see Chapter 2.

12.3.2 Positional isomerism

Positional isomerism means that, in a homologous series of compounds, the same functional group is present but is located in different positions along a chain. This type of isomerism is commonly encountered with alcohols. Figure 12.9 shows the two positional isomers for butanol: butan-1-ol and butan-2-ol. Although butan-1-ol and butan-2-ol are positional isomers, they are still members of the same homologous series (alcohols) and have similar chemical reactivity but different physical properties.

12.3.3 Functional group isomerism

Methoxymethane contains an ether functional group (R–O–R).

With functional group isomerism, the molecular formula is the same, but the functional groups present are different. This can have a major impact on both the chemical reactivity and physical properties of a compound. For example, two compounds have molecular formula C_2H_6O: ethanol and methoxymethane, containing an alcohol and an ether, respectively (Figure 12.10).

HO⤳   OH

butan-1-ol   butan-2-ol

Figure 12.9 Positional isomers of butanol, $C_4H_{10}O$.

ethanol   methoxymethane

Figure 12.10 Functional groups isomers of C_2H_6O.

It is worth noting that when the general formulae are the same, the most common functional group isomer pairs are

- Alcohol/ether

- Cycloalkane/alkene

- Ester/carboxylic acid

- Aldehyde/ketone

12.3.4 *Z* and *E* isomerism (alkenes only)

Z and *E* isomerism is an example of stereoisomerism. Stereoisomers have the same molecular formula but are arranged differently in three-dimensional space. In alkenes, the most important substituents on either side of the double bond can be arranged two ways: either on the same side (*Z*) of the alkene or on opposite sides (*E*) (Figure 12.11). This is the case because the double bond does not allow free rotation; therefore, the two different substituents are fixed in space.

This form of isomerism can also be referred to as *cis* and *trans* isomerism; however, strictly speaking, *cis* and *trans* should only be used if there is a proton at either end of the alkene. The letters *E* and *Z* are derived from German. The *E* stands for *entgegen*, meaning 'opposite', and the *Z* for *zusammen*, meaning 'together'.

> Within organic chemistry, the word 'proton' is often used to indicate a hydrogen atom within a molecule.

> Stereoisomers have the same molecular formula and same connectivity, but the atoms are arranged differently in three-dimensional space.

(*E*)-but-2-ene   (*Z*)-but-2-ene   (*E*)-3,4-dimethylhex-3-ene   (*Z*)-3,4-dimethylhex-3-ene

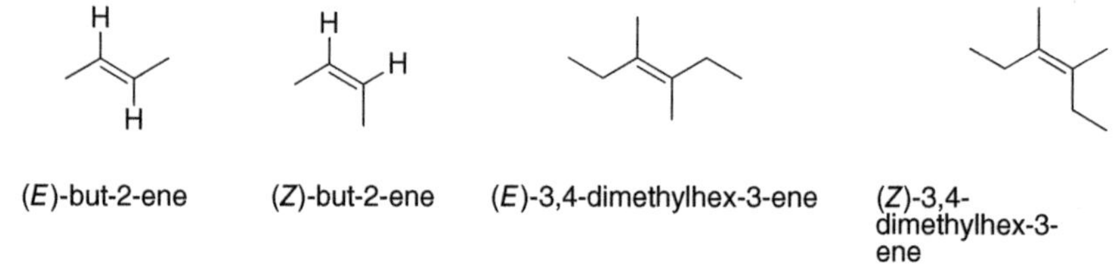

Figure 12.11 The difference *E* and *Z* (*trans* and *cis*) isomers.

Box 12.2 A set of rules called the Cahn–Ingold–Prelog rules can be followed to determine whether an alkene is *E* or *Z*:

1. Locate the alkene, and divide it into two, perpendicular to the double bond.

2. Rank the substituents at either end of the double bond according to atomic number.

3. If the two highest-ranked substituents are on diagonally opposite sides of the double bond, then the stereochemistry is designated (*E*). If the two highest-ranked substituents are on the same side of the double bond, the stereochemistry is designated (*Z*).

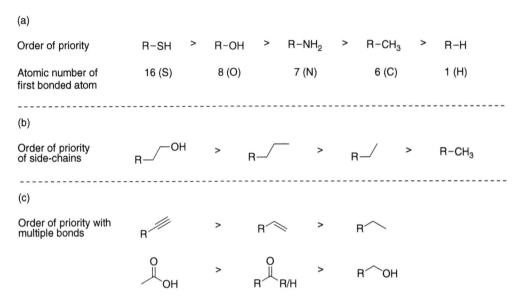

Figure 12.12 A brief summary of the hierarchy of priority when using Cahn–Ingold–Prelog rules. (a) The atomic number of an element determines the priority; (b) atoms are ranked according to the substituents on the side chains; (c) the hierarchy for multiple bonds.

In an *E* alkene, the most important (highest-priority) substituents are on the opposite side of the alkene. In a *Z* alkene, the most important substituents are on the same side of the alkene.

However, before we can use the Cahn–Ingold–Prelog system, some rules need to be discussed:

1. The atomic number of the atoms closest to the stereocentre (or alkene) determines the order in which we consider the substituents. The largest atomic number is given highest priority (Figure 12.12a).

2. If the atoms have the same atomic number, then the atomic number of the next atom away from the stereocentre is examined. If no difference if found, continue to move away from the stereocentre until a difference is found (Figure 12.12b).

3. If the substituent has a double or triple bond (to any element), then it is treated as being bonded to that atom twice or three times, respectively (Figure 12.12c).

Worked Example 12.6   Determine whether the alkene shown, 3-chloro-4-methylhex-3-ene, is either (*E*) or (*Z*):

Solution
Step 1: Divide the alkene in two perpendicular to the double bond:

Step 2: Rank the substituents on either side according to atomic number.

If we take the left-hand end of the alkene first, a chlorine atom and a carbon atom are bonded directly to the carbon atom of the alkene. Chlorine has atomic number 17, and carbon has atomic number 6; 17 is higher than 6, so chlorine has the higher priority (designated 1 in the diagram):

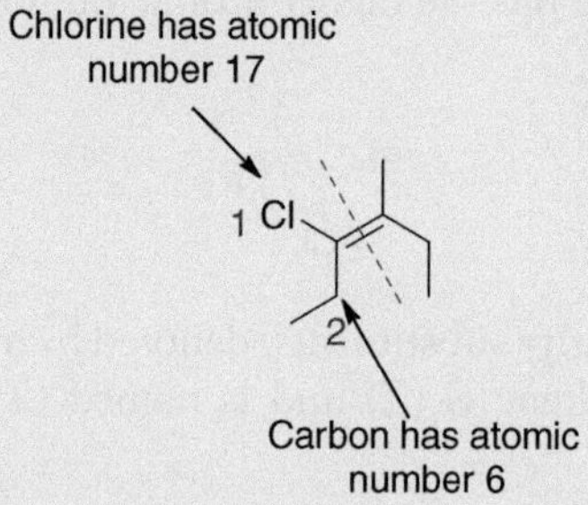

Chlorine has atomic
number 17

Carbon has atomic
number 6

Looking at the other end of the double bond, two carbon atoms are directly bonded to the carbon atom of the alkene, both having atomic number 6. In this case, we need to look further. The carbon atom at the top is bonded to three other atoms, all of which are protons and all of which have an atomic number of 1. The carbon atom at the bottom is bonded to one carbon atom (atomic number 6) and two protons (atomic number 1). The carbon atom at the bottom therefore possesses higher priority because it is bonded to an element that has a higher atomic number:

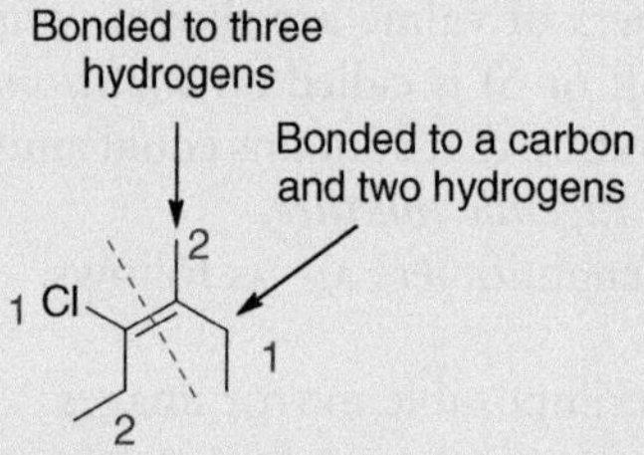

Bonded to three
hydrogens

Bonded to a carbon
and two hydrogens

Finally, we need to consider where each of the higher-priority substituents is physically located around the double bond. The two highest-priority substituents, denoted by a 1, are on diagonally opposite sides; therefore, the alkene is (E), and the overall name of the alkene is (E)-3-chloro-4-methylhex-3-ene.

Worked Example 12.7   Determine whether the alkene shown is (E) or (Z):

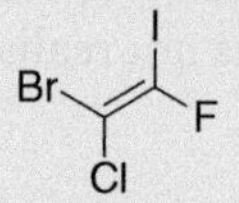

Solution
Step 1: Divide the alkene in two perpendicular to the double bond:

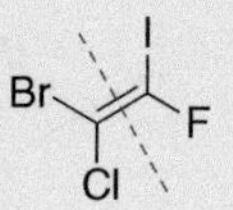

Step 2: Rank the substituents on either side according to atomic number.

If we take the left-hand end of the alkene first, a chlorine atom and a bromine atom are bonded directly to the carbon atom of the alkene. Chlorine has atomic number 17, and bromine has atomic number 35; therefore, bromine has higher priority. Looking at the other end of the double bond, there is an iodine atom with atomic number 53 and a fluorine atom with atomic number 9. Iodine has the higher atomic number, so it takes priority:

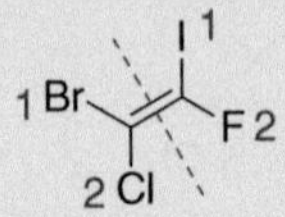

The two highest-priority substituents, denoted by a 1, are on the same side of the alkene, so the alkene is (Z) and is named (Z)-1-bromo-1-chloro-2-fluoro-2-iodoethene.

12.3.5   Chirality

> When a carbon atom bonds to four different substituents, it is called a *chiral carbon* or a *chiral centre*.

The final type of isomerism to be discussed is another form of stereoisomerism; but in this case, it is called *optical isomerism,* and we most commonly observe it when a carbon atom is bonded to four different substituents. Chirality is very common in nature, and amino acids are examples of chiral compounds.

The amino acid valine is a simple example of a chiral compound (Figure 12.13). Two isomers of valine are shown, which are denoted (R) and (S). Each optical isomer (R or S) is called an *enantiomer,* and they are mirror images of each other. A mixture that contains equal amounts of each enantiomer is termed a *racemate* or a *racemic mixture.*

The characteristics of enantiomers are as follows:

> Usually, in a beam of light, the photons that make the light are all moving randomly in different directions. Plane-polarised light occurs when all of the photons within the light source are moving in only one direction, so they oscillate together.

- They have non-superimposable mirror images.

- They rotate plane polarised light in opposite directions.

- They are chemically identical.

Chirality, or at least appreciating that it exists, is extremely important. A famous example of where chirality was not considered is the drug thalidomide. Thalidomide was administered to pregnant women in the late 1950s as a treatment for morning sickness. The drug itself was administered as a mixture of two enantiomers (Figure 12.14). However, one enantiomer was highly toxic and led to birth defects in children. This was not realised until 1961, and unfortunately,

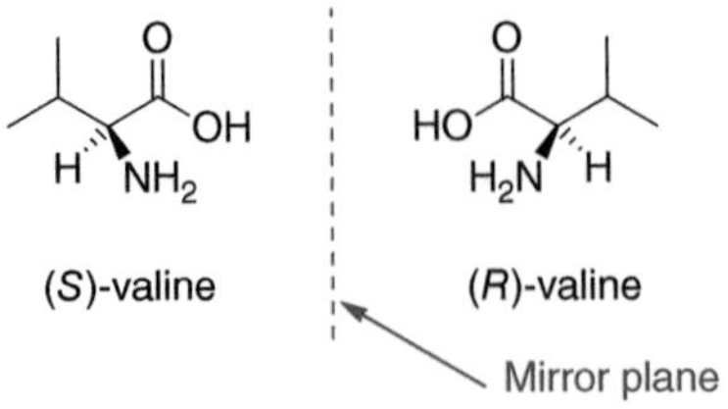

(S)-valine   (R)-valine

Mirror plane

Figure 12.13 The R and S forms of valine.

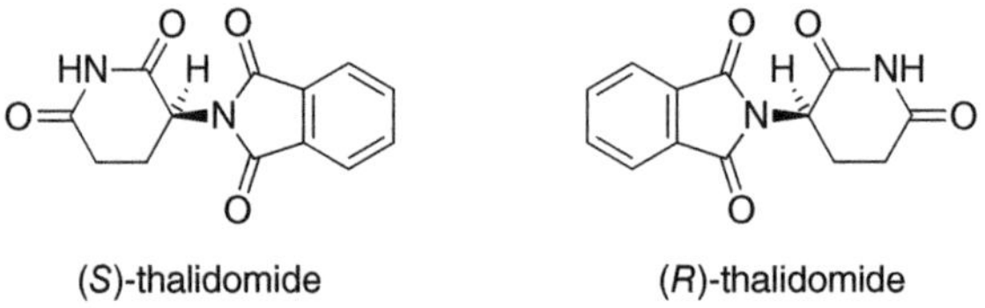

(S)-thalidomide   (R)-thalidomide

Figure 12.14 The R and S forms of thalidomide.

by this time, over 10,000 babies had been born with a thalidomide-related dis-
ability. Unfortunately, even if the two enantiomers had been separated and the
non-toxic enantiomer administered, the body is able to chemically transform
(*racemise*) the enantiomers; so once in the body, the toxic enantiomer would
have been generated anyway.

The Cahn–Ingold–Prelog rules used for alkenes in Section 12.3.4 can also be
used to determine whether a carbon atom has (R) or (S) configuration and are
used in a similar manner as for alkenes. In order for a stereocentre to be designated as having R or S configuration, the central carbon atom must be bonded to
four different substituents.

For a carbon atom to be designated
as having R or S configuration, the
atom in question must be bonded
to four different substituents.

Box 12.3 For chiral centres, the Cahn–Ingold–Prelog rules are used to determine which nomenclature (R or S) should be used:

1. Locate the chiral centre (typically a carbon atom with four different
   substituents).

2. Rank all of the substituents according to decreasing atomic number.
   Three points to note:

   a. Where two atoms are the same, we must look further along the
      chain to determine the higher priority.

   b. If there are different isotopes (e.g. hydrogen H and deuterium D),
      the higher-mass isotope takes priority over a lower-mass isotope:
      i.e. D (A_r = 2) takes priority over H (A_r = 1).

   c. If a substituent contains a double or triple bond, it is treated as being
      bonded to that atom twice or three times, respectively: e.g. C≡C is
      prioritised over C=C, which is prioritised over C—C.

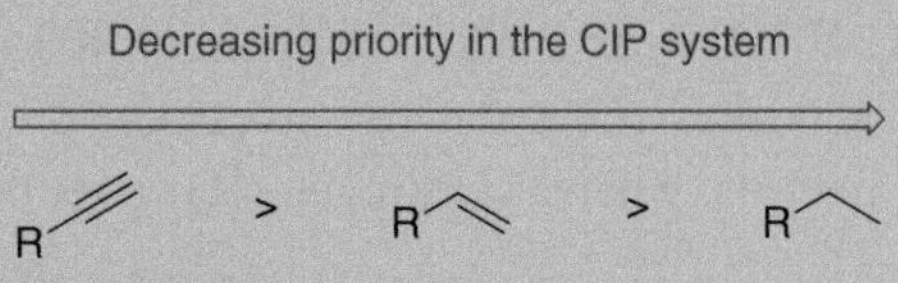

   Decreasing priority in the CIP system

3. The lowest-ranked group should be oriented furthest away from the
   observer.

4. Determine whether the remaining three substituents

   a. *Decrease* in priority in a clockwise direction: this is **R-**

   b. *Decrease* in priority in an anticlockwise direction: **S-**

Worked Example 12.8   Assign the chiral centre in the following molecule as either R or S:

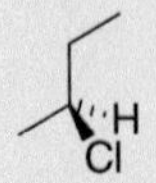

Solution
In this molecule, the chiral centre is bonded to a proton with atomic number 1, a chlorine atom with atomic number 17, and two carbon atoms with atomic number 6. The highest atomic number is Cl, 17, and the lowest H, 1. The two carbon atoms have the same atomic number, 6, so they cannot be differentiated yet. The molecule can therefore be labelled as follows:

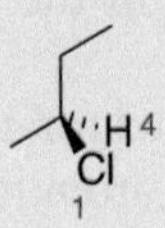

We now need to differentiate the two carbon atoms. The methyl group is bonded to three hydrogen atoms, each with atomic number 1, and the other carbon atom is bonded to two hydrogen atoms (with atomic number 1) and a carbon atom (with atomic number 6). The carbon atom bonded to the other carbon atom takes priority, as it is directly attached to the highest atomic number. The molecule can therefore be labelled as follows:

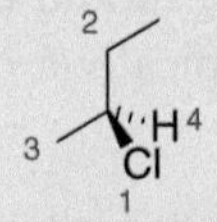

The lowest-priority substituent is facing away from us, so all we now need to do now is determine whether the molecule is R or S. We can do this by drawing an arrow from the highest-priority substituent to the lowest-priority substituent (ignoring the substituent designated as 4):

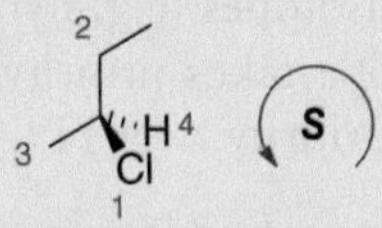

In this case, the arrow turns anticlockwise, so the chiral centre is S.

Worked Example 12.9   Assign the chiral centre in the following molecule as either R or S. (D is deuterium ($^2_1$D) and is an isotope of hydrogen that contains one neutron in the nucleus, whereas hydrogen ($^1_1$H) does not contain any.)

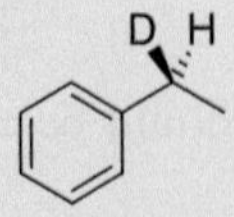

Solution
In this molecule, the chiral centre is bonded to a proton with atomic number 1; a deuterium atom, also with atomic number 1; and two carbon atoms with atomic number 6. The D and the H are isotopes of each other, so we can look

at the atomic mass: D has higher atomic mass than H, so it is higher priority. The two carbon atoms have the same atomic number, 6, so they cannot be differentiated yet. The molecule can therefore be labelled as follows:

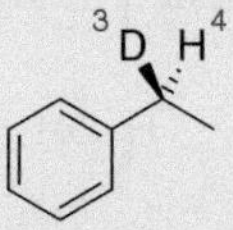

We now need to differentiate the two carbon atoms. The methyl group (to the right) contains three hydrogen atoms, each with atomic number 1, whereas the carbon atom to the left is bonded to two carbon atoms (each with atomic number 6, coloured in blue) However, one of these carbon atoms is part of a double bond; therefore, this carbon atom is counted twice as we have added a 'ghost carbon' (labelled in red). This carbon atom takes priority, as it is directly bonded to the atom with the highest atomic number. The molecule can therefore be labelled as follows:

The lowest-priority substituent is facing away from us, so all we need to do now is determine if the molecule is R or S. We can do this by drawing an arrow from the highest-priority substituent to the lowest-priority substituent (ignoring the substituent designated as 4):

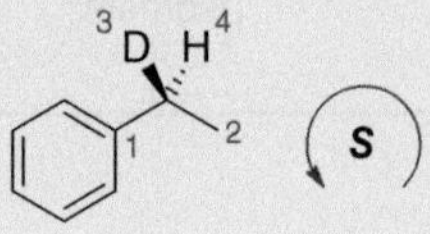

Worked Example 12.10   Assign the chiral centre in the following molecule as either R or S:

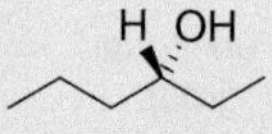

Solution
In this molecule, the chiral centre is bonded to a proton with atomic number 1, an oxygen atom with atomic number 8, and two carbon atoms with atomic number 6. Oxygen is higher priority than hydrogen, so these can be designated. The two carbon atoms have the same atomic number, 6, so they cannot be differentiated yet. The molecule can therefore be labelled as follows:

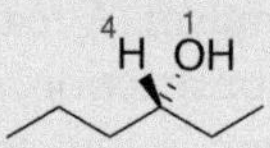

We now need to differentiate the two carbon atoms. The carbon atom to the right is bonded to two hydrogen atoms, each with atomic number 1, and a carbon atom with atomic number 6. The same is true of the carbon atom on the left, so we need to go out one more layer. The second carbon atom of the ethyl group (to the right) is bonded to three hydrogen atoms with atomic number 1. The second carbon atom in the propyl group (to the left) is bonded to one carbon atom and two hydrogen atoms and therefore takes priority. The molecule can be labelled as follows:

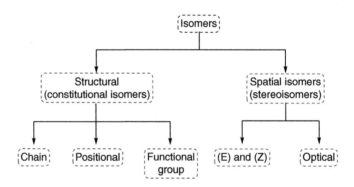

In this case, the lowest-priority substituent is towards us; therefore, the structure needs to be rotated such that it faces away. It can be redrawn and assigned as follows:

12.3.6   Summary of isomerism

This section has discussed a great deal of information about isomerism. Figure 12.15 summarises the different types of isomerism and the key differences.

Figure 12.15 A scheme to summarise the different types of isomerism discussed.

12.4   Drawing reaction mechanisms

When discussing organic chemistry and the reactions that occur, you will often be asked to give the **reaction mechanism**. A reaction mechanism is the proposed pathway through which a process, or reaction, occurs. When drawing a reaction mechanism, you need to be confident in two areas: (i) drawing the structures of any starting materials or intermediates; and (ii) showing the movement of electrons. In this chapter, we look at the basic concepts behind drawing a mechanism and things to look out for. In later chapters, we will consider some specific mechanisms.

A reaction mechanism is the proposed pathway by which a process, or reaction, occurs.

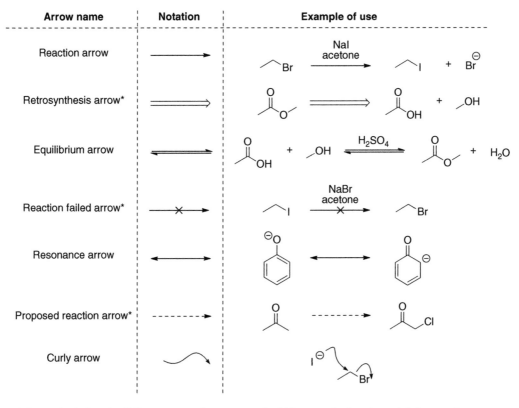

Figure 12.16 A figure depicting the different types of arrows used within organic chemistry. A * denotes that you are unlikely to meet this arrow in a foundation programme.

12.4.1 Types of arrows

A curly arrow shows the *movement of electrons* and is commonly used by organic chemists around the world. Before discussing curly arrows and how to use them, you should familiarise yourself with all of the types of arrows that you will commonly see in organic chemistry (Figure 12.16).

As mentioned earlier, a curly arrow shows the movement of electrons and therefore has to be from an area of high electron density to an area of low electron density. The arrow always points the way that the electrons are moving. For example, a negative charge is an area of high electron density, and a positive charge is an area of low electron density. Figure 12.17a shows this; the iodide has a negative charge and therefore is electron-rich, and the central carbon atom has a positive charge and is therefore electron-poor. Electrons move *from* the electron-rich iodide *to* the electron-poor carbon atom. In this process, the central carbon atom has gained the two electrons from the iodide, and a bond has been made between the carbon atom and the iodine. The fact that a bond has been made is often the part that confuses students the most, so this is shown in terms of electrons in Figure 12.17b. There will be plenty of opportunities to practise using curly arrows throughout the chapters on organic chemistry.

A curly arrow shows the movement of electrons and has to be from an area of high electron density to an area of low electron density.

The ⊖ denotes a pair of electrons. It is important to remember this.

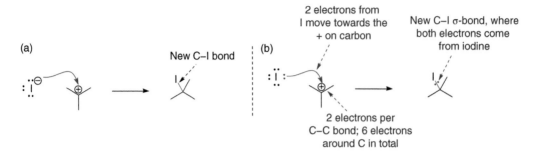

Figure 12.17 (a) Curly arrows showing the attack of iodide to a positive carbon atom; (b) showing how the electrons move to create a new bond. For clarity, the two electrons depicted in red are designated the negative charge although in reality it could be any pair of the eight electrons around the iodine.

12.4.2 Electrophiles, nucleophiles, and radicals

You are likely to encounter the terms *electrophile*, *nucleophile*, and *radical* a great deal, so a good understanding of what they mean will be useful.

Electrophiles

An electrophile is a species that is *electron deficient*. The term *electrophile* is derived from the Greek *electron* ('amber') and *phileo* ('I love'). Electrophiles can have either a positive charge, an empty orbital that can accept electrons, or an uneven charge distribution (dipole). In all cases, electrons must flow towards the electrophilic centre, so a curly arrow must never start at an electrophile. Figure 12.18 shows some commonly encountered electrophiles.

> An electrophile is a species that is electron deficient.

> For a refresher on dipoles, see Chapter 2.

Nucleophiles

A *nucleophile* is a species that is *electron-rich*. The term *nucleophile* is derived from one Greek and one Latin term: *nucleus* ('kernel, core'; Latin) and *phileo* ('I love'; Greek). A nucleophile must have either a negative charge or a lone pair of electrons that it can donate. In all cases, electrons flow from the nucleophile. Figure 12.19 shows some commonly encountered nucleophiles.

> Electrons flow towards an electrophilic centre.

> In BF$_3$ and AlCl$_3$, the central atom only has six electrons that are shared with the three halogen atoms.

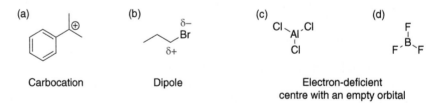

(a) Carbocation   (b) Dipole   (c)   (d) Electron-deficient centre with an empty orbital

> A nucleophile is a species that is electron-rich.

> Electrons flow from a nucleophilic centre.

Figure 12.18 Examples of some electrophiles. (a) A carbocation: in this case, the carbon atom bearing the positive charge is the electrophilic centre; (b) a dipole: in this case, the carbon atom bearing the δ+ positive charge is the electrophilic centre; (c) and (d) both contain a central atom with an empty orbital that can act as an electrophilic centre.

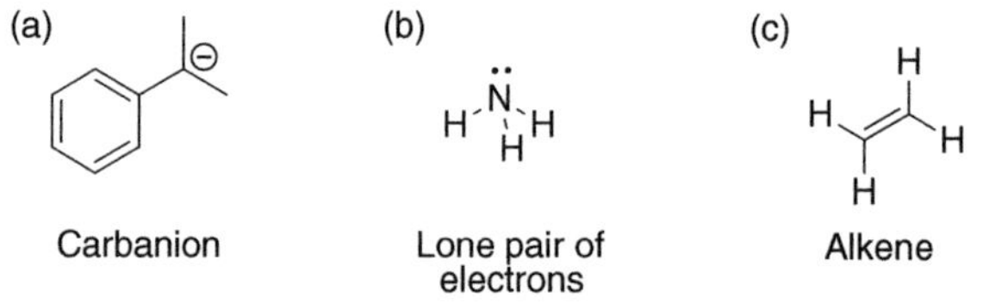

(a) Carbanion

(b) Lone pair of electrons

(c) Alkene

Figure 12.19 Examples of some nucleophiles. (a) A carbanion: in this case, the carbon atom bearing the negative charge is the nucleophilic centre; (b) a lone pair: in this case, the pair of electrons on the nitrogen atom will act as a nucleophile; (c) a double bond: the π bond is electron rich and can broken in order to attack an electrophile.

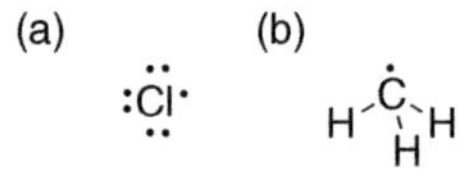

(a) :Cl·

(b) H–C·–H

Figure 12.20 Examples of radical species. Note that the radical is always shown, even on skeletal formulae. (a) A Cl radical, or Cl atom; (b) a methyl radical.

Radicals

A *radical* is a species that has *one unpaired electron*. A radical can be either electrophilic or nucleophilic, but a discussion about this is beyond the scope of a foundation course. Figure 12.20 shows some commonly encountered radical species.

Worked Example 12.11   Classify the following compounds as an electrophile, nucleophile, or radical:

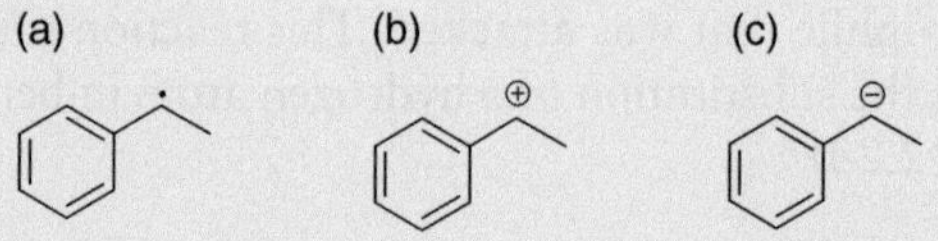

(a)   (b)   (c)

A radical species contains an unpaired electron.

Solution
The species in (a) has a single electron, so it has to be a radical. In (b), there is a positive charge on the carbon atom, which means that the carbon atom has only six electrons surrounding it (two from each σ bond) when it requires eight for a full valence shell. It is therefore electron-deficient and is an electrophile. Finally, (c) has an anion on the carbon atom. This means that the carbon centre has eight electrons around it (six from the three σ bonds and two unpaired on the carbon) but only has three bonds, so it will act as a nucleophile and can make a fourth bond by attacking an electrophile.

Remember that in a skeletal representation the H atoms are not shown. This applies to this example.

12.5   Types of reactions

During your studies using this textbook, you will encounter a variety of different classes of reactions. These will be discussed in more detail in the appropriate chapters, but a summary of the reactions covered is outlined next.

12.5.1   Electrophilic addition (to an alkene)

In this class of reaction, an alkene reacts with an electrophile to generate a new molecule where the alkene is (usually) no longer present. An example is the addition of bromine to pent-1-ene to give 1,2-dibromopentane (Figure 12.21).

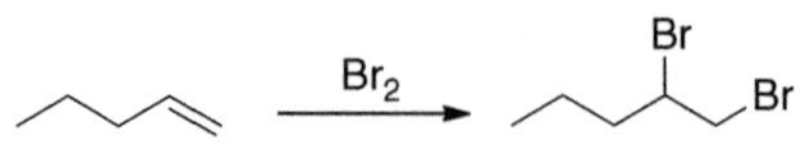

Figure 12.21 Electrophilic addition of bromine to pent-1-ene.

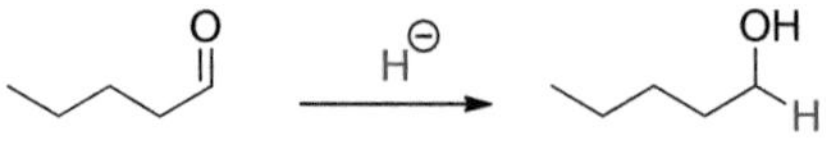

Figure 12.22 Nucleophilic addition of hydride to pentanal.

12.5.2   Nucleophilic addition (to a carbonyl group)

In this class of reaction, a nucleophile reacts with the δ+ carbon atom within a C=O, breaking the carbon-to-oxygen double bond to leave a single C—O bond and a new bond between the nucleophile and the original carbon atom in the C=O. An example is the addition of hydride into pentanal to give pentan-1-ol (Figure 12.22).

12.5.3   Electrophilic aromatic substitution

In this class of reaction, an electron-rich benzene ring (π bond) reacts with an electrophile. After a few further steps that are discussed later in this book, we eventually replace one of the C—H bonds with a new bond between the carbon atom and the electrophile that was attacked. This reaction often requires a catalyst; an example is the substitution of a hydrogen atom in benzene with a chlorine atom (Figure 12.23).

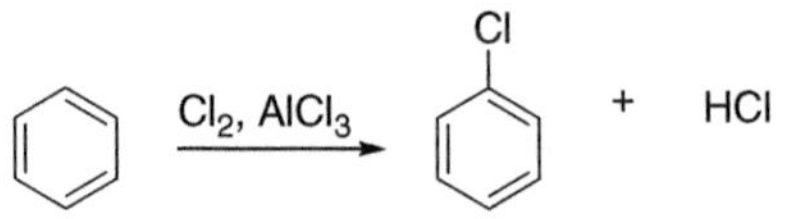

Figure 12.23 Electrophilic aromatic substitution of benzene with chlorine.

12.5.4   Nucleophilic substitution

In this class of reaction, a nucleophile displaces a leaving group in an alkyl chain, so they effectively swap places. There are two subclasses of this type of reaction: S_N1 and S_N2. An example is the substitution of chlorine in 1-chloropropane with iodide to give 1-iodopropane (Figure 12.24).

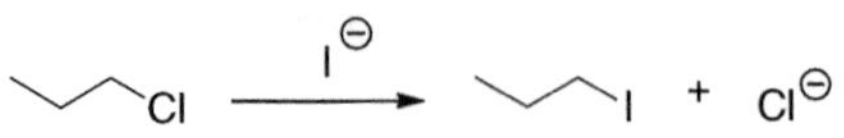

Figure 12.24 Nucleophilic substitution of 1-chloropropane with iodide to give 1-iodopropane.

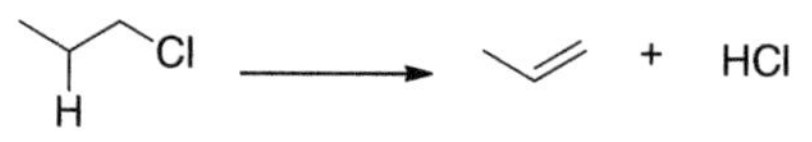

Figure 12.25 Elimination of HCl in 1-chloropropane to give prop-1-ene.

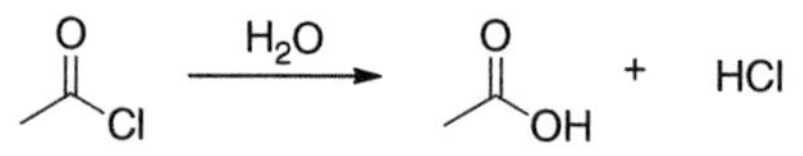

Figure 12.26 Condensation of water with ethanoyl chloride to give ethanoic acid and hydrogen chloride.

12.5.5   Elimination

In this class of reaction, a leaving group is on a carbon atom directly next to a carbon atom that is bonded to a proton. Overall, an alkene is formed, because as the leaving group leaves, the adjacent proton is also removed. There are two subclasses of this type of reaction: E1 and E2. An example is the elimination of hydrogen chloride from 1-chloropropane to give prop-1-ene (Figure 12.25).

12.5.6   Condensation

In this class of reaction, two molecules join together and, in doing so, a small molecule (e.g. water or HCl) is ejected. This type of reaction is commonly seen when making esters and amides. An example is the reaction between ethanoyl chloride and water to give ethanoic acid and hydrogen chloride (Figure 12.26).

Quick-check summary

- There are six types of molecular formula: molecular, empirical, general, structural, display, and skeletal. All are used as different ways of representing a compound and are used for different applications.

- Organic compounds are named according to a set of rules determined by IUPAC. The name has three parts: the parent, the prefix, and the suffix.

- A homologous series is a group of compounds with similar chemical properties.

- Compounds are often named according to the key functional groups present.

- Compounds can exist as different isomers, which can be broadly divided into structural isomers and stereoisomers.

- Chain isomerism, positional isomerism, and functional group isomerism are all types of structural isomerism.

- *E* and *Z* isomerism and optical isomerism (chirality) are both types of stereoisomerism.

- Alkenes exist as *E* or *Z* isomers (also called *cis* and *trans* isomerism), and a chiral centre can exist as an *R* or *S* isomer.

- There are rules that can be used to determine whether an alkene isomer is *E* or *Z*, or a stereocentre is *R* or *S*.

- A reaction mechanism is the proposed pathway through which a process, or reaction, occurs and is often represented using curly arrows.

- A curly arrow shows the movement of electrons.

- An electrophile is an electron-deficient species, a nucleophile is an electron rich species, and a radical contains an unpaired electron.

End-of-chapter questions

1 Name the functional groups highlighted in the following molecules:

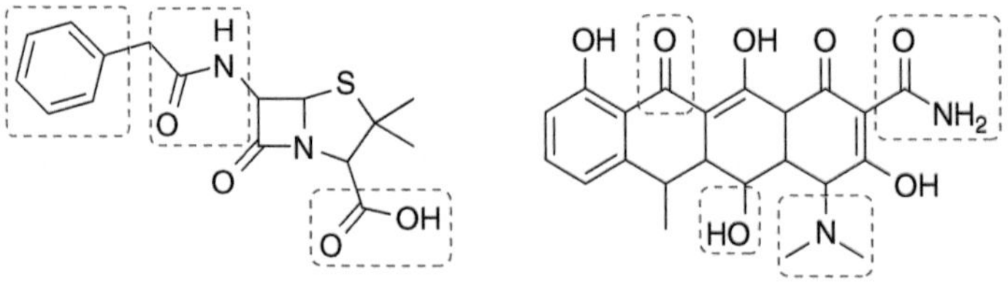

2 Name the following structures according to IUPAC nomenclature rules:

a.

b.

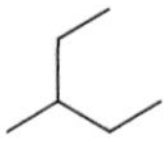

c.

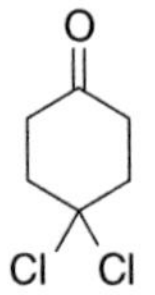

d.

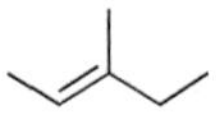

e.

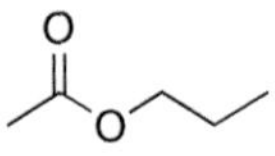

f.

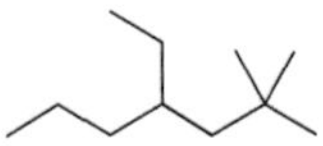

3 Calculate the empirical formula if 16.7 g of a sample only contains 12.7 g iodine and 4.0 g of oxygen.

4 A compound contains 58.50% carbon and 9.87% hydrogen by mass. Calculate the empirical formula.

5 A compound contains 62.04% carbon and 10.41% hydrogen by mass. The molecular mass is 116.16 g mol^{-1}. Calculate the molecular formula.

6 Combustion analysis of 0.100 g of a hydrocarbon compound showed that it contained 0.309 g of CO_2 and 0.142 g of H_2O. Calculate the empirical formula.

7 Draw and name all isomers with molecular formula C_6H_{14}.

8 Draw and name all isomers of bromopentane.

9 Draw all isomers with molecular formula $C_4H_{10}O$.

10 Identify the following alkenes as either *E* or *Z*:

a.

b.

c.

d.

11 Mark all of the chiral centres on the following molecules:

12 Draw the enantiomer of lysine:

13 Are the chiral carbon atoms in the following molecules *R* or *S*?

14 Classify the following molecules as a nucleophile, an electrophile, or a radical species:

a.

b.

c.

d.

e.

f.

g.

13

Alkanes, alkenes, and alkynes

At the end of this chapter, students should be able to:

- Understand and explain the structural difference between alkanes, alkenes, and alkynes

- Suggest a test for the presence of an alkene in a molecule and explain the test in terms of the chemical reaction

- Understand the difference between sigma (σ) and pi (π) bonds in relation to alkanes, alkenes, and alkynes

- Be able to explain and describe the general reactivity of alkanes, alkenes, and alkynes.

13.1 Alkanes: an outline

Alkanes are arguably the simplest molecule that you will encounter during your chemistry studies and are a type of hydrocarbon. Strictly speaking, a hydrocarbon only contains the elements carbon and hydrogen. The carbon atoms can be joined together, making the backbone of the molecule a long chain (also called a *straight-chain alkane*), a long chain with branches, or a ring. An alkane only contains single bonds between the carbon and hydrogen atoms, with each carbon atom making four single bonds. A molecule that contains only single C—C and C—H bonds can also be called *saturated*. All alkanes that are in a straight chain must have molecular formula C_nH_{2n+2}, whereas alkanes that are cyclic have molecular formula C_nH_{2n}. In both cases, *n* is a whole number.

All alkanes that are in a straight chain have molecular formula C_nH_{2n+2}.

The naming of alkanes has already been covered in Chapter 12, so it will not be covered again here. However, in order to succeed in this chapter, you need to be able to name and draw the first 10 alkanes.

For a refresher on naming and drawing alkanes, see Section 12.2

Foundations of Chemistry: An Introductory Course for Science Students, First Edition.
Philippa B. Cranwell and Elizabeth M. Page.
© 2021 John Wiley & Sons Ltd. Published 2021 by John Wiley & Sons Ltd.
Companion website: www.wiley.com/go/Cranwell/Foundations

13.1.1 Alkanes and crude oil

Hydrocarbons are found in crude oil, which is a fossil fuel that was made naturally over millions of years by decomposition of dead plants and animals. Crude oil contains over 150 different hydrocarbons, including straight-chain alkanes, cyclic and branched alkanes, alkenes, and aromatic compounds. When the crude oil is extracted from the ground, these different components need to be separated from each other for application in their varied purposes. The different components are usually separated by fractional distillation, a technique that separates compounds based on their boiling point. The boiling point of a compound is dictated by the strength of the intermolecular forces, with stronger intermolecular forces leading to a higher boiling point.

For a refresher on intermolecular forces, see Chapter 2.

The longer a hydrocarbon chain is, the more intermolecular forces that can exist, and the higher the boiling point. Straight-chain alkanes generally have higher boiling points than their branched analogues; this is also related to structure. A straight-chain molecule can approach another straight-chain molecule more closely, leading to stronger intermolecular forces and therefore a higher boiling point.

> **Worked Example 13.1** Butane and 2-methylpropane have the same molecular formulae (C_4H_{10}); however, their boiling points at room temperature and pressure are –1 and –12 °C, respectively. Explain the difference between their boiling points.
>
> ***Solution***
> The difference in this case is purely due to structure, as both molecules contain the same number and type of atoms. The difference in boiling points is related to the ability of the chains to interact with each other and the possible intermolecular forces. As neither molecule contains a dipole, we only need to consider van der Waals forces (or London forces). Butane has a higher boiling point than 2-methylpropane, so the intermolecular forces holding the butane molecules together must be stronger. If we consider the structure of butane, due to the linear chain, two butane molecules can align reasonably closely and interact with each other quite strongly. However, molecules of 2-methylpropane cannot line up so closely due to the branching, so the intermolecular forces are weaker. This can be seen in the diagram, where the left-hand side are molecules of butane and on the right are 2-methylpropane:
>
>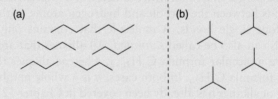

This key principle is crucial for fractional distillation where complex mixtures of crude oil are separated into simpler mixtures based on boiling points using a fractionating column (Figure 13.1). Each fraction contains a mixture of hydrocarbons with similar boiling points. If a pure hydrocarbon is required, the crude-oil fraction after fractional distillation can be distilled further with even greater control.

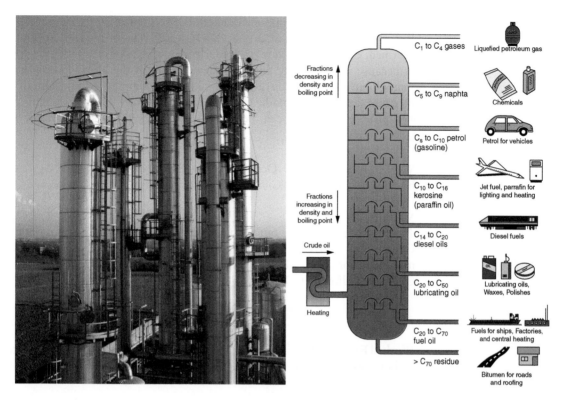

Figure 13.1 (a) A fractionating column. *Source:* Luigi Chiesa, https://commons.wikimedia.org/wiki/File:Colonne_distillazione. jpg, licensed under CC BY 3.0, (b) The fractions. *Source:* Based on http://year10polymers.blogspot.com/2015/07/from-fossil-fuels-to-plastics.html.

Crude oil is pumped into the bottom of the fractionating column (Figure 13.1) and heated to approximately 350 °C. At this temperature, all but the very long and heavy hydrocarbons boil, change state to gas, and evaporate. The very long hydrocarbons left behind (with more than 70 carbon atoms in their chains) are liquid and therefore drain out of the bottom of the fractionating column. These materials, known as *bitumen* or *asphalt*, are used for road surfaces and on roofs.

As the gases rise up the fractionating column, the temperature cools, and they start to condense back to liquids. The exact point at which this happens depends upon the compound in question. Longer hydrocarbon chains start to condense first because they have higher boiling points and require higher temperatures to evaporate.

The shortest, lightest hydrocarbons remain in the gaseous state for longer because they have a lower boiling point, due to weaker intermolecular forces. The different fractions have different uses, which can be seen in Figure 13.1.

> The column is very hot at the bottom, so a lot of energy is available to evaporate the liquids; but as the gas progresses up the column, the temperature cools.

13.1.2 Combustion of alkanes

Short-chain alkanes are often used for fuel; for example, petrol is a commonly used short-chain alkane that usually contains between 7 and 11 carbon atoms in the chain. To burn fully, a plentiful supply of oxygen is needed. If a hydrocarbon chain burns cleanly in oxygen, two products are formed: carbon dioxide, (CO_2), and water, (H_2O). If a compound does not burn completely – for example, due to a lack of oxygen – carbon monoxide, (CO), water, (H_2O), and carbon (soot) can be formed. Carbon monoxide is a colourless, odourless gas that is highly poisonous because it irreversibly binds to haemoglobin in the blood, preventing the haemoglobin binding to oxygen. In the home, incomplete combustion of fuels can cause carbon monoxide to be formed in faulty heating systems. This incomplete combustion can be caused by blocked flues or chimneys, or inadequate ventilation.

> **Worked Example 13.2**   Octane burns in the presence of oxygen to give carbon dioxide and water. Give the chemical equation for this process.
>
> *Solution*
> The first thing to work out is the molecular formula for each compound. Octane is a hydrocarbon containing eight carbon atoms; therefore it has molecular formula C_8H_{18} (C_nH_{2n+2}, where n = 8). Oxygen has formula O_2, carbon dioxide CO_2, and water H_2O:
>
> $$C_8H_{18} + 12.5O_2 \rightarrow 8CO_2 + 9H_2O$$

When balancing an equation, the number of atoms of each element must be the same on the left as it is on the right. For a refresher, see Chapter 0.

> **Worked Example 13.3**   When pentane is burnt in a limited supply of oxygen, carbon monoxide and water are produced. Give the chemical equation for this process.
>
> *Solution*
> As with Worked Example 13.2, the first thing to work out is the molecular formula for each compound. Pentane is a hydrocarbon containing five carbon atoms and therefore has molecular formula C_5H_{12} (C_nH_{2n+2}, where n = 5). Oxygen has formula O_2, carbon monoxide CO, and water H_2O:
>
> $$C_5H_{12} + 5.5O_2 \rightarrow 5CO + 6H_2O$$

13.1.3 Cracking alkanes

A zeolite is an alumoinosilicate species that is often used as a catalyst. They consist mainly of silicon, aluminium and oxygen atoms alongside other trace metals

After crude oil has been subjected to fractional distillation, there are lots of long-chain hydrocarbons but not so many short-chain hydrocarbons. The short-chain hydrocarbons are in higher demand than long-chain hydrocarbons because they are used as fuels and can be used in polymer and plastic production. *Cracking* is one way in which long-chain saturated hydrocarbons are broken into shorter chains or derivatives that are more useful.

Cracking breaks long-chain saturated hydrocarbons into alkanes and alkenes.

Thermal cracking was first used in the 1910s but was improved in the 1930s by using Al_2O_3 and SiO_2 as catalysts; however, there was not much control over the lengths of the shorter chains produced. Nowadays, zeolites are used as catalysts, and the temperature used is usually around 450 °C. When cracking alkanes, there are three possible outcomes, each of which can be controlled by the catalyst used:

- Straight-chain alkanes can break down randomly to shorter alkane(s) and alkene(s).

• Straight-chain alkanes can undergo isomerisation to branched alkanes.

• Straight-chain alkanes can produce cyclic hydrocarbons.

Forming branched and cyclic alkanes from straight-chain alkanes is important because it improves the octane number (ON), which is related to how well a fuel burns. Fuels with a high ON (i.e. near 100) burn cleanly, but fuels with a lower ON burn less efficiently. Branched and cyclic alkanes promote more efficient burning in fuels than their straight-chain counterparts and so are frequently added to fuels.

Worked Example 13.4 Straight-chain octane is cracked into two smaller fragments, one of which is an alkene that contains three carbon atoms. Draw the structures of the products.

Solution
Octane is a hydrocarbon containing eight carbon atoms and therefore has molecular formula C_8H_{18}. Cracking breaks hydrocarbons down into an alkene and an alkane. The alkene contains three carbon atoms; therefore, the alkane product must contain five carbon atoms due to mass balance. The newly formed alkene must always be at the end of the chain (terminal):

The newly formed alkene must always be at the end of the chain because this is the point at which the longer chain is severed.

octane   propene   +   pentane

There are many outcomes for the cracking of octane. Usually, the question will specify the intended products.

Worked Example 13.5 Hexane can be isomerised into 2,2-dimethylbutane or cyclised, giving a cyclic alkane and hydrogen gas. Give an equation for both of these outcomes.

Solution
Hexane is a hydrocarbon containing six carbon atoms and therefore has molecular formula C_6H_{14}. The isomerised product contains four carbon atoms in the longest chain, and there are two methyl groups in the 2-position. When undergoing isomerisation, the molecular formula of the starting material must match that of the product:

hexane   2,2-dimethylbutane
C_6H_{14} C_6H_{14}

When undergoing isomerisation, the numbers and types of each atoms in the original molecule must match that of the products.

When undergoing cyclisation, two hydrogen atoms are formally lost in the form of hydrogen gas, H_2. Again, the number of elements (i.e. atoms of each type) on either side of the equation must balance:

hexane   cyclohexane   + H_2
C_6H_{14} C_6H_{12}

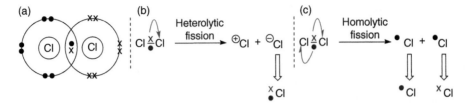

Figure 13.2 (a) The bonding between two atoms of chlorine in the molecule Cl_2; (b) heterolytic fission, where both electrons move to one atom; (c) homolytic fission, where one electron moves to each atom.

A radical contains a single, unpaired electron.

13.1.4   Reactions of alkanes: radicals

The reaction of an alkane with a halogen requires radical species. A radical is formed when an atom has a single, unpaired electron and comes about by covalent bond breaking. Remember, in a covalent bond, there are two electrons: usually one from each bonding partner. When a single covalent bond breaks, the electrons in the bond can move in one of two ways. They can either both move to the same atom or each move to a different atom (Figure 13.2).

Consider a chlorine molecule, Cl_2, where each chlorine atom has contributed one electron to the single covalent bond, allowing each element to gain the octet of electrons that it requires to be stable (Figure 13.2a). When breaking that Cl—Cl bond, there are two options: the electrons can both move to the same atom (as in Figure 13.2b), or one electron can move to each atom (as in Figure 13.2c).

If both electrons move to the same atom, one chlorine atom has gained one more electron than it had originally and therefore has a single negative charge. The other chlorine atom has lost one electron, the one it had used to bond to the other chlorine atom, so it has a single positive charge. In this case, the bond has broken such that there are two different species; therefore, it is termed *heterolytic bond cleavage.*

Homolytic bond cleavage forms radicals; heterolytic bond cleavage forms ionic species.

If each electron in the bond moves to a different atom, the original chlorine atoms have each gained their original electron back and so are not charged. In this case, the bond has broken such that two species each contain a single unpaired electron; this is termed *homolytic bond cleavage.* An atom with a single unpaired electron is called a radical.

Alkanes are generally very unreactive due to the strength of the C—H and C—C bonds. However, one situation in which they will react is with a halogen in the presence of UV light. The UV light is crucial; if this is not present, the alkane and halogen will not react because the reaction cannot be initiated. An example of such a reaction is methane with chlorine gas to give chloromethane and hydrogen chloride:

$$CH_4 + Cl_2 \xrightarrow{UV} CH_3Cl + HCl$$

This reaction is a substitution reaction; overall, one hydrogen atom is swapped with one chlorine atom. The mechanism of this reaction, however, is not so simple. When halogens such as chlorine are exposed to UV light, the single bond breaks homolytically, generating radicals. These radicals are highly reactive and are able to break C—H covalent bonds, which are usually unreactive. You need to know three steps for this:

1. Initiation (radicals are formed)

2. Propagation (radicals continue to be regenerated during the reaction)

3. Termination (radicals react together to make a species where there are no more radicals)

13.1.4.1 Initiation

As stated, *initiation* is where a radical is formed. When considering methane and molecular chlorine, methane only contains strong C—H bonds that cannot easily be broken by UV light, whereas chlorine has a comparatively weak Cl—Cl bond. It is this bond that breaks to form the radicals:

$$Cl_2 \rightarrow 2Cl^{\bullet} \qquad \text{Initiation}$$

13.1.4.2 Propagation

Propagation is where a radical reacts with another non-radical species, generating a new radical. There are two propagation steps. In the first step, the chlorine radical ($Cl^{\bullet}$) reacts with methane to generate HCl and a new methyl radical. It is important to note that the radical generated resides on the carbon atom. In the second step, the methyl radical reacts with another chlorine molecule to generate chloromethane and a chlorine radical:

$$Cl^{\bullet} + CH_4 \rightarrow HCl + {}^{\bullet}CH_3 \qquad \text{Propagation step 1}$$
$$^{\bullet}CH_3 + Cl_2 \rightarrow CH_3Cl + Cl^{\bullet} \qquad \text{Propagation step 2}$$

13.1.4.3 Termination

Termination is when radicals react to make another species, resulting in no more radicals. Usually, this is when two radicals react together. In this reaction, three termination steps are possible as two radicals were formed, $Cl^{\bullet}$ and $^{\bullet}CH_3$:

$$Cl^{\bullet} + Cl^{\bullet} \rightarrow Cl_2 \qquad \text{Termination step 1}$$
$$^{\bullet}CH_3 + Cl^{\bullet} \rightarrow CH_3Cl \qquad \text{Termination step 2}$$
$$^{\bullet}CH_3 + {}^{\bullet}CH_3 \rightarrow C_2H_6 \qquad \text{Termination step 3}$$

Once chloromethane has been formed, it can react by the propagation steps in place of methane, and eventually generate products with two (CH_2Cl_2), three ($CHCl_3$), or even four chlorine atoms (CCl_4).

The reaction of halogens with alkanes is the only radical reaction that you are likely to encounter during your foundation studies. Drawing the mechanism is a little different to other reactions that you encounter because it uses 'fishhook' arrows where only half of the arrow point is drawn rather than the classical double-headed arrows we use for two-electron processes. Drawing this mechanism using curly arrows probably is not necessary for your syllabus, but if you can do it and know where each of the electrons is moving, it will help you in other

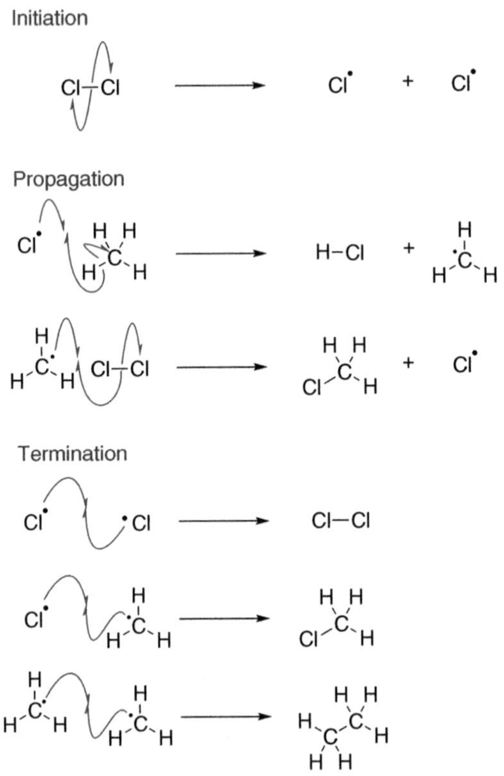

Initiation

Propagation

Termination

Figure 13.3 Curly arrows showing the movement of electrons in radical substitution. Note that the curly arrows used are fishhook arrows with one head; this denotes that only one electron is moving each time.

areas of the course. Figure 13.3 shows the movement of electrons using curly arrows.

In the initiation step, the Cl—Cl bond breaks homolytically, shown by the fishhook arrows moving away from the centre of the bond towards each chlorine atom. This generates two chlorine radicals, each of which has an unpaired electron. In the propagation step, the chlorine radical removes (or *abstracts*) a proton from the methane molecule. In the first propagation step, one of the electrons in the covalent C—H bond moves to join the radical on the chlorine, making a new Cl—H bond, and the other electron in the C—H bond moves back onto the carbon, making a methyl radical. It is important to note that at this point, the radical (electron) cannot move away from this carbon atom, so this is the carbon atom that has to undergo substitution. This is particularly important to note if there is more than one carbon atom in the alkane substrate.

In the second propagation step, the methyl radical reacts with the weakest bond in the system: the Cl—Cl bond. The Cl—Cl bond is again broken homolytically, with the two electrons of the Cl—Cl bond moving so that one moves to bond to the carbon radical, making a new C—Cl bond, and the other moves generating a new chlorine radical. The final steps are the three termination steps, where in all cases the radicals move together to generate a new covalent bond, which is shown by the curly arrows. Three products are possible: chlorine, chloromethane, and ethane.

Box 13.1 Radical golden rules:

1. To make a radical, look for the weakest bond in the system (usually a halogen–halogen bond).

2. Once a radical is made, it is extremely reactive and will readily react with the next-weakest bond in the system (propagation).

3. Radical curly arrows have only half an arrowhead (like a fishhook) as only one electron is moving.

13.2 Alkenes: an outline

An *alkene*, in its simplest form, is a molecule that contains carbon and hydrogen, so it belongs to the hydrocarbon class of compounds. However, it also contains carbon–carbon double bonds. Alkenes are often referred to as *unsaturated* because of these double bonds. Two naturally occurring compounds containing alkenes are shown in Figure 13.4. Linoleic acid is one of two essential fatty acids that the human body cannot synthesise, so has to be ingested through diet. It is found in sunflower oil, nuts, and fatty seeds. It is sometimes referred to as a *polyunsaturated* compound because of the two alkenes. Limonene is another alkene that occurs naturally and is found in the peel of citrus fruit. Limonene contains a chiral centre and exists in both *R* and *S* forms.

Chirality is discussed in Chapter 12.

In analogy with alkanes, the carbon atoms in alkenes can be joined together either in a long chain (also called a *straight-chain alkene*) or a ring. All straight-chain alkenes with one double bond have molecular formula C_nH_{2n}, whereas cyclic alkenes with one double bond have the molecular formula C_nH_{2n-2}. Alkenes can form *E* and *Z* isomers; these are discussed in Section 12.3.4, along with naming them. Alkenes are more reactive than alkanes due to the presence of the double bond.

Alkenes are sometimes referred to as olefins.

All alkenes that are in a straight chain with one double bond have molecular formula C_nH_{2n}.

13.2.1 Bonding in alkenes

Thus far, we have referred to *double bonds* with little regard to what this actually means. A double bond is where two atoms are joined together by two covalent bonds. However, these two bonds, although they are both covalent, are not the same and the electrons in each bond are in a slightly different location. Overall, a double bond is stronger than a single bond because there are two bonds between the atoms, but it is not twice as strong.

Within organic chemistry, there are three types of covalent bonding: *sigma* (σ), *pi* (π), and *dative*. Sigma (σ) and pi (π) bonds are both present in alkenes and are discussed further in this section. Dative covalent bonding was discussed in

The bond energy of a C—C single bond is $+347\,kJ\,mol^{-1}$, whereas the bond energy of a C=C double bond is $+612\,kJ\,mol^{-1}$.

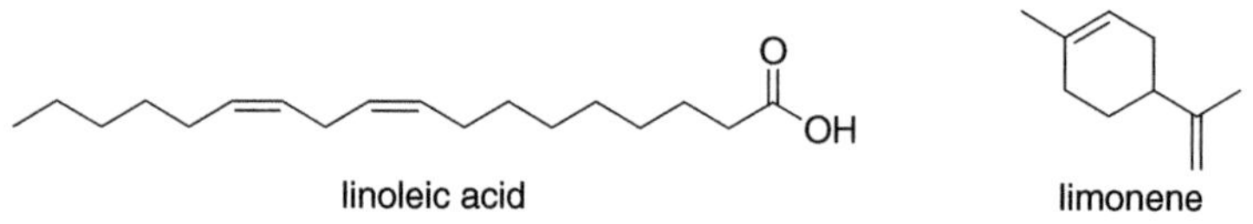

Figure 13.4 Naturally occurring alkenes linoleic acid and limonene.

Chapter 2 and occurs when both electrons in a covalent bond come from one atom. If you need a refresher on covalent bonding in general, see Chapter 2 before starting this section.

13.2.2   Sigma (σ) bonding

A sigma (σ) bond is also known as a *single bond*. In this case, the two electrons that contribute to the bond are directly between the nuclei of the bonding atoms. When there is a single bond between two atoms, the bond is a sigma (σ) bond; a dative covalent bond is a type of sigma (σ) bond. There can be only one sigma (σ) bond between two elements.

> In a sigma bond, the electrons are aligned along the bonding axis, between the two bonding nuclei.

If we consider a single C—H bond, the two electrons that make this bond are situated between the carbon and hydrogen nuclei and are therefore in a sigma (σ) bond. One way of showing these electrons is by using an 'electron cloud' (Figure 13.5a). In a molecule of methane, CH_4, we can see that there are four single bonds, and each is drawn as a single line; therefore, each of the bonds is a sigma (σ) bond (Figure 13.5b). The electron cloud for each of these bonds is shown in Figure 13.5c. The electron cloud also represents a *bonding orbital*: in Figure 13.5c, the four C—H bonding orbitals drawn.

> The notations σ and π are often used on diagrams to show the type of bond present.

In methane, CH_4, the four sigma (σ) bonds are non-polar because carbon and hydrogen have very similar electronegativities and the electrons are therefore evenly distributed through the bond. In bonding situations where the two elements contributing to the sigma (σ) bond have different electronegativities, the electron cloud is distended towards the more electronegative element, causing a dipole. Polar and non-polar bonds were discussed in Section 2.3.2. An example can be seen in chloromethane in Figure 13.6. Uneven distribution of electron density is important when discussing how a molecule may react when subjected to different conditions.

13.2.3   Pi (π) bonding

Pi (π) bonding is often confused by students, possibly from misconceptions in prior study. A pi (π) bond is formed when the electron density of the bond is above and below the plane of the nuclei that are forming the bond. It is *not* a double bond, which has both sigma (σ) and pi (π) components. Pi (π) bonds are made from two p orbitals overlapping sideways on; therefore, a pi (π) bond has no electron density directly between the two nuclei, which explains its comparative weakness compared to a sigma (σ) bond.

> A double bond has both sigma (σ) and pi (π) components.

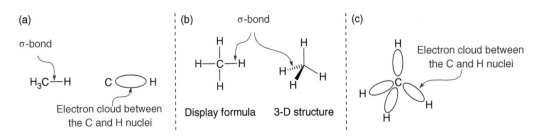

Figure 13.5 (a) A sigma (σ) bond and an electron cloud; (b) the structure of methane; (c) the electron clouds in methane.

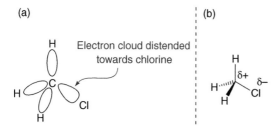

Figure 13.6 The electron clouds in chloromethane and how they are distended towards the electronegative chlorine; (b) the dipole at the C—Cl bond in chloromethane.

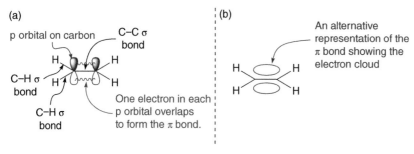

Figure 13.7 (a) The p orbitals on ethene and their overlap to form a π bond; (b) an alternative representation of the π bond showing the electron cloud.

Let's consider ethene, C_2H_4. Ethene contains a double bond; but as stated earlier, there can be only one sigma (σ) bond between two atoms, so the two bonds between the central carbon atoms cannot both be sigma (σ). The other bond therefore has to have electron density above and below the sigma (σ) bond; it is a pi (π) bond. A pi (π) bond is made when two p orbitals overlap sideways on (Figure 13.7).

For a refresher on the shapes of orbitals, see Chapter 1.

Worked Example 13.6   Draw the structure of formaldehyde (CH_2O), and clearly show the sigma (σ) and pi (π) bonds.

Solution
Formaldehyde is the smallest molecule that contains a carbonyl group (C=O). The carbonyl group contains a double bond between the C and O, which has both sigma (σ) and pi (π) components, as well as two further sigma (σ) C—H bonds:

Carbonyls will be covered in more detail in Chapter 14.

(a)

p orbital on carbon   C–O σ bond
C–H σ bond   p orbital on oxygen
C–H σ bond

(b)

An alternative representation of the π bond showing the electron cloud

13.2.4   Testing for alkenes

Alkenes are more reactive than alkanes due to the presence of the carbon–carbon double bond, in particular the pi (π) component of the double bond. When testing for an alkene, it is this difference that is exploited. The standard test for an alkene is to shake it in bromine water. If an alkene is present, the bromine water changes from brown to colourless because the bromine has reacted with the alkene (Figure 13.8).

The alkene reacts with the bromine water because the electrons in the pi (π) bond are more available compared to a sigma (σ) bond, so it is more reactive. The pi (π) bond contains two electrons that are able to attack other reagents (in this case bromine) that are electron-poor; the pi (π) bond acts as a nucleophile and the bromine as an electrophile.

For a refresher on electrophiles and nucleophiles, see Section 12.4.2.

Let us consider the reaction of ethene with bromine to give 1,2-dibromoethane (Figure 13.9). In order for this reaction to occur, the bromine molecule needs to collide with the double bond in ethene. As the bromine molecule approaches, the electrons in the double bond repel the electrons in the Br—Br bond, causing an induced dipole and leaving the bromine furthest from the alkene with a slight negative charge (δ–) and the closest bromine atom with a slight positive charge (δ+). This causes the electrons in the pi (π) bond of the

Figure 13.8 (a) Two test tubes, the left containing a mixture of bromine water and cyclohexene, the right bromine water and cyclohexane at t = 0. (b) After shaking for five minutes. The left contains a mixture of decolourised bromine water (lower layer) and cyclohexene (upper layer), the right bromine water (lower layer) and cyclohexane (upper layer). *Source:* Images reproduced with kind permission of Dr Jenny Eyley.

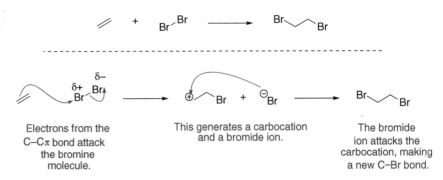

Electrons from the C–Cπ bond attack the bromine molecule.

This generates a carbocation and a bromide ion.

The bromide ion attacks the carbocation, making a new C–Br bond.

Figure 13.9 Reaction of bromine with ethene.

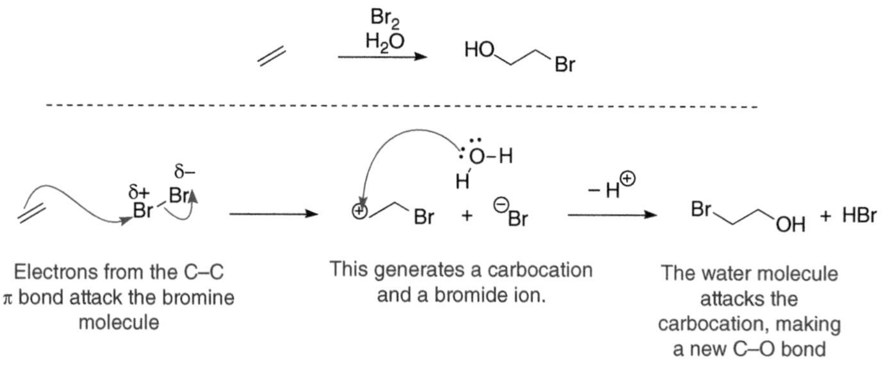

Figure 13.10 Reaction of bromine water with ethene.

alkene to move towards the δ+ bromine atom; the electrons in the sigma (σ) bond remain untouched. This is shown by a curly arrow starting at the alkene (using the pair of electrons in the weaker pi (π) bond) and ending at the δ+ bromine atom, as shown. The electrons from the alkene, moving towards the δ+ bromine atom, start to make a new bond to the bromine atom. The bromine atom can make only one bond in total (otherwise it would have too many electrons and would violate the octet rule); therefore, the bond to the other bromine atom has to break. Both of the electrons in the pi (π) bond are used to make the new C–Br bond, and both of the electrons in the Br–Br bond are given to the leaving bromine atom. This electron movement means that one of the carbon atoms originally in the alkene has only three bonds, and it is positively charged. The bromine atom that has left has an extra electron, so it has a negative charge and forms a bromide ion (Br^-). In a second step, the newly released bromide ion attacks the newly formed positively charged carbon atom to form another C–Br bond.

In reality, when using bromine water, there is much more water present than bromide ions. So the product is actually a bromohydrin (a compound containing an alcohol and a bromine), formed from attack of the carbocation by water, rather than bromide (Figure 13.10).

13.2.5   Reaction of alkenes with electrophiles

The reaction of an alkene with bromine is an example of *electrophilic addition*. Electrophilic addition is where an electrophile is added to a nucleophilic centre. The addition of bromine across an alkene is not the only example of electrophilic addition, and alkenes will react with a range of electrophiles. The example that will be discussed is the reaction of prop-1-ene with HBr (Figure 13.11). In this reaction, two products can be formed, although one of them is preferred.

Figure 13.11 Reaction of prop-1-ene with HBr.

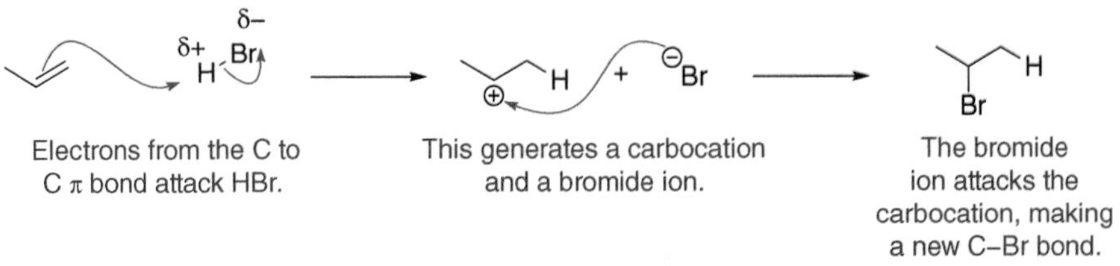

Electrons from the C to C π bond attack HBr.

This generates a carbocation and a bromide ion.

The bromide ion attacks the carbocation, making a new C–Br bond.

Figure 13.12 Curly arrows to show the reaction of prop-1-ene with HBr.

Before a reaction could happen between a bromine molecule and an alkene, the alkene had to induce a dipole in the bromine molecule (Figure 13.9). However, in the case of reacting an alkene with HBr, there is already a permanent dipole (due to the electronegativity difference between bromine and hydrogen atoms), so this is not necessary. When the HBr molecule approaches the prop-1-ene, the electrons in the pi (π) bond move to attack the δ+ hydrogen atom; and because the hydrogen atom can only make one bond, the electrons in the H—Br bond move towards the δ– bromine atom, causing the bond to break and generate Br⁻ (bromide ion). Both of the electrons in the pi (π) bond have been used to attack the HBr, so a carbocation is formed. Following the arrows in Figure 13.12, the carbocation has been formed on the carbon atom that has two other carbon atoms bonded to it (a secondary carbocation) because the curly arrow is drawn such that electrons move from the left to the right. The bromide ion formed by cleavage of the H—Br bond then attacks the carbocation, generating a new C—Br bond. The resulting product is 2-bromopropane.

There is another outcome, however, as the electrons in the alkene did not have to travel such that the hydrogen atom became bonded to the terminal carbon atom; they could move so the new C—H bond is on the central carbon atom (Figure 13.13). This would generate a carbocation at a carbon atom that is only bonded to one other carbon atom (a primary carbocation), which could then be attacked by the bromide ion, generating a new C—Br bond. The resulting product in this case would be 1-bromopropane.

In fact, in our first example (Figure 13.12), the secondary carbocation is more stable than a primary carbocation, as it is lower in energy. The reaction will follow the lower-energy pathway; hence 2-bromopropane is formed preferentially. The secondary carbocation is more stable because it is bonded to two other carbon atoms. Each carbon chain that is bonded to the carbocation is able to push electron density towards the carbocation through the sigma (σ) bonds through a process called *induction* (Figure 13.14). A further stabilisation process is called hyperconjugation and uses neighbouring C—H bonds, but this is beyond the scope of a foundation-level course. However, you should know that the

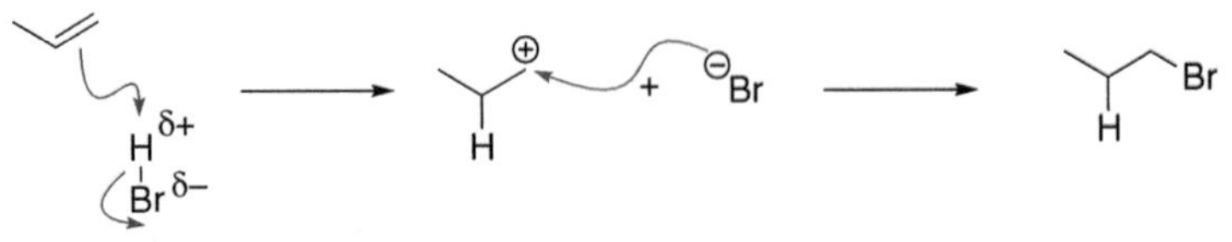

Figure 13.13 Curly arrows to show an alternative reaction of prop-1-ene with HBr.

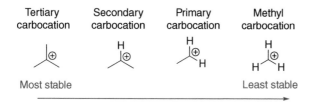

Figure 13.14 Stability of carbocations.

reaction pathway proceeds through the more stable carbocation, and the product that results from this will be the favoured product.

Worked Example 13.7 Give a curly-arrow mechanism for the reaction of but-1-ene with HCl, and predict the dominant product.

Solution
This example is very similar to the explanations already given: an unsymmetrical alkene reacts with an electrophile that already has a dipole. The first thing to ascertain is which partner is the nucleophile and which the electrophile. Here we have an alkene and HCl, the latter having a dipole with a $\delta+$ H and a $\delta-$ Cl. The alkene is able to attack the $\delta+$ H, leaving a carbocation. There are two possibilities: a primary carbocation (less stable) or a secondary carbocation (more stable, comparatively). Attack of the primary carbocation by chloride gives 1-chlorobutane, and the secondary carbocation 2-chlorobutane. In terms of the dominant product, the secondary carbocation in pathway B is more stable; therefore, 2-chlorobutane dominates as the product.

Pathway A:

primary
carbocation

Pathway B

secondary
carbocation

13.2.6   General reactions of alkenes

A further two reactions of alkenes need to be considered: the reaction of an alkene with steam in the presence of phosphoric acid, H_3PO_4, to generate an alcohol and reduction of an alkene with hydrogen in the presence of a metal catalyst to produce an alkane.

The first reaction, to generate an alcohol, proceeds in much the same way as the electrophilic addition; but in this case, the first step is protonation of the

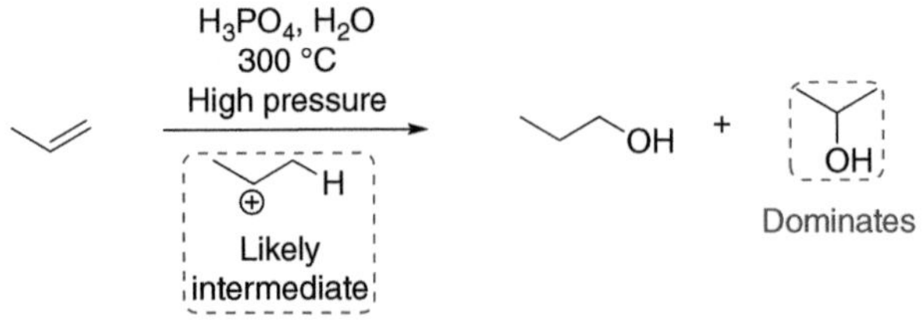

Figure 13.15 Reaction between prop-1-ene, phosphoric acid, and water.

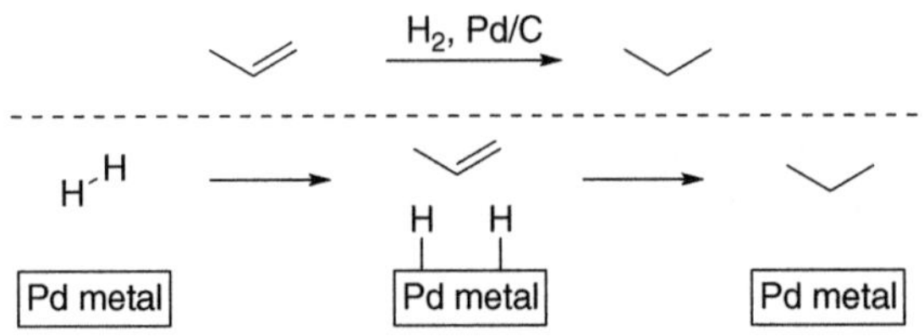

Figure 13.16 Reaction of an alkene with hydrogen in the presence of a metal catalyst.

alkene by the H⁺ present in phosphoric acid and subsequent attack by water (Figure 13.15). This reaction needs to be undertaken at high temperature (300 °C) and high pressure (60–70 atm).

The mechanism of the second reaction is slightly different and uses a metal catalyst, such as palladium, as a means to activate the hydrogen so that the alkene can attack. The mechanism for this is likely to be concerted, where the metal breaks the H–H bond, holding the hydrogen atoms so the alkene can react (Figure 13.16). Converting an alkene to an alkane is also referred to as *hydrogenation*; and because hydrogen is being added to the molecule, the reaction is classed as a reduction.

13.3   Alkynes: an outline

In its simplest form, an *alkyne* is a molecule that contains carbon and hydrogen atoms, so it belongs to the hydrocarbon class of compounds. Alkynes contain carbon–carbon triple bonds. They are also referred to as *unsaturated* compounds because of these triple bonds, and they are quite reactive towards electrophiles. Whereas alkenes contain one pi (π) bond, alkynes contain two; one is above and below the plane, and the second pi (π) bond is found to the left and right of the plane (going in front of and behind the page) (Figure 13.17). Alkynes are linear, with a bond angle of 180° between the two C–H bonds in the case of ethyne.

13.3.1   General reactions of alkynes

Alkynes can react with hydrogen in the presence of a metal catalyst to form either an alkene or an alkane. This is known as *reduction*. To convert an alkyne to an alkane, hydrogen in the presence of palladium on carbon is used. To convert an alkyne to an alkene, hydrogen is required, but the palladium needs to be

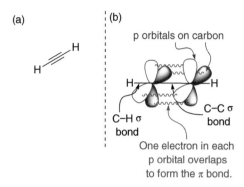

Figure 13.17 (a) A molecule of ethyne; (b) the p orbitals on ethyne and how they overlap to form a π bond.

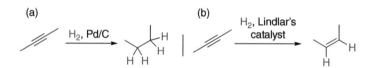

Figure 13.18 (a) Reduction of but-2-yne to propane with hydrogen and palladium; (b) reduction of but-2-yne to propene with hydrogen and a poisoned palladium (Lindlar's catalyst).

'poisoned' so it is less active and does not lead to over-reduction. The usual catalyst for this is Lindlar's catalyst: palladium on calcium carbonate that is poisoned with lead or sulfur. The palladium is poisoned to prevent the product alkene being reduced all the way to the alkane. When forming an alkene, the hydrogens are added to the same side, giving a *cis-* alkene (Figure 13.18).

Quick-check summary

- There can be single, double, or triple bonds between carbon atoms, which lead to the functional groups alkanes, alkenes, and alkynes, respectively.

- Alkanes contain a carbon–carbon single bond, alkenes contain a carbon–carbon double bond, and alkynes contain a carbon–carbon triple bond.

- Structural differences in molecules lead to differences in boiling and melting points.

- Hydrocarbons can be separated according to boiling point by using fractional distillation.

- Alkanes are often used as fuels. When they are burnt, they release carbon dioxide (CO_2) and water (H_2O).

- Long-chain hydrocarbons can be 'cracked' and broken into shorter-chained derivatives. Cracking a hydrocarbon leads to a short alkane and an alkene.

- Subjecting an alkane to UV light in the presence of a halogen leads to halogenation of the alkane.

- Three discrete steps are required for a radical reaction: initiation, propagation, and termination.

- Alkenes can be detected using bromine water as a chemical test. A positive result is when the bromine water changes from orange/brown to colourless.

- Alkanes contain sigma (σ) bonds.

- Alkenes and alkynes contain both pi (π) and sigma (σ) bonds.

- Alkenes can react with electrophiles.

- Alkynes react with hydrogen gas to form either alkanes or alkenes.

End-of-chapter questions

1 What differences would you expect to see in the boiling points of hexane and 2,3-dimethylbutane? Explain your answer.

2 Write a balanced equation for the complete combustion of decane.

3 If 15.0 g of hexane is burned completely, how much CO_2 is made? Give your answer in both grams and moles.

4 Decane can be cracked into pentane and another product. Draw the equation for this process.

5 Draw ethene. Clearly label the σ and π bonds.

6 Draw ethyne. Clearly label the σ and π bonds.

7 Pentane reacts with bromine in the presence of UV light to give a monobrominated product.

   a. Write out the equations for this process.
   b. Draw the mechanism for each step.

8 Draw the product for the reaction of 2-chloropropene with bromine. Give a mechanism for this reaction.

9 Give a mechanism for the reaction of phenylethene with hydrogen bromide, HBr.

10 How would hex-1-yne be reduced to each of the following?
   a. hexane

   b. hex-1-ene

14

Reactivity of selected homologous series

At the end of this chapter, students should be able to:

- Explain the reactivity and bonding of alcohols

- Define and use the terms primary, secondary, and tertiary in relation to alcohols

- Understand that aldehydes, ketones, and carboxylic acids can be formed from alcohols

- Be able to describe and explain the formation and hydrolysis of esters and amides

- Explain the bonding in nitriles and predict their reactivity

- Describe and explain the reactivity of amines

- Be able to suggest synthetic routes to simple organic compounds using functional-group interconversions

14.1 Alcohols

An **alcohol** is a molecule that contains a discrete hydroxyl (OH) group. The OH group is the part of the molecule that dictates the reactivity and many of the properties of this homologous series. The OH group can undergo intermolecular **hydrogen bonding**, which means that alcohols have relatively high melting and boiling points in comparison to their alkane equivalents (Figure 14.1 and Table 14.1). Hydrogen bonding is discussed in more detail in Chapter 2.

In addition, the presence of the OH group means that if another molecule is present that can undergo hydrogen bonding, the alcohol can hydrogen bond to

Foundations of Chemistry: An Introductory Course for Science Students, First Edition.
Philippa B. Cranwell and Elizabeth M. Page.
© 2021 John Wiley & Sons Ltd. Published 2021 by John Wiley & Sons Ltd.
Companion website: www.wiley.com/go/Cranwell/Foundations

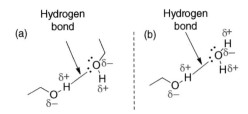

Figure 14.1 (a) Hydrogen bonding between two ethanol molecules; (b) hydrogen bonding between ethanol and water.

Table 14.1 A comparison of boiling points between alkanes and the corresponding alcohols.

Alkane	Boiling point (°C)	Alcohol	Boiling point (°C)
Methane	−161.0	Methanol	64.7
Ethane	−88.0	Ethanol	78.0
Propane	−42.1	Propan-1-ol	97.0
Butane	−0.5	Butan-1-ol	116.0–118.0
Pentane	35.0–36.0	Pentan-1-ol	136.0–138.0

that, too (Figure 14.1b). This has implications for solubility: short-chain alcohols are usually soluble in water and other polar solvents. In fact, the first three alcohols in the homologous series are fully miscible/soluble in water, but their solubility decreases as the hydrocarbon chain length increases. This is because the longer hydrocarbon chain cannot hydrogen bond with water; therefore, the alcohol becomes less soluble in water.

There are three main classifications for alcohols:

- Primary

- Secondary

- Tertiary

This classification relates to the position of the alcohol group in the chain and what else is bonded to the carbon directly adjacent to the OH. This is important because the class of OH group has a bearing on the reactivity.

14.1.1   Primary alcohols

An alcohol is classified as primary when the OH group is bonded to a carbon atom that is bonded to one alkyl chain. Ethanol and propan-1-ol are both examples of primary alcohols (Figure 14.2). Another example is pentan-1-ol, where the carbon atom to which the alcohol is bonded is only bonded to one other alkyl group.

14.1.2   Secondary alcohols

An alcohol is classified as secondary when the OH group is bonded to a carbon atom that is bonded to two alkyl chains. An example of a secondary alcohol is propan-2-ol, where the carbon atom bonded to the OH is bonded to two alkyl chains and one hydrogen atom (Figure 14.3).

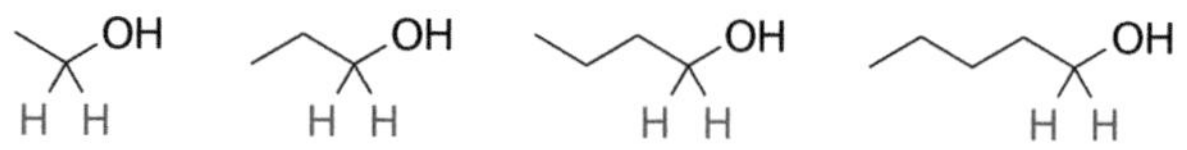

Figure 14.2 Structure of the primary alcohols ethanol, propan-1-ol, butan-1-ol and pentan-1-ol. Note that in a skeletal formula, the hydrogen atoms would not normally be shown; but in this case, selected hydrogen atoms have been added to aid clarity.

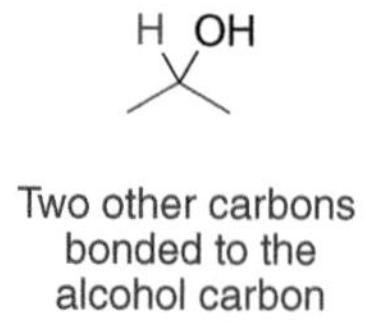

Figure 14.3 Structure of a secondary alcohol, propan-2-ol. Note that in a skeletal formula, the hydrogen atoms would not normally be shown; but in this case, selected hydrogens have been added to aid clarity.

Figure 14.4 Structure of a tertiary alcohol, 2-methylpropan-2-ol.

14.1.3   Tertiary alcohols

An alcohol is classified as tertiary when the OH group is attached to a carbon atom that is bonded to three alkyl chains. An example of a tertiary alcohol is 2-methylpropan-2-ol (Figure 14.4).

14.1.4   Combustion of alcohols

In a similar fashion to alkanes, alcohols burn in an excess of oxygen to form carbon dioxide, CO_2, and water, H_2O:

$$C_2H_5OH + 3O_2 \rightarrow 2CO_2 + 3H_2O$$

In analogy with alkanes, if they are burnt in a deficiency of oxygen, they also generate the dangerous gas carbon monoxide.

14.1.5   Oxidation of alcohols

Both primary and secondary alcohols can be oxidised to a carbonyl-containing compound, whereas tertiary alcohols cannot.

Box 14.1 Within organic chemistry, the use of the term **oxidation** can mean the addition of oxygen; the removal of hydrogen; increasing the number of bonds to oxygen; or the loss of electrons. This relates to the OIL RIG mnemonic, as formally, the oxidation number of the atom in question increases.

14.1.5.1 Primary alcohols

A primary alcohol can be oxidised either to an aldehyde or a carboxylic acid, depending upon the reaction conditions (Figure 14.5). The identity of the oxidation product can be controlled by either varying the reagents used or stopping the reaction at the intermediate aldehyde. It should be noted that the opposite process to oxidation is **reduction**, and it is possible to reduce a carboxylic acid to either an aldehyde or an alcohol by choosing the correct reagent. This will be discussed further in Section 14.2.1.

The usual condition cited in an A-level or foundation-level course for achieving the oxidation of primary alcohols is using acidified potassium dichromate: $K_2Cr_2O_7$ in the presence of H_2SO_4. This is shown for ethanol in Figure 14.6. In the first step, ethanol is oxidised to ethanal. Ethanal can be collected if it is removed during the reaction by distillation, because this will ensure that it cannot react any further (as it has been removed to another vessel). If the ethanal is not collected by distillation but is allowed to heat at reflux in the presence of acidified potassium dichromate, it will be further oxidised to ethanoic acid. In reality, many other conditions for oxidation could be used, as potassium dichromate is not always the most suitable reagent – especially as it can oxidise other functionality present in the molecule (for example alkenes), as well as any other primary or secondary alcohols. It is important to note that in the mechanism of oxidation, the aldehyde must be made first *before* further oxidation to the carboxylic acid can occur.

The oxidation of a primary alcohol to an aldehyde or ketone using acidified potassium dichromate is always accompanied by a colour change from orange to green as the reaction progresses. This is because potassium dichromate contains

It is extremely important that you understand the difference between distillation and reflux. Both are sets of reaction conditions, not chemical processes. *Distillation* allows a volatile product to be collected in a separate flask, whereas *reflux* describes when a compound/solvent is boiled and the vapours are returned to a reaction flask, typically over an extended period. In this example, the chemical process is an 'oxidation', and the conditions are either distillation (to give the aldehyde) or reflux (to give the carboxylic acid).

Figure 14.5 Possible oxidation states of primary alcohols.

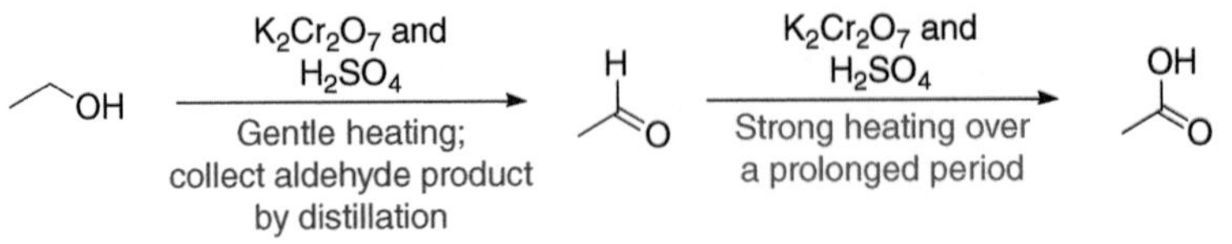

Figure 14.6 Oxidation of ethanol to ethanal or ethanoic acid using acidified potassium dichromate.

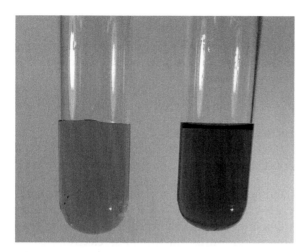

Figure 14.7 The dichromate(VI) ion is reduced to chromium(III) when it is used to oxidise an alcohol. The test tubes show the $K_2Cr_2O_7$ solution before oxidation of the alcohol (orange) and after oxidation of the alcohol (green). *Source:* Image reproduced with kind permission from Dr Jenny Eyley.

chromium(VI), which is orange, and is reduced to a chromium(III) species that is green (Figure 14.7).

When oxidising ethanol to ethanal, the half-equations are as follows:

$$CH_3CH_2OH \rightarrow CH_3CHO + 2e^- + 2H^+ \quad \text{Alcohol oxidation}$$

$$Cr_2O_7^{2-} + 14H^+ + 6e^- \rightarrow 2Cr^{3+} + 7H_2O \quad \text{Chromium reduction}$$

Overall equation:

$$3CH_3CH_2OH + Cr_2O_7^{2-} + 8H^+ \rightarrow 3CH_3CHO + 2Cr^{3+} + 7H_2O$$

14.1.5.2 Secondary alcohols

A secondary alcohol can only be oxidised to a ketone (Figure 14.8). In analogy with a primary alcohol, the conditions for achieving this oxidation are acidified potassium dichromate, $K_2Cr_2O_7$, in the presence of H_2SO_4 with gentle heating. There is also a colour change from orange to green. As with the oxidation of primary alcohols, there many other ways of achieving this that you may learn about during a chemistry degree. The ketone can be reduced back to the secondary alcohol. This will be discussed in Section 14.2.1.

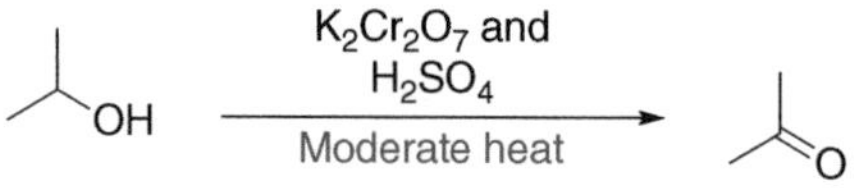

$$\text{(secondary alcohol)} \xrightarrow[\text{Moderate heat}]{\substack{K_2Cr_2O_7 \text{ and} \\ H_2SO_4}} \text{(ketone)}$$

Figure 14.8 Oxidation of a secondary alcohol.

14.1.5.3   Tertiary alcohols

Tertiary alcohols cannot be oxidised to a carbonyl-containing compound and so will not react with an oxidising agent. There will be no colour-change if acidified potassium dichromate is added.

14.2   Aldehydes and ketones

14.2.1   Nucleophilic addition

A C=O functionality is called a carbonyl group. It is similar to an alkene, in that it contains both sigma (σ) and pi (π) components

Both **aldehydes** and **ketones** possess a C=O functionality, with a permanent dipole (Chapter 2) where the carbon has a $\delta+$ charge and the oxygen $\delta-$. This permanent dipole means that the carbon atom, being electron-deficient, is particularly reactive towards nucleophiles, electron-rich species. Many nucleophiles with a negative charge or a lone pair of electrons will attack an aldehyde or a ketone, but the product depends upon the identity of the nucleophile. One common example is the addition of the cyanide ion, NC$^-$, into aldehydes to give a cyanohydrin (Figure 14.9a).

When drawing the mechanism for this reaction, there are two key things to remember (Figure 14.9b). Firstly, KCN is an ionic compound that has a K$^+$ ion and a NC$^-$ ion, which is nucleophilic. Secondly, the carbon atom on the substrate has a $\delta+$ charge. This means that the cyanide ion will act as a nucleophile and attack the $\delta+$ carbon, breaking the C=O bond and generating an oxygen anion. Once the oxygen anion has been formed, it can attack the H$^+$ in the HCl that is present in the reaction mixture, forming an alcohol. When a CN and an OH are both bonded to the same carbon atom, the functional group is called a *cyanohydrin*.

Generally speaking, aldehydes react more quickly with nucleophiles than ketones because the aldehyde carbon atom is more easily accessible, as there is less hindrance of the nucleophile when it attacks. The ketone has two adjacent carbon atoms, which hinder the attack of nucleophiles.

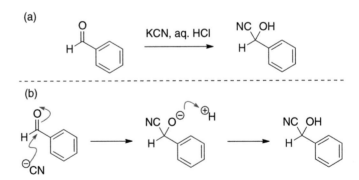

Figure 14.9 Nucleophilic addition into an aldehyde with KCN.

Worked Example 14.1   Draw the mechanism for the reaction between butanal and KCN:

Solution

Butanal contains a C=O that can be attacked by a nucleophile. KCN can act as a nucleophile because it is an ionic compound containing a K^+ ion and an NC^- anion. This means that the NC^- anion can attack the C=O at carbon, generating an oxygen anion that can be quenched with an acid, e.g. HCl:

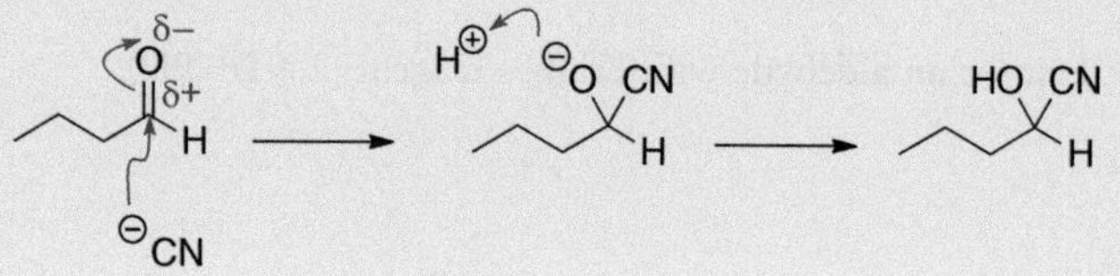

Reduction

As stated in Section 14.1.5, aldehydes and ketones can be formed by the oxidation of primary and secondary alcohols, respectively. They can also be reduced back to the corresponding alcohol using a reducing agent. Within organic chemistry, a reducing agent is often something that can add hydrogen to another molecule. Commonly used reducing agents include hydrides, H^-, such as sodium borohydride, $NaBH_4$, and lithium aluminium hydride, $LiAlH_4$. The addition of a hydride is a form of *nucleophilic addition*; the hydride acts as a nucleophile because it has a negative charge.

For a further discussion about nucleophilic addition into aldehydes or ketones, see Section 14.2.2.

In this book, we draw the hydride ion as H^-. But be aware that if you study chemistry further, H^- may also be used to represent NaH, which is a base and not a reducing agent.

Worked Example 14.2   Draw the product from the reduction of ethanal with sodium borohydride. Provide a reaction mechanism for this transformation.

Solution

The first thing to do when tackling this kind of question is to look for any immediately obvious dipoles. In this case, there is a large dipole on the C=O in the aldehyde, with a δ+ on the carbon atom and a δ− on the oxygen atom. The H^- will therefore attack the δ+ carbon atom, making a bond, which will then break the C to O π bond, as shown. This generates a negatively charged oxygen atom that can then be quenched by addition of an acid:

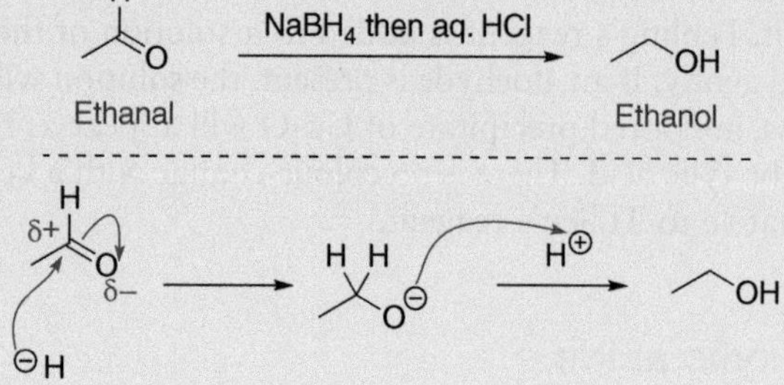

14.2.2   Tests for aldehydes and ketones

Both aldehydes and ketones can be detected by using the reagent 2,4-dinitrophenylhydrazine (2,4-DNPH or 2,4-DNP), which is also known as *Brady's reagent*. During this reaction, Brady's reagent reacts with the C=O present in the

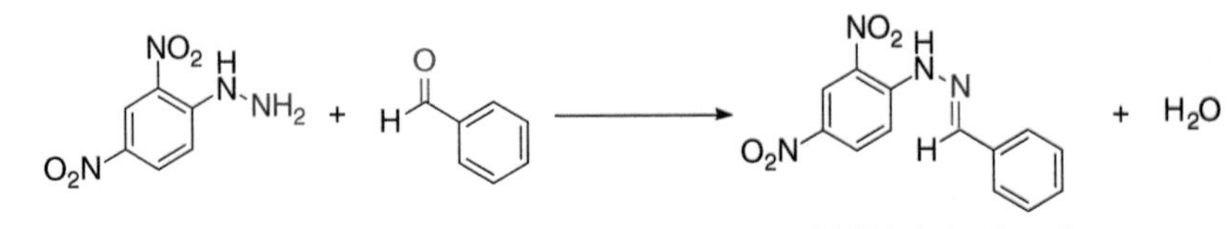

Figure 14.10 Test for an aldehyde with Brady's reagent (2,4-DNPH).

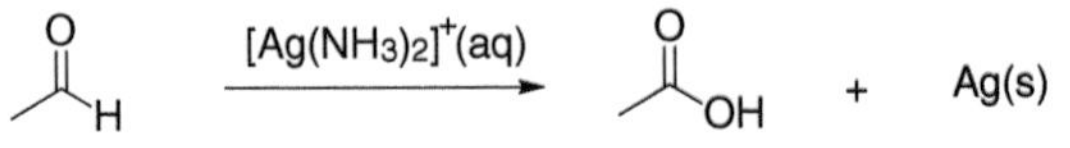

Figure 14.11 Test for an aldehyde with Tollen's reagent (silver mirror test).

aldehyde or ketone, and an organic precipitate is formed (known as a *hydrazone*) (Figure 14.10). There is no reaction with carboxylic acids or esters, because these types of carbonyl-containing compounds are less reactive at the C=O towards nucleophiles.

Unfortunately, Brady's reagent does not distinguish between aldehydes and ketones; it only confirms that they are present. To determine whether an aldehyde or a ketone is present, Tollen's reagent must be used. Tollen's reagent is *ammoniacal silver nitrate* and is prepared by adding aqueous sodium hydroxide to aqueous silver nitrate to give a brown precipitate of silver oxide. Dilute aqueous ammonia is then added until the precipitate just dissolves. This gives the complex $[Ag(NH_3)_2]^+$, which is the active species.

Ketones do not react with Tollen's reagent, but aldehydes do. When an aldehyde is present, a silver mirror is formed; a thin layer of silver will coat the glassware, which is due to the silver ions being reduced to silver metal:

$$Ag^+ (aq) + e^- \rightarrow Ag(s)$$

The aldehyde itself is oxidised to a carboxylic acid (Figure 14.11).

Another test for aldehydes versus ketones is Fehling's solution, which contains a Cu(II) salt. Fehling's reagent is added to a solution of the sample to be tested and heated gently. If an aldehyde is present, the solution will change from blue to colourless, and a red precipitate of Cu_2O will appear as the aldehyde is oxidised to a carboxylic acid. There is no colour-change with a ketone. This test is a useful alternative to Tollen's reagent.

14.3   Carboxylic acids

14.3.1   Preparation and properties of carboxylic acids

Carboxylic acids can be prepared by the total oxidation of a primary alcohol or by oxidation of an aldehyde. Carboxylic acids are found widely in nature, and if they have a short carbon chain, they often smell awful. For example, butanoic acid ($CH_3CH_2CH_2COOH$) smells like parmesan cheese, and hexanoic acid ($CH_3CH_2CH_2CH_2CH_2COOH$) smells like goats!

COOH may also be shorted to CO_2H, so keep an eye out for this.

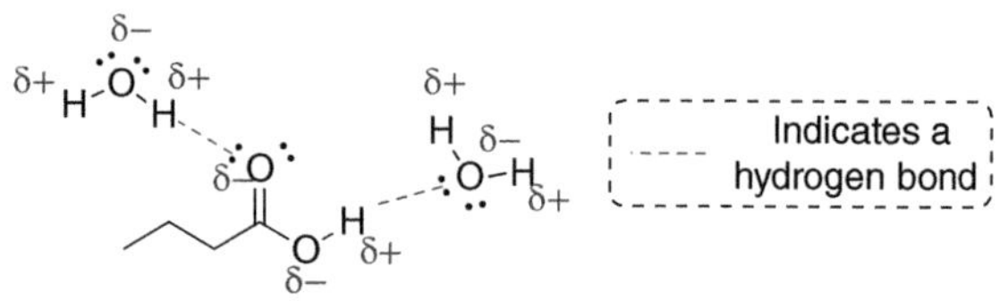

Figure 14.12 Hydrogen bonding between a carboxylic acid and water.

Similar to the alcohols seen previously in Section 14.1, carboxylic acids with a short side chain are generally fairly soluble in water because they can readily form hydrogen bonds with water molecules (Figure 14.12). The longer the alkyl side chain, the less soluble in water they are, because of the increased lipophilicity and reduced opportunities for hydrogen bonding. A very general rule of thumb is that if there are up to five carbon atoms in the carboxylic acid, they will dissolve readily in water; but if there are any more, the carboxylic acid is less soluble. It is possible to increase the solubility of a carboxylic acid in water by deprotonation, i.e. by removing the acidic proton to make a charged species.

14.3.2   Deprotonation of carboxylic acids

As their name implies, carboxylic acids readily give up a proton, H^+. The proton that is easily removed is the one on the OH group (Figure 14.13a). Most bases will remove this proton (Figure 14.13b and c), but some metals will also remove it: for example, sodium or magnesium (Figure 14.13d). When the proton is removed, a *carboxylate salt* is made, which contains an ionic bond. The carboxylate is more soluble in water than the parent carboxylic acid because, having ionic bonding, it dissociates more readily in water.

The acidity of a solution is related to the concentration of H^+ ions in solution (also known as pH). For a refresher, see Chapter 7.

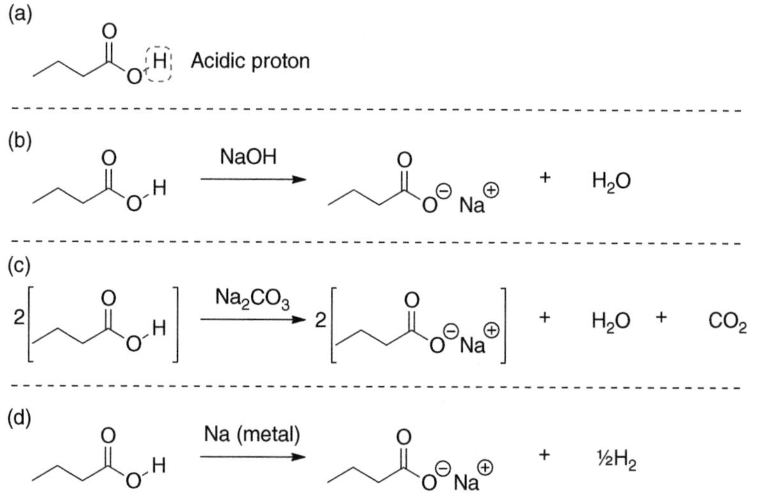

Figure 14.13 (a) The acidic proton in a carboxylic acid; (b) reaction of butanoic acid with sodium hydroxide; (c) reaction of butanoic acid with sodium carbonate; (d) reaction of butanoic acid with sodium metal. A carboxylate salt is formed in each reaction.

Worked Example 14.3   Provide balanced reaction schemes for the reactions of pentanoic acid with KOH, K$_2$CO$_3$, and magnesium metal.

Solution
The first question is reacting a carboxylic acid with a hydroxide base, KOH, which will form a potassium salt and water:

$$CH_3CH_2CH_2CH_2COOH + KOH \rightarrow CH_3CH_2CH_2CH_2COO^-K^+ + H_2O$$

The second example reacts a carboxylic acid with a carbonate base, K$_2$CO$_3$, which generates CO$_2$ and H$_2$O. However, in this case, two equivalents of the carboxylic acid are needed, as there are two basic sites on the carbonate base:

$$2CH_3CH_2CH_2CH_2COOH + K_2CO_3 \rightarrow 2CH_3CH_2CH_2CH_2COO^-K^+ + H_2O + CO_2$$

The third example is the reaction of a carboxylic acid with a metal, Mg, which generates a carboxylate salt and H$_2$. In this case, Mg forms a 2+ ion; therefore, two equivalents of the carboxylic acid are needed, and one equivalent of H$_2$ is released. The two carboxylate species are associated with the one Mg^{2+} ion to balance the charges:

$$2CH_3CH_2CH_2CH_2COOH + Mg \rightarrow [CH_3CH_2CH_2CH_2COO]_2Mg + H_2$$

Magnesium readily makes Mg^{2+} ions because it is in Group 2 of the periodic table. For a refresher, see Chapter 2.

14.3.3   Reduction of carboxylic acids

Carboxylic acids can be reduced to primary alcohols using very strong reducing agents (Figure 14.14). One commonly used reductant is lithium aluminium hydride, LiAlH$_4$.

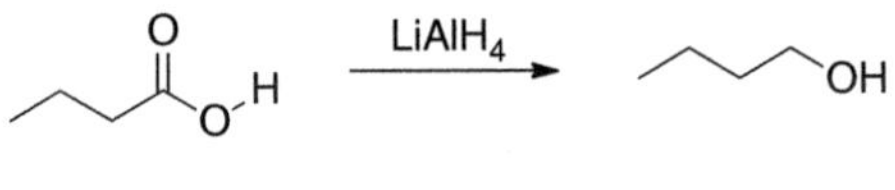

Figure 14.14 Reduction of butanoic acid to butan-1-ol with lithium aluminium hydride.

14.4   Esters

14.4.1   Properties of esters

Esters are another class of compound that contain a carbonyl group, a C=O (Figure 14.15a). When looking at an ester, it can be seen that there are two halves: one half connected to the C=O itself, the acid half, and another joined to the oxygen, the alcohol half. These two halves are important for two reasons. Firstly, they can be used to name the ester; and secondly, they are a clue as to how to form the ester (Figure 14.15b,c).

When naming esters, there are two parts to the name, which correspond to the two halves of the molecule (the alcohol and the acid [Figure 14.15b]). To name an ester, first identify which part of the molecule came from an alcohol, and treat it like a side chain. Secondly, identify which part of the molecule came from a

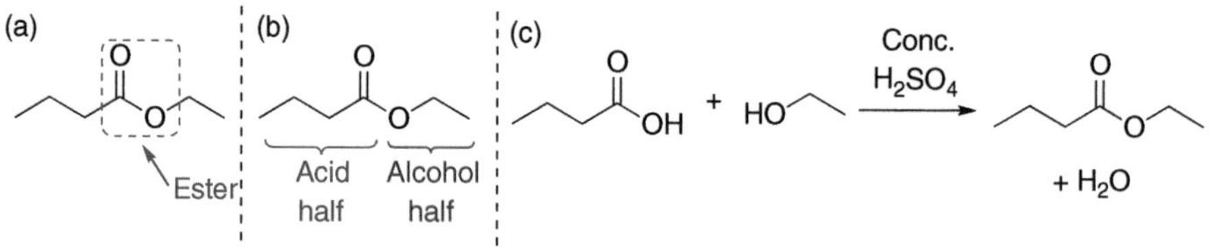

(a) (b) (c)

Figure 14.15 (a) Identification of the ester group; (b) naming an ester; (c) formation of an ester.

carboxylic acid. The prefix from the alcohol is written first, followed by the prefix from the carboxylic acid, but replacing the 'oic acid' part of the carboxylic acid with 'oate'. Remember, the alcohol part comes first, and the carboxylic acid part is second. Therefore, the ester in Figure 14.15a is ethyl butanoate.

For a refresher on naming esters, see Chapter 12.

> **Worked Example 14.4**   Name the following ester: 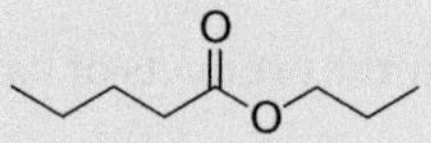
>
>
> **Solution**
> This ester can be divided into two halves, which are shown by the colours. The red half is derived from the carboxylic acid, and the blue half from an alcohol: 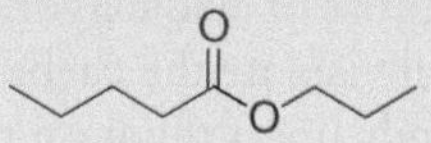
>
>
> The alcohol is propanol, and the carboxylic acid is pentanoic acid. The name of the ester is therefore propyl pentanoate.

14.4.2 Hydrolysis of esters

Water can be used to break esters bonds. This is called *hydrolysis*, and it can be accelerated under acidic or basic conditions, with subtly different processes for each. In both cases, the ester cleavage is across the C—O single bond (Figure 14.16). Under acidic conditions, aqueous HCl is used and will result in the formation of the desired carboxylic acid and the corresponding alcohol. Under basic conditions, the resulting carboxylic acid must be deprotonated to give a carboxylate salt due to the difference in pK_a between the hydroxide base

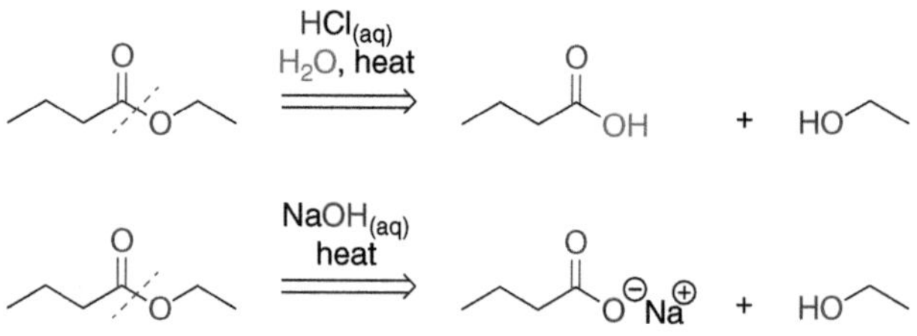

You will notice that in Figure 14.16, an arrow like this ⟹ is used. This arrow means *retrosynthetically* and is used when chemists are showing the products from breaking bonds. Retrosynthesis is a technique used by chemists when planning how to make complex molecules.

Figure 14.16 (a) Cleavage of an ester under acidic conditions; (b) cleavage of an ester under basic conditions.

and the carboxylic acid. pK_a is discussed further in Chapter 7. Under basic conditions, if a carboxylic acid is required, the reaction solution containing the carboxylate salt must be acidified to protonate the salt and give the desired product.

Worked Example 14.5   Draw the products of acid-mediated ester hydrolysis for ethyl ethanoate.

Solution
The structure of ethyl ethanoate is shown, and a line has been drawn on the diagram to show where cleavage under acidic conditions would occur. This allows us to deduce that the ester would be hydrolysed to give ethanol and ethanoic acid.

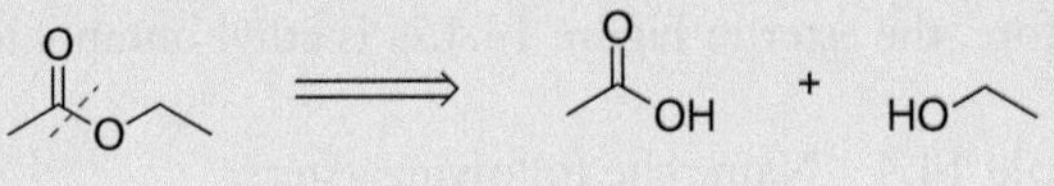

Worked Example 14.6   Draw the products of base-mediated ester hydrolysis for ethyl ethanoate.

Solution
The structure of ethyl ethanoate is shown, and a line has been drawn on the diagram to show where cleavage under basic conditions would occur. The products would be similar to those of acidic hydrolysis, but instead of the carboxylic acid, we would actually isolate the carboxylate salt. The counter-ion for the carboxylate salt depends upon the base used.

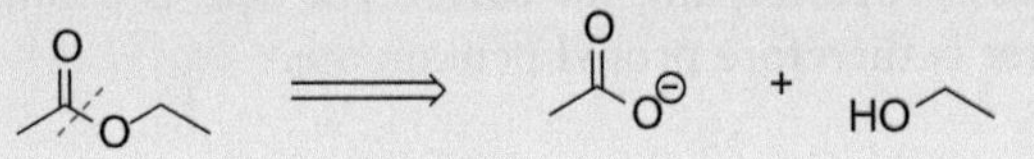

14.5   Amides

14.5.1   Preparation and properties of amides

The chemistry of **amides** discussed at the foundation level is very similar to that of esters. When naming amides, the same rules apply as for esters, but instead of replacing 'oic acid' with 'oate', it is replaced with 'amide'.

Worked Example 14.7   Name the following amide:

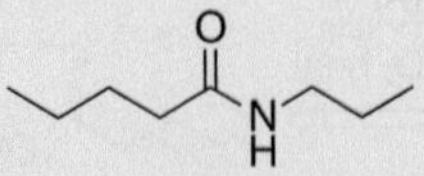

Solution
This amide can be divided into two halves, which are shown by the colours. The red half is derived from the carboxylic acid and the blue half from an amine:

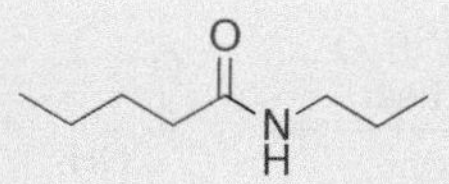

The amine is propylamine, and the carboxylic acid is pentanoic acid. The name of the amide is therefore N-propylpentanamide.

The reason for the N- in the name is to show that the propyl group is bonded to the nitrogen atom.

Amides cannot usually be prepared by reacting an amine with a carboxylic acid because the amine would deprotonate the carboxylic acid to make a salt (Figure 14.17). This is because the lone pair on the nitrogen atom will act as a Brønsted base, removing the acidic proton on the carboxylic acid. This is all related to pK_a, which is discussed in Chapter 7.

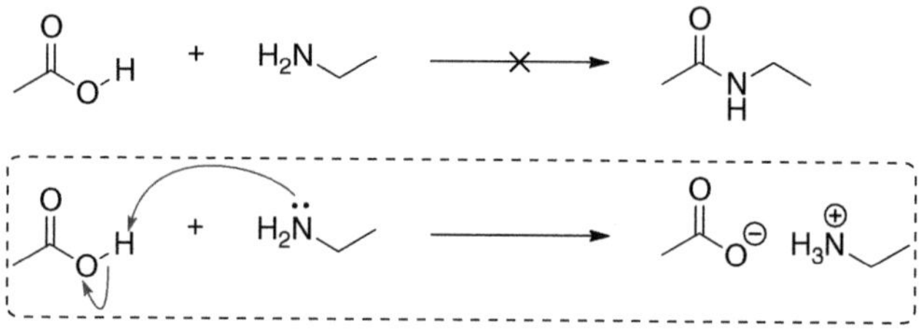

Figure 14.17 Reaction of a carboxylic acid and an amine to make a salt.

To make an amide, a more reactive carboxylic-acid surrogate without an acidic proton must be used. A common method is to turn the carboxylic acid into an acid chloride and combine that with an amine (Figure 14.18a). The acid chloride is very reactive to nucleophiles due to the strong δ+ on the carbon atom in the carbonyl group and therefore readily reacts with the amine. The mechanism proceeds through an additional step where the amine reacts with the C=O, initially forming a tetrahedral intermediate, that then collapses to form a new C=O and eliminates chloride (Figure 14.18b).

14.5.2   Hydrolysis of amides

Amides can be hydrolysed to give a carboxylic acid and an amine. The methods used to achieve this are very similar to those used for esters, and either acids or bases can be used. Amides are typically more robust than esters and usually need

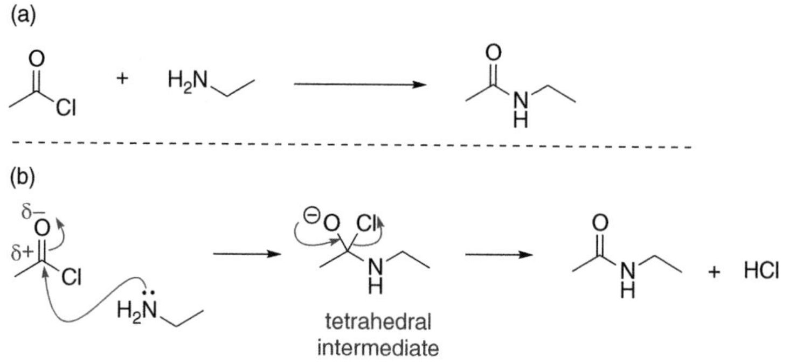

Figure 14.18 (a) Reaction of an acid chloride and an amine to make an amide; (b) the mechanism for amide formation.

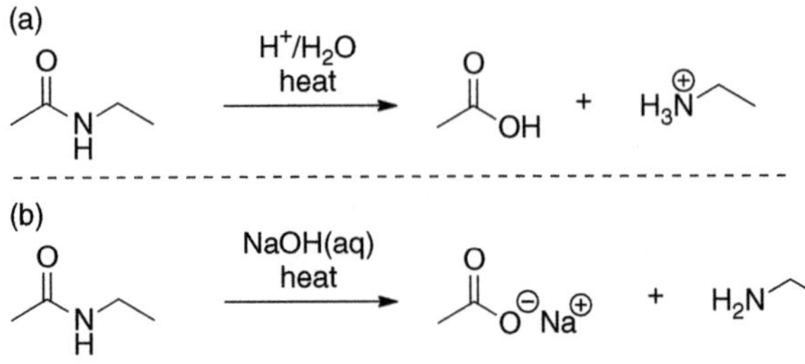

Figure 14.19 (a) Cleavage of an amide under acidic conditions; (b) cleavage of an amide under basic conditions.

to be heated for hydrolysis to occur. When hydrolysing the amide, cleavage is across the C—N single bond (Figure 14.19). When cleavage is completed using acidic conditions, the amine product will likely be protonated by the acid. Conversely, when cleaving amides using basic conditions, the carboxylic acid formed will be deprotonated by the base.

Worked Example 14.8   Give the products of cleavage of N-ethyl propanamide under acidic conditions.

Solution
The structure of N-ethyl propanamide is shown, and a line has been drawn on the diagram to show where cleavage under acidic conditions would occur. The products would be an amine and a carboxylic acid, but the amine would be protonated because of the acidic conditions.

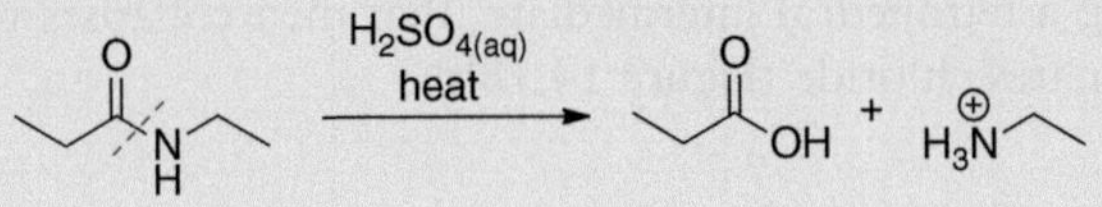

14.6   Amines

14.6.1   Naming amines

Amines are similar in structure (and name!) to amides, and because of this, students often confuse amides and amines. An amide has a C=O group next to the nitrogen, whereas an amine does not (Figure 14.20). There are four different

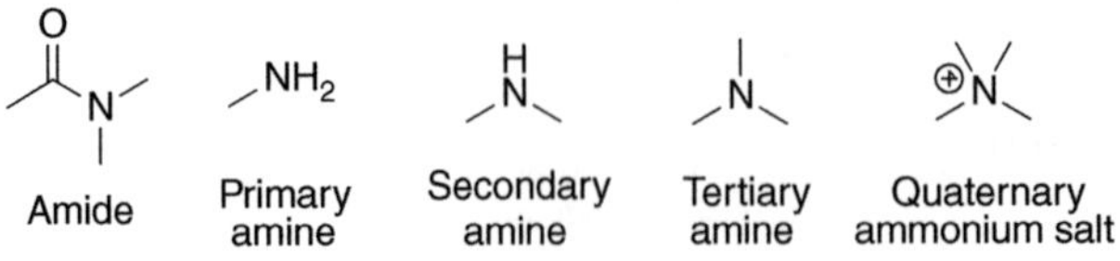

Figure 14.20 An amide and the different classes of amine.

classes of amine: primary, secondary, tertiary, and quaternary. The classes are related to how many carbon atoms are bonded to the central nitrogen atom. In primary amines, the nitrogen atom is bonded to one carbon atom; in secondary, it is bonded to two carbon atoms; and in tertiary, it is bonded to three carbon atoms. This is subtly different to the naming of primary, secondary, and tertiary alcohols and haloalkanes, so be careful. Quaternary amines are named slightly differently in that they are called *quaternary ammonium salts*. This is because a nitrogen atom that is positively charged and bonded to four alkyl chains is called *ammonium*. It is a salt because an anionic counter-ion is associated with the positively charged nitrogen centre. Amines are pyramidal with a lone pair pointing upwards from the nitrogen, whereas quaternary ammonium salts are tetrahedral.

For a refresher on shapes of molecules, see Chapter 2.

When naming amines, if amine is the dominant functional group, the suffix is *amine*. The main rules applied to naming amines are the same as for naming any organic compound; find the longest continuous carbon chain, and work from there.

For a refresher on naming compounds, see Chapter 12.

Worked Example 14.9 Name the amines shown:

Solution

In (a), the longest alkyl chain contains four carbon atoms, so it has the parent name 'butyl'. The dominant functional group is an amine, which is on carbon atom 1, so the overall name is butan-1-amine. *n*-Butylamine and 1-aminobutane are also acceptable names.

Question (b) is a little more difficult, but the same rules apply. The longest chain contains two carbon atoms, so the parent is 'ethan'. The dominant functional group is an amine, bonded to carbon atom 1, so the suffix is 'amine'. The amine itself is bonded to another alkyl chain containing one carbon atom, so it is termed 'methyl'. The overall name is therefore N-methylethan-1-amine.

When studying chemistry it is likely that you will encounter different names for the same compound, so be careful.

Ethyl side-chain

Question (c) is harder still, but as with (b), the same rules apply. The longest chain contains four carbon atoms, so the parent is 'butan'. The dominant functional group is an amine, so the suffix is 'amine'. The amine itself is bonded to two other alkyl chains, each containing one carbon atom, so they are termed 'methyl'. Finally, the amine itself is in position 1, i.e. at the end of the chain, which must be reflected in the name. This amine is therefore named N,N-dimethylbutan-1-amine.

The reason for the N- in the name is to show that the methyl group is bonded to the nitrogen atom and a carbon atom on the ethanamine chain.

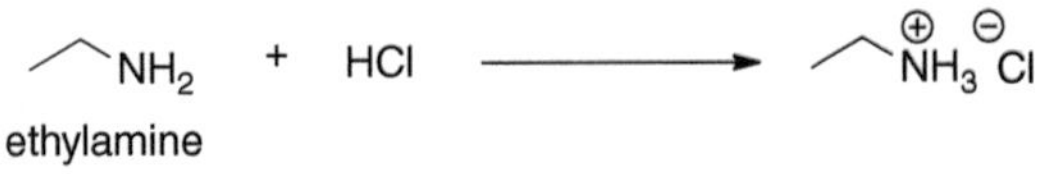

$$\text{CH}_3\text{CH}_2\text{NH}_2 + \text{HCl} \longrightarrow \text{CH}_3\text{CH}_2\overset{\oplus}{\text{NH}_3}\ \overset{\ominus}{\text{Cl}}$$

ethylamine

Figure 14.21 Reaction of ethylamine with hydrochloric acid to give the ethylammonium salt.

14.6.2 Amines as bases

Primary, secondary, and tertiary amines all contain a lone pair of electrons on the nitrogen atom that can act as a Brønsted base by accepting a proton (Figure 14.21). Adding the proton to the nitrogen gives a positive ammonium species that can readily release the proton again; therefore, although amines are basic, they are not particularly so. It is important to remember that when the positively charged ammonium salt is made, there will also be a counter-ion that you should be able to derive from the reaction conditions.

14.6.3 Preparation of alkyl amines

For further information about nucleophilic substitution, see Chapter 16.

Aliphatic amines can be prepared a multitude of ways, but in foundation-level studies, it is usually enough to know that amines are prepared by displacement of a halogen in a halogenoalkane by ammonia or other suitable amine: a nucleophilic substitution reaction. The reason this displacement works is that ammonia, NH_3, has a lone pair that is able to attack electrophiles. In the case of an alkyl halide, this is the δ+ carbon atom bonded to the halogen.

Consider the reaction of chloroethane with ammonia (Figure 14.22). Ammonia attacks the δ+ carbon atom indicated, displacing chloride in an S_N2 reaction pathway and generating ethylamine. During the reaction, hydrochloric acid is formed, which is likely to associate with the amine product until the reaction undergoes a work-up. One drawback of forming amines in this way is the likelihood for over-alkylation of the nitrogen atom, giving a multitude of products. This is something that you are likely to study further if you continue studies in chemistry.

14.7 Nitriles

Nitriles are the final nitrogen-containing functional group that you are likely to encounter during foundation studies. Nitriles contain a triple bond between the carbon and nitrogen atoms, which means that the nitrile functional group is

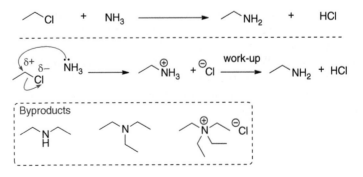

Figure 14.22 Reaction between chloroethane and ammonia to give ethylamine.

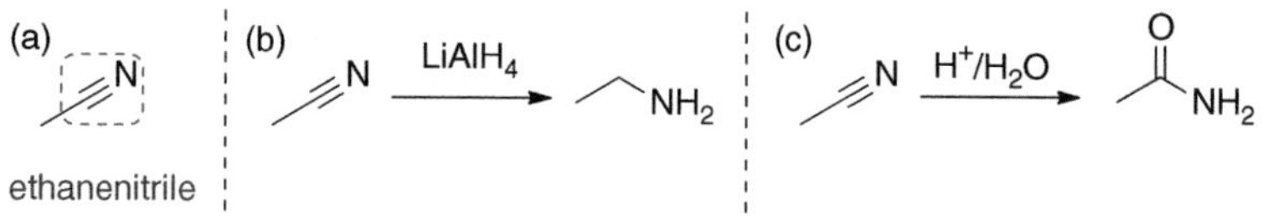

Figure 14.23 (a) A nitrile; (b) reduction of a nitrile to an amine; (c) hydrolysis of a nitrile to an amide.

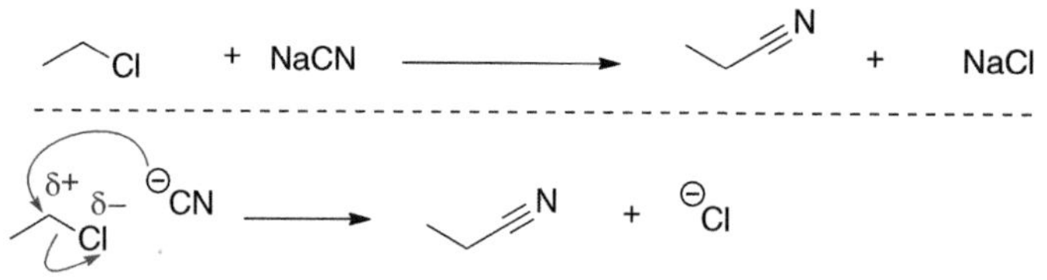

Figure 14.24 Reaction of chloroethane with sodium cyanide to give propanenitrile.

linear (Figure 14.23). Nitriles can undergo a variety of reactions, and the two that are most relevant to foundation studies are reduction, to give a primary amine, and hydrolysis, to give an amide.

14.7.1 Nitrile formation

Nitriles can be formed by displacement of a halogenoalkane using sodium cyanide. It is important to note that when this reaction occurs, the original alkyl chain is increased in length by one carbon atom due to the carbon atom present in $^-$CN (Figure 14.24).

Quick-check summary

- An alcohol is a molecule that contains a discrete OH group.

- Alcohols can undergo hydrogen bonding.

- There are three different classes of alcohol: primary, secondary, and tertiary.

- Alcohols can be burned in oxygen to give carbon dioxide (CO_2) and water (H_2O).

- A primary alcohol can be oxidised to give an aldehyde and/or a carboxylic acid. A secondary alcohol can be oxidised to give a ketone. A tertiary alcohol cannot be oxidised.

- An aldehyde or a carboxylic acid can be reduced to a primary alcohol, and a ketone can be reduced to a secondary alcohol.

- Aldehydes and ketones can undergo nucleophilic addition.

- Aldehydes and ketones can be detected chemically by using the reagent 2,4 dinitrophenylhydrazine (2,4-DNPH or 2,4-DNP), which is also known as Brady's reagent.

- Aldehydes react with Tollen's reagent, to give a silver mirror, or with Fehling's solution, which will change from blue to colourless with formation of a red precipitate. Ketones do not react with either Tollen's reagent or Fehling's solution.

- Aldehydes, ketones, carboxylic acids, esters, and amides all contain a carbon-oxygen double bond.

- In the presence of a base, carboxylic acids are deprotonated to give a salt.

- Esters can be hydrolysed to give a carboxylic acid and an alcohol.

- Amides can be hydrolysed to give a carboxylic acid and an amine.

- Amines contain a carbon–nitrogen bond and can be designated as primary, secondary, or tertiary.

- All amines contain a lone pair of electrons on nitrogen that can act as a Brønsted base and accept a proton.

- A nitrile contains a carbon–nitrogen triple bond. Nitriles can be reduced to give a primary amine.

End-of-chapter questions

1 Explain the difference in boiling point between butane ($-1\,°C$) and butan-1-ol ($118\,°C$) in terms of intermolecular forces.

2 Classify the following alcohols as either primary, secondary, or tertiary:

3 What are the possible organic products of oxidation of the alcohols in question 2?

4 What are the products of reduction of the following molecules? Suggest a reagent, and draw the mechanism.

5 NaCN is added to the following compounds. What are the products? Draw a mechanism for this reaction (assume an acidic work-up).

6 Describe a method for how you would chemically differentiate between acetophenone and benzaldehyde.

7 Describe a chemical test that you could use to determine if a substance is a carboxylic acid.

8 Name the esters shown:

9 What are the products of hydrolysis of the esters in question 9 using
   (a) aqueous NaOH and (b) using aqueous HCl.

10 What reagents would you need to prepare the following compounds?

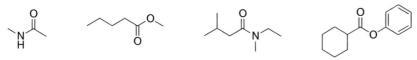

11 Suggest a route to prepare the amines shown:

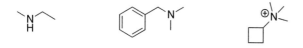

15

The chemistry of aromatic compounds

At the end of this chapter, students should be able to:

- Understand the structure of benzene and other aromatic compounds
- Be able to describe and explain electrophilic aromatic substitution, S_EAr
- Determine if a ring substituent is activating or deactivating
- Compare and contrast the reactivity of benzene, phenol, and aniline with electrophiles

15.1 Benzene

15.1.1 The structure of benzene

Benzene, C_6H_6, is a molecule that can be described as **aromatic**. This is because when benzene and benzene-containing compounds were first isolated, they were noted to smell pleasant, so their name was derived from the Greek 'aroma', meaning spice, which was then used in Latin to mean 'sweet smell'. Benzene is a simple aromatic compound that is often studied in undergraduate courses. It is a colourless liquid with a slightly sweet odour and is fully aromatic. Benzene can be represented in a number of ways depending upon what information needs to be depicted (Figure 15.1).

Each carbon atom in the benzene molecule is trigonal planar in geometry, with a bond angle of 120° between the other two carbon atoms and the hydrogen. In terms of the bonding, there are three σ bonds and one π bond per carbon atom. The π bond is formed by an electron in the p orbital of one carbon atom overlapping with an electron in a p orbital on an adjacent carbon atom. This overlap causes a cyclic system of bonding p orbitals (or π bonds) to be formed, where the electrons are free to move around the ring. The movement of electrons in this manner is called *delocalisation*, and benzene can be said to have a *delocalised electron system above and below the ring*.

For a refresher on the shapes of molecules, please see Chapter 2.

Benzene has a delocalised electron system above and below the ring.

Foundations of Chemistry: An Introductory Course for Science Students, First Edition.
Philippa B. Cranwell and Elizabeth M. Page.
© 2021 John Wiley & Sons Ltd. Published 2021 by John Wiley & Sons Ltd.
Companion website: www.wiley.com/go/Cranwell/Foundations

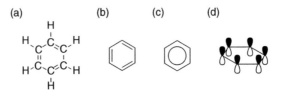

Figure 15.1 (a) Benzene represented using a display formula; (b) benzene represented as the Kekulé structure with alternating single and double bonds; (c) an alternative representation of benzene showing that the carbon–carbon bond lengths are the same; (d) representation of the p orbitals in benzene to show the overlap of the π bonds above and below the ring.

Conjugation is when alkenes are directly next to each other and are only separated by one single bond. In the diagram shown, the compound on the left contains a conjugated system because the alkenes are directly next to each other. However, the compound on the right is not, as there is a CH_2 between each alkene.

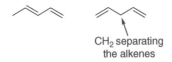

CH₂ separating
the alkenes

In the skeletal and Kekulé representations of benzene (Figure 15.1a,b), it can be seen that there are alternating single and double carbon–carbon bonds. If this were true, then there would be two different bond lengths in benzene: that for a single carbon–carbon bond and that for a double carbon–carbon bond. Experimentally, this is not observed, and in benzene all of the carbon–carbon bonds are the same length. This therefore leads to the representation of benzene where the delocalised π system is represented as a circle in the middle of a cyclohexane (Figure 15.1c). This representation is often used at A-level, but during undergraduate studies structure (b) is usually used. You should become used to using both (b) and (c) interchangeably. Finally, Figure 15.1d shows the p orbitals overlapping above and below the benzene ring to give the conjugated π system. A total of six electrons are in the π system above and below the ring.

In 1931, German chemist Erich Hückel proposed a theory that would allow chemists to determine whether a compound was aromatic by looking at the structure, and would also allow the properties of the compound to be hypothesised. This led to the four criteria for aromaticity, Hückel's rules, which chemists still use today.

Box 15.1
Hückel's rules for aromaticity:
   1. The molecule is cyclic.

   2. The molecule is planar.

   3. The molecule is fully conjugated.

   4. The molecule has $[4n + 2]$ π electrons, where n is an integer.

Worked Example 15.1   Explain, using Hückel's rules and a diagram, why benzene, C_6H_6, can be thought of as aromatic.

Solution
Using Hückel's rules:

1. Benzene is cyclic.

2. The benzene molecule itself is flat, as all of the carbon atoms are trigonal planar.

3. The molecule contains double bonds that are in conjugation (i.e. they can interact with each other).

4. It contains $[4n + 2]$ π electrons.
   Working: there are 6 π electrons.
   $[4n + 2] = 6$
   $n = 1$ (*n* must always be an integer)

Using a diagram:
Benzene is a ring and all of the carbon atoms are trigonal planar. There is an extended system of π electrons that are in p orbitals, so the double bonds are fully conjugated:

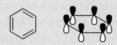

15.1.2   Nomenclature

When naming compounds that contain a benzene ring, some rules need to be applied. When one substituent is directly bonded to the benzene ring, the ring is usually referred to as *benzene* and the substituents are then added to the name, but there are some exceptions. Sometimes benzene-containing compounds are referred to by their trivial name. Some commonly encountered benzene-containing compounds are shown in Figure 15.2.

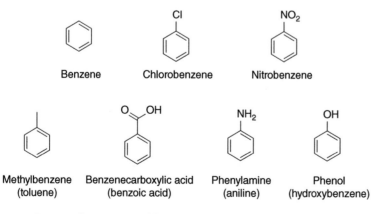

Figure 15.2 Commonly encountered benzene-containing compounds. Trivial names are in brackets.

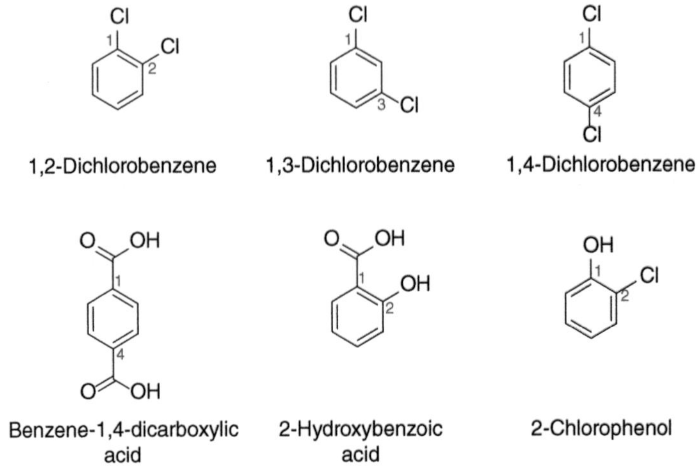

1,2-Dichlorobenzene	1,3-Dichlorobenzene	1,4-Dichlorobenzene

Benzene-1,4-dicarboxylic acid   2-Hydroxybenzoic acid   2-Chlorophenol

Figure 15.3 Examples of benzene-containing compounds containing two substituents.

If a benzene ring has two or more substituents, these need to be indicated using a numbering system. The numbering used is very similar to that discussed in Chapter 12, and in all cases, substituents have the lowest numerical denomination possible. As noted earlier, sometimes benzene-containing compounds have a trivial name, so it is important to be able to use both the IUPAC and trivial names. Some examples are shown in Figure 15.3.

Worked Example 15.2   Name the five compounds drawn below:

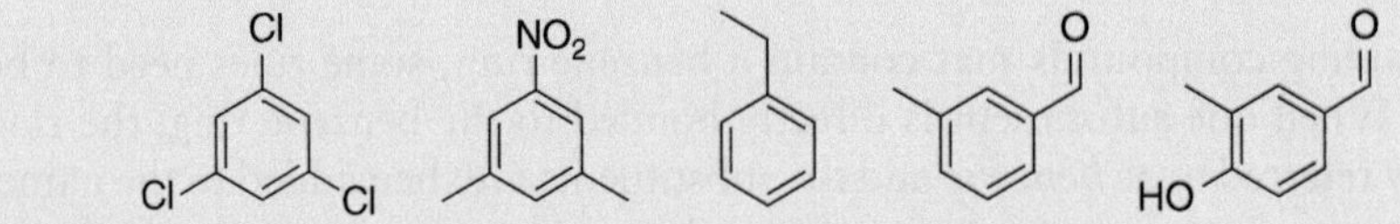

Solution
The first example contains a benzene ring with three chlorine atoms. These chlorine atoms are in positions 1, 3, and 5 in relation to each other. The fact that there are three chlorine atoms means that the prefix 'tri' needs to be added. The name is therefore 1,3,5-trichlorobenzene.

The second example has a nitro group and two methyl groups. The methyl groups take priority over the nitro group; therefore, one of them must be numbered as 1 (See Table 12.4 for a reminder of the functional group hierarchy). The name is therefore 1,3-dimethyl-5-nitrobenzene.

The third example contains one substituent, an ethyl group, on a benzene ring. The name is therefore ethylbenzene.

The fourth structure contains a methyl group and a side chain with an aldehyde. The aldehyde takes priority over the methyl group, so this side chain is numbered as 1. A benzene ring with an aldehyde bonded directly to it is called 'benzaldehyde', so the name is 3-methylbenzaldehyde.

Finally, the last structure contains an alcohol (OH), an aldehyde, and a methyl group. In this case, the benzaldehyde is still the functional group with the highest priority; therefore, the base name is 'benzaldehyde' and the carbon atom bonded to the aldehyde is carbon-1. There is a methyl group in the 3 position, and the alcohol in the 4 position is referred to as 'hydroxy' because this is a prefix, rather than a suffix (when it would be a phenol). When writing the name, the substituents need to be named in alphabetical order, so the name is 4-hydroxy-3-methylbenzaldehyde. If the benzaldehyde were not present, the OH would be dominant, so the name would be 2-methylphenol.

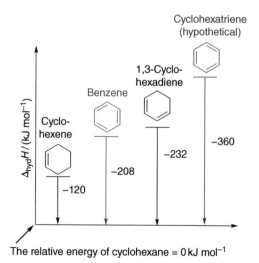

15.1.3   The reactivity of benzene

Benzene is less reactive than you might think. Even though there are effectively three double bonds in the ring, the aromaticity makes the benzene ring more stable than it would be if it were linear, with three double bonds (Figure 15.4). The aromatic stabilisation of benzene was experimentally observed by measuring the energy released by hydrogenation of a double bond in a six-membered ring. Adding hydrogen, H_2, to cyclohexene releases 120 kJ mol^{-1} of energy. In theory, the hydrogenation of benzene should release three times this, as there are three double bonds ($\sim$360 kJ mol^{-1}); but in actual fact, the hydrogenation of benzene releases 208 kJ mol^{-1}, which is 152 kJ mol^{-1} less than expected. This means that benzene is 152 kJ mol^{-1} more stable than would be expected for a six-membered ring containing three double bonds.

This aromatic stability of benzene means that it is less reactive than simple alkenes; therefore, when using benzene in a chemical reaction, a catalyst is often required. The classic example compares the bromination of ethene (Chapter 13) with the bromination of benzene (Figure 15.5). Bromination of ethene with bromine, Br_2, proceeds reasonably rapidly at room temperature. However, the bromination of benzene with bromine requires a catalyst (FeBr$_3$), and sometimes it needs to be heated to further increase the rate of reaction. The other key differences are the products formed. In the case of ethene, both of the bromine atoms in the bromine molecule are added to the alkene (this is *electrophilic addition*), whereas in the case of benzene, only one bromine atom is added, and the benzene loses a proton. This is called *electrophilic aromatic substitution*, S$_E$Ar.

Figure 15.4 Heat of hydrogenation of cyclohexene, benzene, 1,3-cyclohexadiene, and cyclohexatriene (hypothetical) to give cyclohexane.

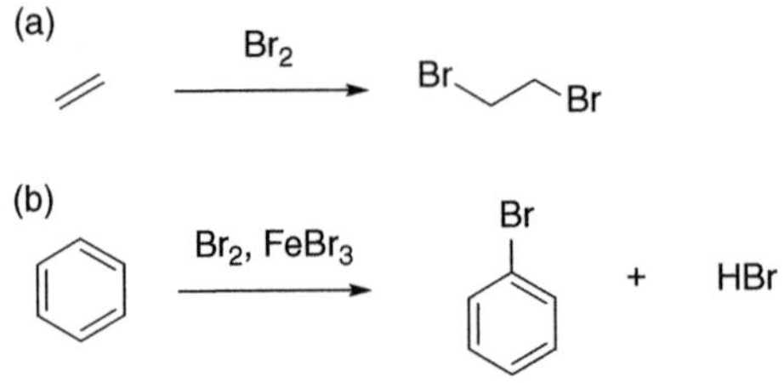

(a)

Br$_2$

Br$\sim$Br

(b)

Br$_2$, FeBr$_3$

Br

+   HBr

Figure 15.5 (a) Bromination of ethene; (b) bromination of benzene, which requires catalytic FeBr$_3$.

15.1.4   Resonance in benzene

You may have noticed in Section 15.1.1 that benzene possesses six overlapping p orbitals that provide the three π bonds, and that there is electron density above and below the benzene ring. This structural feature is important because it means that the electrons can move around the ring freely; this is called **resonance**. To understand resonance, it is important to understand what a double bond is and what it represents.

For a refresher on double bonds, see Chapter 13.

A double bond suggests that the electrons in the π bond are evenly spread between the two atoms that are bonded (Figure 15.6a). Another way of representing this is to draw an extreme form of a π bond, where we imagine that the two electrons of the π bond have been transferred to one atom, thereby creating a negative charge at that end of the bond. To balance the negative charge, a positive charge must be formed on the other atom (Figure 15.6b). In reality, the electrons are not likely to both be on one atom, nor are they likely to be completely in the centre of the two atoms; they will be somewhere between these two extremes. It is often useful to consider both of these cases when thinking about structure and reactivity. Using curly arrows, we can show the electrons moving to give the extreme form (Figure 15.6c).

Due to the three double bonds in the benzene ring system that are all conjugated to each other, we can draw resonance structures and the electrons moving around the ring (Figure 15.7). In Figure 15.7a, the curly arrows show the

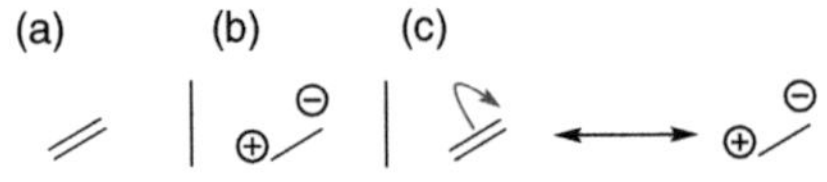

(a)   (b)   (c)

Figure 15.6 (a) A double bond represented by two lines; (b) a representation of an extreme form of a double bond; (c) the curly arrows required to convert (a) into (b).

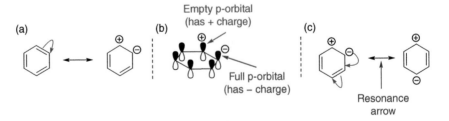

Figure 15.7 (a) Curly arrows to show the electrons moving to the extreme form of the alkene; (b) a representation of the filled/empty p orbitals from the curly arrow; (c) the movement of the negative charge due to adjacent p orbitals.

electrons moving to give an extreme form of the double bond with a positive and a negative charge. Figure 15.7b shows that after drawing the curly arrow in (a), there is now an empty p orbital with a positive charge, and a filled p orbital with a negative charge. The fact that these filled/empty orbitals are the same as the other orbitals in the benzene ring means that the electrons can move around the ring from orbital to orbital; this is resonance. The arrow between resonance structures is double-headed. Resonance is extremely important when discussing the reactivity of benzene rings and other polyconjugated systems because it can be used to explain their relative stability. Structures where a negative or positive charge can be moved over more atoms are more stable.

Worked Example 15.3   Explain, using resonance arguments, why the electrons in 1,3-pentadiene can undergo resonance stabilisation, but those in 1,4-pentadiene cannot:

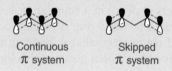

1,3-pentadiene   1,4-pentadiene

Solution
1,3-Pentadiene contains two alkenes that are directly next to each other, and all four of the p orbitals that form the π bond are able to interact with each other. This means that the electrons in these orbitals can move freely over the four carbon atoms that have π bonds to them. However, in 1,4-pentadiene, the two alkenes are not directly adjacent to each other and there is a methylene unit (CH_2) in between. This means that the two π bonds cannot interact, and hence no resonance stabilisation is possible:

Continuous   Skipped
π system   π system

Worked Example 15.4   Draw the resonance structures for the 1,3-pentadiene anion and the 1,3-pentadiene cation:

Solution
Considering the 1,3-pentadiene anion on the left: a negative charge is residing on a carbon *directly adjacent* to a π bond, so the anion can interact with the π bond. This is because the negative charge could reside in a p orbital. In addition, the two π bonds are conjugated to each other; therefore, the anion can move all the way along the chain. Remember that the curly arrows drawn show electron movement, so the arrow starts at the negative charge:

Negative charge
in a p orbital

The 1,3-pentadiene cation has a positive charge on the terminal carbon, so the electrons in the double bond must flow *towards* the positive charge. In this case, the positive charge could reside in a p orbital. Three resonance structures can be drawn:

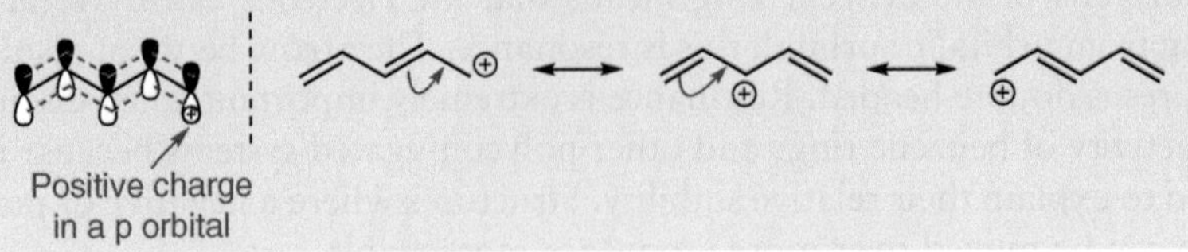

Positive charge
in a p orbital

15.1.5   Substituent effects on reactivity

As discussed in Section 15.1.3, benzene is less reactive than we would expect; however, substituents that are directly attached to the benzene ring can have a marked impact upon the reactivity of the ring system. These effects are related to the electronic impact of the substituents and whether they are electron-withdrawing or electron-donating. Substituents that are electron-withdrawing decrease reactivity of a benzene ring towards electrophiles, and substituents that are electron-donating increase the reactivity of a benzene ring towards electrophiles.

Two types of electronic effect can occur, and they relate to the orbitals through which the electrons are withdrawn or donated. Effects that are through sigma (σ) bonds are called *inductive* effects, and those through pi (π) bonds are called *mesomeric* effects. An example of an inductive effect is the electron-withdrawing nature of fluorine, which can cause a dipole (Figure 15.8a). An example of a mesomeric effect is when a lone pair on a nitrogen atom pushes electron density into a benzene ring (Figure 15.8). It is not normally necessary to know about these effects in detail in a foundation-level course.

Inductive effects are through sigma (σ) bonds, and mesomeric effects are through pi (π) bonds.

When deciding whether a benzene ring is reactive towards electrophiles or not, it is best to look and see exactly what is bonding to the benzene ring. As a general rule, any element or functional group with a lone pair of electrons that can interact with the benzene ring system (except fluorine) is likely to be activating in a mesomeric manner through the pi (π) bonds, because they will push electron density onto the benzene ring, making it more reactive. Conversely, any element that is strongly electronegative, or any group that contains a positive charge/strong dipole next to the benzene ring, is likely to be deactivating: it can pull electron-density out of the benzene ring in either an inductive (through σ bonds) or a mesomeric (through π bonds) manner, so it is less reactive, Figure 15.9.

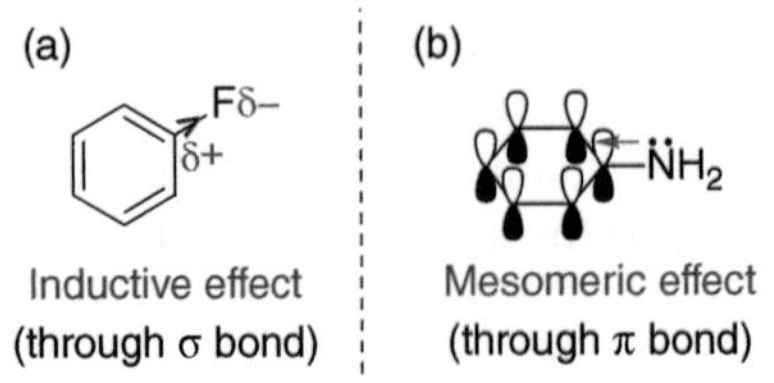

(a)   (b)

Inductive effect   Mesomeric effect
(through σ bond)   (through π bond)

Figure 15.8 (a) Electrons pulled towards electronegative fluorine to create a dipole through the σ bond (inductive effect); (b) electrons pushed towards the benzene ring from the lone pair on nitrogen through the π bond (mesomeric effect).

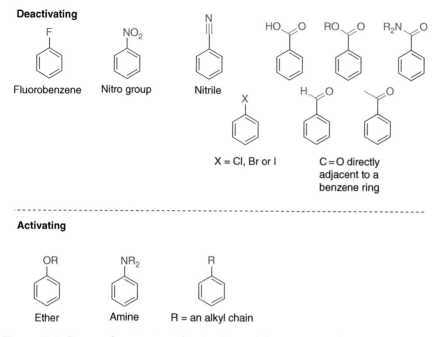

Figure 15.9 Commonly encountered activating and deactivating substituents.

When undergoing reaction with electrophiles, the electronic effects of ring substituents are important and have an impact upon where the electrophile reacts.

15.2 Reactions of benzene with electrophiles

You will encounter four main reactions when studying the reactivity of benzene with electrophiles:

1. Halogenation

2. Friedel–Crafts alkylation

3. Friedel–Crafts acylation

4. Nitration

15.2.1 Halogenation

Benzene can undergo **halogenation** by treatment with a halogen (usually bromine or chlorine) in the presence of a catalyst (usually aluminium(III) chloride or iron(III) bromide) (Figure 15.10). The catalyst is required because, as discussed earlier in this chapter, benzene is much less reactive than an alkene. Therefore, for benzene to react with the halogen, the halogen itself needs to be activated or made more reactive. As a general rule, when undertaking chlorination, use aluminium(III) chloride as a catalyst; and for bromination, use iron(III) bromide. The equivalent reaction with an alkene is discussed in Chapter 13.

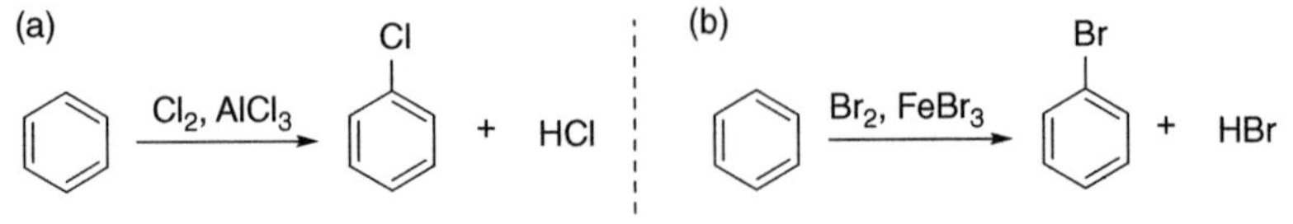

Figure 15.10 (a) Reaction of benzene with chlorine in the presence of AlCl$_3$; (b) reaction of benzene with bromine in the presence of FeBr$_3$.

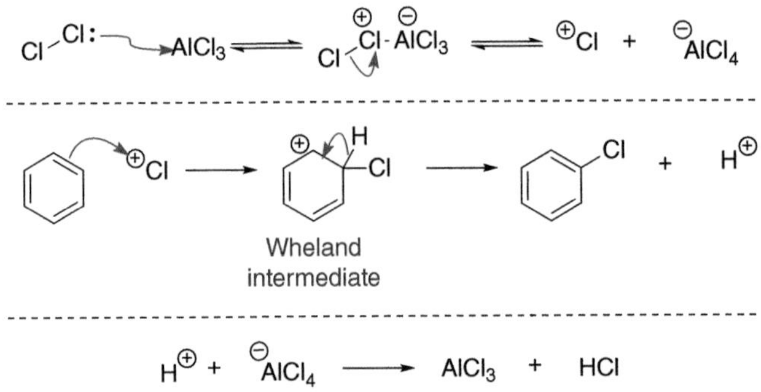

Figure 15.11 Reaction of benzene with chlorine in the presence of aluminium(III) chloride to give chlorobenzene.

The mechanism contains a few steps, but they are all logical. In addition, the mechanistic steps for this reaction are broadly similar to those for the other reactions of benzene with electrophiles.

The first step involves activation of the chlorine molecule, so it is more electrophilic. This occurs by a lone pair of electrons on the chlorine atom attacking the electron-deficient aluminium centre, which is acting as a Lewis acid. This causes the chlorine atom to have a positive charge, which then means that the chlorine–chlorine σ bond breaks towards the positive chlorine, generating a chloronium ion (Cl$^+$) and a tetrachloroaluminate species (AlCl$_4^-$) (Figure 15.11). The Cl$^+$ is a very strong electrophile, so the benzene ring attacks it, generating a carbocation in the process. This carbocation is also called a **Wheland intermediate**. The carbocation is quite unstable because it has lost aromaticity; therefore, it very quickly expels H$^+$, regenerating the benzene ring and generating HCl as a by-product.

15.2.2   Friedel–Crafts alkylation

Friedel–Crafts acylation and **Friedel–Crafts alkylation** are very similar in mechanism to that of halogenation. The only difference is the electrophile that is generated. In Friedel–Crafts alkylation, the starting material is usually an alkyl halide, often an alkyl chloride.

In analogy with the previous mechanism, the chlorine-containing molecule must be activated so that it becomes more electrophilic (Figure 15.12). To do this, the lone pair of electrons on the chlorine atom move to attack the aluminium(III) species, generating a new chlorine to aluminium bond and a chlorine with a positive charge. This weakens the carbon–chlorine bond, which breaks towards the positively charged chlorine atom, giving a carbocation that is a strong electrophile. This carbocation is attacked by the benzene ring, which quenches the positive charge on the carbocation but creates a new positive

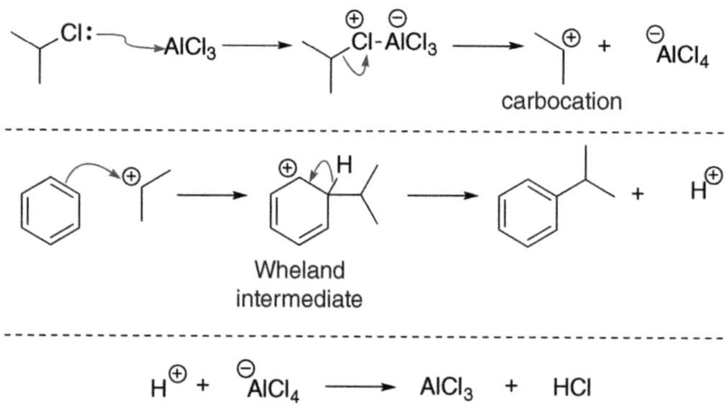

Figure 15.12 Reaction of benzene with 2-chloropropane in the presence of aluminium(III) chloride to give *iso*-propylbenzene.

charge on the benzene ring, forming the Wheland intermediate. Again, the Wheland intermediate is unstable because it is no longer aromatic, so it quickly falls apart and ejects H⁺. The H⁺ reacts with the tetrachloroaluminate, regenerating $AlCl_3$ and making HCl as a by-product.

15.2.3   Friedel–Crafts acylation

In Friedel–Crafts acylation, the starting material is usually an acid chloride, and the catalyst used is aluminium(III) chloride. In analogy with the alkylation, the aluminium(III) interacts with the chlorine on the acid chloride, generating an acylium ion and tetrachloroaluminate (Figure 15.13). Once the acylium ion is formed, attack by the benzene ring to form the Wheland intermediate can occur. Subsequent deprotonation leads to an aromatic product where there is a carbonyl group (C=O) directly adjacent to the benzene ring.

15.2.4   Nitration

The final reaction of benzene that is usually discussed at the foundation level is **nitration**. In this case, nitric acid (HNO_3) and sulfuric acid (H_2SO_4) are premixed and then added to the benzene. Once the electrophile (NO_2^+) has been

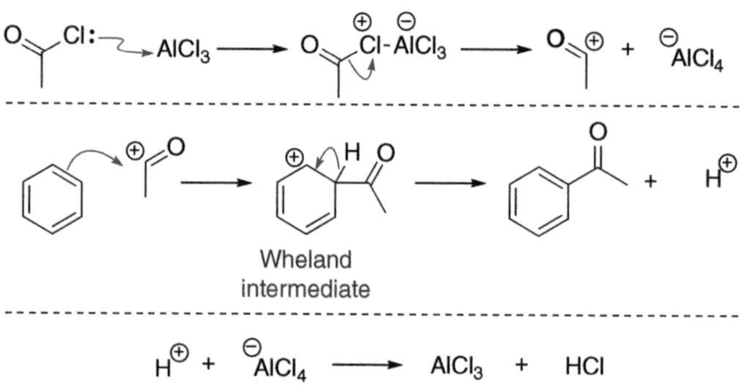

Figure 15.13 Reaction of benzene with ethanoyl chloride in the presence of aluminium(III) chloride to give acetophenone.

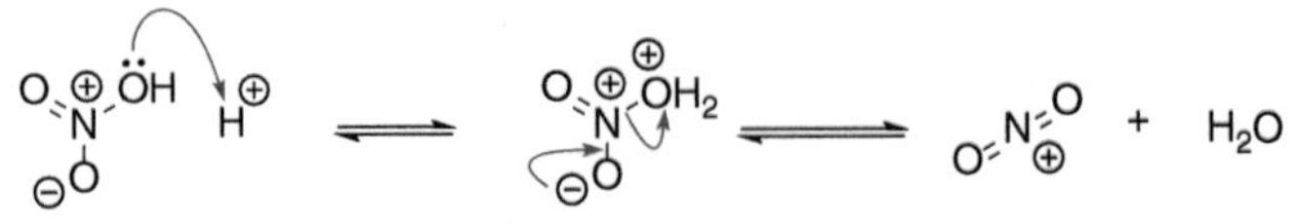

Figure 15.14 Reaction of nitric acid with H^+ in sulfuric acid to generate the nitronium ion, NO_2^+.

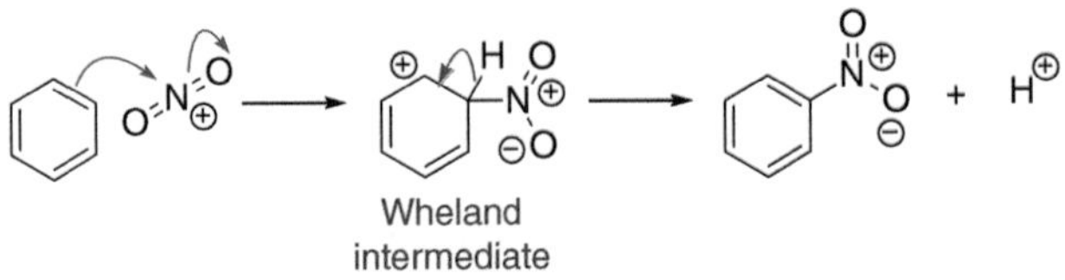

Figure 15.15 Reaction of benzene with the nitronium ion to generate nitrobenzene.

generated, the reaction mechanism is identical to that for the previous three reactions discussed.

In the first step, the nitric acid is protonated by the H^+ present in sulfuric acid, which is very strongly acidic, and then water is eliminated to generate a nitronium ion (Figure 15.14). Once the electrophilic nitronium species is formed, it reacts with the benzene ring to form the Wheland intermediate. This is followed by the elimination of H^+ to give the aromatic product (Figure 15.15).

15.2.5   Substituent effects on position of substitution

As discussed in Section 15.1.5, different substituents can affect the rate of reaction of an aromatic ring with the electrophile. Electron-rich aromatic rings react with electrophiles quickly, and electron-poor aromatic rings react with electrophiles more slowly. This is logical and is shown by the arrows that we use; to react with an electrophile, the benzene ring needs to have electrons that are free to attack the electrophile. If the benzene ring is electron-poor, these electrons are less available; hence the benzene ring is a less reactive species, and the reaction will be slower.

In addition to the speed at which the benzene ring attacks the electrophile, the position that is substituted also varies with electronics. In electron-rich species, substituents are added *ortho* or *para* to the directing substituent; and in electron-poor species, the substituents are added *meta* to the directing substituent (Figure 15.16). An extended discussion about this is beyond the scope of a foundation-level course.

> Electron-rich aromatic rings react with electrophiles quickly, and electron-poor aromatic rings react with electrophiles more slowly.

15.2.6   Reaction of phenol with electrophiles

Phenol has an oxygen atom that has two lone pairs of electrons directly adjacent to the benzene ring. These electrons are able to undergo resonance into the benzene ring through mesomeric effects, rendering phenol electron-rich (Figure 15.17). This is particularly the case in positions 2, 4, and 6, as indicated in Figure 15.17b,c.

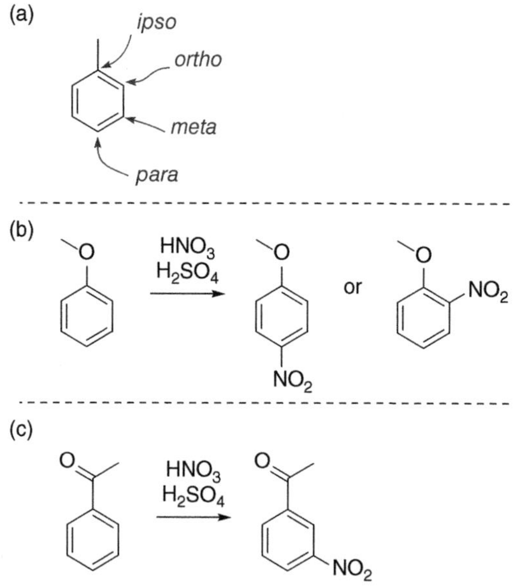

Figure 15.16 (a) The named positions on a benzene ring in relation to a substituent. (b) Reaction of an electron-rich benzene ring with an electrophile gives *ortho* and *para* substitution. (c) Reaction of an electron-poor benzene ring with an electrophile gives *meta* substitution.

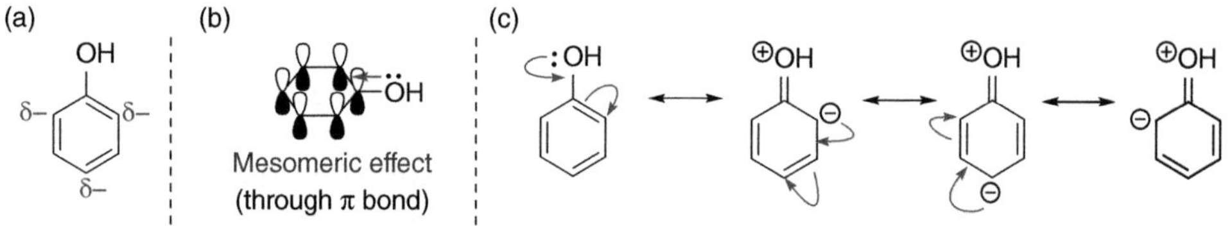

Figure 15.17 (a) Phenol; (b) donation of a lone pair of electrons on oxygen into the π system through mesomeric effects; (c) resonance of the lone pair on the oxygen atom into the benzene ring, leading to electron-rich *ortho* and *para* positions.

Worked Example 15.5   Draw the mechanism and product from the reaction of phenol with nitric acid in the presence of concentrated sulfuric acid:

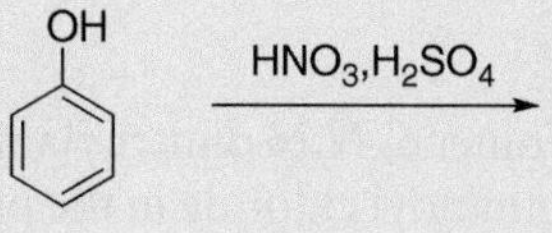

Solution

In this reaction, there are two things to consider: (i) the generation of the electrophile and (ii) the reaction of the aromatic ring with the electrophile. In this case, the electrophile arising from the reaction between nitric acid and concentrated sulfuric acid is the nitronium ion (Section 15.2.4):

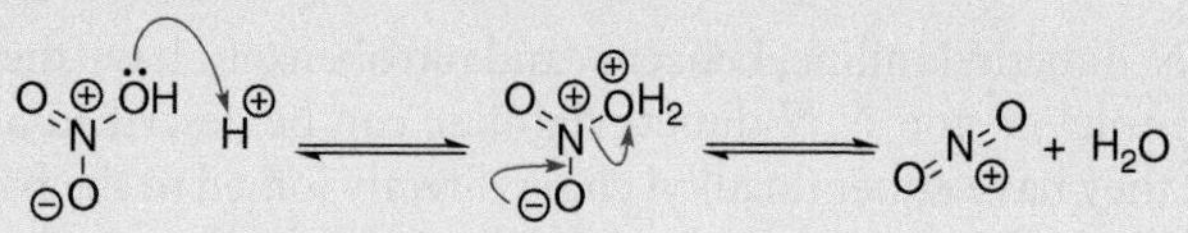

Once the electrophile has been made, attack by the benzene ring can occur. Phenol is electron-rich in the 2, 4, and 6 positions (see Figure 15.17), so the electrophile will be added in those positions preferentially, i.e. *ortho* or *para* to the OH group. In this example, the phenol is added *para*, but the mechanism for *ortho* addition is the same:

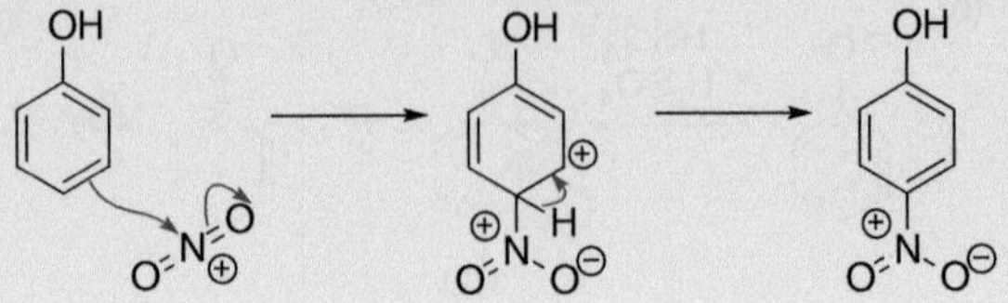

15.2.7   Reaction of toluene with electrophiles

Toluene has a methyl group that is directly attached to the benzene ring, which can push electron density onto the benzene ring so that the ring is electron-rich. This means that toluene will react reasonably quickly with an electrophile and will predominantly undergo electrophilic substitution in the *ortho* or *para* position. It should be noted that the electron-donating effect of the methyl group is lower than that of an oxygen or nitrogen.

15.2.8   Reaction of nitrobenzene with electrophiles

Nitrobenzene has an electron-deficient benzene ring; therefore, reaction with electrophiles is slower than it would be with an electron-rich benzene ring. In addition, due to the electron-poor nature of the benzene ring, substitution occurs only in the *meta* position, in relation to the nitro group (Figure 15.18).

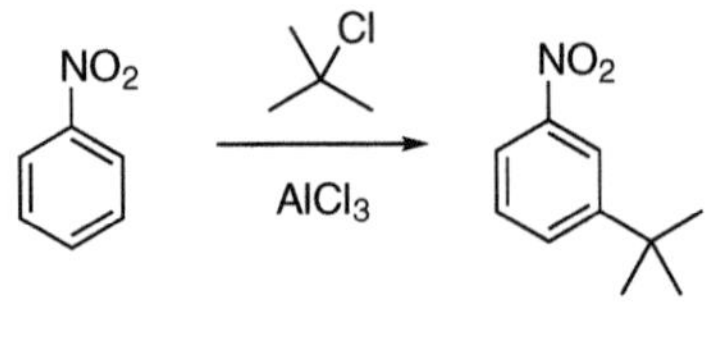

Figure 15.18 Reaction of nitrobenzene with an alkyl chloride in the presence of AlCl$_3$.

Worked Example 15.6   Toluene, *N,N*-dimethylaniline, benzene, and nitrobenzene can all react with ethanoyl chloride in the presence of aluminium(III) chloride to make the corresponding Friedel–Crafts product. Which would you expect to react the fastest, and which would you expect to react the slowest? Draw a mechanism for the reaction of toluene with ethanoyl chloride in the presence of aluminium(III) chloride.

Solution
Toluene, *N,N*-dimethylaniline, benzene, and nitrobenzene have the structures shown. Both toluene and *N,N*-dimethylaniline can be described as electron-rich because they have either an alkyl chain directly joined to the benzene ring (toluene) or a substituent with a lone pair directly joined to the benzene ring

(aniline). The lone pair on nitrogen is more electron-donating, so *N,N*-dimethylaniline is more electron-rich than toluene. Benzene could be described as electron-neutral because it does not have any substituents. Finally, nitrobenzene is electron-poor because the nitro group removes electron density from the benzene ring.

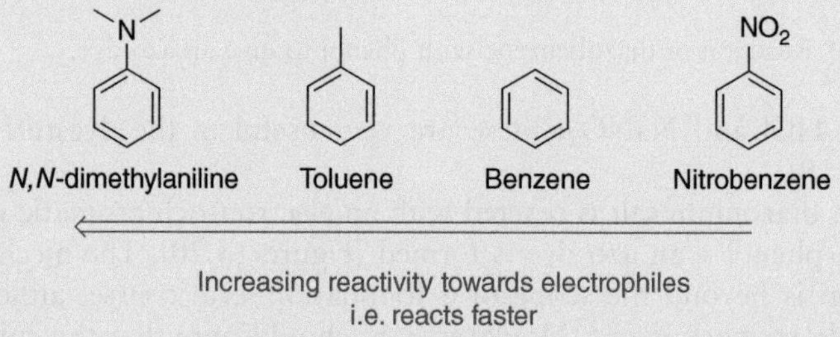

N,N-dimethylaniline   Toluene   Benzene   Nitrobenzene

Increasing reactivity towards electrophiles
i.e. reacts faster

In terms of reactivity with an electrophile, the more electron-rich a benzene ring is, the faster it will react. In this case, *N,N*-dimethylaniline is the most electron-rich, so it will react fastest, followed by toluene, then benzene, and finally nitrobenzene.

The mechanism for the reaction of toluene with ethanoyl chloride is the same as that discussed in Section 15.2.3. The first step is to generate the electrophile, and then the benzene ring can attack it. Toluene is electron-rich, so the substitution could happen either *ortho* or *para* to the methyl group. In this example, *ortho* attack is drawn:

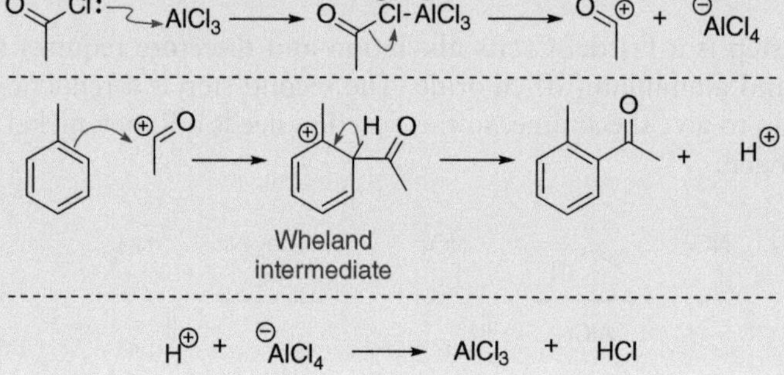

Wheland
intermediate

15.3 Aniline

The final aromatic compound to be discussed is **aniline** (Figure 15.19). Aniline is electron-rich and so will react quickly with electrophiles and undergo substitution at the *ortho* and *para* positions, in relation to the aniline group.

Aniline can be prepared by reduction of nitrobenzene using a Raney-nickel catalyst and hydrogen gas. In addition, anilines can be converted into diazonium

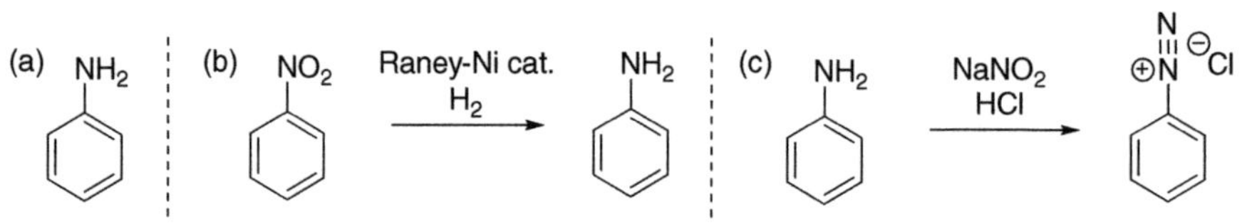

Figure 15.19 (a) Aniline; (b) formation of aniline by reduction of nitrobenzene; (c) formation of diazobenzene from aniline.

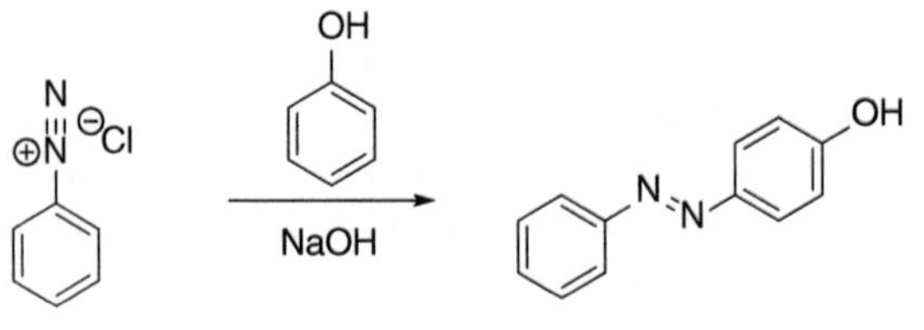

Figure 15.20 Reaction of diazobenzene with phenol to give an azo-dye.

salts using HCl and NaNO$_2$; these are very useful in the dyestuff industry (Figure 15.19).

When a diazonium salt is reacted with an electron-rich aromatic ring – for example, a phenol – an azo dye is formed (Figure 15.20). The mechanism of the reaction is beyond the scope of a foundation-level course, although you may be able to work it out. However, you should note that the substitution has occurred *para* to the OH group in the phenol.

A diazonium salt contains a N ≡ N$^+$ bonded to a benzene ring, whereas an azo-dye contains an N=N between two benzene rings.

Worked Example 15.7   Give the conditions required for the following steps:

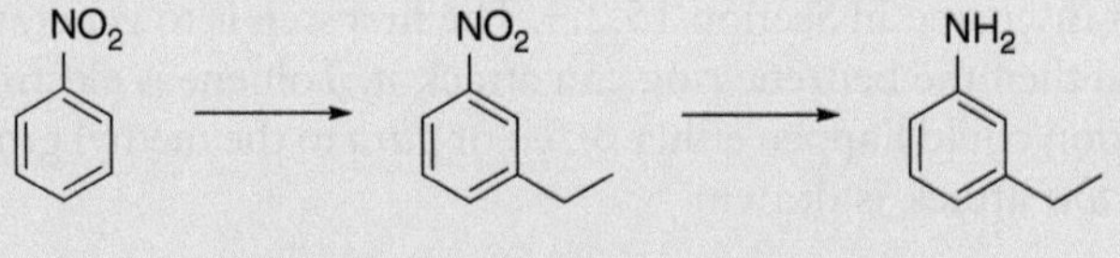

Solution
The first step is a Friedel-Crafts alkylation and therefore requires an alkyl chloride and aluminium(III) chloride. The second step is a reduction of the nitro group to give the aniline, so this reaction needs a Raney-nickel catalyst and hydrogen:

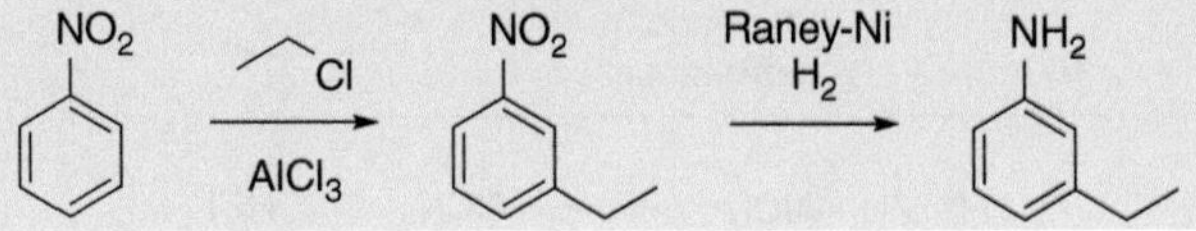

Quick-check summary

- Benzene, C$_6$H$_6$, is an example of an aromatic compound. It is flat and has a cloud of delocalised electrons above and below the ring.

- Hückel's rules can be used to determine whether a compound is aromatic.

- There are rules for naming aromatic ring systems.

- Benzene is not as reactive as might be expected because it is aromatic, and aromaticity adds stability to a compound.

- The electrons in the benzene ring system can move and undergo resonance.

- The substituents on a benzene ring can affect the reactivity towards nucleophiles and electrophiles.

- Electron-rich compounds containing a benzene ring can react with electrophiles, leading to electrophilic aromatic substitution (S_EAr). Halogenation, Friedel–Crafts acylation, Friedel–Crafts alkylation, and nitration are all examples of this.

- Phenol contains a benzene ring that is directly bonded to an OH group, and aniline contains a benzene ring directly bonded to a nitrogen atom. Phenol and aniline are both more reactive towards electrophiles than benzene as they are comparatively electron-rich.

- Nitrobenzene contains a benzene ring that is bonded to a nitro (NO_2) group, and it is less reactive towards electrophiles than benzene as it is comparatively electron-poor.

End-of-chapter questions

1 Draw a molecule of benzene, clearly showing the orbitals that make up key bonding interactions.

2 Name the following compounds:

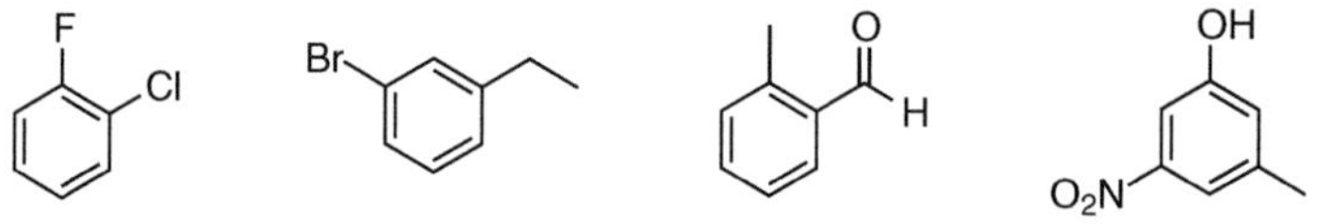

3 Why is nitrobenzene considered electron-poor and phenol considered electron-rich? Use suitable diagrams to support your answer.

4 Draw the mechanism and suggest the product for the following reaction:

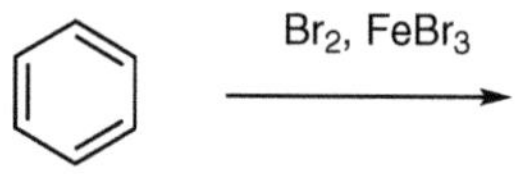

5 Draw the mechanism and suggest the product for the following reaction:

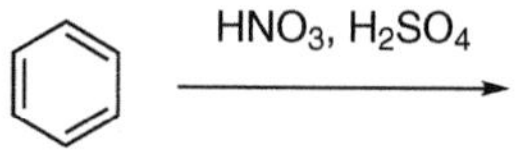

6 Draw the mechanism and suggest the product for the following reaction:

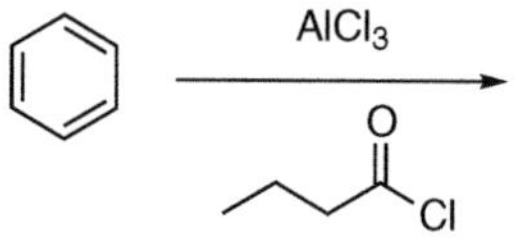

7 How might you prepare the following aromatic compounds from benzene?

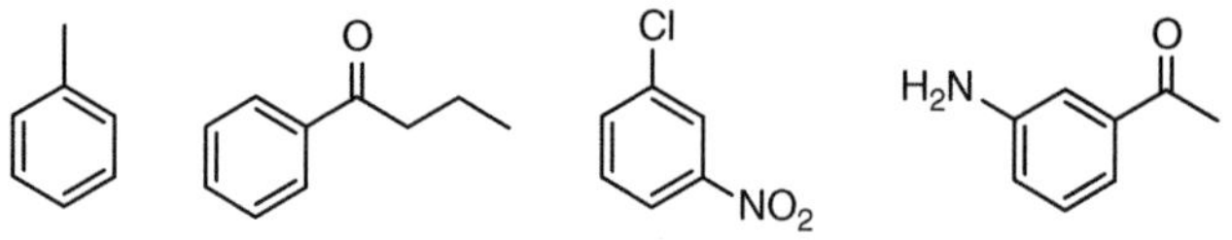

8 What would be the likely comparative rates of reaction of the following aromatic systems with an electrophile?

9 Give conditions required for the transformations shown:

10 What would be the likely product from the reaction shown?

Substitution and elimination reactions

At the end of this chapter, students should be able to:

- Understand what is meant by the terms *substitution* and *elimination*

- Be able to draw the mechanism for an S_N1 or S_N2 reaction pathway

- Be able to draw the mechanism for an E1 or E2 reaction pathway

- Understand and explain the difference between Zaitsev and Hofmann alkenes

16.1 Substitution reactions

A *substitution* reaction is defined by the International Union of Pure and Applied Chemistry (IUPAC) as 'a reaction, elementary or stepwise, in which one atom or group in a molecular entity is replaced by another atom or group'; i.e. one atom or group is swapped for another. Figure 16.1 shows an example of a substitution reaction where the bromine in 2-bromopropane is substituted by hydroxide to give propan-2-ol as the organic product. The OH group comes from sodium hydroxide.

Two different types of substitution for alkyl compounds are likely to be encountered at the foundation level: S_N1 and S_N2. The difference between them is subtle and depends upon the structure of the substrate that is undergoing substitution.

There are two types of substitution that will be discussed: S_N1 and S_N2.

16.1.1 S_N1 reactions

An S_N1 reaction has two steps, the first of which (the rate-determining step) is first order with respect to the starting material i.e. only the starting material is required and the concentration of the attacking species does not affect the rate of the

Foundations of Chemistry: An Introductory Course for Science Students, First Edition.
Philippa B. Cranwell and Elizabeth M. Page.
© 2021 John Wiley & Sons Ltd. Published 2021 by John Wiley & Sons Ltd.
Companion website: www.wiley.com/go/Cranwell/Foundations

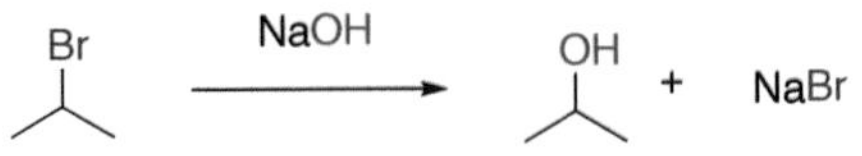

Figure 16.1 Substitution of 2-bromopropane with sodium hydroxide to give propan-2-ol.

For a refresher on carbocation stability and structure, see Chapter 13.

reaction. This is reflected in the term S_N1, which actually means *substitution, nucleophilic, first order*. The reaction proceeds via a discrete carbocation intermediate, which has a positive charge. The fact that a carbocation intermediate is needed means that this intermediate must be stable; therefore, an S_N1 reaction pathway is followed *only* if a secondary or tertiary carbocation can be formed.

An S_N1 reaction has two steps in the mechanism and proceeds via a discrete carbocation intermediate. This reaction pathway is followed only if a secondary or tertiary carbocation can be formed.

Box 16.1 The two steps in an S_N1 reaction are as follows:
1. The leaving group leaves the substrate to form a carbocation.
2. The nucleophile attacks the carbocation to form the product.

The term *substrate* is another word for 'starting material'. It is effectively the compound that is undergoing the chemical reaction.

For example, Figure 16.2 outlines a reaction that undergoes an S_N1 pathway. It can be seen that in the overall reaction, iodine is substituted with cyanide (Figure 16.2a). In terms of the mechanism, two discrete steps are involved: (i) the leaving group (I^-) leaves the substrate to form a carbocation intermediate, which is planar; and (ii) the nucleophile (NC^-) attacks the carbocation to form the product. This can be seen in Figure 16.2b.

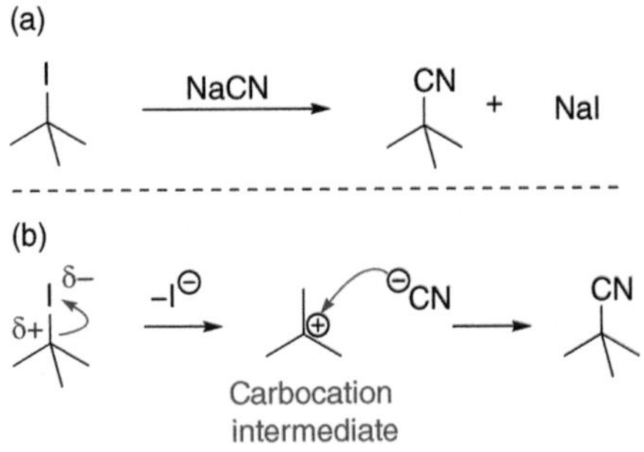

Sodium cyanide contains an ionic bond that can be considered as two closely associated ions: Na^+ and NC^-.

Figure 16.2 (a) Overall reaction for substitution of 2-iodo-2-methylpropane with sodium cyanide to give 2,2-dimethylpropionitrile; (b) the mechanism for this reaction. Note that the Na^+ is a spectator ion and is not involved in the reaction. It is likely to associate with the I^- to make NaI.

Worked Example 16.1   Give the reaction mechanism for the transformation shown, assuming an S_N1 pathway is followed:

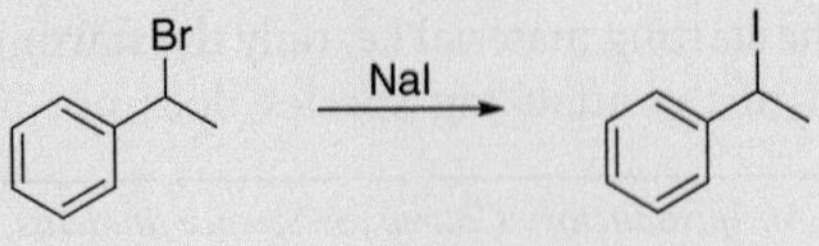

Solution

It can be seen from the scheme that overall, the bromine atom has been substituted with an iodine atom. A good start when trying to draw any reaction mechanism is to draw on any dipoles. In the substrate, there is a dipole between the carbon atom and the bromine atom; the carbon atom is δ+ and the bromine atom δ−. This dipole will help determine where any nucleophile will attack (i.e. identify any electrophiles) and will also show which group is likely to leave.

The second thing to do is to identify a likely nucleophile. In this reaction, NaI is an ionic compound, so it can exist as Na$^+$ and I$^-$ in solution. The I$^-$ will act as the nucleophile because it has a negative charge. The I$^-$ can therefore attack electrophiles, in this case, the δ+ carbon.

Finally, we need to determine if there are any intermediates. The question states that the pathway followed is S$_N$1; therefore, we know that there must be a carbocation intermediate, so the leaving group must leave *before* the nucleophile can attack:

Carbocation
intermediate

16.1.2   S$_N$2 reactions

An **S$_N$2** reaction has only one step, where the nucleophile attacks and the leaving group leaves at the same time. This reaction is first order with respect to both the nucleophile and the substrate and therefore is second order overall. This is reflected in the term S_N2, which actually means *substitution, nucleophilic, second order*. This means that in the all-important rate-determining step, both the nucleophile and substrate must be present. In this case, as the nucleophile attacks, the leaving group leaves; this happens *simultaneously* and is different to S$_N$1, which is a *step-wise* process. For S$_N$2, whether or not this reaction will occur is governed by the steric hindrance near to the leaving group on the substrate (less hindrance is better). Therefore, this reaction pathway will be followed only if the leaving group is on a primary or a secondary position.

Note: in this textbook, we do not consider the transition state.

Box 16.2 In an S$_N$2 reaction, the nucleophile displaces the leaving group in one step.

An S$_N$2 reaction has one step in the mechanism; it is *concerted*. This reaction pathway is followed only if the leaving group is on a primary or secondary carbon atom.

Figure 16.3 outlines a reaction that undergoes an S$_N$2 pathway. In analogy with Figure 16.1, during this reaction, iodine is replaced by cyanide (Figure 16.3a). The mechanism is similar to that shown in Figure 16.1 but occurs in a single step. In this case, the leaving group (I$^-$) leaves at the same time as the nucleophile (NC$^-$) attacks, forming the product. This can be seen in Figure 16.3b.

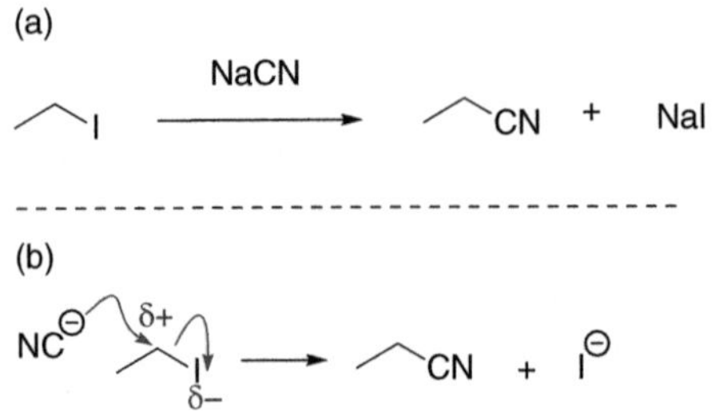

(a)

(b)

Figure 16.3 (a) Overall reaction for substitution of iodoethane with sodium cyanide to give propionitrile; (b) the mechanism for this reaction. Note that the nucleophile attacks from the opposite side of the carbon atom that the leaving group is attached to. The transition state has not been drawn.

Worked Example 16.2   Give the reaction mechanism for the transformation shown, assuming an S_N2 pathway is followed:

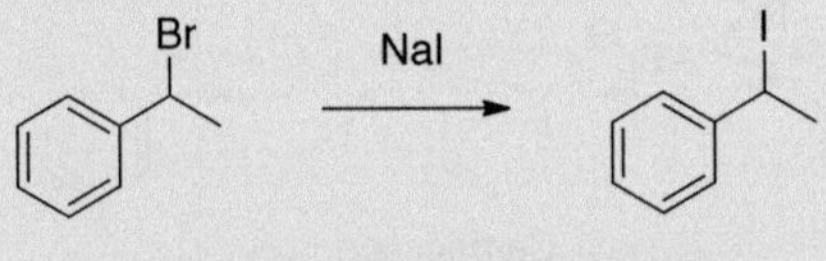

Solution
As with Worked Example 16.1, it can be seen from the scheme that overall, the bromine atom has been substituted with iodide; however, in this case, the question states that the reaction follows an S_N2 pathway. As with the previous example, the best way to attack this question is to draw on any dipoles and identify the nucleophile and the electrophile. In the substrate, there is a dipole on the starting material between the carbon atom (δ+) and bromine atom (δ−), and sodium iodide can exist as two discrete ions: Na$^+$ and I$^-$. The I$^-$ will act as the nucleophile, and bromide will be the leaving group. In an S_N2 reaction, there is no intermediate, and the reaction proceeds in one step:

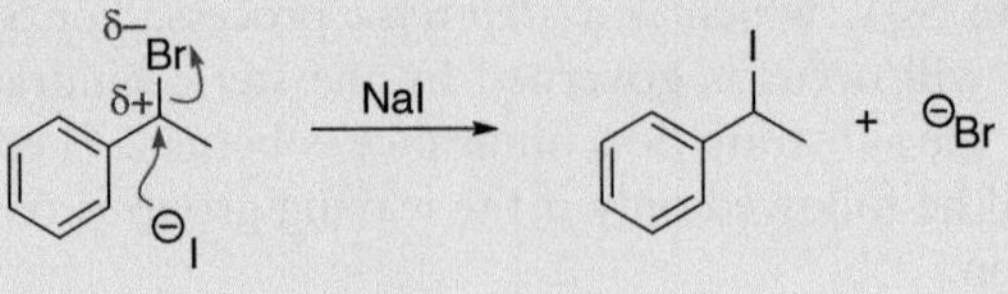

Worked Example 16.3   Consider the reaction shown, and then answer the following questions:

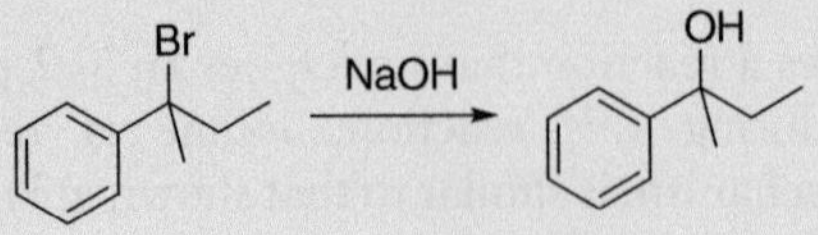

a. Is this reaction likely to follow an S_N1 or S_N2 reaction pathway? Give reasons for your answer.

b. Draw the mechanism for this reaction.

c. What is the inorganic by-product?

Solution
The first part of this question asks if the reaction follows an S_N1 or S_N2 reaction pathway. To determine this, we first need to work out exactly what has happened overall. Looking at the scheme, we can see that the bromine atom has been replaced by an OH; therefore, the bromine atom must be the leaving group and the OH (from NaOH) must be the nucleophile. The carbon atom that the bromine atom is bonded to has three carbon atoms attached, so it is tertiary. The fact that the bromine is bonded to a tertiary carbon means that the reaction pathway must be S_N1. Now that we have established that the reaction is S_N1, a mechanism can be drawn.

In an S_N1 reaction, the leaving group always leaves first, forming a carbocation where it was once bonded. Once the carbocation has been formed, the nucleophile can attack the positive carbon centre, making a new bond. As with all the other examples, it is always best to draw a dipole, determine which species is the nucleophile, determine the likely reaction pathway, and then add curly arrows:

Carbocation
intermediate

Finally, the question asked what the inorganic by-product was likely to be. During the reaction, Br^- was liberated, and an Na^+ ion (from NaOH) is available; therefore, the inorganic product is NaBr.

Worked Example 16.4   Consider the reaction shown, and then answer the following questions, assuming the substrate undergoes nucleophilic substitution:

a. Is this reaction likely to follow an S_N1 or S_N2 reaction pathway? Give reasons for your answer.

b. What is the likely product?

c. Draw the mechanism for this reaction.

Solution

The first part of this question asks if the reaction follows an S_N1 or S_N2 reaction pathway. To determine this, we first need to work out exactly what has happened overall; however, our scheme does not show us the product, so we need to use some chemical intuition. Drawing dipoles onto the substrate gives a $\delta+$ carbon atom and a $\delta-$ bromine atom, so it is likely that the carbon will be attacked by a nucleophile and bromide will leave. Looking more closely at the bromine atom, we can see that it can be designated as primary; therefore, the reaction pathway followed will be S_N2.

The likely product will be formed by displacement of the bromide by iodide, giving 2-(iodoethyl)benzene.

The mechanism will be an S_N2 pathway; therefore, the substitution will occur in one step. The first thing to do is draw on any dipoles, then identify the nucleophile, and finally draw the required curly arrows:

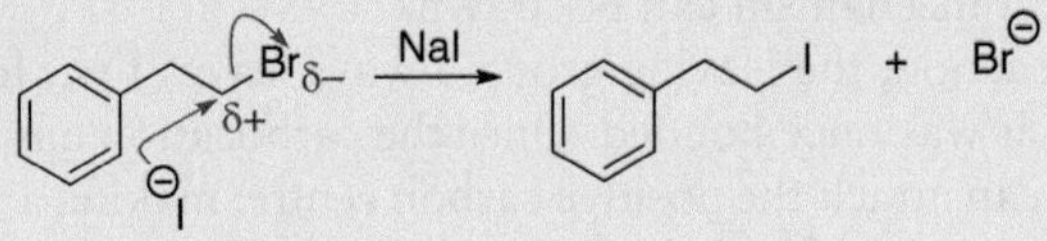

16.2   Elimination reactions

An **elimination** reaction occurs when two groups are lost from two different centres, with concomitant formation of unsaturation in the molecule (i.e. a double bond or triple bond). In this chapter, we will only discuss the formation of alkenes by elimination (Figure 16.4). Three types of elimination are commonly encountered: E1, E2, and $E1_{cb}$. We will only discuss E1 and E2.

In analogy with S_N1 and S_N2, the designation of a reaction as either E1 or E2 depends upon the order (concentration dependence) of the rate-determining step. If the reaction rate is only dependent on the substrate in the rate-determining step, the elimination is **E1**; and if the rate is dependent on both the substrate and the base, the elimination is **E2**.

It is important to remember that in an elimination reaction, the proton removed needs to be on the carbon directly adjacent to the carbon that the leaving group is attached to.

16.2.1   E1 reactions

In an E1 reaction, there are two steps in the mechanism and the reaction proceeds via a discrete carbocation intermediate. This first step, the formation of the carbocation, is identical to that of an S_N1 reaction. As this is an E1 reaction, it is first order, and the overall rate of the reaction relies upon the concentration

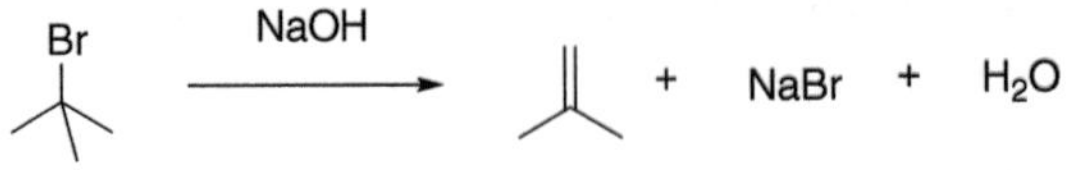

Figure 16.4 An elimination reaction from the treatment of 2-bromo-2-methylpropane with sodium hydroxide to give 2-methylprop-1-ene, NaBr, and water.

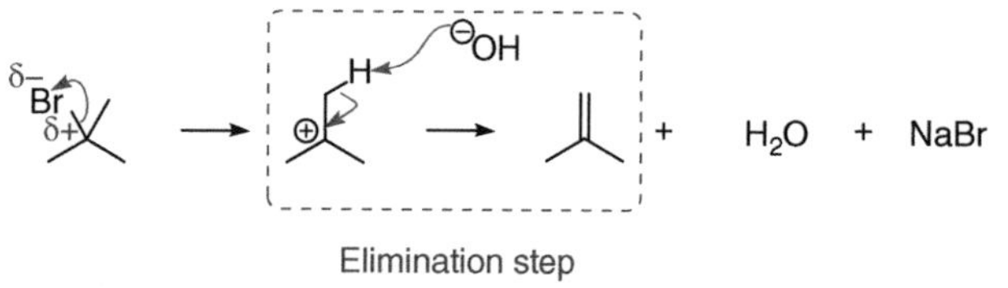

Elimination step

Figure 16.5 The mechanism of an E1 elimination from the treatment of 2-bromo-2-methylpropane with sodium hydroxide to give 2-methylprop-1-ene, NaBr, and water. Note that the H adjacent to the Br has been added to aid clarity; it is not necessary to always draw it.

of one chemical component: in this case, the starting material. The speed at which this reaction proceeds is dependent upon how easy it is for the leaving group to leave and form the carbocation intermediate. The fact that a carbocation intermediate is needed means that this intermediate must be stable; therefore, an E1 reaction pathway is followed *only* if a secondary or tertiary carbocation can be formed.

Once the carbocation is formed, the reaction mechanism deviates from that of S_N1. A proton is removed from the carbon atom adjacent to that of the carbocation, with HO^- acting as a base. The HO^- attacks the proton, and the electrons in the C—H bond then move towards the carbocation forming the double bond, resulting in an alkene (Figure 16.5).

> An E1 reaction pathway is followed *only* if a secondary or tertiary carbocation can be formed.

16.2.2   E2 reactions

An E2 reaction has only one step in the mechanism. In this single *concerted* step, the proton is removed, the alkene is formed, and the leaving group leaves, all at once (Figure 16.6). In terms of bonds breaking and forming, the HO^- acts as a base and removes the proton on the carbon atom adjacent to the carbon atom on which the leaving group is bonded. The electrons in this C—H bond then move and start to push the leaving group out: in this case, the bromine atom. This generates an alkene, but, as stated earlier, all of the electron movement is occurring at the same time.

In analogy with an S_N2 reaction, this reaction pathway will be followed only if the leaving group is primary or secondary. In this reaction mechanism, six electrons are moving at once, so three curly arrows are needed.

> An E2 reaction has one step in the mechanism. This reaction pathway is followed only if the leaving group is on a primary or secondary position.

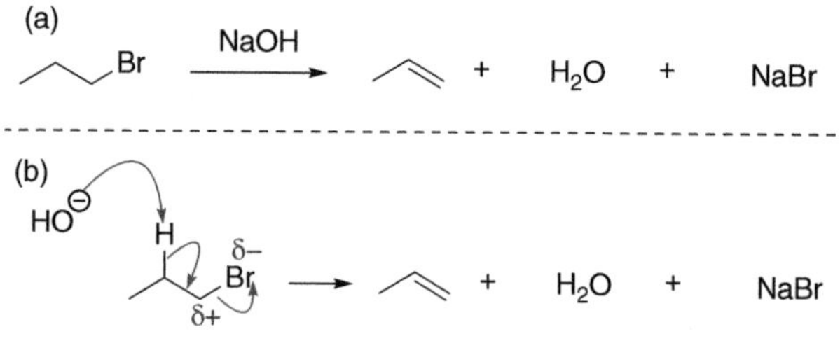

Figure 16.6 (a) Overall reaction between 1-bromopropane and NaOH to give propene; (b) the mechanism for the E2 elimination.

Worked Example 16.5   Consider the following reaction, and then answer the following questions:

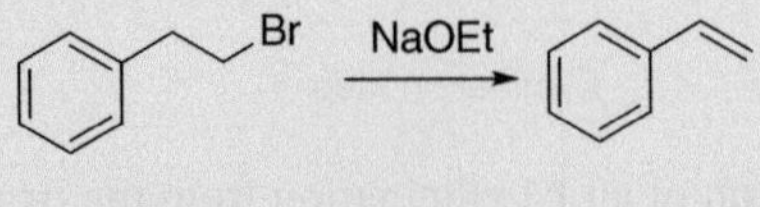

a. Is this reaction likely to follow an E1 or E2 reaction pathway? Give reasons for your answer.

b. Draw the mechanism for this reaction.

Solution
This reaction is likely to proceed via an E2 reaction pathway because the leaving group (bromide) is located on a primary carbon atom.

An E2 reaction is *concerted,* and all the electrons (and therefore curly arrows) move at once; the base removes the proton at the same time as the electrons move to form the double bond, and the bromide leaves. In this reaction, the base is sodium ethoxide, EtONa, a common alternative to NaOH. Like NaOH, EtONa is an ionic solid and exists as EtO$^-$ and Na$^+$ in solution. The key to answering any question about an elimination reaction is firstly to draw on any dipoles, and secondly remember that the proton removed is on the carbon atom adjacent to the carbon atom attached to the leaving group:

Remember that 'Et' stands for 'ethyl'.

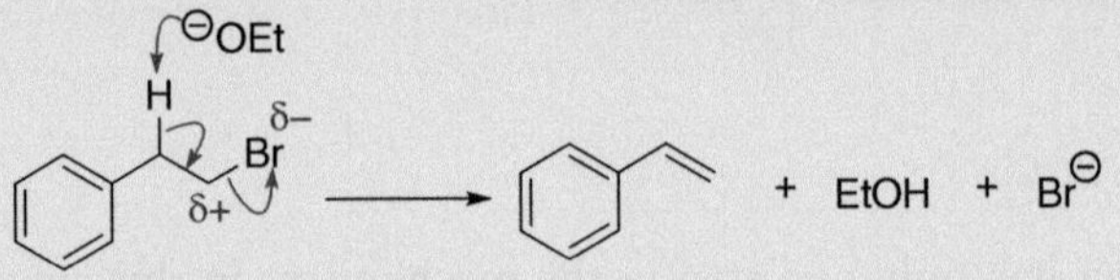

Worked Example 16.6   The following elimination reaction can lead to two different products (shown). Is this reaction likely to be E1 or E2? Draw a mechanism for the formation of each product:

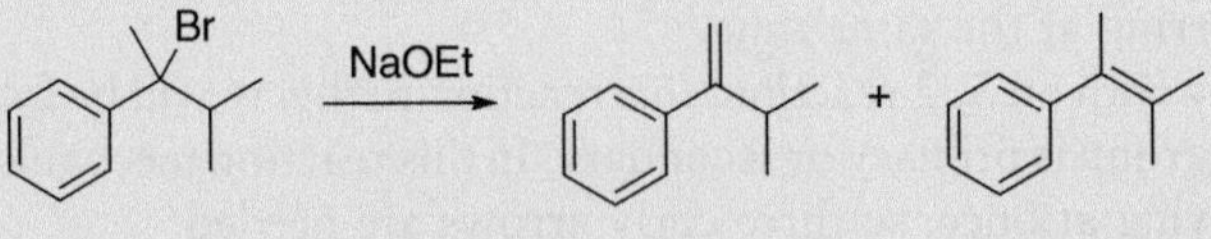

Solution
The bromine atom is bonded to a tertiary carbon centre, so the pathway is likely to proceed via a carbocation intermediate and will therefore be E1. To access both products, a tertiary carbocation will be formed in the first step. The difference is because two different protons can be removed by the base, coloured red and blue in the answer:

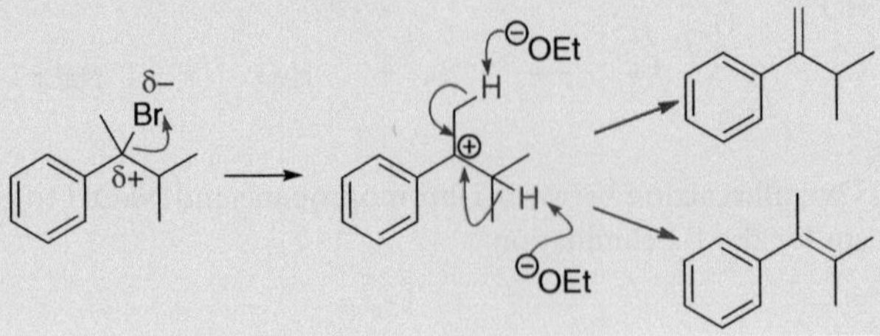

16.2.3   Zaitsev and Hofmann alkenes

When forming an alkene in an elimination reaction, as shown in Worked Example 16.6, sometimes more than one alkene can be formed. Generally speaking, during elimination reactions, the more substituted alkene is formed because it is more stable. This is the *Zaitsev product*. In Worked Example 16.6, this is the product from the red pathway. Sometimes, however, if the bulky base cannot access the proton required for the Zaitsev product, the more-substituted alkene cannot be formed, so the less-substituted alkene is formed. This is called the *Hofmann product*, and in Worked Example 16.6, it is the alkene formed from the blue pathway. E1 reactions are selective for the Zaitsev product, but E2 eliminations may lead to the Hofmann product if the bulky base cannot access the proton required for the Zaitsev product.

16.3   Comparison of substitution and elimination reactions

In this chapter, we have suggested that elimination and substitution reactions are mutually exclusive; however, in the laboratory, we may find that they compete with each other. For example, HO^- can act as both a nucleophile and a base, and it can be confusing when trying to decide between a substitution and an elimination. That said, at the foundation level it is unlikely that you will be asked to decide whether a substitution or elimination is more likely. To be successful at exam questions about substitution or elimination, the following checklist may help:

1. Draw the dipole on the substrate to identify the leaving group.
2. Determine if the leaving group is primary, secondary, or tertiary:
   a. Primary/secondary – S_N2 or E2
   b. Tertiary/secondary – S_N1 or E1

      If the substrate is secondary, you are likely to be told the reaction pathway to follow.
3. Identify the nucleophile/base.
4. Determine from the question if it is an elimination reaction (an alkene will be formed) or a substitution (the nucleophile will bond to the substrate).

Quick-check summary

- A substitution reaction is where one atom or group in a molecule is swapped with another.

- Two types of substitution reactions were discussed: S_N1 and S_N2, each of which follows a slightly different mechanistic pathway.

- In an S_N1 reaction, the leaving group leaves, generating a discrete carbocation intermediate that is then attacked by a nucleophile.

- In an S_N2 reaction, the leaving group is directly displaced by the incoming nucleophile.

- An elimination reaction occurs when two groups are lost from two different centres, which leads to unsaturation within the molecule (i.e. a double or triple bond).

- Two types of elimination reaction were discussed: E1 and E2, each of which follows a slightly different mechanistic pathway.

- In an E1 reaction, the leaving group leaves, generating a discrete carbocation intermediate. The adjacent proton is then removed, leading to an alkene.

- In an E2 reaction, the proton is removed, the alkene is formed, and the leaving group leaves at the same time.

- Elimination can lead to two different classes of alkene: Zaitsev and Hofmann alkenes. A Zaitsev alkene is the more-substituted alkene, and the Hofmann the less-substituted alkene.

End-of-chapter questions

1 Classify the following bromides as primary, secondary, or tertiary:

2 Considering the bromides in question 1, which will undergo S_N1 and which will undergo S_N2 substitution?

3 Considering the bromides in question 1, which will undergo E1 and which will undergo E2 elimination?

4 Draw the likely mechanism for each of the substitution reactions shown:

5   Draw the likely mechanism for each of the elimination reactions shown:

(a) 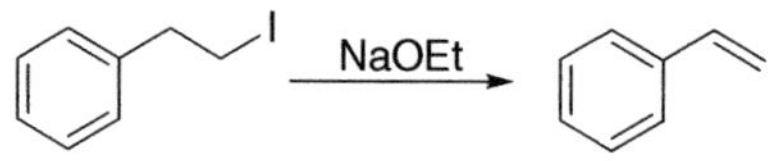 NaOEt

(b) 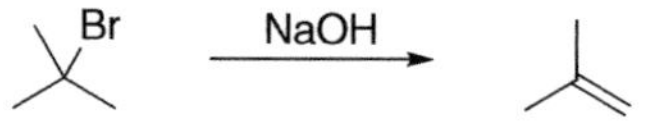 NaOH

(c) 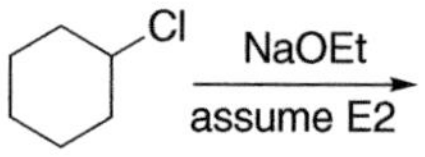 NaOEt / assume E2

(d) 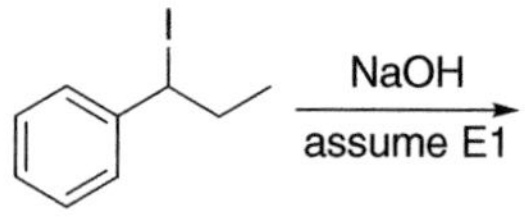 NaOH / assume E1

6   Draw the likely mechanism for the following elimination reaction. Which alkene product is more likely?

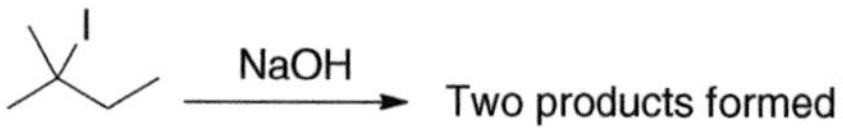

NaOH ⟶ Two products formed

17

Bringing it all together

At the end of this chapter, students should be able to:

- Understand the term *functional group interconversion*
- Draw mechanisms for short synthetic sequences based on material in Chapters 12–16
- Be able to suggest routes to simple target molecules

17.1 Functional group interconversion

Functional group interconversion is where one functional group is converted into another. Often this is simply a reaction occurring: for example, in the reduction of a ketone to an alcohol, the ketone group is converted to an alcohol group (Figure 17.1a). Sometimes the interconversion cannot be performed directly, and it is necessary for a functional group to be converted to something else before a further transformation to the desired functional group. For example, it is not possible to easily convert 2-iodopropane into propan-2-one directly (Figure 17.1b), so it must undergo a functional-group interconversion to an alcohol, which can then be oxidised to the ketone (Figure 17.1c).

17.2 Bringing it all together

The best way to practise using everything you have done in Chapters 12–16 is to make sure that you understand the mechanisms of the reactions in the previous

Foundations of Chemistry: An Introductory Course for Science Students, First Edition.
Philippa B. Cranwell and Elizabeth M. Page.
© 2021 John Wiley & Sons Ltd. Published 2021 by John Wiley & Sons Ltd.
Companion website: www.wiley.com/go/Cranwell/Foundations

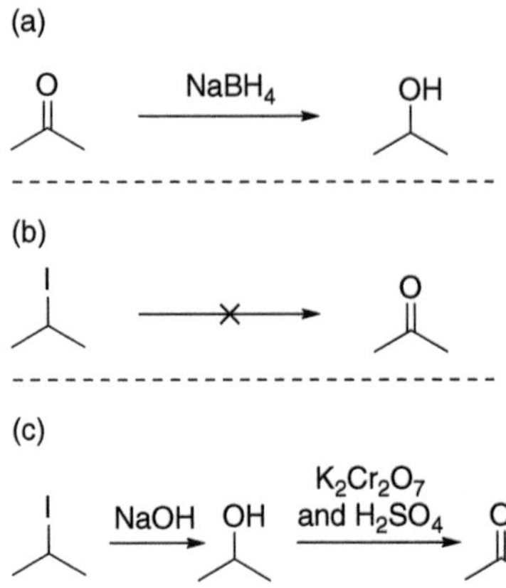

(a)

(b)

(c)

Figure 17.1 (a) Reduction of a ketone to give an alcohol. (b) Conversion of the iodide into the ketone directly is not particularly easy. (c) The alcohol can be formed by a substitution reaction from the iodide, and then the alcohol can be oxidised to the ketone.

chapters and then complete practice questions. This will be showcased by using worked examples that become increasingly more difficult.

Worked Example 17.1   What reagent(s) would be needed to effect transformations A–C?

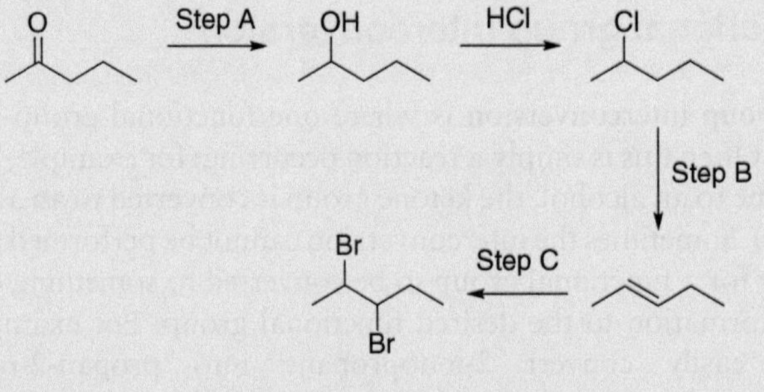

Solution
Step A is a reduction of a ketone to an alcohol, so a source of hydride is required. Acceptable answers would include $NaBH_4$ or $LiAlH_4$. Step B is an elimination reaction, as an alkene is formed by elimination of the chloride, which is a very good leaving group. A suitable reagent may be NaOH, as the hydroxide could act as a base and remove the requisite proton. It is worth noting that the alkene formed is likely to be the more substituted (Zaitsev) alkene. Finally, step C is bromination of an alkene and so requires Br_2.

Worked Example 17.2   Consider the reaction scheme, and answer the questions that follow:

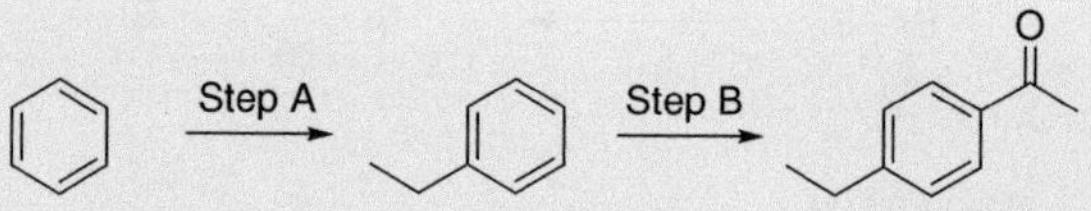

1. What reagent(s) would you need to effect steps A and B?

2. Would you expect ethyl benzene to be more or less reactive towards electrophiles than benzene?

3. How would you chemically test for the ketone present in the final product?

Solution

Steps A and B are both Friedel–Crafts reactions; step A is a Friedel–Crafts alkylation, and step B a Friedel–Crafts acylation. For step A, ethyl chloride with $AlCl_3$ would be suitable; and for step B, ethanyol chloride with $AlCl_3$ would likely be successful.

Ethyl benzene would likely be more reactive towards electrophiles than benzene because the ethyl group is electron-inducing (through inductive effects), so the benzene ring in ethyl benzene is comparatively electron-rich. Finally, the ketone group could be identified using 2,4-DNP; an orange precipitate would be formed. In addition, there would be a negative result with both Tollen's reagent and Fehling's solution.

Worked Example 17.3   Consider the reaction scheme, and answer the questions that follow:

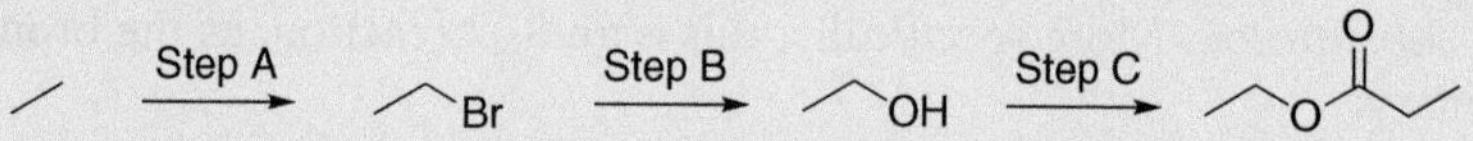

1. What reagent(s) would you need to effect steps A–C?

2. Draw the mechanism for step A.

3. Draw the mechanism for step B. What type of reaction is this?

4. Name the final ester.

Solution

Step A is a radical-mediated halogenation, so UV light and Br_2 are required. Step B is a substitution reaction requiring NaOH, and step C is ester formation between ethanol and either propanoyl chloride or propanoic acid with H_2SO_4 at reflux.

The mechanism for step A is quite complex and requires three steps – initiation, propagation, and termination:

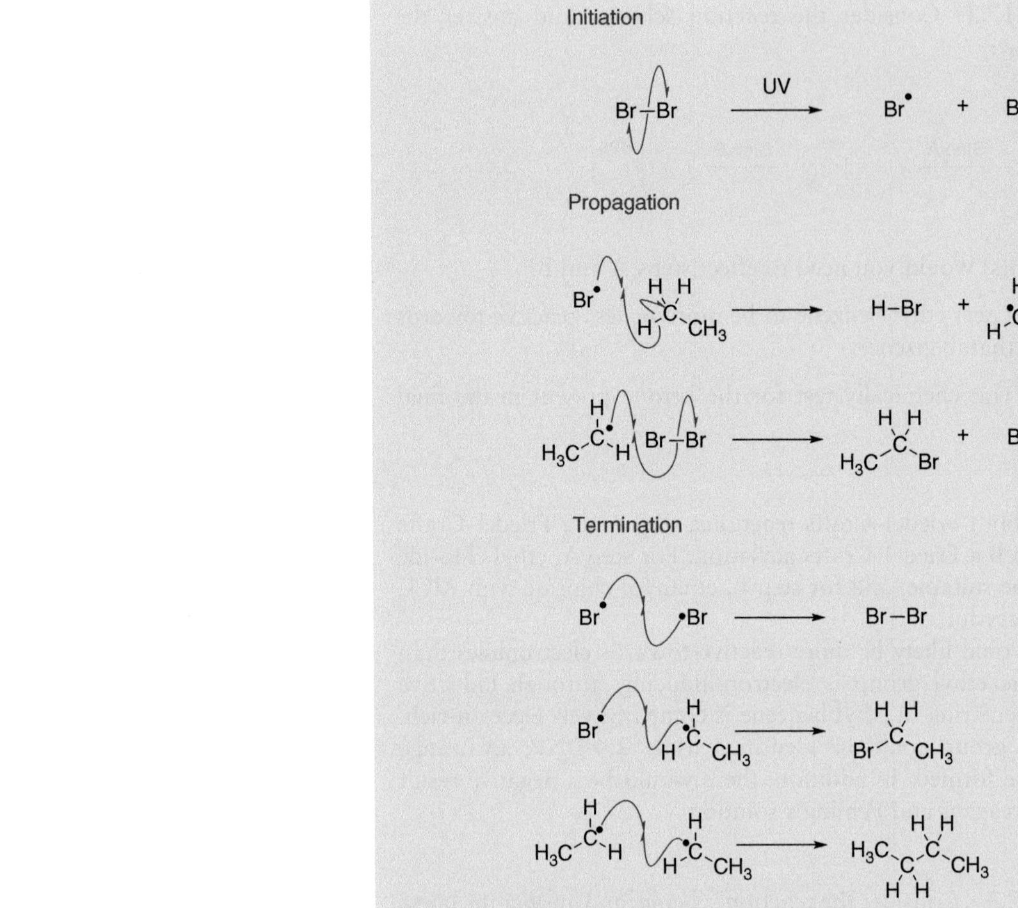

Initiation

Propagation

Termination

The mechanism for step B is slightly easier. The reaction class is *nucleophilic substitution*. More specifically, this is an S_N2 reaction, as the bromide is primary:

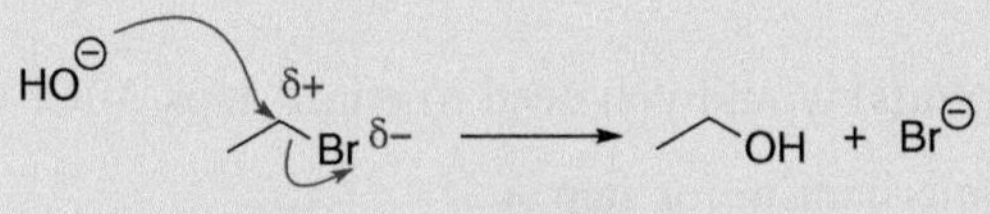

The final ester is called ethyl propanonate.

Worked Example 17.4   Consider the reaction scheme, and answer the questions that follow:

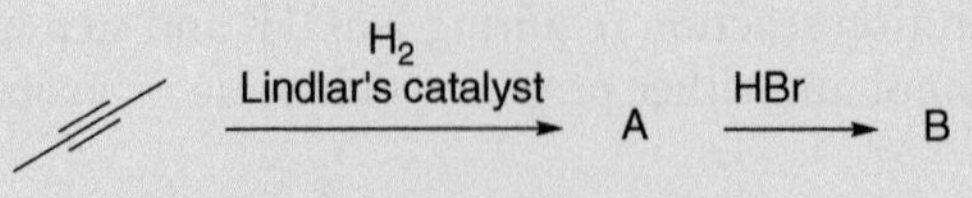

1. What are the structures of compounds A and B?

2. Draw the mechanism for the formation of B. What class of reaction is this?

3. Would you expect B to exhibit isomerism?

Solution

The formation of A is the reduction of an alkyne using Lindlar's catalyst. The catalyst is important because using this Lindlar's catalyst will reduce the alkyne to form the *cis*-alkene. Treatment of the alkene with HBr leads to *electrophilic addition*, where HBr is added across the alkene:

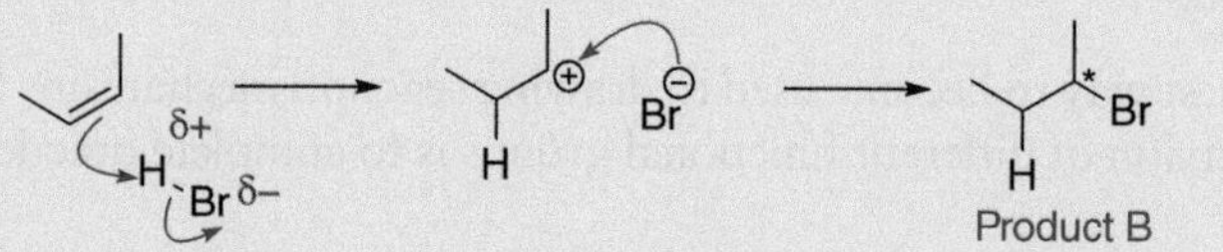

The final product, B, contains a chiral centre (marked with an ∗), so it will exhibit optical isomerism.

Worked Example 17.5 Consider the reaction scheme, and answer the questions that follow:

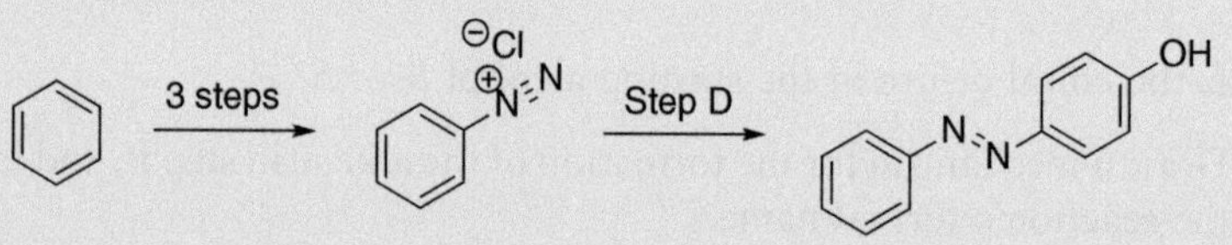

1. Suggest reagents and conditions to prepare the diazonium compound shown from benzene.

2. Suggest reagents, and draw the mechanism, for step D.

Solution

The three steps from benzene to the diazonium compound are nitration (HNO_3/H_2SO_4), reduction (Raney nickel and H_2), and diazotisation ($NaNO_2/HCl$):

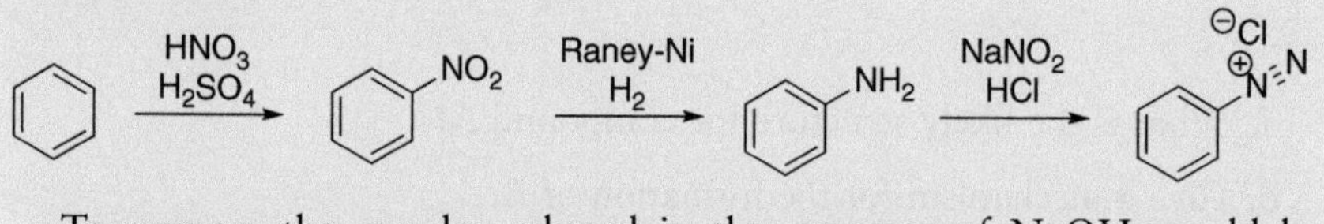

To prepare the azo-dye, phenol in the presence of NaOH would be needed. The mechanism for this is probably above that of a foundation-level course, but it is outlined here:

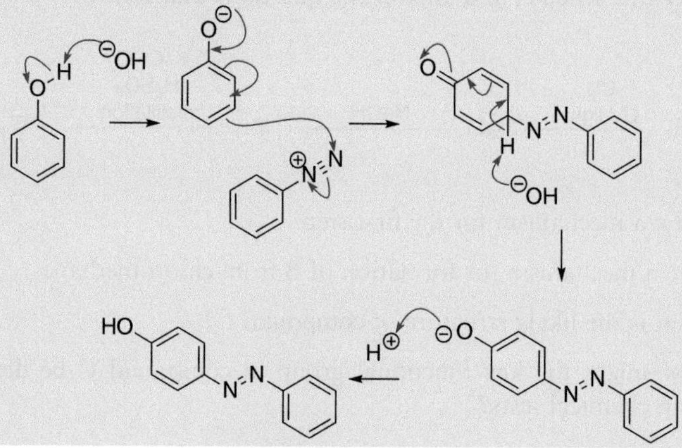

Quick-check summary

- Functional group interconversion is where one functional group is converted into another.

- By amending functional groups, the properties of a molecule can be altered.

- The best way to become used to drawing reaction mechanisms for the interconversion of different functional groups is to complete practice questions.

End-of-chapter questions

1 Consider this scheme, and answer the questions that follow:

a. Is the chiral centre in the starting alcohol R or S?

b. Draw a mechanism for the formation of the alkene in step B, and identify the reaction's formal name.

c. What reagents would be required to convert the alkene into the dibromide?

2 Consider this scheme, and answer the questions that follow:

a. What is the likely structure for compound A?

b. Give a mechanism for the formation of A.

c. Draw the mechanism for reduction of the ketone to the alcohol.

3 Consider this scheme, and answer the questions that follow:

a. Draw a mechanism for the first step.

b. Give a mechanism for formation of B from chloromethane.

c. What is the likely structure of compound C?

d. How might the key functional group in compound C be determined using chemical tests?

4 Consider this scheme, and answer the questions that follow:

a. Draw the bonding in both the alkyne and the benzene ring, showing all relevant orbitals.

b. Draw a likely structure of D.

c. Give a mechanism for conversion of D into E and F. Which structure is more likely as the product?

d. What conditions are required for step C? State the colour change (if any) that would be observed.

e. What might you use as a chemical test for the aldehdye, and what would you observe?

f. What is the likely structure of compound G?

5 Consider this scheme, and answer the questions that follow:

a. Suggest a mechanism for the first step.

b. Suggest a mechanism for the bromination step. The FeBr$_3$ catalyst is not strictly necessary. Can you account for this?

c. What conditions are required for step C? Draw the mechanism.

6 Consider this scheme, and answer the questions that follow:

a. Suggest a mechanism for the first step, and provide the formal name for this reaction.

b. What is the likely product of the second step? Draw a mechanism for this reaction.

c. What is the structure of compound H? Identify any chiral centres.

18

Polymerisation

At the end of this chapter, students should be able to:

- Understand the term *polymerisation*

- Understand the difference between addition and condensation polymerisation and their use in the formation of LDPE and HDPE

- Be able to describe the polymerisation processes to form polyamides and polyesters

18.1 Polymerisation

Polymers are substances that can also be called *macromolecules*. A macromolecule is a molecule that contains a very large number of atoms and has a high molecular mass. Examples include proteins, nucleic acids, and many structures in living tissue. A polymer is made by combining many smaller units (also called **monomers**) to make one larger structure (Figure 18.1).

The two types of **polymerisation** that you are likely to encounter during a foundation-level course are **addition polymerisation** and **condensation polymerisation**. Addition polymerisation is where monomers are added together; it usually occurs with alkene monomer units. Condensation polymerisation occurs when, during the polymerisation process, a small molecule (usually water) is produced as a by-product.

18.1.1 Addition polymerisation

As stated earlier, addition polymerisation occurs when alkene monomers react together to form a polymer. The length of the polymer chain depends upon how many monomer units are present; n monomer units leads to a polymer chain

Foundations of Chemistry: An Introductory Course for Science Students, First Edition.
Philippa B. Cranwell and Elizabeth M. Page.
© 2021 John Wiley & Sons Ltd. Published 2021 by John Wiley & Sons Ltd.
Companion website: www.wiley.com/go/Cranwell/Foundations

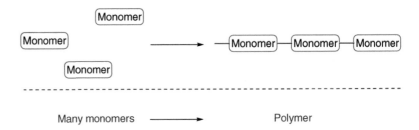

Figure 18.1 Combination of many monomers to give a polymer.

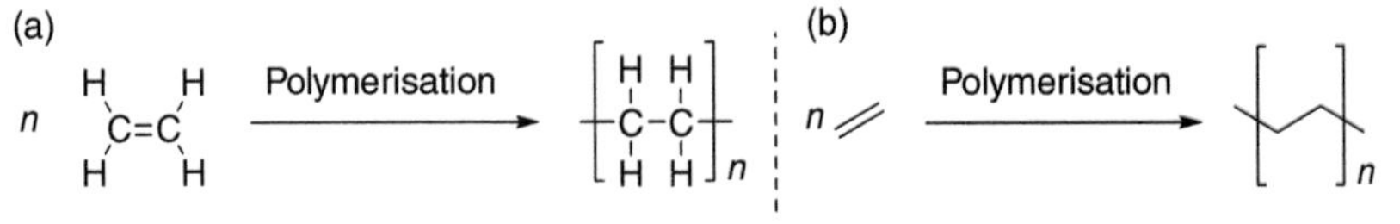

Figure 18.2 (a) Combination of many ethene monomers to give polyethene. The repeat unit is shown; (b) Combination of many ethene monomers to give polyethene using skeletal formulae.

with n repeat units. When drawing polymers, their sheer size means that it is not possible to draw the entire structure; therefore, polymers are usually represented by the smallest repeating structure. For example, polymerisation of ethene leads to the polymer polyethene (Figure 18.2). The smallest repeat unit of the polymer is a structure that contains two carbon atoms and four hydrogen atoms, and this is shown within the square brackets in Figure 18.2.

When determining the structure of an addition polymer, it is helpful to think of the alkenes as gates: one opens up and joins to another one, which then has to open up to continue the polymerisation process.

Worked Example 18.1 Give the structure of the polymer formed from the monomer propene.

Solution
Propene is a hydrocarbon containing three carbon atoms and an alkene. The addition polymerisation must occur between alkenes, so the polymer must contain the propene units. The polymer is called polypropene or polypropylene. In the worked example, both skeletal and display formulae are shown, but you should use skeletal formulae as much as you can:

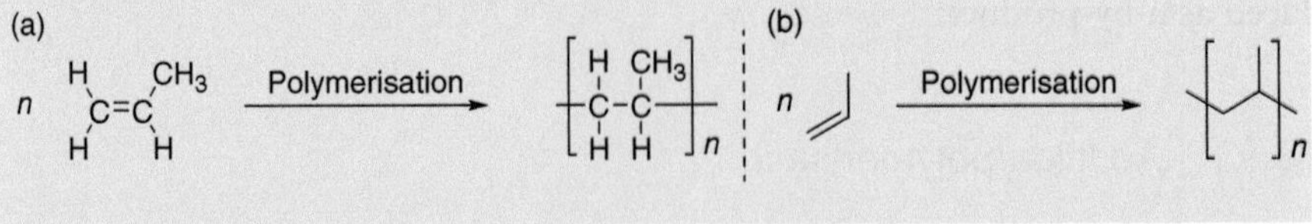

Worked Example 18.2 Give the structure of the polymer formed from a combination of the two monomers propene and chloroethene.

Solution
Both propene and chloroethene contain an alkene, so they will undergo addition polymerisation. When completing this question, we will assume that they

react in an orderly manner and so are alternating with each other. The product is shown, and the monomer units contributing to the polymer are shown in differing colours for clarity:

As well as knowing how to form polymers from monomers, it is also important to be able to suggest the monomers that were used to form the polymer. In this case, all that needs to be identified is the repeat unit.

Worked Example 18.3 Identify the monomer required to form the polymer shown:

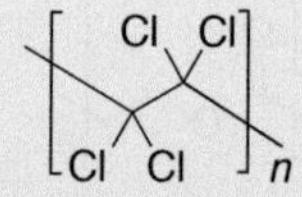

Solution
In this example, the repeat unit contains two carbon atoms and four chlorine atoms. The monomer needs to contain the same composition of atoms and is therefore tetrachloroethene ($Cl_2C=CCl_2$).

18.1.2   LDPE and HDPE

LDPE and HDPE are two different types of polyethene (also called polyethylene): **low-density polyethylene (LDPE)** and **high-density polyethylene (HDPE)**. To favour one over the other, different conditions for polymerisation are used. LDPE polymers are formed when ethene is polymerised at high temperature (approx. 200 °C) and high pressure (2000 atm). LDPE contains chains with a lot of branching, so the polymer chains cannot align closely. This means that the polymer structure is highly irregular and the material is not crystalline. The melting point is lower, so the structure formed is more flexible. LDPE is mainly used for materials such as plastic bags and flexible plastic sheets.

HDPE is formed at lower temperatures (approx. 60 °C) and low pressure (10–80 atm), and a catalyst is used to facilitate the reaction. HDPE does not contain branches; therefore, the layers of polymer are able to align closely to each other. This leads to a highly ordered, crystalline structure that is strong and has a high melting point. HDPE is used to make more rigid materials, such as milk bottles, washing-up bowls, and pipes.

18.1.3   Condensation polymerisation

As stated earlier, condensation polymerisation occurs when monomers react together to form a polymer, and a small molecule, e.g. H_2O or HCl, is formed as a by-product during the process. The two most commonly encountered condensation polymers are polyesters (often used for clothing)

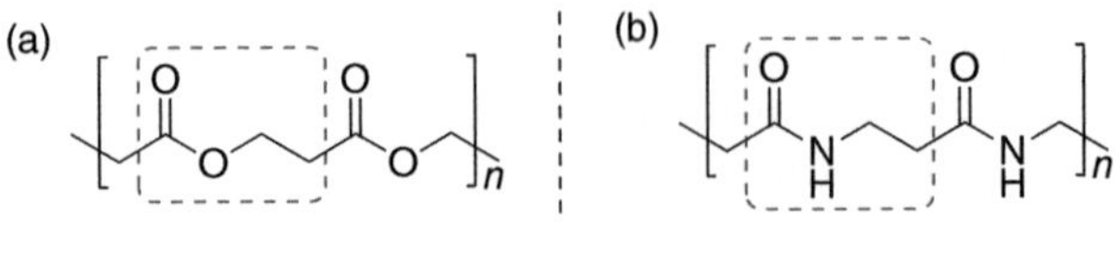

Figure 18.3 (a) A hypothetical polyester. The repeat unit is highlighted; (b) A hypothetical polyamide. The repeat unit is highlighted.

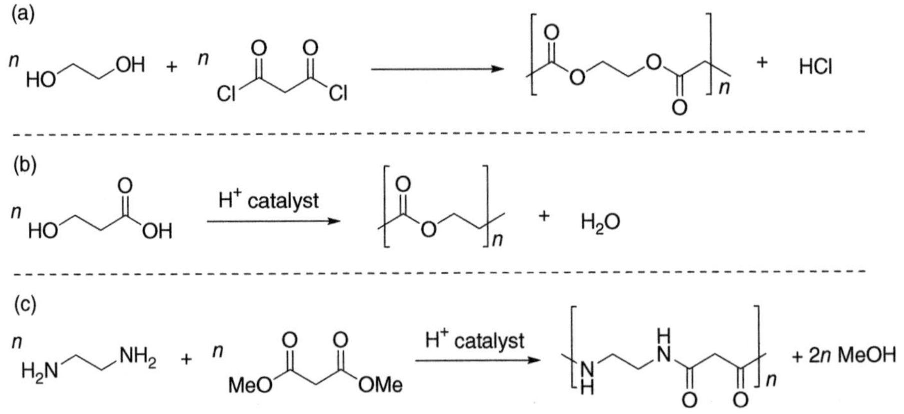

Figure 18.4 (a) Combination of a 1,2-diol and a diacyl chloride to give a polyester; (b) Combination of an alcohol and a carboxylic acid to give a polyester, where the electrophile and nucleophile are both contained in the monomer; (c) Combination of a diamine and a diester to give a polyamide.

and polyamides (such as nylon). Polyesters, as their name suggests, contain many ester linkages, and polyamides contain many amide linkages (Figure 18.3).

The formation of polyesters and polyamides usually requires two different monomer units. To make a polyester, an alcohol and an acid chloride, or an alcohol and a carboxylic acid, can be used, provided an acid catalyst is present in the latter reaction. The reagents required are the same as used to make a simple ester. To make a polyamide, an amine and an acid chloride, an amine and a carboxylic acid, or an amine and an ester can be used (Figure 18.4).

Worked Example 18.4   Give the structure of the polymer formed from the combination of the two monomers shown:

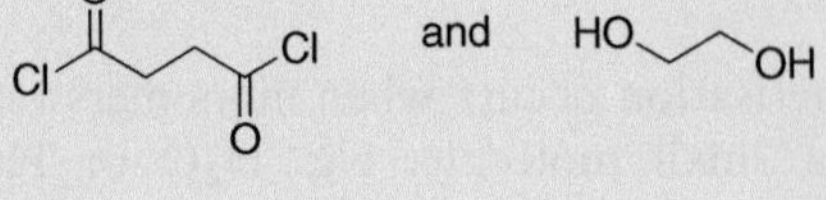

Solution

The monomers in this example are an acid chloride and an alcohol. Reacting these together will form a polyester:

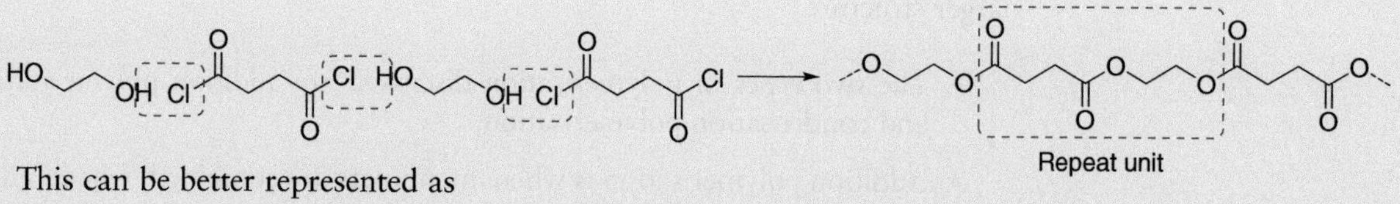

Repeat unit

This can be better represented as

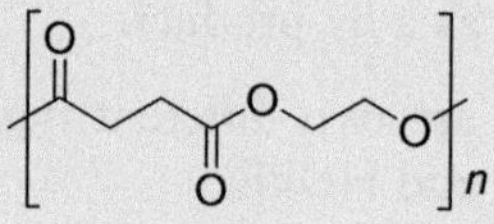

Worked Example 18.5   Give the structure of the monomers required to form the polymer (Kevlar) shown:

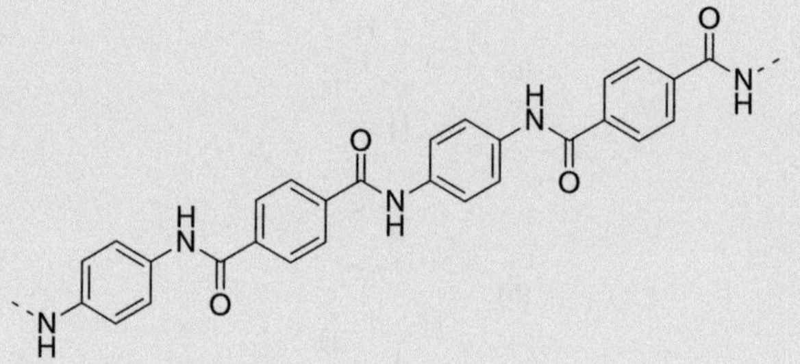

Solution

This polymer is a polyamide; therefore, the monomers required are an amine and either a carboxylic acid or an acid chloride. The logical place to cleave the structure is across the amide bond, which would give one fragment containing an amine and another fragment containing a carbonyl group:

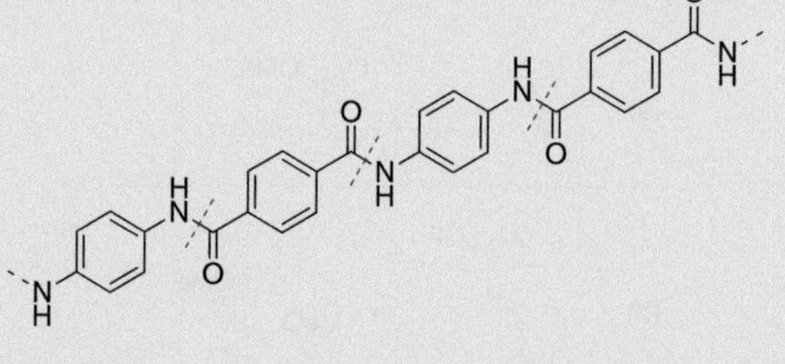

The two likely monomers are benzene-1,4-diamine and either benzene-1,4-dicarboxylic acid (terephthalic acid) or benzene-1,4-dicarbonyl chloride (terephthaloyl chloride):

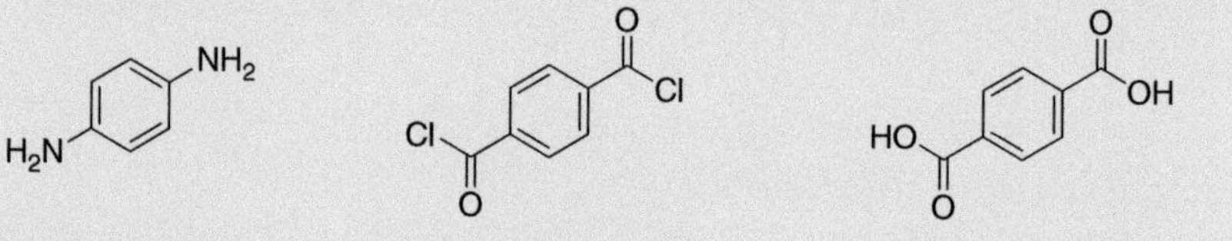

benzene-1,4-diamine   benzene-1,4-dicarbonyl   benzene-1,4-dicarboxylic
                                       chloride   acid

Quick-check summary

A polymer is made by combining many smaller units (monomers) to make one larger structure.

- The two types of polymerisation discussed are addition polymerisation and condensation polymerisation.
- Addition polymerisation is where monomers are combined, whereas during condensation polymerisation, a small molecule (usually water) is produced as a by-product.
- There are two different types of polyethene (also called polyethylene): LDPE and HDPE.

End-of-chapter questions

1 Draw three repeat units for the polymerisation of the monomers shown

(a)

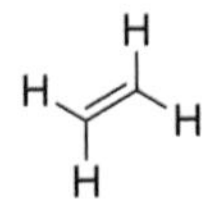

(b)

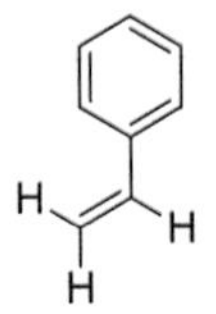

(c)

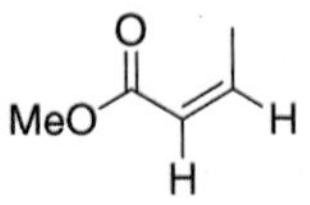

2 Draw three repeat units for the following polymerisations (Ph is phenyl):

(a)

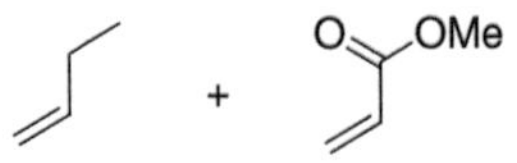

(b)

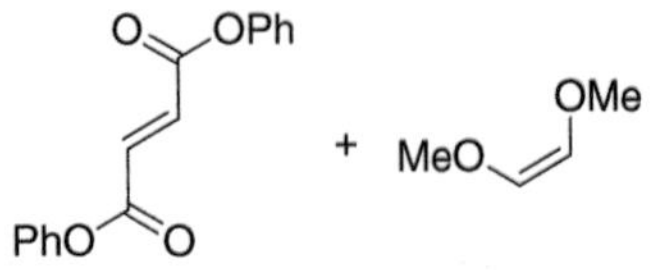

(c)

3 What are the likely monomers required for preparation of these polymers?

(a)

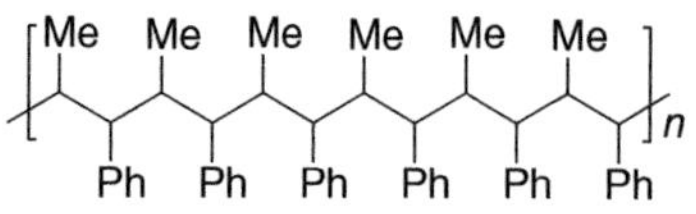

(b)

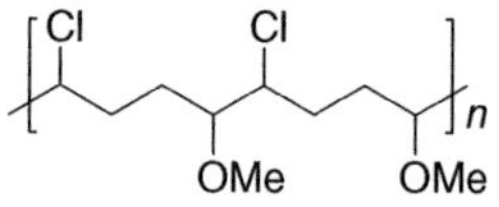

(c)

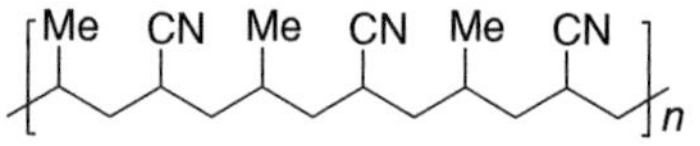

4 Draw two repeat units for the following polymerisations:

(a) n

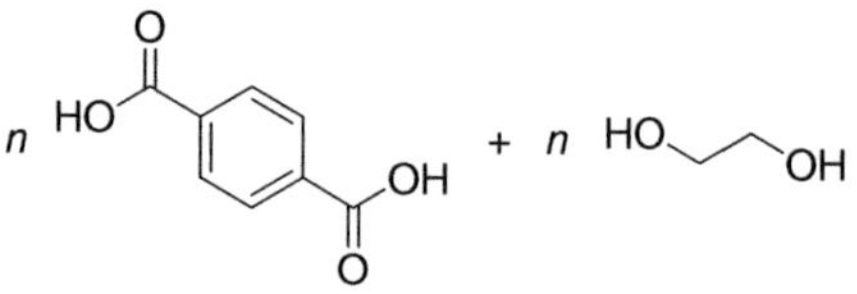

(b) n

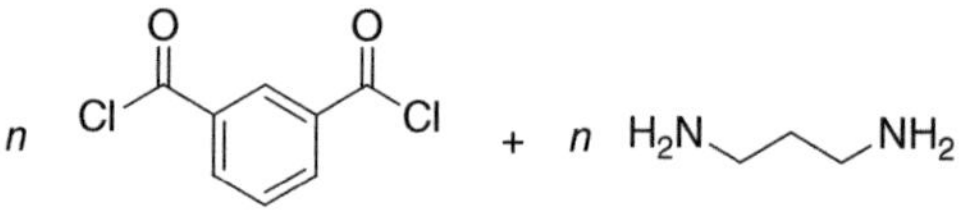

(c) n

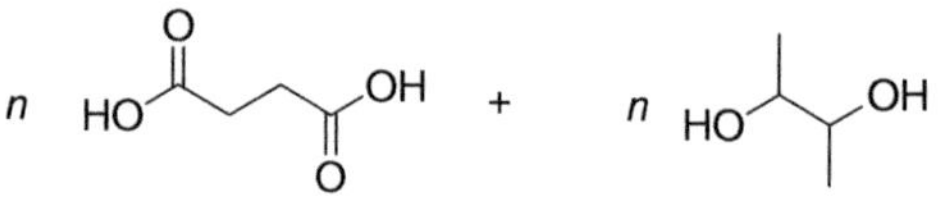

(d) n

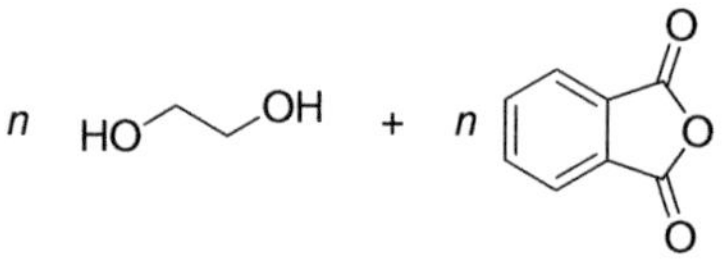

5 What are the likely monomers required to prepare these polymers?

(a)

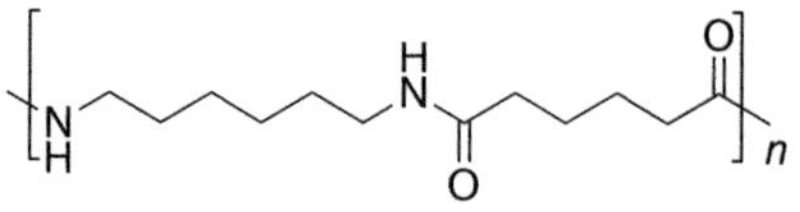

(b)

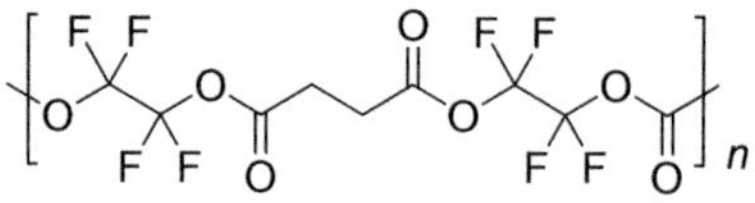

19

Spectroscopy

At the end of this chapter, students should be able to:

- Understand the processes behind mass spectrometry (MS), and be able to interpret simple data

- Understand the concept of infrared spectroscopy (IR), and be able to interpret simple data

- Interpret simple nuclear magnetic resonance (NMR) spectra

- Use MS, IR, and NMR data together to determine molecular structures

19.1 Mass spectrometry

Mass spectrometry is an analytical technique that allows us to determine the molecular mass of a compound.

19.1.1 How a mass spectrometer works

A mass spectrometer measures the mass of an atom or a molecule by measuring the mass-to-charge ratio of ions, referred to as the m/z (where m is the mass and z is the charge on the ion). A large range of mass spectrometers is available for use, but the way that they collect data is broadly the same and is outlined here:

1. The sample is introduced to the mass spectrometer via the sample inlet.

2. The molecules are converted into ions (either positive or negative) in the gas phase by an ionisation source. There are a few ways that this can

The *molar mass* is the mass of one mole of a substance, while the *molecular mass* is the mass of one molecule of a substance. A mole, mol, is the amount of substance that contains exactly 6.022×10^{23} particles such as atoms, molecules, electrons or ions. For further information See Chapter 3.2.1.

Foundations of Chemistry: An Introductory Course for Science Students, First Edition.
Philippa B. Cranwell and Elizabeth M. Page.
© 2021 John Wiley & Sons Ltd. Published 2021 by John Wiley & Sons Ltd.
Companion website: www.wiley.com/go/Cranwell/Foundations

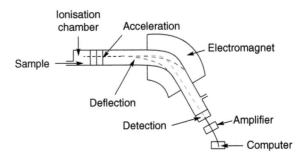

Figure 19.1 A simple schematic of a mass spectrometer.

happen, but the most common are *electron impact ionisation* and *electrospray ionisation*.

3. The ions are propelled towards the mass analyser by using a magnetic field.

4. The ions are separated according to their *m/z* ratio. As a general rule, lighter ions are deflected more than heavier ions, and more highly charged ions (2+ versus 1+) are also deflected more.

5. The ions are detected, and a mass spectrum is generated that plots the mass of an ion against the relative abundance.

This process is outlined in Figure 19.1.

Electron impact ionisation (EI) is where the molecule to be analysed is bombarded by a stream of high-energy electrons. It is a reasonably aggressive form of ionisation that causes the molecule to be broken apart at weak bonds to give more stable, smaller, fragments. The drawback of this approach is that it is not normally possible to obtain the mass of an intact ion; however, the fragments and resulting spectrum generated can act as a 'fingerprint', which can aid identification.

Electrospray ionisation (ESI) is much less aggressive and is referred to as a *soft* ionisation technique. This method allows the whole molecular ion to be detected. ESI has both positive (ESI+) and negative (ESI–) modes, and the charge of the ion depends upon the functional group(s) present. Positive ion mode is best when working with amines, or functional groups that can be protonated, and gives an *m/z* peak at the mass of the molecule (M) plus one $(M + 1)^+$. Negative ion mode is best when working with carboxylic acids, or functional groups that can be deprotonated, and gives an *m/z* peak at the mass of the molecule minus one $(M - 1)^-$.

19.1.2   Using the data from the mass spectrum

There are two ways in which you are likely to be asked to interpret a mass spectrum: either to calculate the relative atomic mass (RAM) of an element or to calculate the relative molecular mass of a compound (RMM).

Worked Example 19.1   The mass spectrum of magnesium showed three peaks due to the isotopes of magnesium. The first was at m/z 24 with an abundance of 79%; the second was at m/z 25 with an abundance of 10%; and the third was at m/z 26 with an abundance of 11%. What is the relative atomic mass of magnesium?

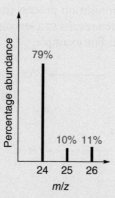

Solution
The question and spectrum shown give us enough information to answer the question. All we need to do is take into account how much each m/z contributes to the overall molecular mass:

$$\text{Relative atomic mass} = \frac{(24 \times 79.0) + (25 \times 10.0) + (26 \times 11.0)}{100} = 24.3$$

An alternative calculation would be:

$$\text{Relative atomic mass} = (24 \times 0.790) + (25 \times 0.10) + (26 \times 0.110) = 24.3$$

This calculation already takes into account the percentage; therefore, it does not need to be divided by 100.

19.1.3   Mass spectrometry in organic chemistry

Using mass spectrometry within organic chemistry is slightly more complicated but very similar in what the data tells us. The type of mass spectrometry used (i.e. EI or ESI) affects the ions observed during the ionisation process. EI leads to many smaller fragments; but in ESI, the molecule usually stays intact. In this textbook, we only discuss mass spectrometry using EI as the method for ionisation; but if you study chemistry further, you will likely encounter other ionisation methods.

As stated earlier, EI utilises a beam of electrons that bombard the molecule. This knocks an electron off the parent molecule, creating a positive ion (also called the *parent ion*). As one electron is removed, one unpaired electron will be left behind somewhere on the molecule, which is represented by •. It is the positive ion that is detected by the instrument. In this example, ethanol is ionised using EI:

$$C_2H_5OH + e^- \longrightarrow [C_2H_5OH]^{\ddagger} + 2e^-$$

Molecular
ion

However, this is only one possible outcome of many. The ionisation process can also lead to fragmentation of the radical ethanol ion, generating *fragment ions*. This occurs because the ionisation process causes energy to be transferred to the molecular ion, which often results in a strong enough molecular vibration to cause weak bonds to break. For example:

$$[C_2H_5OH]^{\ddagger} \longrightarrow {}^{\bullet}CH_3 + [CH_2OH]^+$$

Molecular   Fragment
ion   ions

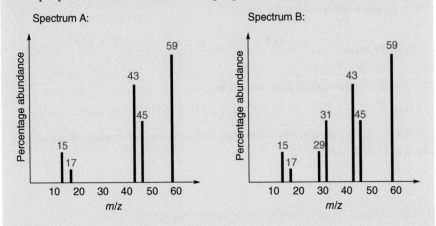

Worked Example 19.2   The mass spectra of propan-1-ol and propan-2-ol are shown. Using your chemical knowledge, determine which spectrum is from propan-1-ol and which is from propan-2-ol.

Solution

The trick with these questions is to draw the structure for both propan-1-ol and propan-2-ol and then work backwards to see the possible fragments that could be made. Propan-1-ol and propan-2-ol are structural isomers, so they have the same relative molecular mass and therefore would have the same parent ion peak (60 *m/z*).

If we consider propan-1-ol first, in addition to losing the H from the OH group to give a peak at 59 *m/z*, the structure can fragment as shown (assuming that the fragment ions do not further fragment):

$$\begin{array}{c}\text{H H H}\\|\ |\ |\\\text{H-C-C-C}\!\!\downarrow\!\!\text{O-H}\\|\ |\ |\\\text{H H H}\\60\end{array}\Longrightarrow\left[\begin{array}{c}\text{H H H}\\|\ |\ |\\\text{H-C-C-C}\\|\ |\ |\\\text{H H H}\end{array}\right]^{+}\text{and}\ \begin{array}{c}\cdot\text{O-H}\\17\end{array}$$

Fragmentation so that the radical and cation are on the opposite fragments is also possible.

$$\begin{array}{c}\text{H H H}\\|\ |\ |\\\text{H-C-C}\!\!\downarrow\!\!\text{C-O-H}\\|\ |\ |\\\text{H H}\\60\end{array}\Longrightarrow\left[\begin{array}{c}\text{H H}\\|\ |\\\text{H-C-C}\\|\ |\\\text{H H}\end{array}\right]^{+}\text{and}\ \begin{array}{c}\text{H}\\|\\\cdot\text{C-O-H}\\|\\\text{H}\\31\end{array}$$

Radical ions will not be observed.

$$\begin{array}{c}\text{H H H}\\|\ |\ |\\\text{H-C}\!\!\downarrow\!\!\text{C-C-O-H}\\|\ |\ |\\\text{H H H}\\60\end{array}\Longrightarrow\left[\begin{array}{c}\text{H}\\|\\\text{H-C}\\|\\\text{H}\end{array}\right]^{+}\text{and}\ \begin{array}{c}\text{H H}\\|\ |\\\cdot\text{C-C-O-H}\\|\ |\\\text{H H}\\45\end{array}$$

In the case of propan-2-ol, in addition to losing the H from the OH group to give a peak at 59 *m/z*, the fragmentation is

$$\begin{array}{c}\text{H}\\|\\\text{H O H}\\|\ |\ |\\\text{H-C-C-C-H}\\|\ |\ |\\\text{H H H}\\60\end{array}\Longrightarrow\left[\begin{array}{c}\text{H}\quad\text{H}\\|\quad\ |\\\text{H-C-C-C-H}\\|\ |\ |\\\text{H H H}\end{array}\right]^{+}\text{and}\ \begin{array}{c}\cdot\text{O-H}\\17\end{array}$$

Radical ions will not be observed.

$$\begin{array}{c}\text{H}\\|\\\text{H O H}\\|\ |\ |\\\text{H-C-C}\!\!\downarrow\!\!\text{C-H}\\|\ |\ |\\\text{H H H}\\60\end{array}\Longrightarrow\left[\begin{array}{c}\text{H OH}\\|\ |\\\text{H-C-C}\\|\ |\\\text{H H}\end{array}\right]^{+}\text{and}\ \begin{array}{c}\text{H}\\|\\\cdot\text{C-H}\\|\\\text{H}\\15\end{array}$$

By comparing the spectra given and the possible fragmentation patterns, it is possible to make the assignment. If we consider spectrum A, there are five peaks at 59, 45, 43, 17, and 15 *m/z*, which corresponds to the fragmentation pattern for propan-2-ol.

In spectrum B, there are peaks at 59, 45, 43, 31, 29, 17, and 15 *m/z*. Many of these can be formed by propan-2-ol, but the peaks at 29 and 31 cannot, as they would require even further fragmentation of smaller fragments. This means that spectrum B is propan-1-ol.

19.2 Infrared spectroscopy (IR)

Where mass spectrometry allows us to determine the molecular weight of a compound, **infrared spectroscopy** allows us to determine which bonds are present in a compound. IR works by irradiating the molecule with infrared light, which causes the bonds to stretch and bend (Figure 19.2). We can observe which radiation is absorbed and therefore infer which bonds and functional groups are present.

Different bonds bend and stretch at different frequencies, so it is possible to infer which bonds are present by the frequency of light absorbed (Table 19.1).

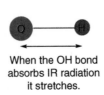

When the OH bond
absorbs IR radiation
it stretches.

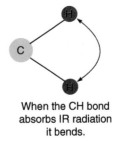

When the CH bond
absorbs IR radiation
it bends.

Figure 19.2 Stretching and bending vibrations from absorbing infrared radiation.

Table 19.1 Characteristic absorptions of some functional groups commonly encountered in a foundation-level course.

Bond	Functional Group	Wavenumber/cm^{-1}	Notes
C–O	Alcohol, ester	1060–1300	Often quite hard to pick out
C=C	Alkene	1610–1690	Not always very strong, so can be difficult to see
C=O	Aldehydes, ketones, carboxylic acids, esters, amides	1650–1750	Very strong peak that is usually easy to see
C≡N	Nitrile	2220–2260	Usually quite weak but is often the only peak in this region
C≡C	Alkyne	2100–2260	Not always very strong and so can be difficult to see
C–H	Non-aromatic compounds	2850–3300	Quite weak, but very sharp
C–H	Aromatic compounds	3000–3150	Quite weak, but very sharp
N–H	Amines, amides	3100–3600	Usually quite strong; not as broad as an O–H
O–H	Alcohols, phenols, carboxylic acids	3000–3600	Can be strong or weak, but usually broad and covers a large range

To have a strong peak in the infrared region, the bond that is vibrating or bending needs to have a change in dipole moment, i.e. a dipole.

An infrared spectrum has a large number of troughs, which are called *peaks*. Each peak represents the absorbance of energy by the molecule from infrared radiation. This absorbance causes the vibration or bend of a particular bond in a molecule. For example, the IR spectrum of propan-1-ol shows a characteristic O–H peak between 3500 and 3100 cm^{-1}, C–H peaks between 2950 and 2800 cm^{-1}, and C–O peaks at approximately 1300 cm^{-1} (Figure 19.3).

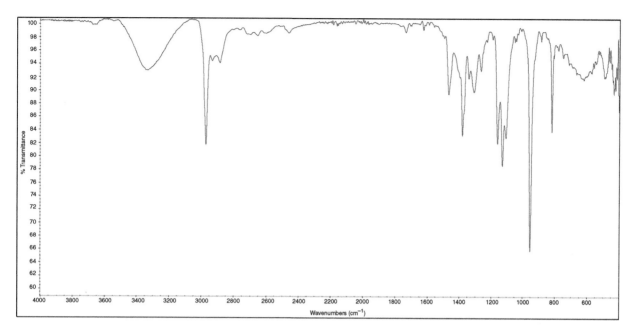

Figure 19.3 The infrared spectrum of propan-1-ol with the characteristic O—H peak at $3350\,cm^{-1}$.

Worked Example 19.3 The IR spectra of methanol, ethyl ethanoate, and benzoic acid and are shown. Determine which spectrum corresponds to each compound.

Spectrum A

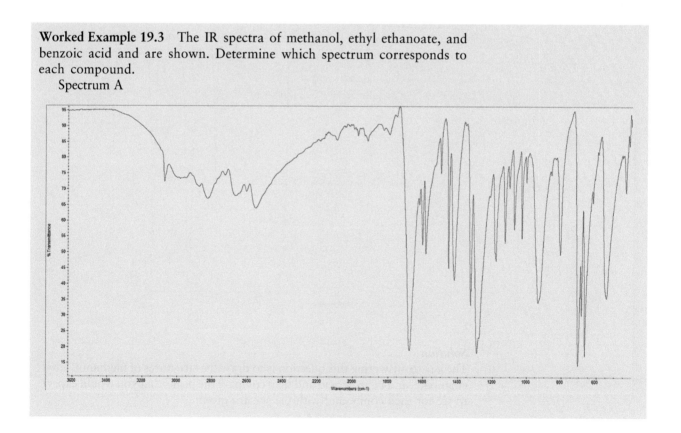

Spectrum B

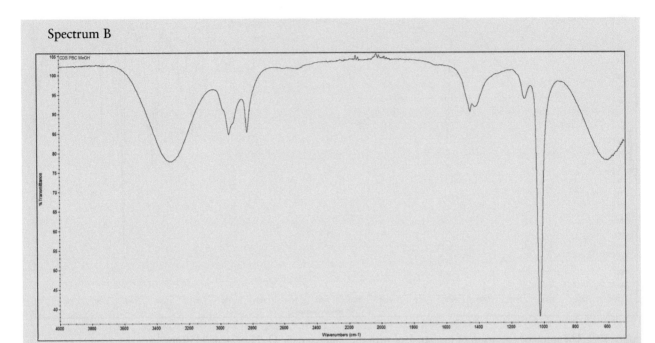

Spectrum C

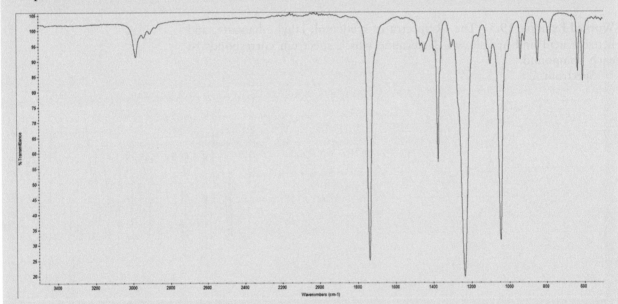

Solution

The key to answering this question is to draw the structures of methanol, ethyl ethanoate, and benzoic acid and then compare the peaks that you would expect to see for each compound with the spectra given:

Methanol   Ethyl ethanoate   Benzoic acid

Methanol, CH_3OH, contains an OH group, a C—O bond, and three C—H bonds. The OH group gives a peak between 3000 and 3600 cm^{-1}, the C—O gives a peak between 1060 and 1300 cm^{-1}, and the C—H bonds peaks between 2850 and 3300 cm^{-1}. The only spectrum that shows these peaks is spectrum B.

Ethyl ethanoate contains an ester; therefore, there will be a C=O peak between 1650 and 1750 cm^{-1}, a C—O peak between 1060 and 1300 cm^{-1}, and C—H peaks between 2850 and 3300 cm^{-1}. There will be no O—H peak between 3000 and 3600 cm^{-1}. The only spectrum that fits these peaks is spectrum C.

Finally, benzoic acid will show a C=O peak between 1650 and 1750 cm^{-1}, a C—O peak between 1060 and 1300 cm^{-1}, an OH peak between 3000 and 3600 cm^{-1}, and C—H peaks between 3000 and 3150 cm^{-1}. The spectrum that is consistent with these observations is spectrum A. The reason for the OH peak being so broad and containing so many other peaks is likely to be due to intermolecular hydrogen bonding.

19.3 Nuclear magnetic resonance spectroscopy (NMR)

Where mass spectrometry allows us to determine the molecular weight of a compound, and infrared spectroscopy allows us to determine which bonds are present in a compound, **nuclear magnetic resonance** (NMR) spectroscopy allows us to look at the finer detail within the compound and determine the environment in which a proton (H) lies. A number of different NMR techniques and nuclei are available, and the further through your studies in chemistry you progress, the more you will see. In this book, we only look at NMR spectra obtained for protons: ^1H NMR.

NMR spectroscopy works by looking at the nucleus of an atom. For example, in proton NMR, the NMR spectrometer analyses the proton nuclei. These nuclei contain magnetic moments and effectively behave like small bar magnets. This means that when the sample is placed in a magnetic field, the nuclei will either be in alignment with the field or not. Application of electromagnetic radiation causes transitions in these nuclei bar magnets, which can be detected by the NMR spectrometer. The computer processes these nuclear transitions into data or a spectrum. The environment in which that particular atom resides can then be interpreted.

19.3.1 The NMR spectrum

The spectrum obtained from a ^1H (proton) NMR spectrum runs from 0 to 12 ppm, and where a resonance (peak) occurs along this spectrum depends upon the environment in which the proton resides. It should be noted that 0 ppm is to the right and 12 ppm is to the left as we are looking at the spectrum. Figure 19.4 shows a representative NMR spectrum with different aspects annotated, and Table 19.2 shows chemical shifts for some commonly encountered proton environments. This information can also be represented visually (Figure 19.5).

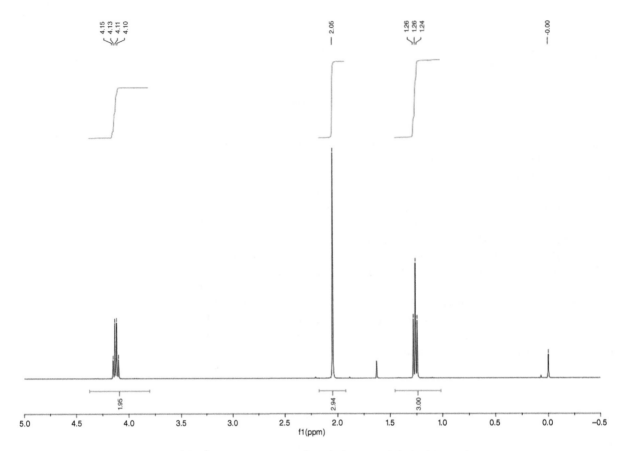

Figure 19.4 A illustrative section of the ¹H NMR spectrum for ethyl acetate (ethyl ethanoate).

Table 19.2 The chemical shifts for different ¹H environments.

Proton environment	Typical chemical shift (δ)	Proton environment	Typical chemical shift (δ)
H₃C–H (methane structure shown)	0.7–1.6	R₂C=CHR (alkene) Note that the alkene can be either *cis*- or *trans*-.	4.5–6.0
R₂CH₂ / R₃CH	1.4–2.3	C₆H₅–H (benzene)	6.4–8.2
CH₂ adjacent to C=O C–H adjacent to a C=O in ester, amide, aldehyde, ketone, or carboxylic acid	2.0–2.7	RCHO (aldehyde)	9.4–10.0
CH₂ adjacent to N (amine)	2.3–2.9	CH₂ adjacent to O–H	0.5–4.5*

Table 19.2 *(continued)*

Proton environment	Typical chemical shift (δ)	Proton environment	Typical chemical shift (δ)
⬡ H H / R (benzyl CH₂)	2.3–3.0	⬡–O–H (phenol)	4.5–10.0*
R–CH₂–O–R, R–C(=O)–O–CH₂–	3.3–4.8	R–CH₂–N(H)–H (amine)	1.0–5.0*
R–CH₂–Cl, R–CH₂–Br	3.0–4.2	R–C(=O)–N(H)–R (amide)	5.0–12.0*
		R–C(=O)–O–H (carboxylic acid)	9.0–15.0*

An * denotes that the proton shift can be very broad and may lie outside the range given. In addition, these protons are easily exchanged (labile), and if D₂O is added to the NMR sample, they will no longer be observed.

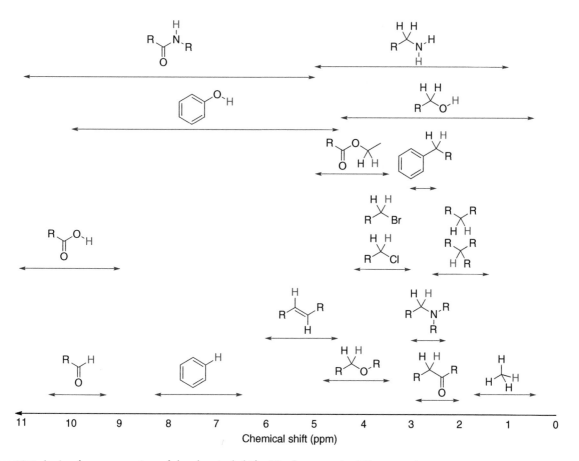

Figure 19.5 A visual representation of the chemical shifts (δ) of protons in different environments.

When looking at a spectrum, there are four things to note:

1. *The number of peaks indicates the number of proton environments in a molecule.* This is the same as saying, 'How many different types of proton are there?'

Worked Example 19.4   How many proton environments are expected in the molecules shown (protons have been added to aid clarity)?

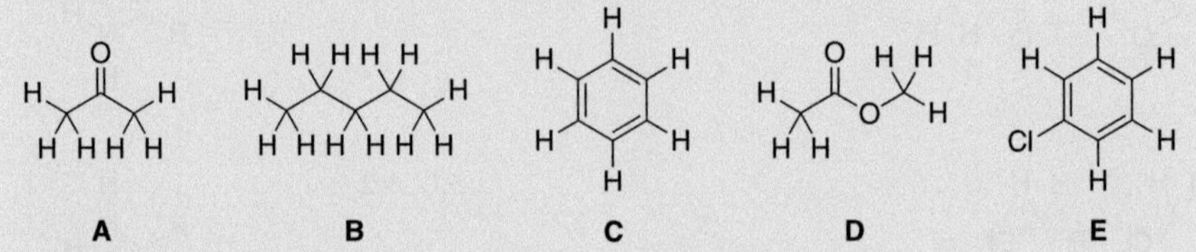

Solution
In structure A, there are six protons, each belonging to a CH_3 group. There are two CH_3 groups, but they are the same as each other because they are both directly next to a C=O. These six protons are therefore all in the same environment, labelled as H-1:

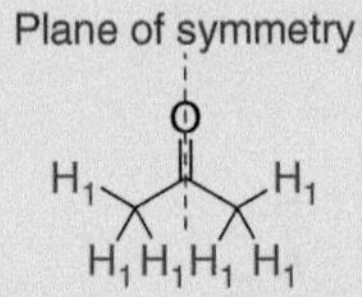

In structure B, there are 12 protons and 3 different proton environments. Taking this slowly, it can be seen that there are two CH_3 groups, but they are the same as each other because they are both at the end of the chain. They are next to a CH_2, and the molecule is symmetrical: H-1. The protons labelled H-2 are the same as each other: they are both next to a CH_3 and also a CH_2. Finally, the central CH_2 is different to H-1 and H-2 as it has a CH_2 group on either side of it. These two protons are therefore in a third proton environment: H-3.

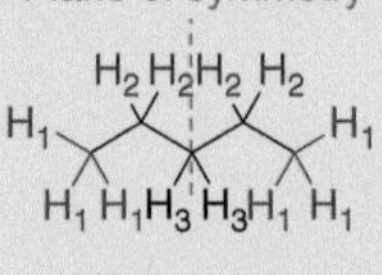

In structure C, there are six protons that are all in the same proton environment. The protons in this structure are all aromatic, all have one other proton on either side, and are indistinguishable when rotating the molecule, so they are all chemically identical: H-1.

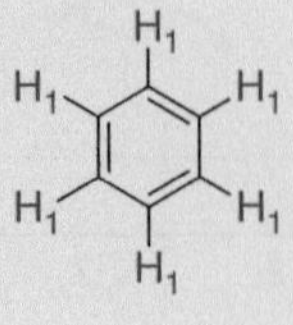

In structure D, there are six protons, each belonging to a CH_3. There are two CH_3 groups, but they are different to each other; one is directly next to a C=O, and the other is bonded to an oxygen. These six protons are therefore in two different environments, labelled as H-1 and H-2:

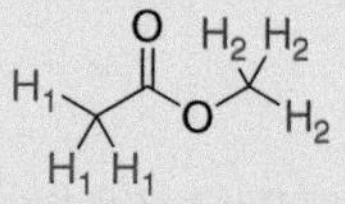

Finally, structure E contains three proton environments. The protons numbered H-1 both have a chlorine atom on one side and a proton on the other. The protons labelled H-2 both have a proton either side, and the proton labelled H-3 has two protons labelled H-2 on either side. H-3 is not the same as H-2, as H-2 are in closer proximity to the chlorine atom than H-3 (H-2 are *meta* to the chlorine, whereas H-3 is *para*):

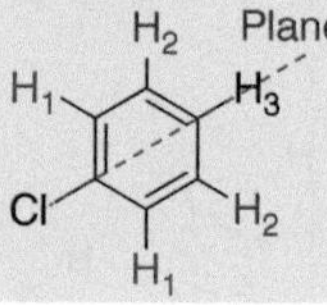

2. *The peak position depends upon the proton environment.* For example, if we consider Table 19.2 and Figure 19.5, it can be clearly seen that the location of the proton in a molecule affects the chemical shift (peak position).

Worked Example 19.5   Using Table 19.2, what are the likely chemical shifts of the protons in these structures?

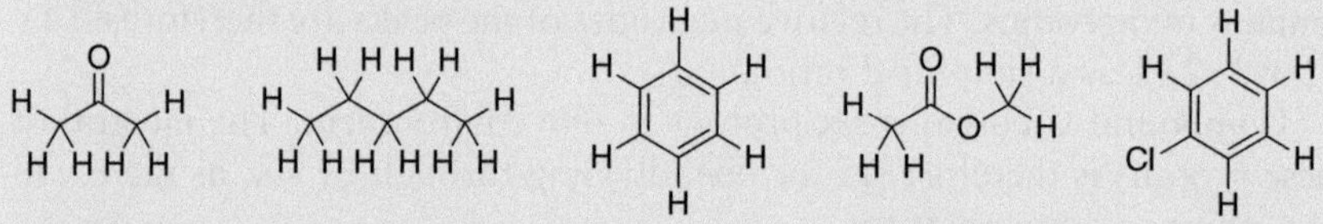

Solution

In example A, all six protons are in the same environment; therefore, there will be one peak at a chemical shift of 2.0–2.7 ppm, which is the shift for a proton directly adjacent to C=O.

In example B, there are three proton environments, but they are all very similar. There will be three different peaks, but all will be located between 0.7 and 1.6 ppm, the shift for protons in an alkyl chain.

In example C, there is one proton environment. Therefore, there will be one peak between 6.4 and 8.2 ppm, the shift for protons directly bonded to a benzene ring.

In example D, there are two proton environments, so there will be two peaks: one at 2.0–2.7 ppm, the shift for a proton directly adjacent to C=O,

The presence of the chlorine atom on the aromatic ring affects where the other proton resonances are, and we can distinguish where they are likely to be in relation to each other in terms of chemical shift. However, this is likely to be outside the scope of a foundation-level course.

and the other at 3.3–4.8 ppm, the shift for protons in an alkyl chain bonded to an oxygen.

In example E, there are three proton environments, and all are reasonably similar. There will be three peaks between 6.4 and 8.2 ppm, the shift for protons directly bonded to a benzene ring.

3. *The area under each peak relates to the number of contributing nuclei.* This is a very useful piece of information because it means that if two resonances are near to each other, the relative intensities (also called the *integrals*) can give more information about the protons. For example, a CH_3 and a CH_2 may have very similar chemical shifts, but the CH_3 group will integrate to three protons, and the CH_2 group will integrate to two protons.

Worked Example 19.6   Give the relative intensities (integrals) of the protons in the structures shown:

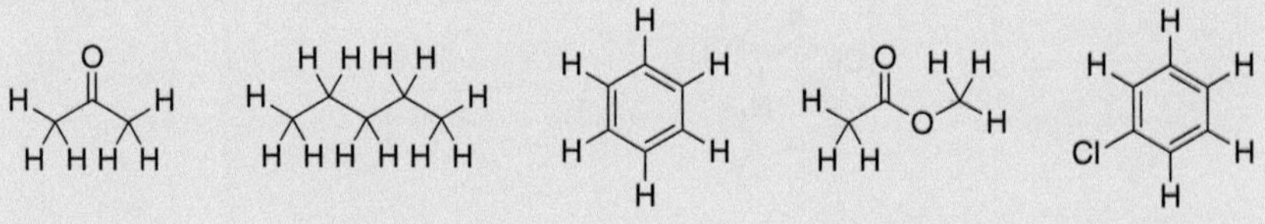

A   **B**   **C**   **D**   **E**

Solution

As discussed in the previous examples, compound A contains six protons in one environment. The integral of these protons is therefore six (or one, dividing through by six, as there will be nothing to compare it to).

Compound B has three different proton environments. The environment labelled H-1 contains six protons in total, H-2 contains four protons, and H-3 contains two protons. The relative intensities of the peaks are therefore $6:4:2$ (or $3:2:1$ as a simplified ratio).

Compound C contains six protons in one environment. The integral of these protons is therefore six (or one, dividing through by six, as there will be nothing to compare it to).

Compound D contains six protons in two different proton environments. Each proton environment will integrate to three protons, so the ratio is $3:3$ (or $1:1$ as a simplified ratio).

Compound E contains five protons in three different environments. In H-1, there are two protons; in H-2, there are also two protons; and in H-3, there is one proton. The ratio is therefore $2:2:1$.

4. *Peaks have a fine-splitting structure that is related to the presence of neighbouring nuclei.* For example, if two protons are on adjacent carbon atoms but in different chemical environments, they will be 'aware' of each

Table 19.3 A table summarising the most commonly encountered peak (resonance) multiplicities and their key characteristics.

Multiplicity name	Number of lines	Relative intensity of lines	Example
Singlet	1	1	
Doublet	2	1 : 1	
Triplet	3	1 : 2 : 1	
Quartet	4	1 : 3 : 3 : 1	

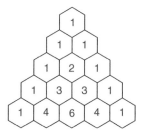

Figure 19.6 Pascal's triangle showing the relative intensities of the peaks (resonances).

other, which will lead to splitting of the resonances. This is known as the $(n + 1)$ rule, where n is the number of neighbouring protons. Protons in the same chemical environment do not lead to splitting. For example, the resonance of a proton that is adjacent to a different environment containing two protons will be split into three lines (also called a *triplet*) because it is $2 + 1 = 3$. Other multiplicities are also seen (Table 19.3).

The relative intensities of the lines are based on Pascal's triangle (Figure 19.6).

In this textbook, we use the term *chemical environment*, which is not strictly correct as NMR also depends upon magnetic environments. You will learn about magnetic environments versus chemical environments if you progress further in your chemistry studies.

Worked Example 19.7   Give the number of proton environments, chemical shift, relative intensities, and splitting patterns for the following compounds:

A   **B**   **C**   **D**

Solution
Compound A has six protons in one environment that have chemical shift 2.0–2.7 ppm. The protons on the two CH_3 groups are in the same environment, so no splitting is observed, because there are no protons on the adjacent

carbon atom and the peak is a singlet that integrates to six protons. As there is nothing to compare these six protons with, the integration could also be equal to one:

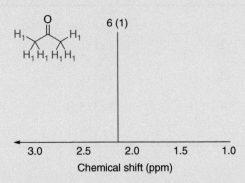

Compound B contains 10 protons in 4 environments. Protons H-1 will be in the range 3.3 to 4.8 ppm, will integrate to three, and will show no splitting, as there are no protons bonded to any adjacent carbon atoms. Protons H-2 and H-3 will be in the range 6.4 to 8.2 ppm as they are all directly bonded to an aromatic ring. There are two protons in environment H-2, so this will integrate to two. The protons on H-2 are each next to one other proton in a different environment, H-3, so this will be a doublet (1 + 1 = 2). There are two protons in environment H-3, so they will integrate to two. These protons are next to one proton in a different environment, H-2, so this will also be a doublet (1 + 1 = 2). Finally, H-4 will be in the range 2.3 to 3.0 ppm, will integrate to three protons and will be a singlet, as there are no other protons on adjacent carbon atoms:

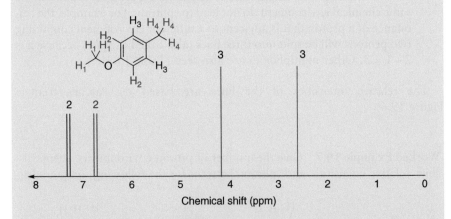

Compound C contains 10 protons in 2 different environments in the range 0.7 to 1.6 ppm. Each set of protons labelled as H-1 is adjacent to two other protons (H-2), so the protons in environment H-1 will be split into three (2 + 1 = 3): a triplet. The protons in environment H-2 are next to three other protons (H-1) and so will be split into four (3 + 1 = 4): a quartet. Remember, protons that are in the same environment cannot 'see' each other. A representation of an NMR spectrum containing this information is shown. Numbers above the peaks represent how many protons are in each environment:

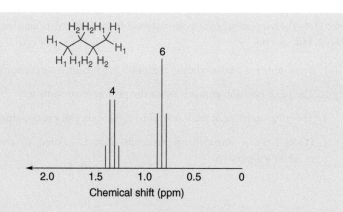

Compound D contains nine protons in four proton environments. Protons H-1 will be in the range 0.7 to 1.6 ppm and integrate to three protons; protons H-2 will be in the range 2.3 to 3.0 ppm and integrate to two protons; protons H-3 will be in the range 6.4 to 8.2 ppm and integrate to two protons; and protons H-4 will be in the range 6.4 to 8.2 ppm and integrate to two protons. Protons H-1 are adjacent to two protons (H-2), so this will be split into three lines (2 + 1 = 3): a triplet. These lines will have relative intensity 1 : 2 : 1 due to Pascal's triangle (Figure 19.6). Protons H-2 are adjacent to three protons (H-1), so this will be split into four lines, a quartet (3 + 1 = 4), with relative intensities 1 : 3 : 3 : 1. The protons in environment H-3 are not on adjacent carbon atoms to those in H-2, so they will not impact the splitting. Each proton in environment H-3 is adjacent to one proton (H-4), so this will be split into two lines (1 + 1 = 2), a doublet, with relative intensity 1 : 1. Each proton in environment H-4 is adjacent to one proton (H-3), so this will be split into two lines (1 + 1 = 2), a doublet, with relative intensity 1 : 1. A representation of an NMR spectrum containing this information is shown. Numbers above the peaks represent how many protons are in each environment:

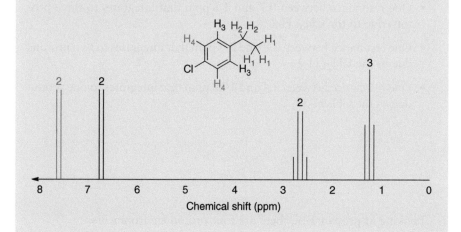

There are a number of pitfalls that you are likely to fall into. Therefore, when using NMR data:

- Do not confuse integrals and splitting (n + 1 rule).
- Check the symmetry in the molecule, and then make a judgement about the splitting patterns you would expect to see. Remember, proton environments are only equivalent if they are related by symmetry or rotation.

If two environments of protons either side of the proton under study are equivalent, add up the number of protons and *then* apply the n + 1 rule.

The best way to understand NMR spectroscopy is to complete some examples. These additional worked examples will clarify some of the information outlined.

Box 19.1 To summarise, four points should be considered when discussing proton (^{1}H) NMR spectra:

1. The number of peaks indicates the number of proton environments in a molecule.

2. The peak position depends upon the proton environment.

3. The area under each peak relates to the number of contributing nuclei.

4. Peaks have a fine-splitting structure that is related to the presence of neighbouring nuclei.

Worked Example 19.8 Consider the structures, and then answer the following questions:

1. How many proton environments are in each structure?

2. What are the likely chemical shifts for each proton environment?

Solution
Considering propanal, the easiest way to tell how many proton environments there are is to draw on all of the protons and then compare them. It can be seen that there are six protons in three different environments; therefore, three resonances would be seen.
In this case, there would be

- One resonance between 0.7 and 1.6 ppm that integrates to three protons due to the CH_3; H-1

- One resonance between 2.0 and 2.7 ppm that integrates to two protons due to the CH_2; H-2

- One resonance between 9.4 and 10.0 ppm that integrates to one proton due to the CH; H-3

Looking at propan-1-ol, there are four proton environments:

- One resonance between 0.7 and 1.6 ppm that integrates to three protons due to the CH_3; H-1

- One resonance between 1.4 and 2.3 ppm that integrates to two protons due to the CH_2; H-2

- One resonance between 3.3 and 4.8 ppm that integrates to two protons due to the CH_2; H-3

- One resonance between 4.5 and 10.0 ppm that integrates to one proton due to the OH

If we consider 2,4-dimethylbenzene, there are two proton environments. This is because the two methyl groups are equivalent, the two protons on either side of the ring are equivalent due to symmetry, and the two protons adjacent to the methyl groups on the ring are also equivalent:

This means that there will be two resonances:

- One resonance between 2.3 and 3.0 ppm that integrates to six protons due to the two CH_3 groups on the benzene ring; H-1

- One resonance between 6.4 and 8.2 ppm that integrates to four protons due to the four equivalent CH groups; H-2

Finally, propan-2-ol also has a plane of symmetry down the centre of the molecule, which means that the two CH_3 groups (H-1) are equivalent to each other. There will be three resonances in total:

- One resonance between 1.4 and 2.3 ppm that integrates to six protons due to the two CH_3 groups; H-1

- One resonance between 3.3 and 4.8 ppm that integrates to one proton due to the CH group; H-2

- One resonance between 0.5 and 4.5 ppm that integrates to one proton due to the OH group; H-3

Worked Example 19.9 What splitting patterns will be observed in the ^{1}H NMR for these structures?

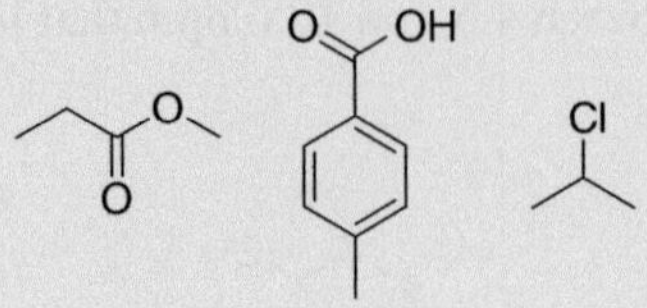

Solution

When answering this question, the best approach is to work out how many proton environments there are and then calculate the likely splitting patterns.

Methyl propanoate contains three proton environments, labelled H-1, H-2, and H-3:

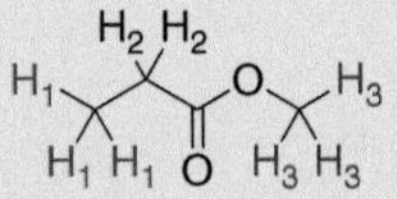

The protons in H-1 and H-2 are close to each other and so will be 'aware' of each other. The protons in environment H-1 will 'see' two protons next to them in environment H-2 and will therefore be split into three lines: a triplet. The protons in environment H-2 will 'see' three protons next to them in environment H-1 and will therefore be split into four lines: a quartet.

Overall, there will be

The number of lines that a signal is split into is one more than the number of adjacent protons.

- One resonance between 1.4 and 2.3 ppm that integrates to three protons due to the CH_3 group; H-1. This resonance will be split into three lines to give a triplet.

- One resonance between 2.0 and 2.7 ppm that integrates to two protons due to the CH_2 group; H-2. This resonance will be split into four lines to give a quartet.

- One resonance between 3.3 and 4.8 ppm that integrates to three protons due to the CH_3 group; H-3. This resonance will not be split as there are no adjacent protons and so will be a singlet.

4-Methylbenzoic acid contains four proton environments and has a plane of symmetry, so the protons on either side of the benzene ring are equivalent. The proton in environment H-1 is far away from H-2, H-3, and H-4, so the resonance will not be split; therefore the signal will be a singlet that integrates to one proton. The same is true of the protons in environment H-4; therefore, the resonance will not be split, so the signal will be a singlet that integrates to three protons. The protons in H-2 and H-3 are close to each other and so will be 'aware' of each other. *Each* proton in environment H-2 will 'see' one proton next to it in environment H-3. Therefore, the resonance for H-2 will be split into two lines: a doublet. *Each* proton in environment H-3 will 'see' one proton next to it in environment H-2. Therefore, the resonance for H-3 will be split into two lines: another doublet.

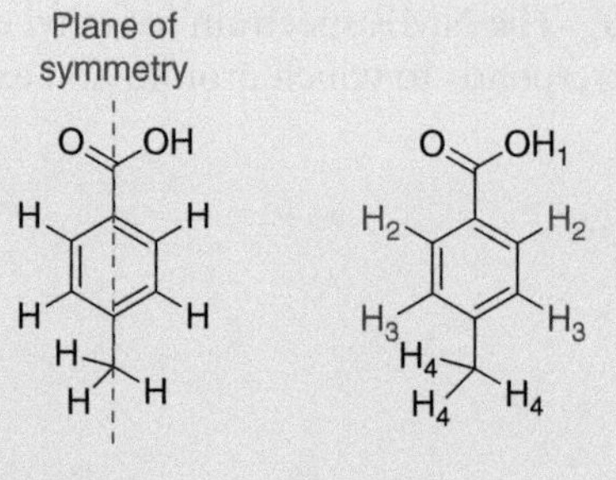

Overall, there will be

- One resonance between 9.0 and 15.0 ppm that integrates to one proton due to the COOH group; H-1. This resonance will not be split, as there are no adjacent protons, and will therefore be a singlet.

- One resonance between 6.4 and 8.2 ppm that integrates to two protons due to the two CH bonds; H-2. This resonance will be split into two lines to give a doublet.

- One resonance between 6.4 and 8.2 ppm that integrates to two protons due to the two CH bonds; H-3. This resonance will be split into two lines to give a doublet.

- One resonance between 2.3 and 3.0 ppm that integrates to three protons due to the CH_3 group; H-4. This resonance will not be split, as there are no adjacent protons, and will therefore be a singlet.

Finally, 2-chloropropane contains two proton environments because it contains a plane of symmetry, so both of the CH_3 groups are equivalent. The six protons in environment H-1 will be split into two, giving a doublet, due to the proton in environment H-2. The proton in environment H-2 will be split into seven lines, giving a septet, due to the six protons in environment H-1:

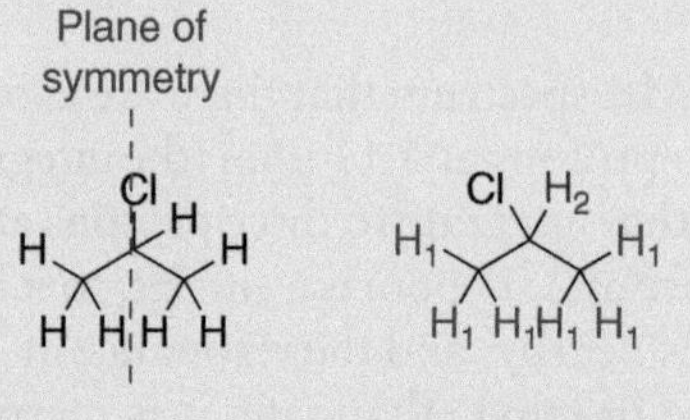

Overall, there will be

- One resonance between 1.4 and 2.3 ppm that integrates to six protons due to the two CH_3 groups; H-1. This resonance will be split into two lines due to the one adjacent proton, so will be a doublet.

- One resonance between 3.0 and 4.2 ppm that integrates to one proton due to the CH bond; H-2. This resonance will be split into seven lines to give a septet.

Worked Example 19.10   The NMR spectrum for ethyl ethanoate is shown. Identify which peak corresponds to which proton, and explain any coupling patterns observed.

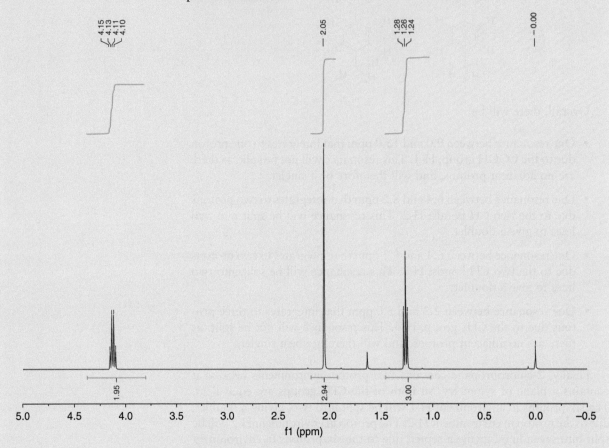

Solution

It can be seen from the NMR spectrum that there are three peaks present, one of which integrates to two protons at 4.11 ppm (the integral is under the peak) and is a quartet, and two that integrate to three protons at 2.05 and 1.26 ppm, respectively. The resonance at 2.05 ppm is a singlet, so it is safe to assume that there are no other protons nearby, and the resonance at 1.26 ppm is a triplet. Considering the structure of ethyl ethanoate, it is reasonable that there are three proton environments, and these are marked on the structure:

The resonance at 2.05 is characteristic of a CH_3 adjacent to a carbonyl group and also nowhere near any other protons and so will not be split; therefore, the singlet resonance at 2.05 is assigned to H-1. The resonance at 4.11 is characteristic of a CH_2 adjacent to an oxygen. In addition, it is split into four (a quartet) and so will be adjacent to three protons; therefore, the resonance at 4.11 is assigned to H-2. The resonance at 1.26 is characteristic of a CH_3 adjacent to another alkyl group; it is split into three (a triplet) and therefore will be adjacent to two protons, so the resonance at 1.26 is assigned to H-3.

19.3.2 Confirming the identity of O—H and N—H peaks

It is possible to confirm the identity of O—H and N—H peaks by using D_2O (deuterium oxide). A detailed explanation of this is likely to be above the level of a foundation course, but taking an NMR sample and shaking it with a drop of D_2O allows the O—H and N—H peaks to exchange with the D in D_2O to make O—D and N—D instead. Deuterium cannot be observed in ^1H NMR, so any protons that have switched will no longer be visible in the NMR spectrum.

19.4 Bringing it all together

Now that we have an understanding of mass spectrometry, IR, and NMR, we can start to combine this information to determine the structures of unknown compounds. This is best learned through practice; therefore, there will be worked examples and then some further questions at the end of the chapter.

Worked Example 19.11   Compound A is heated in the presence of acidified potassium dichromate solution to give compound B. The IR and mass spectrometry data for compounds A and B are given, along with some physical data. Determine the molecular formula of compound B, and propose structures for A and B based upon the spectroscopic data.

Compound A: colourless liquid; boiling point 97 °C; C_3H_8O

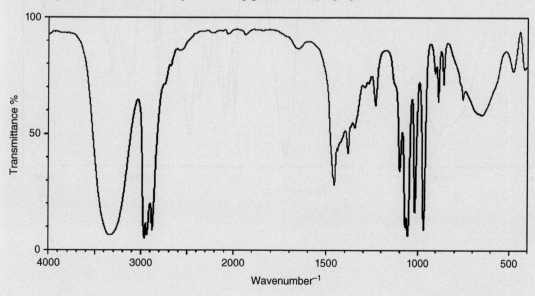

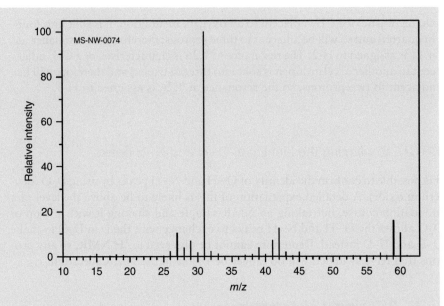

Compound B: colourless liquid; boiling point 141 °C; contains 48.65% C and 8.11% hydrogen

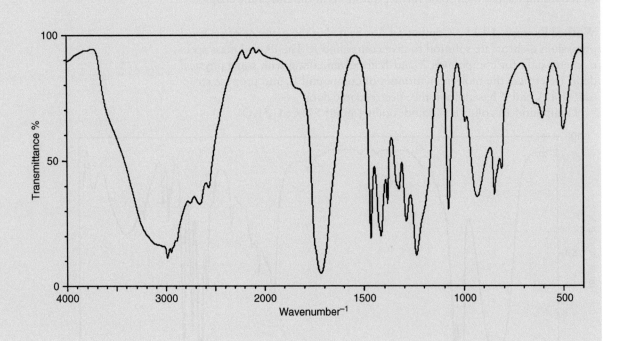

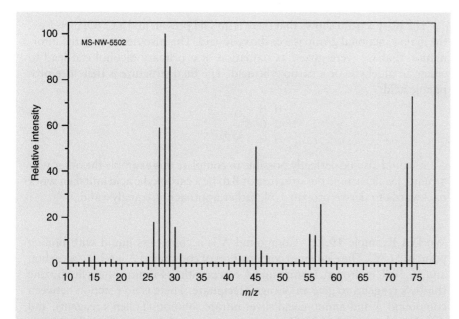

Solution

The molecular formula of compound A is C_3H_8O, which has relative molecular mass = $60 \, g \, mol^{-1}$. This ties in with the mass spectrum, where the parent ion shows a peak at m/z 60. In addition, a fragment can be removed that has m/z value $31 \, g \, mol^{-1}$, which could be CH_2OH. Considering the IR spectrum, compound A contains an OH group and C—H bonds and so must be an alcohol. A possible structure for C_3H_8O is propan-1-ol:

$$
\begin{array}{ccc}
\text{H} & \text{H} & \text{H} \\
| & | & | \\
\text{H--C--C--C--OH} \\
| & | & | \\
\text{H} & \text{H} & \text{H}
\end{array}
$$

Considering compound B, the empirical formula can be calculated. We know that oxygen is present firstly because the starting material contains oxygen, but also because acidified potassium dichromate solution is an oxidant.

Element	% Carbon	% Hydrogen	% Oxygen
Percentage composition	48.65	8.11	(100-(48.65 + 8.11)) = 43.24
A_r	12	1	16
$\%/A_r$	4.05	8.11	2.70
Divide by the smallest number	= 4.05/2.70 = 1.5	= 8.11/2.70 = 3.0	= 2.70/2.70 = 1.0
Multiply by 2	3	6	2

The molecular formula is therefore $C_3H_6O_2$, which has relative molecular mass = $74 \, g \, mol^{-1}$. This molecular weight corresponds to the mass spectrum provided.

The IR spectrum shows that there is an OH present and a C=O; therefore, the main functional group is a carboxylic acid. This also ties in with the information that we were given, as oxidation of a primary alcohol can lead to either an aldehyde or a carboxylic acid. The final structure is therefore propanoic acid:

$$H-\underset{\underset{H}{|}}{\overset{\overset{H}{|}}{C}}-\underset{\underset{H}{|}}{\overset{\overset{H}{|}}{C}}-C\overset{\diagup O}{\underset{\diagdown O-H}{}}$$

It would also be perfectly possible to complete this example the other way around, i.e. determine the structure of B to be a carboxylic acid and then work backwards to derive propan-1-ol. Either approach is equally valid.

Worked Example 19.12   Compound A is a colourless liquid with boiling point of 56 °C. The elemental composition of compound A is 62.01% carbon and 10.40% hydrogen. Compound A reacts with 2,4-dinitrophenylhydrazine (Brady's reagent) to give an orange precipitate. There is no reaction between compound A and ammoniacal silver nitrate solution (Tollen's reagent), and an ethereal solution of $LiAlH_4$ reduces compound A to an alcohol.

The NMR spectrum contains only one peak that integrates to six protons at 2.1 ppm. The IR spectrum contains peaks at 3000 and 1750 cm^{-1}. The mass spectrum contains a peak at 58 m/z.

Solution
As with the previous example, the empirical formula is the easiest place to start. We can assume that the missing element is oxygen.

Element	%Carbon	%Hydrogen	%Oxygen
Percentage composition	62.01	10.40	100 – (62.01 + 10.40) = 27.59
A_r	12	1	16
$\%/A_r$	5.16	10.40	1.72
Divide by smallest number	= 5.16/1.72 = 3.0	= 10.40/1.72 = 6.0	= 1.72/1.72 = 1.0

The molecular formula is therefore C_3H_6O, which has relative molecular mass = 58 g mol^{-1}. This molecular mass corresponds to the mass spectrum data provided. The IR spectrum contains peaks at 3000 and 1750 cm^{-1}; the former relates to C—H bonds and the latter to a C=O containing functional group. The fact that compound A reacts with Brady's reagent suggests that there is an aldehyde or ketone present. The lack of reaction with ammoniacal silver nitrate solution means that it is not an aldehyde and so must be a ketone. This also ties in with the reduction of the ketone with $LiAlH_4$ to give an alcohol. Finally, the one peak integrating to six protons at 2.1 ppm in the ^1H NMR suggests that there is only one proton environment and all six protons are equivalent. The compound is therefore propan-2-one (acetone):

$$H-\underset{\underset{H}{|}}{\overset{\overset{H}{|}}{C}}-\overset{\overset{O}{\|}}{C}-\underset{\underset{H}{|}}{\overset{\overset{H}{|}}{C}}-H$$

Quick-check summary

- Mass spectrometry is an analytical technique that allows us to determine the molecular mass of a compound.

- The data that can be gained from a mass spectrum are the relative atomic mass (A_r) or the relative molecular mass (M_r).

- Electrospray ionisation (ESI) and electron impact ionisation (EI) are commonly used ionisation techniques.

- Infrared spectroscopy allows us to determine the key bonds present in a compound.

- Nuclear magnetic resonance (NMR) spectroscopy allows us to look at the finer detail of the molecular structure.

- Proton (^{1}H) NMR allows for the determination of the proton environments in a molecule and therefore derivatisation of a molecular structure.

- A proton NMR spectrum shows four key pieces of information:

  ○ The number of peaks indicates the number of proton environments in a molecule.

  ○ The peak position depends upon the proton environment.

  ○ The area under each peak relates to the number of contributing nuclei.

  ○ Peaks have a fine-splitting structure that is related to the presence of neighbouring nuclei.

- Mass, IR, and NMR data can all be combined to determine molecular structures.

End-of-chapter questions

1 Uranium exists as six isotopes. One isotope, ^{232}U, is only produced synthetically; ^{233}U and ^{236}U are only abundant in trace amounts. However, ^{234}U, ^{235}U, and ^{238}U exist as 0.005%, 0.720%, and 99.274% abundance, respectively. What is the A_r of uranium?

2 Copper has 2 stable isotopes and 27 radioisotopes. ^{63}Cu has 69.15% abundance, and ^{65}Cu has 30.85% abundance. What is the A_r of copper?

3 Consider the following mass spectra. Which compound contains chlorine, and which contains bromine? Give reasons for your answer.

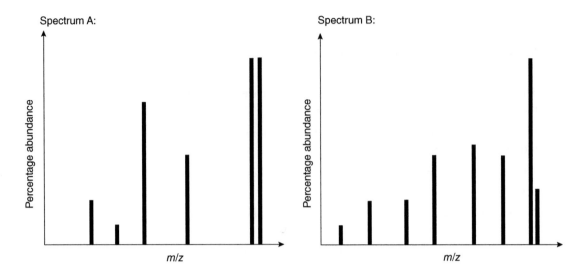

Spectrum A:

Spectrum B:

4 Draw the likely fragmentation patterns that you would expect to see for benzyl alcohol, C_7H_8O.

5 What would be the key differences between the IR spectra for

a. Propanol, propanenitrile, and propanamide

b. Acetone and dichloromethane

6 What key peaks would be present in the IR spectrum of citronellal?

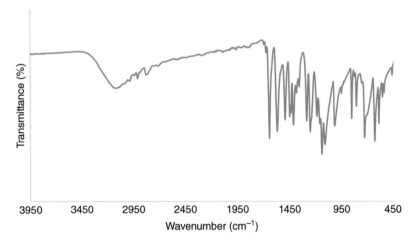

7 The IR spectrum of vanillin is shown. Label the key peaks.

Structure of vanillin:

8 Describe the NMR spectra that you would expect to see for each of the three isomers of C_3H_8O.

9 What differences would you expect to see between the NMR spectra of these compounds?

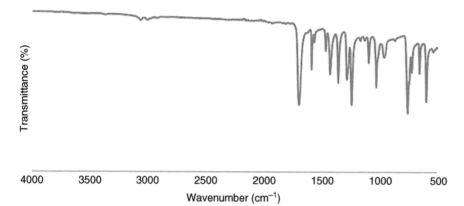

10 A compound is a colourless solid with molecular formula C_8H_7BrO. Using the following IR and NMR data, suggest the structure.

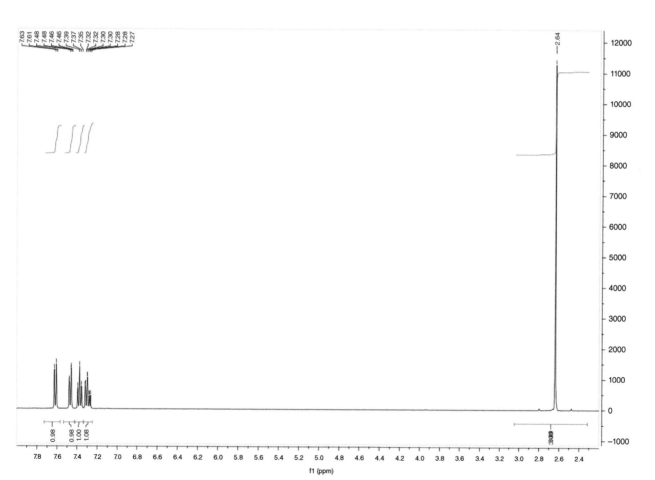

11 Compound A is a carbohydrate found in some common foods and contains C 54.53% and H 9.15%. It is a colourless liquid with a boiling point of 155 °C. Using these data and the IR, NMR, and mass spectrometry data presented, deduce the structure of A. Ignore the peak at 7.26 ppm in the NMR spectrum.

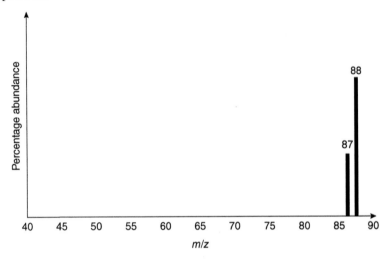

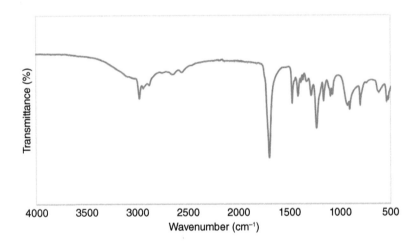

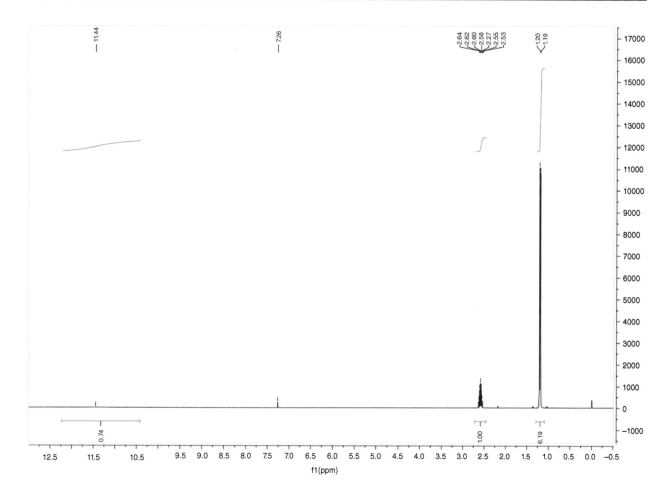

12 Compound B is a colourless liquid with a fishy odour and contains C 79.29%, N 11.56%, and H 9.14%. It has a boiling point of 141 °C. Using these data, the NMR, and the mass spectrometry data, deduce the structure of B. Ignore the peak at 0.0 ppm in the NMR spectrum.

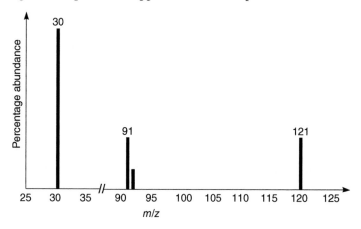

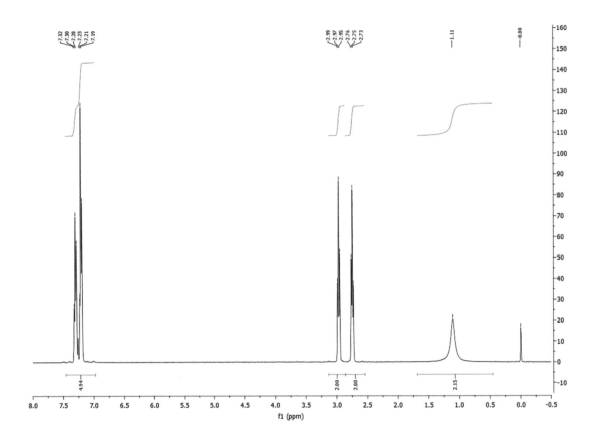

13 Compound C is a colourless liquid with a boiling point of 71 °C. It contains C 26.11%, I 68.96%, and H 4.93%. It reacts on heating with an ethanolic solution of silver nitrate to give a pale-yellow precipitate, soluble in concentrated NH_3 solution. If C is treated with alcoholic KOH, a gas is evolved that decolourizes a solution of bromine in hexane. Using these data and the following IR, NMR, and mass spectrometry data, deduce the structure of C. Ignore the peaks in the NMR spectrum at 0.0 and 1.6 ppm.

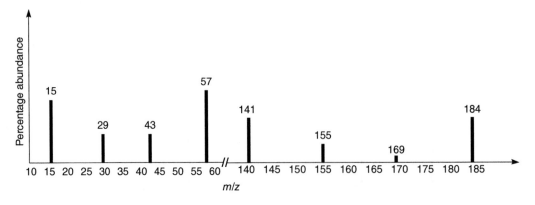

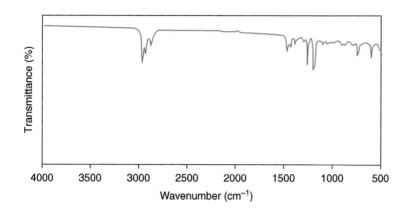

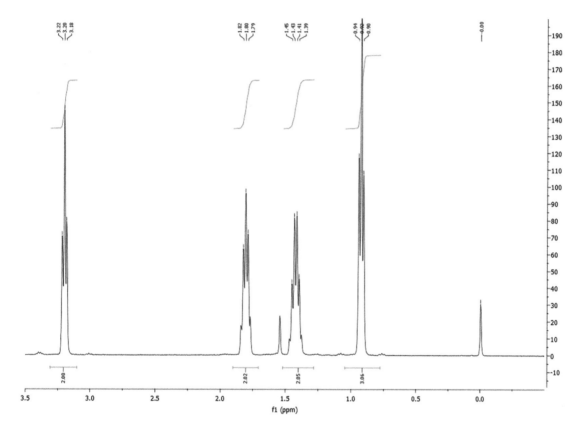

Appendix

Table of constants and other useful information

Physical constants

Name	Symbol	Value
Avogadro constant	N_A	$6.022 \times 10^{23}\ \text{mol}^{-1}$
Ideal gas constant	R	$8.314\ \text{J}\,\text{K}^{-1}\,\text{mol}^{-1}$
Faraday constant	F	$9.6485\ \text{C}\,\text{mol}^{-1}$
Molar gas volume	V_m (ideal) (273 K, 1 atm)	$22.3\ \text{dm}^3\,\text{mol}^{-1}$

SI base units

Quantity	Unit name	Symbol
Length	metre	m
Mass	kilogram	kg
Time	second	s
Electric current	ampere	A
Temperature	kelvin	K
Amount of substance	mole	mol

Foundations of Chemistry: An Introductory Course for Science Students, First Edition.
Philippa B. Cranwell and Elizabeth M. Page.
© 2021 John Wiley & Sons Ltd. Published 2021 by John Wiley & Sons Ltd.
Companion website: www.wiley.com/go/Cranwell/Foundations

Derived units

Quantity	Unit name	Symbol and definition
Area	square metre	m^2
Volume	cubic metre	m^3
Velocity	metre per second	$m\,s^{-1}$
Acceleration	metre per second squared	$m\,s^{-2}$
Density	kilogram per cubic metre	$kg\,m^{-3}$
Concentration	mole per cubic metre	$mol\,m^{-3}$
Energy	joule	$J = kg\,m^2\,s^{-2}$
Force	newton	$N = J\,m^{-1} = kg\,m\,s^{-2}$
Pressure	pascal	$Pa = N\,m^{-2}$
Electric charge	coulomb	$C = A\,s$

Multiples of units and prefixes

Multiple	Prefix	Symbol		Multiple	Prefix	symbol
10^{-1}	deci	d		10^{1}	deca	da
10^{-2}	centi	c		10^{2}	hecto	h
10^{-3}	milli	m		10^{3}	kilo	k
10^{-6}	micro	μ		10^{6}	mega	M
10^{-9}	nano	n		10^{9}	giga	G
10^{-12}	pico	p		10^{12}	tera	T

Short end-of-chapter answers

Chapter 0

1 a. $m\,s^{-2}$.

   b. $kg\,m\,s^{-2}$

   c. $kg\,m^2\,s^{-2}$

2 a. 8.49×10^{-3}

   b. 8.4265×10^{4}

   c. 3.54×10^{2}

   d. 2.18×10^{-5}

3 a. 0.023

   b. 280

   c. 0.000495

   d. 575 900

4 a. 6.300 km

   b. 1.540×10^{-3} s

   c. 4.56×10^{-1} g

   d. 6.395×10^{2} kJ

   e. 6.500 μs

   f. 2.50×10^{-1} dm^3

   g. 5×10^{2} cm^3

Foundations of Chemistry: An Introductory Course for Science Students, First Edition.
Philippa B. Cranwell and Elizabeth M. Page.
© 2021 John Wiley & Sons Ltd. Published 2021 by John Wiley & Sons Ltd.
Companion website: www.wiley.com/go/Cranwell/Foundations

5 a. 3

  b. 5

  c. 4

  d. 3

6 a. $1.36\,m$

  b. $48.2\,cm^2$

  c. $1.1 \times 10^{-1}\,m\,s^{-1}$

  d. $6\,MJ$

7 a. $9.9 \times 10^{-2}\,kg$

  b. $41\,s$

  c. $2.956 \times 10^1\,g$

8 a. KNO_3

  b. $CaCl_2$

  c. Na_2SO_4

  d. $(NH_4)_3PO_4$

  e. $Al_2(SO_4)_3$

9 a. $N_2 + 3H_2 \rightarrow 2NH_3$

  b. $4Fe + 3O_2 \rightarrow 2Fe_2O_3$

  c. $C_5H_{12} + 8O_2 \rightarrow 5CO_2 + 6H_2O$

  d. $2NaOH + H_2SO_4 \rightarrow Na_2SO_4 + 2H_2O$

  e. $2Li + 2H_2O \rightarrow 2LiOH + H_2$

10 a. $2H_2(g) + O_2(g) \rightarrow 2H_2O(l)$

  b. $C(gr) + O_2(g) \rightarrow CO_2(g)$

  c. $CaCO_3(s) + 2HCl(aq) \rightarrow CaCl_2(aq) + CO_2(g) + H_2O(l)$

  d. $AgNO_3(aq) + NaCl(aq) \rightarrow AgCl(s) + NaNO_3(aq)$

  e. $4LiNO_3(s) \rightarrow 2Li_2O(s) + O_2(g) + 4NO_2(g)$

Chapter 1

1 1 mass unit; +1

  1 mass unit; 0

  0.00055 mass units; –1.

2 a. 12e, 12p, 12n

   b. 32e, 32p, 42n

   c. 47e, 47p, 60n

   d. 36e, 40p, 50n

   e. 54e, 53p, 74n

3 See definition in chapter

4 65.45

5 See definition in chapter

6 $^{222}_{86}\text{Rn} \rightarrow \, ^{218}_{84}\text{Po} + \, ^4_2\text{He}$

7 See web site

8 See web site

9 a. $1s^2 \, 2s^2 \, 2p^6 \, 3s^2 \, 3p^4$ or [Ne] $3s^2 \, 3p^4$

   b. $1s^2 \, 2s^2 \, 2p^6 \, 3s^2 \, 3p^6 \, 4s^2 \, 3d^{10} \, 4p^6 \, 5s^2$ or [Kr] $5s^2$

   c. $1s^2 \, 2s^2 \, 2p^6 \, 3s^2 \, 3p^6 \, 4s^2 \, 3d^{10} \, 4p^6$ or [Kr]

   d. $1s^2 \, 2s^2 \, 2p^6 \, 3s^2 \, 3p^6 \, 4s^2 \, 3d^7$ or [Ar] $4s^2 \, 3d^7$

   e. $1s^2 \, 2s^2 \, 2p^6 \, 3s^2 \, 3p^6 \, 4s^2 \, 3d^5$ or [Ar] $4s^2 \, 3d^5$

   f. $1s^2 \, 2s^2 \, 2p^6$ or [Ne]

10 a. C

   b. P

   c. Kr

   d. Cr

Chapter 2

1 See web site for diagram

2 Ionic bonding
   Covalent bonding
   Metallic bonding

3 a. Metallic

   b. Ionic

   c. Covalent

   d. Ionic

   e. Metallic

f. Covalent

g. Covalent

4 See web site for diagrams

5 a. No

b. No

c. No

d. No

e. Yes.

f. Yes.

g. No

6 a. Permanent dipole–permanent dipole; Hydrogen bonding

b. Instantaneous dipole–induced dipole (London forces)

c. Permanent dipole–permanent dipole

d. Permanent dipole–permanent dipole; Hydrogen bonding

e. Instantaneous dipole–induced dipole (London forces)

f. Instantaneous dipole–induced dipole (London forces)

7 See website for diagrams

8 a. Trigonal planar

b. Tetrahedral

c. Trigonal pyramidal

d. Angular

e. Linear

Chapter 3

1 a. 20 g

b. 80 g

c. 25 g

2 a. 1.8 kg

b. 7.4 kg

3 73%

4 C_4H_8O, molecular formula the same

5 95%

6 0.15 mol dm^{-3}

7 6.34 g dm^{-3}

8 K (39.1 g mol^{-1})

9 88%

10 a. 1.4 dm^3.

   b. 8.4 g.

Chapter 4

1 A: simple molecular; B: high high; C: ionic; D: metallic

2 a. Diamond: giant covalent/giant molecular; carbon dioxide: simple molecular/simple covalent

   b. See web site for explanation

   c. See web site for explanation

3 See web site

4 2.84 kg

5 a. 1.20 dm^3. b. 1.72 g

6 a. 5.59 dm^3; b. see web site for explanation

7 a. Xe; b. decane; c. pentane; d. 1,1-dichloroethene: see web site

8 a. 298 K; b. –173 °C; c. 268 K; d. –263 °C

9 a. 80 K.; b. 8 K; c. see chapter and web site

10 a. 4.26×10^{-3} m^3; b. 0.989 atm

Chapter 5

1 a. Cu reduced from +2 to 0. H oxidised from 0 to +1

   b. Mg oxidised from 0 to +2. C reduced from +4 (although covalent) to 0

    c. Ag reduced from +1 to 0. Cu oxidised from 0 to +2

    d. Cl reduced from 0 to –1. I oxidised from –1 to 0

2   a. +1; b. +6; c. -2; d. +7; e. +5; f. +6

3   a. $NaClO$; b. Fe_2O_3; c. KNO_3; d. Na_2SO_3; e. PCl_5

4   a. $Ni(s) + 2Fe^{3+}(aq) \rightarrow Ni^{2+}(aq) + 2Fe^{2+}(aq)$

    b. $5Br^-(aq) + BrO_3^-(aq) + 6H^+(aq) \rightarrow 3Br_2(aq) + 3H_2O(l)$

    c. $6Fe^{2+}(aq) + Cr_2O_7^{2-}(aq) + 14H^+(aq) \rightarrow 6Fe^{3+}(aq) + 2Cr^{3+}(aq) + 7H_2O(l)$

    d. $2MnO_4^-(aq) + 16H^+(aq) + 5C_2O_4^{2-}(aq) \rightarrow 2Mn^{2+}(aq) + 8H_2O(l)$
      $+ 10CO_2(g)$

5   a. $Cu^{2+} + 2I^- \rightarrow CuI + \frac{1}{2}I_2$

    b. $2MnO_4^- + 6H^+ + 5H_2O_2 \rightarrow 2Mn^{2+} + 8H_2O + 5O_2$

6   a. disproportionation; b. coproportionation; c. disproportionation

7   86.6%

8   24.8%

Chapter 6

1 $-85\ kJ\ mol^{-1}$

2   a. $+212.0\ kJ\ mol^{-1}$; b. $+146\ kJ\ mol^{-1}$

3 $-1165\ kJ\ mol^{-1}$

4 $3770\ kJ\ mol^{-1}$, $52.4\ kJ\ g^{-1}$

5   a. $E(Br—Br)$; b. $\Delta_{at}H^{\ominus}(Br_2(l))$; c. $\Delta_{EA}H^{\ominus}(Br(g))$; d. $\Delta_f H^{\ominus}(KBr(s))$;
    e. $\Delta_{lat}H^{\ominus}(KBr(s))$; f. $\Delta_{at}H^{\ominus}(K(s))$

6 $-2010\ kJ\ mol^{-1}$

7 $-5320\ kJ$

8 $-647\ kJ\ mol^{-1}$

9   a. positive; b. negative; c. negative; d. positive

10   a. $+174.2\ J\ mol^{-1}\ K^{-1}$; b. $-0.238\ kJ\ mol^{-1}\ K^{-1}$; c. $-63.8\ J\ mol^{-1}\ K^{-1}$; d. not
     spontaneous

11   a. $+65\ kJ\ mol^{-1}$; b. $+13.7\ J\ K^{-1}\ mol^{-1}$; c. $-65\ J\ K^{-1}\ mol^{-1}$; d. not likely

12 $1100\ K$

Chapter 7

1 0.36 atm

2 a. true; b. false; c. false; d. true

3 a. $0.2 \, mol^{-2} \, dm^6$; b. (i) shift to left, decrease, (ii) shift to right, no change, (iii) shift to right, no change

4 0.04

5 a. $2.5 \, mol^{-1} \, dm^3$; b. $9.6 \times 10^{-3} \, kPa^{-1}$

6 0.62 atm

7 a. See chapter definition; b. $1.00 \times 10^{-3} \, mol$; c. $32.0 \times 10^{-9} \, mol^3 \, dm^{-9}$; d. 0.022 g e., see web site for explanation

8 a. 4; b. 10; c. 3.4; d. 11.6; e. 4.3

9 a. 2.71; b. 4.42; c. 4.3

10 red: strong acid and strong base; blue: weak acid and strong base; red: phenolphthalein/methyl orange/methyl red; blue: phenolphthalein, c. red: about pH = 7; blue: about pH 9

Chapter 8

1 a. $Rate_1 = -\dfrac{\Delta[N_2O_5]}{\Delta t}$, $Rate_2 = \dfrac{\Delta[NO_2]}{\Delta t}$, $Rate_3 = \dfrac{\Delta[O_2]}{\Delta t}$;
   b. $-\dfrac{1}{2}\dfrac{\Delta[N_2O_5]}{\Delta t} = \dfrac{1}{4}\dfrac{\Delta[NO_2]}{\Delta t} = \dfrac{\Delta[O_2]}{\Delta t}$; c. 2.4 mol

2 a. Heterogeneous; b. see Chapter 8, Figure 8.6 or web site for diagram

3 a. 1st order in $S_2O_8^{2-}$, 1st order in I^-, 2nd order overall. b. 2nd order in BrO_3^-, 2nd order overall. c. 1st order in NO, 1st order in Cl_2, 2nd order overall.

4 a. 2nd; b. $k[CH_3CHO]^2$; c. $3.74 \times 10^{-2} \, mol^{-1} \, dm^3 \, s^{-1}$; d. more than one step

5 1st order, 340 s

6 C.

7 a. $k[NO_2]^2$; b. $k \times 0.04 \, mol^2 \, dm^{-6}$, $k \times 0.16 \, mol^2 \, dm^{-6}$; c. reduce by one quarter; d. as answer c

8 See web site for plot of $1/[C_4H_6]$ against time; 2nd order 0.0612 $mol^{-1} \, dm^3 \, s$

9 Rate $= k[NO_2]^2$

10 a. $2.3 \times 10^3 \, mol^{-1} \, dm^3 \, s^{-1}$; b. at least one; c. see web site for answer

11 $274 \, kJ \, mol^{-1}$

Chapter 9

1 i. a. $Ag^+(aq)/Ag(s)$; b. Pb; c. $Ag^+(aq)/Ag(s)$; d. $Pb(s)|Pb^{2+}(aq) \parallel Ag^+(aq)|Ag(s)$; e. +0.93 V

   ii. a. $Co^{2+}(aq)/Co(s)$; b. Zn; c. $Co^{2+}(aq)/Co(s)$; d. $Zn(s)|Zn^{2+}(aq)\parallel Co^{2+}(aq)|Co(s)$; e. +0.48 V

   iii. a. $Cu^{2+}(aq)/Cu(s)$; b. Fe; c. $Cu^{2+}(aq)/Cu(s)$; d. $Fe(s)|Fe^{2+}(aq)\parallel Cu^{2+}(aq)|Cu(s)$; e. +0.78 V

2 –0.22 V

3 Both $Fe^{2+}(aq)$ and $Sn^{2+}(aq)$ will reduce $VO_2^+(aq)$

4 a. Electrons flow to I_2/I^- electrode; b. electrons flow to H^+/H_2 electrode

5 a. 450 C; b. $2H^+(aq) + 2e^- \rightleftharpoons H_2(g)$; c. 52 cm^3

6 a. 1.88×10^{-2} mol; b. 3600 C; c. 2.25×10^{22} electrons; d. 6.00×10^{23} mol^{-1}

7 $Cu^+(aq)$ ions disproportionate to $Cu(s)$ and $Cu^{2+}(aq)$

8 a. see web site; b. +0.15 V so reaction would probably not take place

9 a. See web site; b. +1.05 V

10 a. $Pb(s)$; b. $PbO_2(s)$; c. H_2SO_4; d. see web site; e. +2.05 V

Chapter 10

1 a. i. $1s^2 2s^2 2p^6 3s^2 3p^6 4s^1$, ii. $2s^2 2p^6 3s^2 3p^4$, iii. $1s^2 2s^2 2p^6 3s^2 3p^1$, iv. $1s^2 2s^2 2p^6 3s^2 3p^6 3d^{10} 4s^2 4p^5$

   b. i. K^+, ii. S^{2-}, iii. Al^{3+}, iv. Br^-

2 a. (i) Cl, (ii) Na; b. (i) Na, (ii) K; c. (i) Fe^{3+} (ii) Fe^{2+}

3 a. Br; b. Mg; c. Ne.

4 See web site for explanation

5 a. H^+; b. H^-.

6 See web site

7 See chapter for explanation

8 See chapter

9 a. $1s^2 2s^2 2p^2$; b. $1s^2 2s^2 2p^6 3s^2 3p^2$; c. and d. see web site

10 See web site

Chapter 11

1 a. See chapter for definition; b. $Li(g) \rightarrow Li^+(g) + e^-$; c. see web site; d. ionic

2 a. $1s^2 2s^2 2p^6 3s^2 3p^6 4s^2$; b. see web site; c. $Ca(s) + 2H_2O(l) \rightarrow Ca(OH)_2(aq) + H_2(g)$; d. K; e. Sr

3 a. $1s^2 2s^2 2p^6 3s^2 3p^5$; b. see chapter; c. see chapter; d. $AgNO_3$ followed by conc. NH_3; e., f., and g. see web site

4 a. A = V, B = Ca; b. and c. see web site; d. $VO_3^- + 5$, VO^{2+} +4, $[V(H_2O)_6]^{2+} = +2$

5 a. See chapter for definition; b. see web site

6 a. Cu: $[Ar]3d^{10}4s^1$, Cu^{2+}: $[Ar]3d^9$; Zn: $[Ar]3d^{10}4s^2$; Zn^{2+}: $[Ar]3d^{10}$; b. and c. see web site

7 a. See web site; b. Fe: $[Ar]3d^6 4s^2$, Fe^{2+}: $[Ar]3d^6$, Fe^{3+}: $[Ar]3d^5$; c. hexaaquairon(II), hexacyanoferrate(III); d. and e. see web site

8 See web site

9 a. See web site; b. X^- –1; XO_3^- + 5; X_2 0; c. X^- oxidised from –1 to 0; XO_3^- reduced from +5 to 0; d. see web site; e. see web site.

10 a. IF: I = +1, F = –1; IF_5: I = +5, F = –1; I_2: I = 0; b. disproportionation; c. see web site; d. –365 kJ mol^{-1}

11 +0.75 V

12 a. See web site; b. see web site; c. +13 kJ mol^{-1}; d. and e. see web site

Chapter 12

1 a. Benzene ring, amide, carboxylic acid; b. ketone, alcohol, amine, amide

2 a. pentane

  b. 3-methylpentane

  c. 4,4-dichlorocyclohexanone

  d. (E)-3-methyl-pent-2-ene

  e. Propyl ethanoate

  f. 4-ethyl-2,2-dimethylheptane

3 I_2O_5

4 $C_5H_{10}O_2$

5 $C_6H_{12}O_2$

6 C_4H_9

7   Hexane, 2-methylpentane, 3-methylpentane, 2,3-dimethylbutane, 2,2-dimethylbutane

8   1-bromopentane, 2-bromopentane, 3-bromopentane, 2-bromo-3-methyl-butane, 2-bromo-2methylbutane.

9   Butan-1-ol; butan-2-ol; 2-methylpropan-2-ol; 2-methylpropan-1-ol; 1-methoxypropane; ethoxyethane; 2-methoxypropane

10 a. *E*, b. *Z*, c. *E*, d. *Z*

11 See web site

12

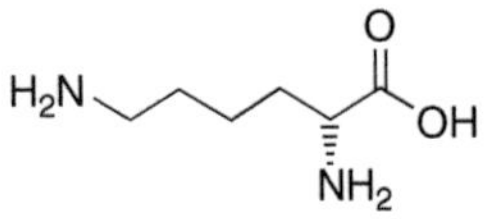

13 a. *S*, b. *S*, c. *R*, d. *S*

14 a. Electrophile; b. Nucleophile; c. Radical; d. Nucleophile; e. Radical; f. Electrophile

Chapter 13

1   Hexane would have a higher boiling point than 2,3-dimethylbutane – stronger intermolecular forces (London forces/van der Waals).

2   $C_{10}H_{12} + 13O_2 \rightarrow 10CO_2 + 6H_2O$

3   1.04 mol CO_2 is 46.05 g CO_2.

4   $C_{10}H_{22} \rightarrow C_5H_{12} + C_5H_{10}$

5   See web site

6   See web site

7   a. $Br_2 \rightarrow 2Br^{\cdot}$ Initiation

$Br^{\cdot} + C_5H_{12} \rightarrow HBr + {}^{\cdot}C_5H_{11}$ Propagation step 1

${}^{\cdot}C_5H_{11} + Br_2 \rightarrow C_5H_{11}Br + Br^{\cdot}$ Propagation step 2

$Br^{\cdot} + Br^{\cdot} \rightarrow Br_2$ Termination step 1

${}^{\cdot}C_5H_{11} + Br^{\cdot} \rightarrow C_5H_{11}Br$ Termination step 2
${}^{\cdot}C_5H_{11} + {}^{\cdot}C_5H_{11} \rightarrow C_{10}H_{22}$ Termination step 3

b. See web site

8   See web site

9   See web site

10 a. Hydrogen in the presence of a palladium (or nickel) catalyst

b. Hydrogen in the presence of a poisoned palladium catalyst (Lindlars)

Chapter 14

1 Butan-1-ol has a higher boiling point. Butane – van der Waals (or London) forces; butan-1-ol hydrogen bonding.

2 Left to right: primary, secondary, tertiary, secondary

3 The tertiary alcohol cannot be oxidised. The primary alcohol can either become an aldehyde or a carboxylic acid. The secondary alcohols can be oxidised to ketones.

4 Suitable reagents would be $LiAlH_4$ or $NaBH_4$. 4-Chloropentan-1-ol, cyclohexanol.

5 See web site.

6 Acetophenone is a ketone, and benzaldehyde is an aldehyde. Both would show a positive test with 2,4-DNP (an orange precipitate). Benzaldehyde will give a positive test with Tollen's reagent (silver mirror), will also be oxidised using potassium permanagante with a colour change from orange to green, and will give a red precipitate with Fehling's solution.

7 Using a carbonate base e.g. $NaHCO_3$ with an acidic solution would give carbon dioxide gas, which would be bubbled through lime water.

8 Left to right: methyl ethanoate; propyl 3-methylbutanoate; ethyl hexanoate

9 Left to right: a. methyl methanoate: NaOH – methanol and sodium ethanoate. Aqueous HCl – methanol and ethanoic acid. b. Ethyl 3-methylbutanoate: NaOH – propanol and sodium 3-methylbutanoate. Aqueous HCl – propanol and 3-methylbutanoic acid. c. Ethyl hexanoate: NaOH – ethanol and sodium hexanoate. Aqueous HCl – ethanol and hexanoic acid.

10 Left to right: a. methylamine and ethanoyl chloride; b. methanol and pentanoyl chloride *or* methanol and pentanoic acid with H_2SO_4; c. ethylmethylamine and 3-methylbutanoyl chloride; d. phenol and cyclohexylacetyl chloride.

11 Left to right: a. methylamine and chloroethane *or* ethylamine and chloromethane; b. benzyl chloride and dimethylamine *or* benzylamine and chloromethane (two equivalents); c. cyclobutylamine and chloromethane (three equivalents). In all cases, the chloroalkane could also be a bromoalkene or an iodoalkane.

Chapter 15

1 See web site

2 Left to right: 1-chloro-2-fluorobenzene; 1-bromo-3-ethylbenzene; 2-methylbenzaldehyde; 3-methyl-5-nitrophenol

3 The nitro group is electron withdrawing; the alcohol group on phenol is electron donating.

4 Bromobenzene

5   Nitrobenzene

6   1-Phenylbutan-1-one

7   Left to right: a. Benzene, chloromethane, and $AlCl_3$. b. Benzene, butanoyl chloride, and $AlCl_3$. c. Step a: benzene, HNO_3/H_2SO_4; step b: chlorine and $AlCl_3$. d. Step a: benzene, ethanoyl chloride, and $AlCl_3$; step b: HNO_3/H_2SO_4; step c: Raney-Ni and H_2.

8   Most reactive to least reactive: toluene > benzene >1-phenylbutan-1-one > nitrobenzene

9   Step a.: Benzene, Cl_2, $AlCl_3$; step b.: ethanoyl chloride and $AlCl_3$; step c.: HNO_3/H_2SO_4; step d.: Raney-Ni and H_2

10 N,N-Dimethyl-4-(phenyldiazenyl)aniline

Chapter 16

1   Left to right: secondary; secondary; primary; tertiary

2   Left to right: S_N1 or S_N2; S_N1 or S_N2; S_N2; S_N1

3   Left to right: E1 or E2; E1 or E2; E2; E1

4   a. S_N2; b. S_N1

5   a. E2; b. E1

6   2-Methylbut-2-ene formed preferentially

Chapter 17

1   a. S

    b. E2 (elimination)

    c. Br_2

2   a. 1-(3-nitrophenyl)ethan-1-one; b. and c. see web site

3   a. and b. See web site; c. formaldehyde or formic acid; d. depends on answer to c.

4   a. See web site; b. styrene; c. see website; d. acidified potassium dichromate; e. Tollen's reagent or add acidified potassium dichromate; f. ethyl benzoate

5   a. and b. See web site; c. HNO_3 and H_2SO_4

6   a. S_N2 (substitution); b. bromination at every alkene; c. see web site

Chapter 18

See web site for all

Chapter 19

1 RAM = 237.98

2 RAM = 63.62

3 Spectrum A contains bromine as there is a peak at [M + 2] that is in a $1:1$ ratio with that of the [M] peak. Spectrum B contains chlorine as there is a peak at [M + 2] that is in a $1:3$ ratio with that of the [M] peak.

4 C_7H_8O M_r is 108. Likely peaks include [M] peak (108); the others are 107 (−H), 91 (−OH), 77 (C_6H_5).

5 a. Propanol would contain an OH peak around 3000 cm^{-1}. Propanenitrile would contain a C≡N peak around 3000 cm^{-1}. Propanamide will contain C=O and N−H peaks at around 1700 and 3000 cm^{-1} respectively.

   b. Acetone would contain a C=O peak at around 1700 cm^{-1}, but dichloromethane would not.

6 C=O, C=C, C−H (at around 1700, 1650, 3000 cm^{-1}, respectively)

7 The OH peak between 3450 and 2950 cm^{-1}, the aldehyde peak at approximately 1700 cm^{-1}, and the C−O peak at 1300 cm^{-1}

8 Propan-1-ol, propan-2-ol, methoxyethane. For a full explanation, see web site.

9 Spectrum A contains three proton environments; spectrum B contains three proton environments; spectrum C contains three proton environments. For a full explanation, see web site.

10 2′-bromoacetophenone. For a full explanation, see web site.

11 Molecular formula: $C_4H_8O_2$; 2-methylpropanoic acid. For a full explanation, see web site.

12 Molecular formula: $C_8H_{11}N$ (121 g mol^{-1}); phenylethylamine (PhCH$_2$CH$_2$NH$_2$)

13 Molecular formula: C_4H_9I; 1-iodobutane

Index

Foundations of Chemistry: An Introductory Course for Science Students, First Edition.
Philippa B. Cranwell and Elizabeth M. Page.
© 2021 John Wiley & Sons Ltd. Published 2021 by John Wiley & Sons Ltd.
Companion website: www.wiley.com/go/Cranwell/Foundations

Printed and bound by CPI Group (UK) Ltd, Croydon, CR0 4YY

27/10/2024

14580770-0001